Functional Tissue Engineering

Springer
New York
Berlin
Heidelberg
Hong Kong
London
Milan
Paris
Tokyo

Farshid Guilak   David L. Butler
Steven A. Goldstein   David J. Mooney
Editors

Functional Tissue Engineering

With 133 Figures

Springer

Farshid Guilak
Orthopaedic Research Laboratories
Department of Surgery
375 MSRB, Box 3093
Duke University Medical Center
Durham, NC 22710
USA
guilak@duke.edu

David L. Butler
Department of Biomedical Engineering
University of Cincinnati
2901 Campus Drive
Cincinnati, OH 45221-0048
USA
david.butler@uc.edu

Steven A. Goldstein
Orthopaedic Research Laboratories
University of Michigan
Room G-161, 400 North Ingalls
Ann Arbor, MI 48105-0486
USA
stevegld@umich.edu

David J. Mooney
Department of Biologic and
   Materials Sciences
University of Michigan
1011 North University Avenue
Ann Arbor, MI 48109-1078
USA
mooney@umich.edu

Library of Congress Cataloging-in-Publication Data
Functional tissue engineering/editors, Farshid Guilak . . . [et al.].
     p. cm.
   Includes bibliographical references and index.
   ISBN 0-387-95553-4 (alk. paper)
   1. Biomedical engineering. 2. Biomechanics. 3. Tissue culture. 4. Animal cell
biotechnology. I. Guilak, Farshid.
R856 .F86 2003
610′.28—dc21   2002026659

ISBN 0-387-95553-4   Printed on acid-free paper.

9 8 7 6 5 4 3 2 1   SPIN 10887292

www.springer-ny.com

Springer-Verlag   New York Berlin Heidelberg
A member of BertelsmannSpringer Science+Business Media GmbH

Preface

Tissue engineering is an exciting new field at the interface of engineering and biology that uses implanted cells, scaffolds, DNA, proteins, protein fragments, and inductive molecules to repair or replace injured or diseased tissues and organs. Tremendous progress in biological and biomaterial aspects of this field have been accomplished to date, and several engineered tissues are now being used clinically. However, tissue engineers face major challenges in repairing or replacing tissues that serve a predominantly biomechanical function.

To meet this challenge, the United States National Committee on Biomechanics in 1998 adopted a new paradigm termed *functional tissue engineering* (FTE) to emphasize the importance of biomechanical considerations in the design and development of cell and matrix-based implants for soft and hard tissue repair. Functional tissue engineering represents a relevant and exciting new discipline in the field of tissue engineering. Since many tissues, such as those of the musculoskeletal, cardiovascular, and dental systems, are accustomed to being mechanically challenged, tissue-engineered constructs used to replace these tissues after injury or disease must certainly do the same. Of course, tissue engineers must also attempt to return normal biological activity in order for the construct to truly integrate with the surrounding tissues. Thus, the term *functional* can have many meanings, such as restoration of metabolic function. The primary focus of this text is on the role of biomechanical function in tissue engineering.

To more clearly delineate the important aspects of functional tissue engineering, the editors and a steering committee (Van C. Mow, Columbia University; Robert M. Nerem, Georgia Institute of Technology; Robert L. Spilker, Rensselaer Polytechnic Institute; Michael V. Sefton, University of Toronto; Savio L.Y. Woo, University of Pittsburgh) organized a Functional Tissue Engineering Workshop in September 2000 that serves as the basis for this text. The FTE conference was fortunate to attract a select group of biomedical engineers, biologists, and clinicians who contributed the chapters appearing in the book. From its inception, the organizers had four primary objectives in running the conference. These objectives were: (1) to increase awareness among tissue engineers about the importance of restoring function when engineering constructs; (2) to identify the critical structural and mechanical requirements needed for each construct; (3) to provide a catalyst for the development of functional criteria in the design, manufacture, and optimization of tissue engineered constructs; and (4) to develop a teaching and reference text for students and investigators in the field of tissue engineering.

Stemming from these objectives were more specific goals of the conference. These goals are briefly described below and are the basis for the structure of the remainder of the book.

Primary Goals of Functional Tissue Engineering

In Vivo and *In Vitro* Responses of Native Tissue

1. **Establish *in vivo* load and deformation histories for native tissues under a wide range of activities of daily living.** These histories provide benchmarks against which tissue-engineered repairs can be compared.

2. **Establish safety factors and relevant biomechanical properties for native tissues under a wide range of activities of daily living.** For example, safety factors or ratios of failure force (determined from *in vitro* failure tests) to peak *in vivo* force need to be established in the normal operating ranges of a native tissue for different activities of daily living. Identify and prioritize structural and material properties determined within these *in vivo* ranges so as to provide useful design parameters for tissue-engineered repairs and replacements. These safety factors and biomechanical properties are likely to change with activity level and tissue location.

3. **Understand the biomechanical properties and structure-function relationships for native tissues.** Develop correlates among the biomechanical properties, tissue structure, and biochemistry of normal tissues.

4. **Determine the regulatory processes (including humoral and neurogenic factors) that cells experience *in vivo* as they interact with the extracellular matrix.** Understanding the critical interplay between biological signaling and mechanical forces *in vivo* is important to designing effective tissue engineered implants.

Functional Requirements and Design Parameters

5. **Identify a minimum set of qualification tests for each tissue type.** Select only those tests that are necessary and sufficient for a successful repair.

6. **Develop accurate computational models and corresponding computer simulation tools.** These models and tools should incorporate the functional demands of a native tissue/organ (mechanical properties and anatomies) to determine baselines so as to compare conditions in repair/regenerated tissue/organ in situ. The data should be subject specific data to characterize the local functional environment (mechanical, electrical, chemical) of tissues at scales ranging from systems, tissues, cells, and molecules. These data sets will provide validated predictions of clinically relevant, functional parameters.

7. **Optimize stress, strain, and strain rate characteristics of engineered constructs in culture and their biological, chemical, and morphological properties.** Design, fabricate, and test cell-matrix constructs in culture that can

tolerate expected *in vivo* loads and deformations (Goal 1) and that possess desirable biomechanical properties (Goal 2). If necessary, develop means to protect the constructs and to prepare the surfaces to improve integration into surrounding tissue (which itself may not be normal). Understand structure-function relationships for these constructs by also characterizing relationships between their biological, chemical, and morphologic properties.

8. **Verify *in vivo* load and deformation regimes for engineered tissues.** Determine *in vivo* force and/or deformation histories for already designed tissue engineered repairs for a wide range of activities of daily living. Compare these results with similar histories for native tissues.

Intrinsic Properties of Matrices, Biophysical Factors on Cells and Matrices, and Bioreactors/*In Vitro* Factors

9. **Define carriers to promote functional biological characteristics to enhance tissues and differentiation and incorporation *in vivo*.** Matrices are needed that can withstand the large *in vivo* forces that tissues must often withstand.

10. **Identify bioactive soluble stimulators of tissue regeneration (proteins or their genes)** and study these in the context of appropriate matrices and mechanical environment for each tissue specific application.

11. **Establish how changing mechanical, chemical, and biological regulating factors influence cellular activity in bioreactors and *in vivo* so as to guide tissue repair to be more like normal.** Determine how mechanical stimulation of cell-matrix implants modulates engineered tissue structure and function. Define biological indicators that promote structural enhancement both *ex vivo* and *in vivo*. Characterize bioreactor explant conditions (e.g., stress, strain, pressure, flow velocity, electric potential, and currents) to determine the specificity of mechanical signal transduction mechanism(s), thus enabling the engineering of tissue constructs to be optimized for *in vivo* use.

Assessment of Repair Function

12. **Develop minimally invasive implantation procedures for implants.** Formulate methods and protocols for testing the tissues to be engineered using surrogate markers (biological, imaging, or biomechanical).

13. **Solve implant size issues by exceeding functional limitations of diffusion.** Address problems of vascularization/perfusion.

14. **Explain how host genetic background affects adaptive responses to tissue engineered products, including functional incorporation and remodeling.** Effectively integrate tissue-engineered tissues into adjacent tissues. Correlate the same parameters in a diseased state at the time of reconstruction so as to understand the environment in which the tissue is implanted.

Clinical Evaluation

15. **Set clinical standards to determine whether the repairs and replacements are safe and efficacious after surgery, recognizing the potential impact of patient**

selection on the success of tissue engineering. The clinical tools to evaluate success (clinical, functional, anatomical) must be defined. Develop minimally invasive imaging methods (e.g., PET, FTIR, MRI) to validate biomechanical measurements, to assess remodeling, and to track the fate of implants after surgery. Select quantitative outcome measures to assess product performance in humans. Establish whether the goal of functional tissue engineering is to regenerate normal tissue or simply repair damaged or diseased tissue to meet expected demands.

16. **Establish appropriate (and inappropriate) patient rehabilitation regimes.** Surgical and postsurgical (rehabilitation) principles need to be defined to enhance tissue differentiation, incorporation, and durability.

Obviously, other issues must be considered that are also critical to the success of engineered constructs. These issues include electromechanical coupling (i.e., in muscle and other tissues), vaso-activity, electrical conduction cell-cell communication, cell source, quality control/manufacturing/storage/sterilization (e.g., shelf life, storage, handling procedures must be realistic and feasible). Furthermore, it will be necessary to define the relevance of animal models to the human condition in both health and disease and to correlate these parameters in human tissue to those measured in animal models.

In closing, the conference was truly fortunate to have the financial support of multiple organizations. These organizations included the Whitaker Foundation, the National Science Foundation, Georgia Tech/Emory Center for the Engineering of Living Tissues, and numerous companies including Advanced Tissue Sciences, BioMet, Orthologic, Orquest, Selective Genetics, and Zimmer. The conference was also fortunate to have active participation from representatives of the National Science Foundation (Sohi Rastegar) and the National Institutes of Health, including James Panagis and Harold Slavkin. We hope that this conference and text will set the tone for both research and training in tissue engineering for the next decade and will serve as the paradigm for academic/industrial interactions in this field. The conference is also leading to a synergy between biomaterials, biology, and biomechanics in the next generation of engineered tissue replacements.

Durham, North Carolina, USA   Farshid Guilak
Cincinnati, Ohio, USA   David L. Butler
Ann Arbor, Michigan, USA   Steven A. Goldstein
Ann Arbor, Michigan, USA   David J. Mooney

Contents

Contributors

Steven D. Abramowitch
Musculoskeletal Research Center
Department of Orthopaedic Surgery
University of Pittsburgh
210 Lothrop St., E1641 BST
P.O. Box 71199
Pittsburgh, PA 15213, USA

Robert E. Akins
Department of Biomedical Research
A.I. duPont Hospital for Children
Nemours Children's Clinic
1600 Rockland Road
Wilmington, DE 19803, USA

Kai-Nan An
Department of Orthopaedic Surgery
Mayo Clinic
Rochester, MN 55905, USA

Thomas Andriacchi
Stanford University
Departments of Mechanical Engineering and
    Functional Restoration
Division of Biomechanical Engineering
Stanford, CA 94305-3030, USA

Gerard A. Ateshian
Departments of Mechanical Engineering and
    Biomedical Engineering
Columbia University
New York, NY 10027, USA

Kyriacos A. Athanasiou
Rice University
Department of Bioengineering-MS 142
P.O. Box 1892
Houston, TX 77251-1892, USA

François A. Auger
Laboratoire d'Organogénèse Expérimentale (LOEX)
Centre hospitalier affilié universitaire de Québec
Québec (Québec) G1S 4L8
Canada

Hani Awad
Orthopaedic Research Laboratories
Department of Surgery
Duke University Medical Center
Durham, NC 27710, USA

Albert J. Banes
Department of Orthopaedics
253 Burnette Womack Bldg. CB#7055
University of North Carolina
Chapel Hill, NC 27599-7055, USA

A. Boyd
Departments of Orthopaedics and Radiology
601 Elmwood Avenue, Box 665
University of Rochester
Rochester, NY 14642, USA

S. Bukata
Departments of Orthopaedics and Radiology
601 Elmwood Avenue, Box 665
University of Rochester
Rochester, NY 14642, USA

David L. Butler
Department of Biomedical Engineering
2901 Campus Dr.
University of Cincinnati
Cincinnati, OH 45221-0048, USA

Arnold I. Caplan
Skeletal Research Center
Department of Biology
Case Western Reserve University
Cleveland, OH 44106, USA

Stephen C. Cowin
The New York Center for Biomedical Engineering
    and The Department of Mechanical
    Engineering
The City College of New York
138th Street and Convent Avenue
New York, NY 10031, USA

Robert G. Dennis
Departments of Mechanical and Biomedical
    Engineering
Institute of Gerontology
University of Michigan
Ann Arbor, MI 48109, USA

Michael A. DiMicco
Department of Bioengineering
University of California at San Diego
La Jolla, CA 92093-0412, USA

Matthew Dressler
Department of Biomedical Engineering
2901 Campus Drive
University of Cincinnati
Cincinnati, OH 45221-0048, USA

Angel O. Duty
School of Mechanical Engineering
Georgia Tech/Emory Center for the Engineering of
    Living Tissues
Georgia Institute of Technology
Atlanta, GA 30332, USA

John A. Faulkner
Department of Biomedical Engineering
Institute of Gerontology
300 North Ingalls Building, University of Michigan
Ann Arbor, MI 48109-2007, USA

Lisa E. Freed
Harvard-M.I.T. Division of Health Sciences and
    Technology
Massachusetts Institute of Technology
E25-342, 45 Carleton Street
Cambridge, MA 02139, USA

Shunichi Fukuda
Department of Bioengineering
The Whitaker Institute for Biomedical
    Engineering
University of California San Diego
La Jolla, CA 92093-0412, USA

L. Germain
Laboratoire d'Organogénèse Expérimentale
    (LOEX)
CHA/Hôpital du Saint-Sacrement, 1050
chemin Sainte-Foy et Départment de Chirurgie,
    Université Laval
Québec (Québec)
Canada

Steven A. Goldstein
Orthopaedic Research Laboratories
University of Michigan
Ann Arbor, MI 48105-0486, USA

G. Grenier
Laboratoire d'Organogénèse Expérimentale (LOEX)
CHA/Hôpital du Saint-Sacrement, 1050
chemin Sainte-Foy et Départment de Chirurgie,
    Université Laval
Québec (Québec)
Canada

Farshid Guilak
Departments of Surgery, Biomedical Engineering,
    and Mechanical Engineering and Materials
    Science
Duke University Medical Center
Box 3093, 375 MSRB
Durham, NC 27710, USA

Robert E. Guldberg
School of Mechanical Engineering
Georgia Tech/Emory Center for the Engineering of
    Living Tissues
Georgia Institute of Technology
Altanta, GA 30332, USA

Hai-Chao Han
George W. Woodruff School of Mechanical
    Engineering
Georgia Institute of Technology
Atlanta, GA 30332-0405, USA

Anne Hoger
Department of Mechanical and Aerospace
    Engineering
University of California, San Diego
9500 Gilman Drive
La Jolla, CA 92093-0411, USA

J.C.Y. Hu
Department of Bioengineering-MS 142
PO Box 1892
Rice University
Houston, TX 77251-1892, USA

Jay D. Humphrey
Biomedical Engineering
233 Zachry Engineering Center
Texas A&M University
College Station, TX 77843-3120, USA

Clark T. Hung
Department of Biomedical Engineering
Columbia University
New York, NY 10027, USA

Brett C. Isenberg
Departments of Biomedical Engineering and
    Chemical Engineering and Materials Science
University of Minnesota
Minneapolis, MN 55455, USA

Stephen M. Klisch
Department of Bioengineering
University of California at San Diego
La Jolla, CA 92093-0412, USA

Paul E. Kosnik
Cell Based Delivery, Inc.
1 Richmond Square, Suite 215E
Providence, RI 02906, USA

David N. Ku
George W. Woodruff School of Mechanical
    Engineering
Georgia Institute of Technology
Atlanta, GA 30332-0405, USA

Jack L. Lewis
Department of Orthopaedic Surgery
University of Minnesota
420 Delaware St. S.E.
Minneapolis, MN 55455, USA

Jay R. Lieberman
UCLA Department of
    Orthopedic Surgery
UCLA School of Medicine
Los Angeles, CA 90095, USA

John C. Loh
Musculoskeletal Research Center
Department of Orthopaedic Surgery
University of Pittsburgh
210 Lothrop St., E1641 BST
P.O. Box 71199
Pittsburgh, PA 15213, USA

J.M. Looney
Departments of Orthopaedics and Radiology
601 Elmwood Avenue, Box 665
University of Rochester
Rochester, NY 14642, USA

J. Monu
Departments of Orthopaedics and Radiology
601 Elmwood Avenue, Box 665
University of Rochester
Rochester, NY 14642, USA

David J. Mooney
Departments of Biologic and Materials Sciences,
    Biomedical Engineering, and Chemical
    Engineering
University of Michigan
Room 5213 Dental
1011 North University Avenue
Ann Arbor, MI 48109-1078, USA

Volker Musahl
Musculoskeletal Research Center
Department of Orthopaedic Surgery
University of Pittsburgh
210 Lothrop St., E1641 BST
P.O. Box 71199
Pittsburgh, PA 15213, USA

Robert M. Nerem
Georgia Tech/Emory Center for the Engineering of
    Living Tissues
Parker H. Petit Institute for Bioengineering and
    Bioscience
Georgia Institute of Technology
Atlanta, GA 30332-0363, USA

Regis J. O'Keefe
Department of Orthopaedics
University of Rochester
Rochester, NY 14642, USA

K. Parker
Departments of Orthopaedics and
    Radiology
601 Elmwood Avenue, Box 665
University of Rochester
Rochester, NY 14642, USA

Brett Peterson
UCLA Department of Orthopedic
    Surgery
UCLA School of Medicine
Los Angeles, CA 90095, USA

M. Rémy-Zolghadri
Laboratoire d'Organogénèse Expérimentale
    (LOEX)
CHA/Hôpital du Saint-Sacrement, 1050
chemin Sainte-Foy et Départment de Chirurgie,
    Université Laval
Québec (Québec)
Canada

Maria A. Rupnick
Department of Medicine
Brigham and Women's Hospital
45 Francis Street
Boston, MA 02115, USA

Michael S. Sacks
Department of Bioengineering
Room 749 Benedum Hall
3700 Ohara St.
University of Pittsburgh
Pittsburgh, PA 15261, USA

Robert L. Sah
Department of Bioengineering
The Whitaker Institute for Biomedical Engineering
University of California at San Diego
La Jolla, CA 92093, USA

Dirk Schaefer
Department of Surgery
University of Basel
Basel, Switzerland

Geert Schmid-Schönbein
Department of Bioengineering
The Whitaker Institute for Biomedical Engineering
University of California San Diego
La Jolla, CA 92093-0412, USA

Edith Richmond Schwartz
Biomed Consultants
1401 17th Street, NW
Washington, DC 20036, USA

E.M. Schwarz
Departments of Orthopaedics and Radiology
601 Elmwood Avenue, Box 665
University of Rochester
Rochester, NY 14642, USA

Michael V. Sefton
Institute of Biomaterials and Biomedical
    Engineering
University of Toronto
Toronto M5S 3G9
Canada

G.S. Seo
Departments of Orthopaedics and Radiology
601 Elmwood Avenue, Box 665
University of Rochester
Rochester, NY 14642, USA

Lori A. Setton
Departments of Biomedical Engineering, Surgery,
    and Mechanical Engineering and Materials
    Science
Duke University
Durham, NC 27708, USA

Craig A. Simmons
Departments of Biologic and Materials Sciences and
    Biomedical Engineering
University of Michigan
Ann Arbor, MI 48109, USA

Jan P. Stegemann
Georgia Tech/Emory Center for the Engineering of
    Living Tissues
Parker H. Petit Institute for Bioengineering and
    Bioscience
Georgia Institute of Technology
Atlanta, GA 30332-0363, USA

J. Tamez-Pena
Departments of Orthopaedics and Radiology
601 Elmwood Avenue, Box 665
University of Rochester
Rochester, NY 14642, USA

S. Totterman
Departments of Orthopaedics and Radiology
601 Elmwood Avenue, Box 665
University of Rochester
Rochester, NY 14642, USA

Robert T. Tranquillo
Departments of Biomedical Engineering and
    Chemical Engineering and Materials Science
University of Minnesota
Minneapolis, MN 55455, USA

Herman H. Vandenburgh
Cell Based Delivery, Inc.
Providence, RI 02906, USA

Gordana Vunjak-Novakovic
Harvard-MIT Division of Health and Sciences and
    Technology
Massachusetts Institute of Technology
77 Massachusetts Ave. E25-330
Cambridge, MA 02139, USA

James H.-C. Wang
Musculoskeletal Research Center
Department of Orthopaedic Surgery
University of Pittsburgh
210 Lothrop St., E1641 BST
P.O. Box 71199
Pittsburgh, PA 15213, USA

Savio L.-Y. Woo
Musculoskeletal Research Center
Department of Orthopaedic Surgery
University of Pittsburgh
210 Lothrop St., E1641 BST
P.O. Box 71199
Pittsburgh, PA 15213, USA

Part I

The Functional Properties of Native Tissues

1

How Does Nature Build a Tissue?

Stephen C. Cowin

Introduction

The interest here is in the well-formulated, highly successful plans for the construction of each tissue; the clocklike, highly stable methods of tissue assembly; and the ability of the organism to adapt to different inputs, including trauma, during the construction process of the tissue. If we understand how tissue is built, we will also learn how it grows, how it is repaired, and how we might engineer it. The few facts that I have been able to find in the literature concerning tissue construction processes demonstrate its great sophistication, so that I have come to envy those who will some day have the pleasure of describing the entire construction processes. I think that it will be clear to the reader from the few facts assembled here that what we have taken thus far in this direction are baby steps. Great discoveries of enormous practical use to man are to be expected. Thus, we seek the nature of the components of a biological tissue composite, the cellular processes that produce these constituents, the assembly of the constituents into a hierarchical structure, and the behavior of the tissue composite structure in the adaptation to its mechanical environment. The subsequent five sections deal with the questions: From where do the blueprints for a tissue structure come? From where do the materials of construction come? What is the structural assembly process for a tissue? How are they coordinated so that the proper development of an organism occurs?

Genetic and Epigenetic Influence in Tissue Development

The functional form of biological structures that must operate under specific environmental force systems is strongly influenced by the force systems to which the structures are subjected. As examples, consider the following body substructures: the tendon, ligament, and muscle that carry tension; articular cartilage, bone, and teeth that are subjected to repeated compressive impact loads; membranes (cells, mesentery, skin) that protect one environment from another; and cylinders (long bones, intestine, blood vessels) that are designed for structural support or the transport of fluids. This theme is summarized by the following quote from Trelstad and Silver (1981):

"The processes of morphogenesis, growth, repair, adaptation and aging are all reflected in tissues by the processes of either changing the ratio of their constituents or the properties of their constituents. The capacity of cells in tissues to produce composite extracellular matrices that assemble into multiple diverse forms reflects a successful stratagem of multicellular organisms to segregate cells into functional units of tissues and organs able to contend with the forces of gravity and work in unison to transmit the forces necessary for movement. The shape and size of higher organisms are defined by spaces, partitions, and unique forms of the matrix. In the embryo, the extracellular matrix is the scaffolding that helps determine tissue patterns, and in the adult, it serves to stabilize these same patterns. The mineralized matrices of the bones and teeth are stiff, hard structures, whereas the nonmineralized cartilages are flexible and compressible and serve as joint cushions during compression and translation of joints. The ropelike organization of tendons and ligaments provides them with the capacity to withstand large forces without stretching more than a few percent of their length, making movement possible. Elastic blood vessels transiently store pulse pressures generated by the heart and thus ensure a relatively continuous flow of blood. The stretchable, tough, and tight-fitting skin is a collagenous shield that serves to keep undesired materials out and desired materials in, whereas its counterpart covering the eye, the cornea, is a transparent lattice of collagen fibrils serving both barrier and optical functions."

The biological systems of interest can adapt their structure, in the individual (epigenetically) as well as in the species (genetically), to accommodate a changed mechanical load environment. Numerous studies of biological systems in the last century and a half have demonstrated that slow evolutionary engineering has refined the properties of the biological structures, including mechanical properties. The results of this process are encoded in the genes (DNA). However, the biological structures of an individual may also adapt to a changed mechanical load environment as illustrated by the bodies of weight lifters or body builders. In general, the form of a biological structure is due to both genetic and epigenetic information, but how much of each is not known. The key genetic factor that lies at the base of structural adaptation is likely the sensitivity of cells of the biological structure to physical forces. The general form of the biological structure is recorded in the genes, but during early growth it is stimulated by mechanical factors in the embryo and *finally* refined by the individual after birth. For example, the skin forming the soles of the feet is already thickened in the human fetus, but our common experience shows that walking barefoot or wearing poorly fitting footwear will contribute to the thickness and distribution of this tissue on the foot bottom. An example of development precisely driven by the genes has been observed in the small worm, *Caenorhabditis elegans*. It has been shown that of the 1090 cells formed during the development of the main body of an individual worm, 131 always die (Coen, 1999). The programmed cell deaths (apoptosis) are not random, but involve exactly the same cells in every

individual. Genetic precision is also observed in genetic mutations. Huntington's disease, the gene mutation that killed the folk singer Woody Guthrie, is an example. Its grim mathematical precision was described by Ridley (1999) thus:

"The age at which the madness will appear depends strictly and implacably on the number of repetitions of the DNA sequence CAG (cytosine, guanine, thymine) in one place in one gene. If you have thirty-nine, you have a ninety per cent probability of dementia by the age of seventy-five and will, on average, get the first symptoms at sixty-six; if forty, on average you will succumb at fifty-nine; if forty-one, at fifty-four; if forty-two, at thirty-seven; and so on until those who have fifty repetitions of the 'word' will lose their minds at roughly twenty-seven years of age."

Epigenetic Processes of Tissue Pattern Formation

Many different mechanisms have been proposed for the spontaneous (perhaps partially) epigenetic development of patterns in living systems (Held, 1992). These include chemically reacting and diffusing systems, mechanical instabilities, and different combinations of these mechanisms. Alan Turing, famous for creating, in 1936, the prototype plan for digital computers and, during World War II, for his effort in deciphering the German Enigma cipher, is also renowned for the chemical reaction mechanism hypothesis. In his 1952 paper entitled, "The Chemical Basis of Morphogenesis," he hypothesized that the diffusing patterns of reacting chemicals can form steady-state heterogeneous spatial patterns, and this phenomenon could be a biological mechanism of pattern formation (Turing, 1952). The hypothesized chemicals are called morphogens. Let $a(x, t)$ and $n(x, t)$ be two scalar valued functions of the position x and t representing two different chemical species. The species $a(x, t)$ is the activator and the species $n(x, t)$ is the inhibitor, $n(x, t)$ tending to reverse the advance of $a(x, t)$, or undo the effect of $a(x, t)$. Thus, $a(x, t)$ drives the spatially distributed chemical reacting system faster and $n(x, t)$ causes it to slow down. Turing's coupled system of reaction-diffusion equations are written as

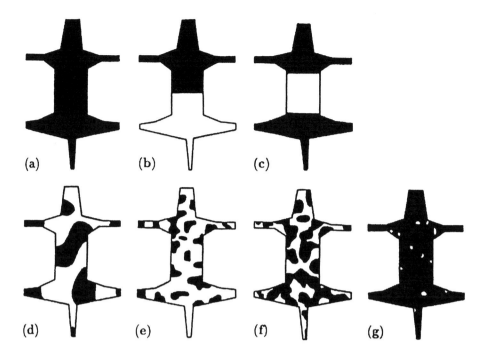

FIGURE 1.1. The patterns on the material surface of the epidermis formed by the reaction-diffusion mechanism. These results are model predictions and the differences (*a*) to (*g*) are associated with scale factor that is a measure of relative animal size. From Murray (1993) with permission.

$$\frac{\partial a}{\partial t} = f(a,n) + D_a\nabla^2 a, \qquad \frac{\partial n}{\partial t} = g(a,n) + D_n\nabla^2 n$$

where D_a and D_n represent diffusion coefficients. As Murray (1993) points out, Turing's idea is a simple but profound one. He said that if, in the absence of diffusion (effectively $D_a = D_n = 0$), $a(x, t)$ and $n(x, t)$ tend to a linearly stable uniform steady state then, under certain conditions, spatially inhomogeneous patterns can evolve by diffusion driven instability if $D_a \neq D_n$. Diffusion is usually considered a stabilizing process, which is why Turing's reaction-diffusion model was such a novel concept (Murray, 1993). The development and application of the Turing system to biological and mathematical problems is summarized by Murray (1993). One particularly graphic use of the Turing system described by Murray is his own application of the reaction-diffusion equations to mammalian coat patterns (Fig. 1.1).

A second mechanism for the morphogenetic event of pattern formation in tissue development is mechanical instability, or buckling. Buckling mechanisms are favored by some over the morphogen approach of Turing for a number of reasons (Harris 1984; Harris et al., 1984). First, mechanical instabilities can create physical structures directly, in one step, in contrast to the two or more steps that would be required with morphogens if positional information first had to be specified by morphogen gradients and then only secondarily implemented in physical form. Second, physical forces act at much longer range (and more quickly) than can diffusing chemicals. There are other difficulties with the morphogen approach, notably the difficulty in the identification of morphogens (Murray, 1993) and the fact that quite different morphogen systems can produce the same pattern (Halken, 1978). Harris (1984), and also to some extent Halken (1978), argue that the mechanical activities of cells can themselves accomplish the morphogenetic functions usually attributed to the diffusion and reactions of chemical morphogens. (For example, Belintsev et al. (1987) employ a model that involves a process of quasi-diffusion of cell polarization.) Systems of cells may also accomplish this by exerting traction forces by which they can propel themselves and rearrange extracellular materials, in particular collagen. Because the compression and alignment created by these cellular forces can, in turn, affect cell behavior, positive feedback cycles of several kinds arise, and these cycles are capable of spontaneously generating regular geometric patterns of cells and matrix. A fascinating axis-symmetric shell buckling pattern formation hypothesis for the patterns associated with leaf and petal morphogenesis has been developed by Green (1999) and his co-workers (Green et al., 1996; Dumais and Steele, 2000). This work is briefly described and illustrated in Cowin (2000).

Two possible models for self-organization of structure in developing biological systems have been considered above, the traditional activator-inhibitor models (chemical or not) based on systems of parabolic equations, and mechanical instability models. There is another set of models in which diffusion-like equations are combined with equations describing mechanical activity. An example of this sort is given by the model of mesenchymal morphogenesis (Murray et al., 1983), where an active stress exerted by cells is taken into account. The mesenchyme is part of the embryonic mesoderm, consisting of loosely packed, unspecialized cells set in a gelatinous ground substance, from which connective tissue, bone, cartilage, and the circulatory and lymphatic systems develop. Another example is the model developed by Belintsev et al. (1987) for epithelial morphogenesis, which describes the spontaneous self-organization in a sheet of actively polarizable (columnarizable) cells. The model of Belintsev et al. involves a process of quasi-diffusion of cell polarization. Stein (1994) presents models which describe the instability in systems with both diffusion and stress dependence of active stress rates. The Stein (1994) models corresponds to models of the Belintsev type when diffusion of a certain agent is assumed and a law for active stresses reduced to a finite dependence on the concentration of this agent and the total stress is accepted.

Melikhov et al. (1983) presented the results of the analysis that showed the possibility of instabilities, including a short-wave instability in the system of two phases with different elastic and growth properties (growth rates depend on stresses in different ways) growing together. The diffusion of one component was involved, but it was later shown that a three-phase system of this sort does not need diffusion for short-wave instability (A.A. Stein, personal communication, February 9, 2000). Stein also points out that, although the mechanochemical models are often interpreted in activator-inhibitor terms, they usually cannot be reduced to pure parabolic systems, as in the Melikhov et al. (1983) example. Some examples of self-organization in systems with no diffusion-like processes involved can be constructed (Stein and Logvenkov, 1993).

The influences of stress in animal development is surveyed by Beloussov (1997) and in plant organs by Hejnowicz and Sievers (1997). The growth rate of a tissue depends upon its mechanical environment both when it is developing morphology and when it is simply increasing the size of existing morphological structures. The developmental growth of most structural tissues is enhanced by the use of those tissues and retarded by their disuse. Models of this type are special cases of some of the above models that deal with the formation of structure. As Hejnowicz and Sievers (1997) point out, plants exhibit stress dependence in their growth rate. Some aspects of this are considered in Stein et al. (1997). Plant cells can create structural patterns by cell deformations without the relative motion of the cells. Differential growth in plants is definitely regulated chemically and also, in some situations, mechanically. Differential growth is the main technique plants use in tissue building and is thought by some to be more probable than the use of mechanical instabilities.

Structural Materials for the Tissue

The production of the structural materials for a tissue is one of the best-understood topics considered in this chapter. Cell and molecular biologists have documented the manufacturing of most of the tissue constituents by cells within or

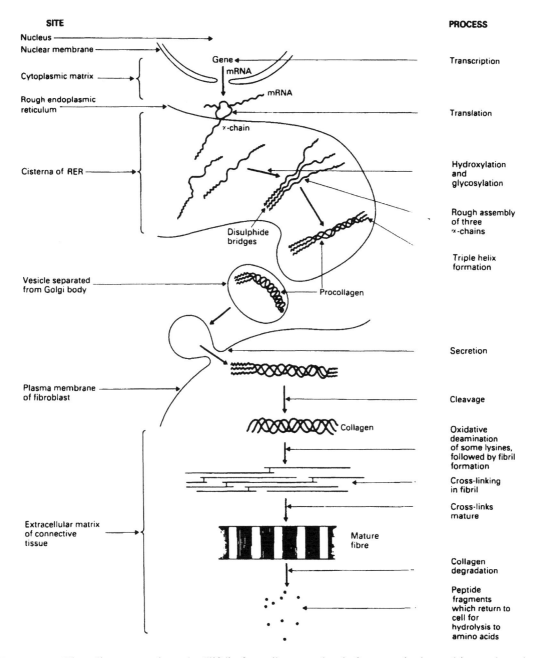

SITE   **PROCESS**

Nucleus

Nuclear membrane

Gene   mRNA   Transcription

Cytoplasmic matrix

mRNA

Rough endoplasmic reticulum   Translation

α-chain

Cisterna of RER   Hydroxylation and glycosylation

Rough assembly of three α-chains

Disulphide bridges

Triple helix formation

Vesicle separated from Golgi body   Procollagen

Secretion

Plasma membrane of fibroblast   Cleavage

Collagen   Oxidative deamination of some lysines, followed by fibril formation

Cross-linking in fibril

Cross-links mature

Extracellular matrix of connective tissue   Mature fibre

Collagen degradation

Peptide fragments which return to cell for hydrolysis to amino acids

FIGURE 1.2. Flow diagram to show the "life" of a collagen molecule from synthesis to ultimate degradation back to amino acids. There is some controversy about whether collagen molecules are actually secreted through the Golgi apparatus. It is shown as occurring here but may not. From Woodhead-Galloway (1980) with permission.

near the tissue. The cells not only manufacture these tissue constituents, but they also maintain the tissue and adapt the tissue structure to changed environments, including mechanical load environments.

The manufacture of proteins begins with the process of "transcription" in the nucleus of a cell. (This paragraph generally follows the structure of the upper two-thirds of Figure 1.2, the portion describing the development of collagen within the cell.) In this process a long mRNA (m for messenger) segment is formed in a pattern determined by the amino acid sequence found in a DNA segment. The mRNA contains the information required to sequence the amino acids in a protein. The mRNA passes out of the nucleus and into the cytoplasm of the cell. Protein synthesis requires transfer RNA (tRNA) and ribosomal RNA (rRNA) in addition to mRNA, and it occurs at a ribosome on the rough endoplasmic reticulum. In the case of collagen, the process of "translation" is the synthesis of polypeptide a chains at the ribosome using the information contained in the mRNA molecules. The flattened sacks of the endoplasmic reticulum are called the cisternae; it is here that the a chains are released from the ribosome and form the triple helix of collagen. The helical structure (the procollagen) travels from the cisterna of the endoplasmic reticulum to the Golgi apparatus where the procollagen is packaged and condensed into a dense membrane-bound granule. The cytoskeletal system moves these granules to the cell membrane, where there is fusion between the external membrane of the cell and the granule, so that the granule contents are released to the extracellular space. At this point, the terminal portions of the procollagen are enzymatically removed and the collagen molecules align themselves to form a collagen fibril. This last step is an extracellular event but occurs at the cell surface of the collagen producing cells (e.g., the fibroblasts or osteoblasts). This is illustrated in Figure 1.2 as is the rest of the life cycle of a collagen molecule.

In the chick tendon, Trelstad and Hayashi (1979) suggest that the cells produce relatively short fibrils of type I collagen within cytoplasmic recesses that form as part of the tendon cell membrane, so allowing the fibrils to add to existing bundles that are themselves held in grooves in the surface of the fibroblasts. The tendon cell surface is infolded at many sites, and such infoldings or recesses contain collagen fibrils. The chick tendon cell discharges its collagenous secretory product, presumably as a procollagen, into these recesses, where the discharged material then apparently adds to the end of a growing fibril. For tendon formation, such an arrangement would provide means for the cell to control the length and orientation of the tendon bundles as they form. It has been suggested that the orientation of the formed microfibrils in tendon assemble as smectic liquid crystals (Figure 3). See also Trelstad (1984).

Not all the structural materials are made by the cells. Some, such as water and minerals, are delivered to the tissue from the vascular system. The mineral apatite combines with the collagen in bone to form a hard, stiff composite that is very different from the collagen itself in mechanical properties. The collagen-apatite system of bone is not the only model of mechanical stiffening and strengthening in nature. It is a scheme widely employed in living systems in which mechanical integrity is required. Different minerals are employed. While vertebrate bone and teeth are mineralized with calcium phosphates, certain invertebrate structure such as shells of clams and coral are mineralized with calcium carbonates. Even though the exoskeleton of insects and crustaceans can be purely organic (chitin, etc.), in many larger species such as crabs and lobsters, it is encrusted with calcium carbonates. Finally, the supporting structure of many plants often contain amorphous silicon oxide (silica) in addition to cellulose (Schmidt-Nielsen, 1983). Some sponges secrete silicon oxide spicules and others make spicules with calcium carbonates.

Assembly of the Tissue Structure

It is of great interest to understand how a tissue develops from nothing, or from an amorphous form having no structure and orientation, to the structure that we know it to be, say, the mature tendon. Although this topic is of great interest we are a long way from a complete knowledge of these processes. In this section the few understood aspects of these developmental processes are described as well as a few of the hypotheses that have been made to try to fill in the gaps. In biology, the word *morphogenesis* is used to refer either to (i) the structural changes observed in tissues as an embryo develops or (ii) the underlying mechanisms that are responsible for the structural changes (Bard, 1990). Both the underlying mechanisms for the structural changes and the changes in structure are of interest here. The unanswered scientific question is the relationship between the information stored in the genome and the resulting tissue structure. It is assumed that tissue structure is the result of specific interactions between the cells and their environment in a sequence leading to a structural phenotype. Once the phenotype has been established, there is no further interaction with the genome except for regeneration and repair.

The unanswered scientific question concerning morphogenesis can be divided into the questions of what initiates a morphogenetic event, what is the morphogenetic process for building the tissue, and what stops the morphogenetic process when it is complete? Concerning initiation of the process, there are known examples in which the initiator is a molecule produced by another tissue. It is possible that the cells have an internal clock that initiates their activity. It is also possible that the trigger is some event in the physical environment of the cell, such as mechanical instability. These three suggestions are not exhaustive of the possibilities, which also include combinations of the mentioned possibilities occurring.

The remainder of this section deals with some of these unanswered questions. It is divided into five subsections that deal with self-assembly at the molecular level, developmental embryology, assembly by cellular activity, supramolecular assembly, and the animal analogue of the plant protein expansin.

Self-Assembly at the Molecular Level

The proteins that cells produce are configured before or after they leave the cell to adopt a shape appropriate for their assimilation into a tissue. It has been established that cells of one type (say fibroblasts) that occur in different tissues are aware of the tissue of which they are a part. The term *molecular self-assembly* is employed to describe the assembly of large, highly ordered structures from substantial numbers of protein molecules (Jarvis, 1992). Molecular self-assembly of proteins is one of the better-understood aspects of tissue development. Self-assembly of the collagen molecule is illustrated in Figure 1.2. Protein subunits fit together like children's agglomeration toys in which similar pieces are fitted together. A classic example of self-assembly is the DNA double helix. When two pre-existing complementary DNA chains

form a double helix, that is self-assembly. The folding of a single protein chain is an example of an assembly process that must occur in one precise kinematic path, of many that are possible, if the result of the folding is to be useful to the resulting tissue. The spontaneous folding of a peptide is molecular self-assembly. When a template is used in the formation of a structure, it is called template assembly. Template assembly occurs when the DNA double helix is formed by polymerization of one chain upon another.

Developmental Embryology

The British embryologist C.H. Waddington used the analogy of a railway switching yard for describing the progressive assignment of embryonic cells in the development of an animal (Harris, 1994). Just as box cars can be shunted into different sidings by throwing switches one way or the other, a series of genetic switches control whether a given cell turns on the set of genes appropriate for being, for example, a skin cell, or those appropriate for being a nerve cell. The development of an individual from a single cell involves many processes that have not yet been modeled. In a pioneering work, Odell et al. (1981) created a mechanical model for the morphogenetic folding of embryonic epithelia based on hypothesized mechanical properties of the cellular cytoskeleton. The interest in this model stems from the fact that embryonic transitions might be explained in terms of local mechanical events rather than in terms of genetic information. Similar models in which the contractile forces generated by cells give rise to patterns in developing biological structures have been studied by Oster et al. (1983), Manoussaki et al. (1996), and, for large deformations, Taber (2000). An alternative to model the Odell et al. (1981) was proposed by Belintsev et al. (1987). In the model of Odell et al. (1981), tangential tensions make the mechanochemical activity spread, whereas, in the Belintsev et al. (1987) model, similar tensions restrict its spreading. The Belintsev et al. (1987) model places more reliance on chemical diffusion effects and less reliance on mechanical effects than the Odell et al. (1981) model and it involves a Turing instability. The interesting model of Odell et al. (1981) is further discussed by Oster et al. (1983), Belintsev et al. (1987), Bard (1990), Harrison (1993), and Taber (1995, 1998, 2000).

The embryology of tendon formation is not well studied. Initially, one finds "patches" of mesenchymal unspecialized cells set in a gelatinous ground substance (Bard, 1990). Similar areas of packed undifferentiated cells appear in the early stages of the development of most connective tissues, for example, bone, cartilage, and the circulatory and lymphatic systems. The signal for such cellular packing or condensation of cells is unknown. The forming tissue converts itself into the one-dimensional structure, a tendon precursor, but is forming simultaneously with embryonic bone and muscle. It is speculated that fibroblasts, aligned by tractional forces, lay down bundles of aligned collagen fibrils that associate to form fine tendons.

Assembly by Cellular Activity

It is convenient to think of steps in the morphogenetic process for building a tissue as being caused by tissue-created forces that are self-equilibrated, that is to say that the tissue itself is both the activator or creator of the force as well as the reactor or recipient of the force. These morphogenetic forces may involve both the cells and the extracellular matrix and can lead to the rearranging, moving, enlarging, folding, condensing, and shaping of tissue. The cell machinery used in generating morphogenetic forces includes cell division, cell sensitivity to stretch, voltage and contact with other chemicals, the cell adhesion molecules, and the cytoskeletal ability for microfilament assembly. Cells contain actin and myosin, variant forms of the same proteins that are the key proteins for the development of force in muscle. These proteins are used in the development of microfilaments, which can be used by the cell to resist deformation, change cell shape, influence cell movement or to develop an external force. A number of actions that cells can accomplish are listed below. Is it possible that these cell actions can accomplish tissue assembly?

Cell Adhesion

Cells adhere to surfaces by forces that are three to four orders of magnitude larger than their own weight (Cowin, 1998). The cells can control this adhesion, and some cells can detach from a surface at their own signal. Cells have been observed moving toward more adhesive surfaces.

Cell Muscle

Harris et al. (1980) reported a graphic demonstration of the tension field created by the fibroblasts. Fibroblasts were placed on a thin sheet of silicone rubber to which they adhered tightly. Although Harris et al. (1980) illustrated that after 48 hours the cells had contracted and caused the silicone rubber to wrinkle, the contraction and wrinkling process often began within a few minutes of cell placement on the thin sheet of silicone rubber (A.K. Harris, personal communication, November 22, 1999). In another experiment, Harris et al. (1981) showed how cell systems can create a line of tension, lines extending many times the length of one cell, to orient collagen fibers. Two groups of chicken heart fibroblasts were planted about 1–2 cm apart on a thick gel of reprecipitated collagen. When the growing cells spread and clustered, they created a field of tension between the two explants. The resulting tensile stress stretched the collagen fibers into alignment.

Cell Locomotion

A cell is capable of self-propelled movement using different systems to propel itself at speeds measured in microns per hour. It can use its adhesion capability and its ability to form (polymerize) and dissolve (depolymerize) microfilaments as a means of locomotion. It has been suggested that cells might organize the energy from Brownian motion as motors for different activities including propulsion (Peskin et al., 1993).

Cell Chemotaxis

Cells are attracted to certain chemicals and repulsed by others. They can follow gradients in concentration of a chemical.

Cell Contact Guidance

Cells are known to guide by man-made straight edges and by long fibers; this is called contact guidance.

(a) Smectic   (b) Nematic   (c) Cholesteric

FIGURE 1.3. Structure of liquid crystals made of rodlike structures such as molecules, microfibrils, or fibrils. Smectic liquid crystals are characterized by rods that lie parallel to one another in layers of equal thickness; they diffuse freely within the layers but not between the layers. Nematic liquid crystals are also characterized by rods that lie parallel to one another, but the layered structure of the smectic does not exist. The ordering is purely orientational and the distribution of the rods in the direction of the rod axis is random. Cholesteric liquid crystals lie parallel to one another in each plane, but each plane is rotated by a constant angle from the next plane of crystals. From Neville (1993) with permission.

Cell Sensors

Cells have many sensors and different cells have different sensors. Cells are sensitive to stretch, voltage and contact with chemicals. For example, hair cells in the ear can pick up very low energy mechanical signals.

It appears reasonable that these cell capabilities can be modeled and simulations with these modeled capabilities might be able to demonstrate the possibility of cells accomplishing certain steps in tissue assembly.

Supramolecular Assembly

The question of supramolecular assembly is not on the same firm ground as molecular self-assembly; however, the ideas involved are ones of deep engineering interest and have a great deal of future potential. The idea is equivalent to self-assembly at a higher structural level. The concept is a useful working hypothesis for individual stages in the construction of supramolecular structures. For connective tissue the most interesting analogy for supramolecular assembly is the liquid crystal analog. The term *liquid crystal* should initially strike one as an oxymoron; a crystal is a well-ordered solid characterized by regular stacking in a three-dimensional lattice with strong attractive forces holding it together, and a liquid has little obvious structure as it takes the shape of its container and changes shape with the changing shape of containers it occupies. A liquid crystal is a mesomorphic state (intermediate form) between a crystal and a liquid. Some organic materials do not show a single transition from liquid to crystal, but have a series of intermediate transitions involving new phases including the liquid crystal phase. In their defined range of temperature and concentration, liquid crystals are rodlike molecules oriented in one, two, or three dimensions. Their orientation is sometimes controllable by electromagnetic fields, and this has led to one liquid crystal (methoxybenzylidenebutylaniline) being used extensively in liquid crystal displays (LCDs).

Liquid crystals are classified as smectic, nematic, or cholesteric (Fig. 1.3). Rods that lie parallel to one another in layers of equal thickness characterize smectic liquid crystals. Smectic liquid crystals have the highest degree of order. Nematic liquid crystals are also characterized by rods that lie parallel to one another, but the layered structure of the smectic does not exist and the nematic have the lowest degree of order. The ordering is purely orientational and the distribution of the rods in the direction of the rod axis is random. Cholesteric liquid crystals lie parallel to one another in each plane, but each plane is rotated by a constant angle from the next plane of crystals. The cholesteric is a chiral form of the nematic phase, chiral describing a structural characteristic of a molecule that prevents it from being superposed upon its mirror image.

When liquid crystals were discovered in 1888 by F. Reinitzer (Mackay, 1999), they quickly became strong candidates for the mechanism by which nature forms living structures from homogeneous multi chemical soups (Haeckel, 1917). This is because liquid crystals display a "striking form of self-organization in which directional order appears spontaneously in a homogeneous liquid, not incrementally, as in the growth of the familiar crystals layer by layer at the surface, but simultaneously throughout a substantial volume" (Mackay, 1999). The analogy with liquid crystal structure is developed by Bouligand (1972), Wainwright et al. (1976), Bouligand et al. (1985), and Neville (1993).

The terms *biological analogues of liquid crystals* and *pseudomorphoses* are used to describe those biological systems that appear to obey liquid crystalline geometries but have lost their liquid character. Examples of birefringent biological materials that can be genuine liquids are biological membranes and certain gland secretions. Nonfluid analogues are cell walls and skeletal structures of plants and animals. The fibrillar material is aligned in a way geometrically similar to the arrangement of molecules in cholesteric liquid crystals; however, the length of the polymer, its crystallization, and the presence of cross-links destroy the liquid character that must exist following the initial secretion of the organic components by the cells. The state of a forming or reforming tissue is called here the *morphosis state* and it is thought to have the characteristics of both a solid and a fluid state, enough fluidity for a structure to self-assemble and enough solidity to maintain the load carrying capacity of the structure that is reforming. A morphosis state of a tissue is thought to exist during morphogenesis and during the process of tissue remodeling. In the case of bone, a morphosis state probably exists at some time during the formation of a secondary osteon and during the process of tooth movement.

Two major types of twists are found in liquid crystals and their biological analogues and are defined by the disposition of the fibrillar elements either in parallel planes (planar twist) or coaxial cylinders (cylindrical twist) (Fig. 1.4).The coaxial cylinders or cylindrical twist are also described as helicoidal, a term that will be em-

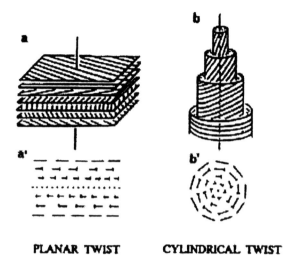

PLANAR TWIST   CYLINDRICAL TWIST

FIGURE 1.4. (*a*) In a planar twist, equidistant straight lines are drawn on horizontal planes, and the direction of the lines rotates regularly from plane to plane. (*a*9) In the conventional notation for a cholesteric geometry applied to a planar twist, lines represent molecules longitudinal to the drawing plane and dots represent molecules perpendicular to it; molecules in oblique position are represented by nails whose points are directed toward the observer. (*b*) In a cylindrical twist, equidistant helices are drawn on a series of coaxial cylinders, and the angle of the helices rotates regularly from one cylinder to the next. (*b*9) Conventional representation of a cholesteric geometry applied to a cylindrical twist. From Giraud-Guille (1996) with permission.

ployed here. Collagen in the secondary osteons of bone tissue is observed to be in a helicoidal pattern. The developments of typical liquid crystal patterns have been associated with flowing liquid crystals. Giraud-Guille (1992, 1996) described the assembly of collagen molecules in typical liquid crystalline phases in highly concentrated solutions after sonication for several weeks. In this work, it was shown that intact 300-nm-long collagen molecules form cholesteric liquid crystalline domains. The sonication does not alter the triple-helical structure of the collagen fragments. The development of typical liquid crystal patterns in situations not involving flow has been suggested by Rey (1996).

The origin of the cholesteric twist is thought to be a geometric effect. Under sonication or Brownian motion, in the morphosis state, the spontaneous alignment of elongated molecules at high concentration originates mainly from steric constraints involving the spatial configuration of the molecules. The rod-shaped molecules do not interpenetrate, and the excluded volume is reduced when the molecules lie parallel; consider a stack of forks. At high concentrations geometric factors associated with the chirality of the molecules govern the change in orientation between layers. A constant characteristic of biological polymers forming cholesteric liquid crystals is that they consist of layered helicoids. The contact between the layered helicoidal domains of assemblies of asymmetric molecules determines the orientation of one layer to its neighbor. To minimize steric hindrance, helical polymers have first aligned in layers, thus gaining maximum space. Each layer creates parallel and equidistant grooves, oblique to the helical axis, where a second layer of molecules can be deposited. This oblique packing generates a twisted plywood system over successive layers. It is generally thought that the coalignment of cytoskeletal elements that occurs in extracellular fibrillar material is accomplished by mechanical forces. However, the nature of these forces has not been determined.

The Animal Analogue of the Plant Protein Expansin

Expansin is a land plant protein that loosens a plant wall structure so that the plant cell wall may expand and provide space for growth (Cosgrove, 2000). In plants, the cell is a structural element and the cell wall is the major component of the cell. The cell wall determines cell shape, glues cells together, provides mechanical strength and stiffness, and acts as a barrier against pathogens. Secreted by growing cells, the primary wall is a polymeric network of crystalline cellulose microfibrils embedded in a hydrophilic matrix. The polysaccharides of the growing plant cell wall are mostly separate long-chained polymers that form a cohesive network through noncovalent lateral associations and physical entanglements. Many animal tissues have a similar topological structure of lateral associations and physical entanglements of fibers, although the fibers are of different proteins. The

interest here in expansin is that there is likely an analogue in animal tissues that loosens the network structure so that more fibers may be effectively added and the tissue may develop by expanding and strengthening.

Plant cells originate mainly in small pockets of dividing cells, called meristems, that are located at the growing points of shoots, roots, and other organs. Meristematic cells are small (approximately 5 mm) and densely packed with cytoplasm. When cells are displaced from the meristem, they undergo a prolonged phase of enlargement and differentiation during which cell volume greatly increases. For example, a water-conducting cell of the xylem may be a million times larger in volume than its meristematic initial. Such cell enlargement is achieved economically by filling the cell with a large central vacuole containing water and solutes. The physical control of this process resides in the ability of the cell wall to undergo turgor-driven expansion. The growing cell wall possesses a remarkable combination of strength and flexibility, enabling it to withstand the large mechanical forces that arise from the cell turgor pressure, while at the same time permitting a controlled creep that distends the wall and creates space for the enlarging protoplast. The stiff cellulose microfibrils themselves are effectively inextensible. Wall expansion occurs by slippage or rearrangement of the matrix polymers that coat the microfibrils and hold them in place. Expansin's unique physical effects on plant cell walls include rapid induction of wall extension and stimulation of stress relaxation. McQueen-Mason and Cosgrove (1994) proposed a mechanism in which expansins weaken the noncovalent binding between wall polysaccharides, thereby allowing turgor-driven polymer creep. In this scheme, expansin makes use of the mechanical strain energy in the wall to catalyze an inchworm-like movement of the wall polymers. Expansin movement may be confined to lateral diffusion along the surface of the cellulose microfibril, as has been observed for other polysaccharide-binding proteins. Such contained diffusion would enable expansin to search the microfibril surface, locally loosening its attachment to the matrix, and allowing chain movement and stress relaxation.

Expansins offer potential applications for bioengineering of cell walls. The animal analogue of expansin would be of use in tissue engineering.

Tissue Building by Nature and by Man

The purpose of this section is to relate nature's process of building a tissue to the way of creating tissues presently evolving in laboratories. A very short summary of nature's way follows. Day one of the creation of tissue in the body occurs when egg and sperm meet. On days two and three, the cell divides and then divides again and the cells begin cell-to-cell signaling. On day four, the cells begin to migrate to different parts of the embryo. Genes direct this migration and indicate to the cells what they are to do in their new specific anatomical location. Being in their new location also changes the cells; their cell-to-cell signaling pattern is different because their neighboring cells are different and the specialization of their cell type begins to increase. On day five, the blastula, a spherical ensemble of cells, forms. The interior of the blastula contains a mass of cells that are the source of stem cells used in some laboratory tissue development. On the sixth to the ninth day, the blastula attaches to the uterine wall. The highly vascularized placenta will then begin to form; the placenta can exchange solutes with the maternal circulation. By day 14, the inner cell mass of the blastula gives rise to three cell layers that will form all the body's tissues and organs: the ectoderm (hair, neurons, skin), the mesoderm (heart muscle, blood, bone), and the endoderm (intestinal lining, bladder, pancreas). The specific tissue construction processes then begin with input from the genes and the cell environment.

Creating a tissue in a laboratory cannot be done in exactly the same way; in fact, it is presently done in a very different way. If stem cells are employed in the tissue building, they may be taken from the blastula; if more specialized cells are employed they will be obtained at a later time in the natural tissue development process. The cells are then placed in a nutrient-rich culture. The cell culture is subjected to growth factors and mechanical environments similar to those the mature tissue would experience. There are of number of chapters in this text that describe this process in detail for specific tissues.

The Need for Mechanics in the Study of Tissue Development

In 1994, the biologist Albert Harris wrote

"Systems of interacting forces and stimuli don't have to be very complicated before the unaided human intuition can no longer predict accurately what the net result should be. At this point computer simulations, or other mathematical models, become necessary. Without the aid of mechanicians, and others skilled in simulation and modeling, developmental biology will remain a prisoner of our inadequate and conflicting physical intuitions and metaphors."

In 1990, the geneticist Jonathan Bard wrote

"I . . . assert that the process of tissue formation is in many ways the cellular equivalent of molecular self-assembly and that the appropriate language in which to analyze morphogenesis is that of the differential equation. . . ." (Bard, 1990).

In 1993, the physical chemist Lionel G. Harrison (1993) wrote

"In developmental biology I found something different, and immensely exciting: a field with a Great Unknown, and no firmly established conceptual basis. To pursue it is like trying to account for the rainbow in the fourteenth century, to do celestial mechanics before Newton, or to pursue quantum theory in the 1890s. There are many ideas around. Some of them are elaborately developed, and some will eventually be recognized as the correct concepts, but none has reached that status yet" (Harrison, 1993).

The book of Beloussov (1998) and the review article of Taber (1995) are suggested as places to learn about these problems. Beloussov suggests the foundation for the analysis of a number of problems in developmental biology. The major interest of Taber is in cardiac development (see Taber, 1998).

A subject often advances by a critical experiment that distinguishes between two or more alternative hypotheses for a specific phenomenon. It is thought that there are not yet enough hypotheses for the influence of mechanical phenomena in developmental biology, consistent or inconsistent, to drive critical experimentation. There is a need for innovative mechanics in the construction of hypotheses for tissue development phenomena.

Acknowledgments

This work was supported by National Institutes of Health grant number 1RO1AR44211-01 and by grant number 668383 from the PSC-CUNY Research Award Program of the City University of New York.

References

Bard, J. 1990. *Morphogenesis.* Developmental and Cell Biology Series, Vol. 23, Cambridge University Press, Cambridge.

Belintsev, B.N., Beloussov, L.V., Zaraisky, A.G. 1987. Model of pattern formation in epithelial morphogenesis, *J. Theor. Biol.* 129:369–394.

Beloussov, L.V. 1997. Mechanical stresses in animal development: Patterns and morphogenetical role. In: *Dynamics of Cell and Tissue Motion.* Alt, W., Deutsch, A., Dunn, G., eds. Birkhauser, Boston, 221–228.

Beloussov, L.V. 1998. *The Dynamic Architecture of the Development of Organisms.* Kluwer Academic Publishers, Dordrecht.

Bouligand, Y. 1972. Twisted fibrous arrangements in biological materials and cholesteric mesophases, *Tissue Cell* 4:189–217.

Bouligand, Y., Denefle, J.-P., Lechaire, J.-P. Maillard, M. 1985. Twisted architectures in cell-free assembled collagen gels: Study of collagen substrates used for cultures, *Biol. Cell* 54:143–162.

Coen, E. 1999. *The Art of Genes: How Organisms Make Themselves.* Oxford University Press, Oxford.

Cosgrove, D.J. 2000. Loosening of plant cell walls by expansins. *Nature* 407:321–362.

Cowin, S.C. 1998. On mechanosensation in bone under microgravity. *Bone* 22:119S–125S.

Cowin, S.C. 2000. How is a tissue built? J. Biomech. Eng. 122:1–17.

Dumais, J., Steele, C.R. 2000. New evidence for the role of mechanical forces in the shoot apical meristem. *J. Plant Growth Regul.* 19:7–18.

Giraud-Guille, M.M. 1992. Liquid crystallinity in condensed type I collagen solutions: A clue to the packing of collagen in extracellular matrices, *J. Mol. Biol.* 224:861–873.

Giraud-Guille, M.M. 1996. Twisted liquid crystalline supramolecular arrangements in morphogenesis. *Int. Rev. Cytology* 166:59–101.

Green, P.B. 1999. Expression of pattern in plants: Combining molecular and calculus-based biophysical paradigms. *Am. J. Botany* 86:1059–1076.

Green, P.B., Steele, C.R., Rennich, S.C. 1996. Phyllotactic patterns: A biophysical mechanism for their origin. *Ann. Botany* 77:515–527.

Haeckel, E. 1917. *Kristallseelen—Studien fiber das anorganische Leben.* Alfred Kroner Verlag, Leibzig.

Halken, H. 1978. *Synergetics.* Springer-Verlag, New York.

Harris, A.K. 1984. Cell traction and the generation of anatomical structure. *Lecture Notes in Biomathematics* 55:104–22.

Harris, A.K. 1994. Multicellular mechanics in the creation of anatomical structures. In: *Biomechanics of Active Movement and Division of Cells.* N. Akkas, ed. Springer-Verlag, New York pp. 87–129.

Harris, A.K., Wild, P., Stopak, D. 1980. Silicone rubber substrata: A new wrinkle in the study of cell locomotion. *Science* 208:177–179.

Harris, A.K., Stopak, D., Wild, P. 1981. Fibroblast traction as a mechanism for collagen morphogenesis. *Nature* 290:249–251.

Harris, A.K., Stopak, D., and Warner, P. 1984. Generation of spatially periodic patterns by mechanical instability: a mechanical alternative to the Turing model. *J. Embryol. Exp. Morph.* 80:1–20.

Harrison, L.G. 1993. *Kinetic Theory of Living Patterns.* Developmental and Cell Biology Series, Vol. 28, Cambridge University Press, Cambridge.

Hejnowicz, Z., Sievers, A. 1997. Tissue stresses in plant organs: Their origin and importance for movements. In: *Dynamics of Cell and Tissue Motion.* Alt, W., Deutsch, A., Dunn, G., eds. Birkhauser, Boston, pp. 235–242.

Held, L.I. Jr. 1992. *Models for Embryonic Periodicity.* Monographs in Developmental Biology, Vol. 24. Karger.

Jarvis, M.C. 1992. Self-assembly of plant cell walls. *Plant, Cell and Environment* 15:1–5.

Mackay, A.L. 1999. Crystal souls (a translation of Haeckel (1917)), *FORMA* 14:1–146.

Manoussaki, D., Lubkin, S.R., Vernon, R.B., Murray, J.D. 1996. A mechanical model for the formation of vascular networks *in vitro. Acta Biotheor* 44:271–282.

McQueen-Mason, S., Cosgrove, D.J. 1994. Disruption of hydrogen bonding between wall polymers by proteins that induce plant wall extension. *Proc. Natl. Acad. Sci. USA* 91:6574–6578.

Melikhov, A.V., Regirer, S.A., Stein or Shtein, A.A. 1983. Mechanical stresses as a factor in morphogenesis. *Sov. Phys. Dokl.* 28:636–638.

Murray, J.D. 1993. *Mathematical Biology.* Springer-Verlag, New York.

Murray, J.D., Oster, G.F., Harris, A.K. 1983. A mechanical model for mesenchymal morphogenesis. *J. Math. Biol.* 17:125–129.

Neville, A.C. 1993. *Biology of Fibrous Composites.* Cambridge University Press, Cambridge.

Odell, G.M., Oster, G.F., Alberch, P., Burnside, B. 1981. The mechanical basis of morphogenesis. I. Epithelial folding and invagination. *Dev. Biol.* 85:446–462.

Oster, G.F., Murray, J.D., Harris, A.K. 1983. Mechanical aspects of mesenchymal morphogenesis. *J. Embryol. Exp. Morphol.* 78:83–125.

Peskin, C.S., Odell, G.M., Oster, G.F. 1993. Cellular motions and thermal fluctuations: The Brownian ratchet. *Biophysical J.* 65:316–324.

Rey, A.D. 1996. Phenomenological theory of textured mesophase polymers in weak flows. *Macromol. Theory Simul.* 5:863–876.

Ridley, M. 1999. Genome. Harper Collins, New York.

Schmidt-Nielsen, K. 1983. *Animal Physiology: Adaptation and Environment,* 3rd Ed., Cambridge University Press, Cambridge pp. 162–163.

Stein, A.A. 1994. Self-organization in biological systems as a result of interaction between active and passive mechanical stresses: Mathematical model. In: *Biomechanics of Active Movement and Division of Cells.* N. Akkas, ed. NATO ASI Series, Vol. H 84, Springer-Verlag, New York, pp. 459–464.

Stein or Shtein, A.A., Logvenkov, S.A. 1993. Spatial self-organization of a layer of biological material growing on a substrate. *Phys. Dokl.* 38:75–78.

Stein, A.A., Rutz, M., Zieschang H. 1997. Mechanical forces and signal transduction in growth and bending of plant roots. In: *Dynamics of Cell and Tissue Motion.* Alt W., Deutsch A., Dunn G., eds. Birkhauser, Boston pp. 255–265.

Taber, L. 1995. Biomechanics of growth, remodeling and morphogenesis. *Appl. Mech. Rev.* 48:487–545.

Taber, L. 1998. Mechanical aspects of cardiac development. *Prog. Biophys. Mol. Biol.* 69:237–255.

Taber, L. 2000. Pattern formation in a non-linear membrane model for epithelial morphogenesis. *Acta Biotheor.* 48:47–63.

Trelstad, R.L. 1984. *The Role of Extracellular Matrix in Development.* Alan R. Liss, New York.

Trelstad, R.L., Hayashi, K. 1979. Tendon fibrillogenesis: Intracellular collagen subassemblies and cell surface charges associated with fibril growth. *Dev. Biol.* 7:228–242.

Trelstad, R.L., Silver, F.H. 1981. Matrix assembly. In: *Cell Biology of the Extracellular Matrix.* E.D. Hay, ed. Plenum Press, New York, pp. 179–216.

Turing, A.M. 1952. The chemical basis of morphogenesis. *Philos. Trans. R. Soc. Lond. B. Biol. Sci.* 237:37–72.

Wainwright, S.A., Biggs, W.D., Currey, J.D., Gosline, J.M. 1976. *Mechanical Design in Organisms.* Edward Arnold.

Woodhead-Galloway, J. 1980. *Collagen: The Anatomy of a Protein.* Studies in Biology No. 117. Arnold, London.

2

Ligament Healing:
Present Status and the Future of
Functional Tissue Engineering

Savio L-Y. Woo, Steven D. Abramowitch, John C. Loh, Volker Musahl,
and James H-C. Wang

Introduction

Ligaments are bands of dense connective tissue
that mediate normal joint movement and share
in the transmission of forces with other articular
tissues in order to provide joint stability. Rupture of ligaments upsets this balance between
mobility and stability, resulting in abnormal
joint kinematics and damage to other tissues
around the joint that may lead to morbidity and
pain. Ligamentous injuries are common, particularly in sports and sports-related activities. A
recent European study revealed that 1 of 4 ski-
ing injuries occurred in ligaments of the knee [1].
In the United States, 62% of acute injuries in
children and adolescents between the ages of 5
and 17 years involved ligament sprains, thus
making it the most common injury during sports
and recreation [2, 3].

The biochemical constituents and biomechanical properties of ligaments make them well designed to function in the unique environment of
a diarthodial joint. For example, the knee is a
versatile structure that serves in primary weight
bearing and mobility. It is composed of five
major components: the femorotibial joint and
its accompanying cruciate ligaments, the patellofemoral joint, capsuloligamentous restraints,
and menisci. The anterior cruciate ligament
(ACL) and posterior cruciate ligament (PCL),
which are located within the joint space surrounded by synovial fluid, restrain the anterior
and posterior tibial displacement with respect to
the femur, and also perform functions such as
maintaining the rotational stability of the knee.
The capsuloligamentous structures include the
medial collateral ligament (MCL), lateral collateral ligament (LCL), posteromedial/lateral
complex, and joint capsule, and also serve collectively to stabilize the knee. Finally, the medial
and lateral menisci, which are intraarticular
semilunar cartilaginous structures, bear a large
portion of the total load transmitted across
the knee and assist in maintaining knee stability
[4, 5].

Within the knee, the potential for healing
varies from ligament to ligament. Injuries to the
midsubstance of the ACL fail to heal, frequently
requiring replacement with an autograft. Injuries
to the MCL, however, generally heal sufficiently
well such that nonsurgical management has become the treatment of choice [6–8]. The ability
of a ligament to heal is influenced by multiple
factors including geometric or anatomic complexity, environment, nutrition, and function as
well as other intrinsic and cellular factors. In the
MCL, the process of healing is well understood
and can be divided into four overlapping phases:
hemorrhagic, inflammatory, proliferating, and
remodeling/maturation. Although the structural
properties of the femur-MCL-tibia complex
(FMTC) are restored within weeks, the entire

healing process of the ligament substance is not complete, even after one year. The mechanical properties of the healed MCL (i.e., stress-strain curve) remain inferior to those for the normal MCL [7–9]. Biochemically, the healing ligament exhibits elevated type III and V collagen and elevated proteoglycans [10]. Histologically, an elevated number of vessels, fat cells, voids, and increased water content are observed [11]. As a result, many questions remain on how to improve these deficiencies.

Functional tissue engineering (FTE) offers an attractive approach to enhance ligament healing and provides a possibility of producing tissue that is biomechanically, biochemically, and histomorphologically similar to normal tissue. If successful, these techniques may be applied to other ligaments that do not heal [12]. There are several ways to functionally tissue engineer a healing ligament. The basic concept rests on the manipulation of cellular and biochemical mediators to affect protein synthesis and to improve tissue formation and remodeling. Ultimately, the process is expected to lead to a restoration of mechanical properties. There are several available approaches on the horizon, that is, the use of growth factors, gene transfer technology to deliver genetic material, stem cell therapy, and the use of tissue scaffolding as well as external mechanical factors. Each approach, or combinations thereof, offers an opportunity to enhance the healing process.

In this chapter, the properties of normal ligaments, including their anatomical, biological, biochemical and mechanical properties, as well as the changes that occur following injury, will be described. Subsequently, sample FTE methods and preliminary findings will be presented. The MCL will be used as a model for ligamentous healing due to its uniform cross-sectional area, large aspect ratio, and propensity for healing. Although the field of FTE is still in its infancy, it has captured the imagination of many investigators and offers exciting frontiers that promise new and improved solutions to the challenging problems in tissue healing. Thus, this chapter will describe the relationship between biochemistry, histological structure, and mechanical behavior of normal ligaments, and provide insight as to how these properties might be

applied to FTE. Finally, future directions in this area of research will be suggested.

Properties of Normal Ligaments

Anatomy and Morphology

Anatomically, ligaments connect bones and are designed to maintain stability during normal joint motion, while tendons connect muscles to bones and transmit muscle forces to generate movement. Ligaments contain fibroblasts that are interspersed in parallel bundles of extracellular matrix composed mainly of collagen. Collagen is organized into a cascade of levels, with procollagen assembled into microfibrils, which in turn aggregate to form subfibrils. Fibrils are composed of multiple subfibrils and are the elemental constituent of collagen fibers [13]. Finally, groups of collagen fibers form fascicular units. Up to 20 of these fascicular units are bound together to form fasciculi that are up to several millimeters in diameter. Hundreds of fasciculi agglomerate to form a ligament [13] (Fig. 2.1).

Ligament insertions to bone are classified as direct and indirect. For direct insertions (e.g., the femoral insertion of MCL, the tibial insertion of the ACL), fibers attach directly into the bone and the transition of ligament to bone occurs in four zones: ligament, fibrocartilage, mineralized fibrocartilage, and bone [14]. For an indirect insertion (e.g., the tibial insertion of MCL), superficial fibers are attached to periosteum while the deeper fibers are directly attached to the bone at acute angles (so called Sharpey's fibers) [14]. The tibial insertion of the MCL is designed to cross the epiphyseal plate so that it can be lengthened in synchrony as the bone grows between the plate and the joint. The human MCL is approximately 80 mm in length and runs from the medial femoral epicondyle distally and anteriorly to the postero-medial margin of the metaphysis of the tibia. The LCL originates from the lateral femoral epicondyle and passes postero-distally to the top of the fibular head.

Although morphologically intraarticular, the cruciate ligaments are surrounded by a synovial layer, making them like extraarticular structures. For functional reconstruction of ligaments, it is

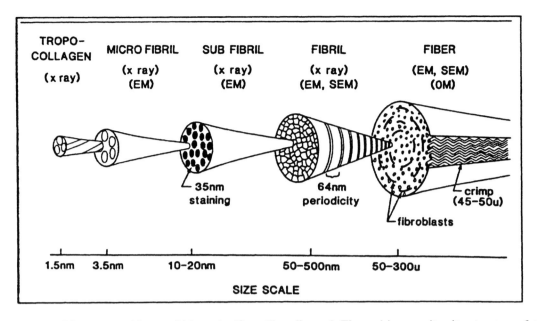

FIGURE 2.1. Ligament architectural hierarchy (from Kastelic et al. The multicomposite ultrastructure of tendons. Connect Tissue Res 1978; 6:11–23)

important to know the exact insertion sites of the ligaments in order to accurately replicate the biomechanical properties. The postero-convex femoral attachment of the ACL lies lateral to the vertical surface of the intercondylar fossa and its broad tibial attachment is located between the intercondylar spines of the tibia. The ACL consists of two bundles, an antero-medial (AM) bundle and a postero-lateral (PL) bundle. The AM bundle is thought to be important as a restraint to antero-posterior translation of the knee, while the PL bundle is thought to be an important restraint to rotational moments about the knee [15]. The PCL is also composed of two bundles, an antero-lateral (AL) and a postero-medial (PM) bundle. Additionally, there are ligaments of minor importance anterior and posterior to the PCL: the anterior meniscofemoral ligament (MFL; i.e., ligamentum Humphrey) and the posterior meniscofemoral ligament (i.e., ligamentum Wrisberg) [16].

A uniform microvascular system originating from the insertion sites of the ligament provides for nutrition of the cell population as well as maintenance of the ligament in terms of matrix synthesis and repair [17]. For the MCL, the blood supply originates from the medial superior genicular artery, while for the LCL, the blood supply originates from the lateral superior genicular artery [18, 19]. For the ACL and PCL, the blood supply is provided by the middle genicular artery.

Biochemical Constituents

The biochemical constituents of ligaments consist of fibroblasts surrounded by aggregates of collagen, elastin, proteoglycans, glycolipids, and water [20]. Between 65% and 70% of a ligament's total weight is composed of water. Type I collagen is the major constituent (70–80% dry weight) and is primarily responsible for a ligament's tensile strength. Type III collagen (8% dry weight) is another component that is elevated during ligament healing, and type V collagen (12% dry weight) has been associated with the regulation of collagen fiber diameter [21, 22]. Type XII collagen (<1% dry weight) provides for lubrication between collagen fibers [10].

Variations in the concentrations or orientations of these basic constituents lead to an incredibly diverse array of mechanical behaviors.

Each ligament's composition, therefore, is directly correlated with its function. Thus, detailed knowledge about a ligament's microstructure and microstructural properties is integral in understanding a ligament's response to load.

Tensile Properties of Normal Ligaments

The kinetic response of a joint to internal and external loads is governed by bone geometry as well as the anatomical location, morphology, and chemical composition of ligaments and other connective tissues contained within or around the joint. To understand the contribution of an individual ligament to joint kinetics, the transfer of load through a ligament's insertion, midsubstance, to opposing insertion must be considered. Physiologically, these bone-ligament complexes have been designed to transfer load uniaxially along the longitudinal direction of the ligament. Thus, tensile testing of a bone-ligament-bone complex (e.g., femur-MCL-tibia complex, or FMTC) is performed to determine the structural properties. The resulting load-elongation curve reveals a nonlinear, concave, upward behavior. Parameters obtained from this curve include stiffness, ultimate load, ultimate elongation, and energy absorbed at failure. From the same test, information about the mechanical properties (quality) of the ligament substance can also be obtained. This is done by normalizing the force by the cross-sectional area of the ligament and the change in elongation by the initial length of a defined region of the ligament midsubstance, defined as stress and strain, respectively [23, 24]. From the stress-strain curve, the elastic modulus, ultimate tensile strength, ultimate strain, and strain energy density of the ligament substance can be determined. Readers are encouraged to study chapter 24 of *The Orthopaedic Basic Science Book*, published by the American Academy of Orthopaedic Surgeons (AAOS), for more detailed information on the terminology [20].

Mechanical Properties of Knee Ligaments

The geometry, aspect ratio, and alignment of the MCL make it suitable for uniaxial tensile testing.

The orientation of the MCL between the tibia and femur allows the collagen fibers within the long axis of the ligament to be aligned along the direction of loading. A recently developed finite element model of the MCL confirmed that loading in this manner (i.e., loading along the longitudinal axis of the MCL) yields a relatively homogenous stress and strain distribution in the the midsubstance near the joint line [25]. Using surface markers to track strain in this region, the elastic modulus of an intact MCL has been determined for a number of animal models (e.g., rabbit, canine, goat) with mean values reported to range between 526 and 726 Mpa [8, 26, 27]. Similarly, dumbell-shaped specimens that are cut from ligaments demonstrate similar mechanical behaviors [28].

Recent studies have used these dumbell-shaped specimens to analyze the anisotropic behavior of the human MCL [28]. The following hyperelastic strain energy equation describes this behavior as transversely isotropic,

$$W(I_1 I_2, \lambda) = F_1(I_1, I_2) + F_2(\lambda) + F_3(I_1, I_2, \lambda), \quad (1)$$

where I_1 I_2 are invariants of the right Cauchy stretch tensor and λ is the stretch along the collagen fiber direction [28]. For a uniaxial tensile test, F_1 is described by a two-coefficient Mooney-Rivlin material model representing the behavior of the ground substance,

$$F_1 = \frac{1}{2}[C_1(I_1 - 3) + C_2(I_2 - 3)], \quad (2)$$

where C_1 and C_2 are constants. F_2 represents the strain energy of the collagen fibers, and is described by separate exponential and linear functions of λ for the toe and linear regions of the stress versus stretch response, respectively. F_3 represents an interaction term accounting for shear coupling, and is assumed to be zero. The Cauchy stress, T, can then be written as,

$$T = 2\{(W_1 + I_1 W_2)B - W_2 B^2\} + \lambda W_\lambda \, a \otimes a + \rho 1, \quad (3)$$

where B is the left deformation tensor and W_1, W_2, and W_λ are the partial derivatives of strain energy with respect to I_1, I_2, and λ, respectively. The unit vector field, a, represents the fiber direction in the deformed state, and ρ the hydrostatic pressure required to enforce incompressibility.

The properties of the tissue, orientation of collagen fibers, and the dimensions of the specimen, however, must be considered to ensure that the dumbell-shaped specimens that are cut from ligaments represent the intact specimen. Tests of dumbell-shaped specimens cut from the human MCL demonstrate that longitudinal specimens display an elastic modulus of 332.2 ± 58.3 MPa and a tensile strength of 38.6 ± 4.8 Mpa. In the transverse direction, values were an order of magnitude lower, that is, an elastic modulus of 11.0 ± 0.9 MPa and tensile strength of 1.7 ± 0.5 MPa [28] (Fig. 2.2).

It was found that this constitutive model can fit both the longitudinal and transverse data equally well, making it a good model for descriptions of the three-dimensional behavior of the MCL.

As mentioned earlier, the human ACL and PCL consist of functionally distinct bundles arranged in a nonuniform geometric configuration [16, 29–31]. The complex geometry of the ACL and PCL make it necessary to model mechancial properties of the whole ligament substance using finite element analysis techniques.

In order to obtain the mechanical properties of the ACL and PCL, it is necessary to separate their bundles such that a specimen with a more uniform cross-sectional area can be obtained. In a study performed at our center, the mechanical properties of the MCL and ACL were found to be different. The elastic modulus of the MCL, which was determined between 4% and 7% strain, was 1120 ± 153 Mpa, or more than twice that of either portion of the ACL, which was determined to be 516 ± 64 Mpa and 516 ± 69 MPa for the medial and lateral portions, respectively [32]. The properties of the bundles within the ACL, however, were nonuniform [33]. The AM bundle exhibited a larger elastic modulus, tensile strength, and strain energy density than the PL bundle. However, it is likely that the properties of both bundles are important for normal joint function. The mechanical properties of the AL and PM bundles of the PCL, as well as the MFL were measured separately under uniaxial tension [30]. It was found that the elastic modulus of the AL bundle (294 ± 115 Mpa) was almost twice that of the PM bundle (150 ± 69 Mpa), supporting the previously published findings of Race and Amis [34]. The elastic modulus of the MFL was not significantly

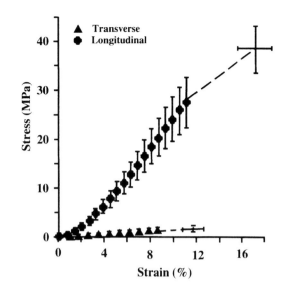

FIGURE 2.2. Stress-strain curves for human MCLs longitudinal ($n = 9$) and transverse ($n = 7$) to the collagen fiber direction.

different from that of the AL bundle. Therefore, knowledge about the properties of these bundles illustrate the importance in considering both bundles when replacing the ACL and PCL during surgical reconstruction.

Mechanical Properties of Ligaments in other Joints

Because of the large range of motion and flexiblty required of the glenohumeral joint, the mechanical behavior of ligaments that surround this joint is substantially different from the behavior of ligaments that surrond the knee. For example, the inferior glenohumeral ligament has a lower elastic modulus of 42 Mpa as well as a tensile strength of only 5 Mpa, that is, a little over 10% of those for the MCL [35]. Conversely, the interosseous ligament (IOL) of the forearm transmits load between the radius and ulna and has a relatively high elastic modulus of 528 ± 82 MPa [36]. Similarly, the anterior longitudinal ligament of the spine has an elastic modulus that is comparable to that of the IOL [37–39]. From these few examples, it is evident that ligaments are highly specialized tissues that govern complex joint function. Engineering ligaments, therefore, must take into consideration not only their complex geometry, but also the diversity of their biomechanical function.

Viscoelastic Properties of Normal Ligaments

Ligaments are also known to possess time and history dependent viscoelastic properties. As a ligament is elongated, complex interactions of collagen, water, and ground substance exhibit creep or stress relaxation behaviors. In our laboratory, the Quasi-Linear Viscoelastic (QLV) theory has been successfully used for modeling the viscoelastic behavior of bovine articular cartilage, canine medial collateral ligaments, porcine anterior cruciate ligaments, and the patellar tendons of human cadavers [40–43]. This theory assumes that the nonlinear elastic response and a separate time-dependent component can be combined with a convolution integral formula to result in a one-dimensional general viscoelastic model expressed as follows:

$$\sigma[\varepsilon(t);t] = G(t) * \sigma^{\varepsilon}(\varepsilon) \qquad (4)$$

The elastic response is a strain-dependent function. One of the representations can be written as follows:

$$\sigma^{\varepsilon}(\varepsilon) = A(e^{B\varepsilon} - 1) \qquad (5)$$

Using Fung's generalized relaxation function based on the assumption of a continuous relaxation spectrum, the time-dependent reduced relaxation function, G(t) [44], takes the form

$$G(t) = \frac{[1 + C\{E_1(t/\tau_2) - E_1(t/\tau_1)\}]}{[1 + C * Ln(\tau_2/\tau_1)]} \qquad (6)$$

where E_1 is the exponential integral and C, τ_1, and τ_2 are constants with $\tau_1 \ll \tau_2$. Assuming $G(t_0) = 1$, $\delta G(t)/\delta(t) = $ constant, and $G(\infty) = $ constant, the QLV theory can be applied to a ligament during uniaxial tensile tests in one dimension for small strains ($<5\%$).

The results from modeling the canine MCL using this theory are shown in Figure 2.3. Based on curve fitting of $\sigma^{\varepsilon}(\varepsilon)$ and G(t) using both the loading and relaxation portions of data, respectively, the parameters describing viscoelastic behavior were obtained [42]. These parameters were then employed to predict the peak and val-

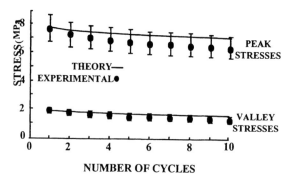

FIGURE 2.3. Theoretically predicted peak and valley stresses using QLV theory compared to experimental results of a canine MCL ($n = 8$) subjected to cyclic loading [42].

ley stress values of a cyclic stress relaxation experiment of canine FMTCs (Fig. 2.3). Recent studies have shown that QLV theory works well for describing the viscoelastic properties of the ACL and patellar tendon [45, 46]. These investigations have also applied QLV theory to cyclic testing and have successfully predicted the time-dependent stress response between two levels of strain.

Recently, our research center has used an even more robust viscoelastic theory, namely the single integral finite strain (SIFS) theory to model ligaments and tendons [47]. This theory incorporates structural assumptions of microstructural change and continuum concepts of fading memory to create a general nonlinear, three-dimensional description of a tissue's nonlinear viscoelastic behavior. By truncating an integral series representation of viscoelastic behavior, and assuming finite deformation, the specific constitutive equation is written as:

$$T = -pI + C_0\{[1 + \mu I(t)]B(t) - \mu B^2(t)\} - (C_0 - C_\infty)$$
$$\int_0^t G(t-s)\{[1 + \mu I(s)]B(t) - \mu F(t)C(s)F^T(t)\} ds, \qquad (7)$$

where T is the Cauchy stress, p is the hydrostatic pressure arising to enforce incompressibility, I is the identity tensor, B is the left Cauchy-Green strain tensor, G(t) is the time-dependent relaxation function, C_0 is the instantaneous modulus, and $I(s) = \text{tr } C$, where C is the right Cauchy-

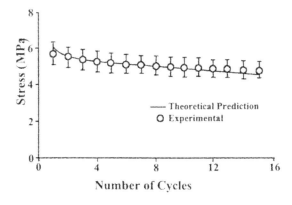

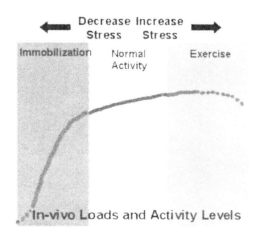

FIGURE 2.4. Theoretically predicted peak stresses using SIFS theory compared with experimental results of a human patellar tendon ($n = 15$, 64–93 years) subjected to cyclic loading [41].

FIGURE 2.5. *In vivo* loads and activity levels to homeostatic response [14].

Green strain tensor. By mathematically combining the constitutive equations for different levels of strain the viscoelastic behavior of ligaments and tendons can be modeled for any magnitude of strain. This theory was used to model the viscoelastic behavior of a human patellar tendon, and was found to adequately predict stresses for a variety of strain histories [41] (Fig. 2.4).

Biological and Experimental Factors Influencing the Mechanical Properties of Ligaments

Many biological factors, such as species, skeletal maturity, age, immobilization, and exercise, can have a profound effect on the mechanical properties of ligaments. Immobilization is known to significantly compromise the biomechanical properties of ligaments [14, 48–51]. After 9 weeks of immobilization, the structural properties of rabbit FMTCs exhibited marked changes, including an 82% decrease in energy absorbed to failure and a 69% decrease in ultimate load at failure. These decreases occurred due to osteoclastic activity that resulted in subperiosteal bone resorption within the insertion sites, as well as microstructural changes in the ligament substance itself. Histologic evaluation revealed marked disruption of the deep fibers that insert into the bone. Additionally, Newton et al. showed that the mechanical properties of the immobilized rabbit ACL were lower then those of nonimmobilized rabbits [52].

Remobilization can reverse the adverse effects of immobilization, but at a much slower rate. It was found that after 9 weeks of immobilization, a remobilization period of up to one year is required to restore normal ligament function [14]. Studies on primates also revealed that cast immobilization for 8 weeks required 52 weeks of remobilization before the structural properties of the FATC returned to values similar to control groups [50].

Exercise can enhance the mechanical properties of the ligament substance and the structural properties of the FMTC; however those effects are not analogous to the effects of immobilization. [53–55]. Exercise training for 1 hour per day at 5.5 km/hour plus ½ hour at 8 km/hour every other day were studied in miniature swine [56]. After 12 months, only a 14% increase in linear stiffness and a 38% increase in ultimate load/body weight were demonstrated. Mechanical properties of the ligament substance also changed slightly with ultimate tensile strength increasing by approximately 20% over matched sedentary controls. Based on the results of these and other related studies, a highly nonlinear representation of the relationship between different levels of stress and ligament properties is depicted in Figure 2.5. The normal range of physiological activities is represented by the middle of the curve. Immobilization results in a rapid reduction in tissue properties and mass and is represented by the steep slope at the beginning of the curve. In contrast, long-term exercise and the associated increase of stress and motion resulted in a slight increase in mechanical properties as compared with those observed in normal physiological activities.

Age has also been shown to cause significant changes to ligaments. A study of the effects of specimen age on the structural properties of the human FATC demonstrated a significant decrease in the stiffness and ultimate load, with values for older specimens (60–97 years) only 74% and 30% of those for younger specimens (22–35 years) [12]. A study by Noyes and Grood noted similar results after testing specimens that were 16–18 years of age and comparing them to specimens that were age 60 years or older [57]. Studies in the rabbit model have revealed changes in the structural properties of the FMTC during aging. Stiffness and ultimate load increased dramatically from 6 to 12 months of age, after which differences between the age groups diminished up to four years of age [58]. For skeletally immature rabbits, failure of the FMTC during tensile testing occurred at the tibial insertion site, suggesting that the tibial insertion site is the weakest link. Yet, for mature animals, failures consistently occurred at the ligament midsubstance [59]. Every ligament, however, is a unique structure in itself and therefore it is not possible to extrapolate age-related changes from one ligament to the other, that is, from age-related changes in the ACL to those in the PCL or other ligaments.

Due to its complexity, a large range of experimental methods have been employed by investigators in the measurement of the mechanical properties of ligaments. Difficulties encountered during testing include gripping of ligaments, strain measurement, definition of the initial length, determination of the cross-sectional area, specimen orientation, and so on. Furthermore, other factors, including temperature, dehydration, freezing and sterilization techniques used during testing, can also cause different outcomes in the experimental data [12]. The readers are encouraged to refer to the chapter by Woo et al. for further details [60].

Ligament Contributions to Joint Kinematics

Although understanding the response of an individual ligament in uniaxial tension has provided much insight, understanding how loads are distributed in a ligament during joint articulation requires knowledge of joint kinematics. As the knee joint can move in six degrees of freedom (DOF)—three translations and three rotations—it is convenient to use three defined axes—the femoral shaft axis, the epicondylar axis, and a floating anterior-posterior axis perpendicular to these two axes—to describe its motion [61]. Translation along these three axes will lead to distraction/compression, medial-lateral translation, and anterior-posterior translation, respectively. Rotations about these three axes will lead to internal-external rotation, flexion-extension, and varus-valgus rotation, respectively.

The in situ force of the knee is an experimentally measured quantity that represents forces existing in the knee at a given position in response to an externally applied load. These forces can generally be measured directly via a multi-DOF load cell rigidly fixed to the femur or tibia. In situ force in a knee ligament (or graft) is the force carried by the ligament (or graft) within an intact joint in response to a given load applied to the knee. To measure the in situ forces in ligaments, various devices, such as buckle transducers, implantable force and pressure transducers and load cells, have been used [62–64].

A robotic/universal force-moment sensor (UFS) testing system was developed and used to determine knee kinematics in multiple DOF as well as the in situ force in ligaments without attaching devices to the ligament [65]. Instead, a 6-DOF kinematic linkage and UFS are attached to the cadaveric knee and the UFS measures forces and moments along and about its orthogonal axes. The line of action and magnitude of an external force can be reconstructed from these data. The robotic manipulator obtains the complex motion of a joint specimen in response to any specified loading condition and then reproduces this motion after the specimen has been altered (e.g., removal of the ACL). The UFS records forces and moments in both cases. The ability of the robot to duplicate the motion

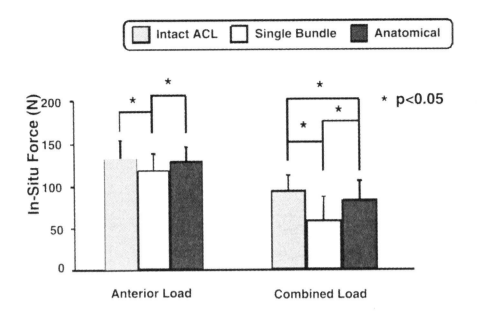

Figure 2.6. In situ forces of the intact ACL, single bundle ACL reconstruction, and double bundle (anatomical) reconstruction at 30° of knee flexion [15].

exactly after modifying the specimen enables the principle of superposition to be applied. Thus, the magnitude, direction and point of application of the in situ forces of the removed tissue can be determined [31, 66–69]. With this technology, the ability exists to evaluate cadaveric knees in conditions of ACL or PCL deficiency [70, 71]. Reconstruction techniques can be preformed and the data can be compared with those for the intact knee [72]. Thus, intraspecimen variability is minimized and the statistical power is increased. The potential also exists to determine knee kinematics *in vivo* and to replay them on the robotic/UFS testing system in order to determine the in situ forces in knee ligaments for those *in vivo* loading conditions.

Human cadaveric knees were tested using the robotic/UFS testing system. In these tests, the magnitude of in situ force in the ACL in response to a 110 N anterior tibial load varied from a high of 111 ± 15N at 15° of knee flexion to a low of 71 ± 30N at 90° of knee flexion [31]. Further, the study showed that the force distribution between the AM and PL bundles of the ACL change with knee flexion angle. The magnitude of force in the PL bundle parallels that for the ACL, while the magnitude of force in the AM bundle remains relatively constant through-

out the range of motion. This observation suggests that the ACL is not a simple ligament, but that both the AM and PL bundles have important functions. Following these findings, an anatomical reconstruction of the ACL using two bundles was studied. It was found that the in situ forces in the graft reached $97 \pm 9\%$ of the intact ACL at 0° and 30° of knee flexion. Similar results were also obtained for a "combined rotatory" load (5 N-m internal-external and 10 N-m varus-valgus load) applied at 30° [15] (Fig. 2.6).

Similar studies of the in situ forces in the PCL demonstrated an increase from 36 ± 13N at full extension to 112 ± 29N at 90° of flexion in response to a 110 N posterior tibial load [67]. Unlike the ACL, the in situ forces in the AL and PM bundles were not significantly different, suggesting that both bundles play a role under certain loading conditions. In a separate study, it was found that hamstring muscle loads increased the in situ forces in the PCL, but the forces were reduced by the addition of quadriceps muscle loads [73]. These results indicate that rehabilitation after reconstruction of the PCL should avoid excessive isolated hamstrings activation especially at high knee flexion angles in order to avoid the likelihood of overloading the graft.

Healing of Knee Ligaments

Ligament injury causes a disruption of normal joint function that, without proper treatment, can lead to injury of other soft tissue structures and eventual osteoarthritis. As mentioned earlier, basic science and clinical studies have shown that a grade III injury (complete tear) of the MCL heals spontaneously after injury and thus can be treated conservatively [6, 8, 74, 75]. Injury models based on rabbit and canine MCLs have been used in experimental studies aimed at understanding the process of ligament healing. These models include surgically transecting the MCL [75] as well as rupturing the MCL to simulate a grade III tear [8].

The results of these experiments have revealed that the quality, rate, and composition of the healing MCL depend on treatment modality. Immobilization after ligament injury was shown to lead to a greater percentage of disorganized collagen fibrils, decreased structural properties of the FMTC, decreased mechanical properties of the ligament substance, and slower recovery of the resorbed insertion sites [14]. Operative treatment (suturing the transected or ruptured MCL) had no long-term biomechanical advantage compared with nonoperative treatment [75]. These findings support clinical observations that isolated grade III MCL injuries should be treated conservatively with active rehabilitation.

Nevertheless, the healing of a completely torn MCL is a long, dynamic process that consists of three overlapping phases [8, 74, 76]. The inflammatory stage has been demonstrated to start immediately after injury and lasts for a few weeks. It was marked by hematoma formation and increased vascular and cellular reactions. After approximately 2 weeks, new blood vessels formed and over a period of weeks, a network of immature collagen fibers were produced by disorganized groupings of fibroblasts marking the events of the reparative phase. Overlapping with the reparative phase was the remodeling phase, which continued for years after injury. The reparative phase was marked by alignment of collagen fibers along the long axis of the ligament and increased collagen matrix maturation [74].

Biomechanical data correlates strongly with the histmorphological and biochemical changes

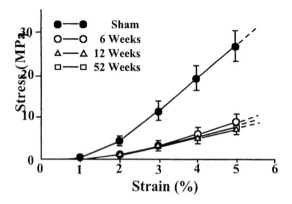

FIGURE 2.7. Stress-strain curves representing the mechanical properties of the medial collateral ligament substance for sham-operated and healing MCLs at time periods of 6 ($n = 6$), 12 ($n = 6$), and 52 ($n = 4$) weeks [77].

of the healing MCL. Using the rabbit model, studies have shown that whereas the structural properties of the healing FMTC returned to near normal levels within a few months, the mechanical properties of the healing ligament's midsubstance remained substantially inferior to those of the intact ligament up to one year after injury [8, 77] (Fig 2.7). The elastic modulus and tensile strength of healing MCLs remained significantly inferior to controls. The cross-sectional area of the healed MCL was found to increase with time. Thus, even after a long period of healing, injured MCLs contained a greater quantity of lesser quality of tissue compared with normal ligaments.

The level of injury has been shown to have a significant effect on ligament healing. Studies that examined the effects of gap size between the torn ends of the injured MCL demonstrated that increased gap size is associated with inferior biomechanical properties for the healing MCL [78]. MCL ruptures with concurrent ACL injury resulted in inferior results compared to isolated MCL injuries [79]. Although repaired and non-repaired FMTCs benefited from ACL reconstruction, structural properties of the FMTC, knee function, and mechanical properties of the healing MCL all remained worse than for isolated MCL injuries, suggesting that the gross joint instability may disrupt the healing process of the MCL. Additionally, no long-term advantages were found between groups with the MCL repairs. Thus, most clinicians reconstruct the ACL in a combined MCL/ACL injury while using nonoperative treatment for the MCL.

New Approaches to Improve Healing

As noted previously, healing ligaments are hist-morphologically and biochemically different from normal ligaments. Novel biological and bioengineering techniques, including application of growth factors, gene transfer, cell therapy, and the use of scaffolding materials and mechanical factors, offer the potential to improve the quality of healing ligaments. These techniques are based on the manipulation of cellular and biochemical mediators to increase expression of specific proteins (e.g., growth factors) or suppress the synthesis of endogenous proteins [80].

Growth Factors

Growth factors are small proteins that are known to induce a variety of cellular responses, including cell proliferation, chemotaxis, and matrix synthesis. The induction of these biological effects is mediated by binding of growth factors to their specific receptors on cell surfaces. Previous studies have shown that expression of many growth factors and their receptors is altered in healing ligaments and tendons. This indicates that a complex interaction exists between growth factors and their receptors, and that this interaction plays a role in ligament and tendon healing [81–85, 86]. *In vitro* studies at our research center demonstrated that transforming growth factor (TGF)-β1 increases collagen synthesis 1.5-fold over controls in both MCL and ACL fibroblasts [87]. *In vivo* studies also showed that treatment of platelet-derived growth factor (PDGF)-BB effects the biomechanical properties of the rabbit FMTCs. Specifically, this growth factor increased the ultimate load, energy absorbed to failure, and ultimate elongation of the rabbit FMTCs, which, on average, were 1.6, 2.4, and 1.6 times greater than those without treatment of the growth factor [88]. This effect was seen at

two weeks after injury and was still present up to six weeks after injury. However, the structural properties of the healing ligament with the growth factor treatment were inferior to uninjured controls.

Gene Transfer

Gene therapy can be defined as the delivery of genetic material (genes) into cells to alter cellular function. Using gene transfer technology, it is possible to increase cellular production of specific proteins (e.g., growth factors) that are functionally important for tissue healing. Gene transfer requires gene carriers including viral vectors and liposomes [89]. Two approaches, ex vivo and *in vivo*, are often used for gene transfer. With the ex vivo approach, genes are placed into cells in culture and the cells later transferred back to the host tissues of interest. With an *in vivo* approach, genes are transferred to cells in host tissues by direct injection of a carrier (e.g., a viral vector) to the host tissues. Hildebrand and co-workers have examined the feasibility of gene transfer to ligaments and tendons [88]. Using adenoviral vectors with both ex vivo and *in vivo* techniques, it was shown that LacZ, a marker gene, was successfully introduced and expressed in the rabbit MCL and ACL. Therapeutic genes, such as PDGF-BB and hemopoietic growth factor (HGF), have also been successfully introduced and expressed in the patellar tendon [89, 90]. In addition, the gene of focal adhesion kinase (pp125 FAK) was overexpressed in flexor tendon using adenovirus [91]. Recently, adeno-associated virus (AAV) has been used for gene delivery to cells from a variety of tissues, such as muscle [92]. Compared with other virus vectors, AAV has several outstanding advantages for gene delivery. For example, AAV can transduce both dividing and nondividing cells without causing a cellular immune response, a problem for other types of viral vectors such as adeno virus [93]. Because of these advantages, AAV has become a promising gene delivery system for enhancement of ligament and tendon healing.

Other techniques have also been used to deliver therapeutic genes to healing rabbit MCLs. For example, antisense decorin oligodeoxynucleotides were introduced into the ligaments using HVJ-liposomes [89, 94]. It was found that decorin mRNA expression and protein synthesis were significantly decreased. Decorin, a small leucine-rich proteoglycan, plays an important role in collagen fibrillogenesis [95, 96].

Cell Therapy

It has been established that mesenchymal progenitor cells (MPCs), from both humans and animals, can differentiate into a variety of cell types, including fibroblasts, osteoblasts, and endothelial cells [97]. These cells are responsible for tissue wound healing since they are the major types of cells producing matrix proteins such as collagens, proteoglycans, cytokines, and tissue enzymes, which are essential for wound healing. Bone marrow, blood, muscle, or adipose tissue are potential sources of MPCs for cell therapy. For cell therapy, MPCs are isolated from the bone marrow, cultured, and finally transplanted to host tissues. MPCs appear to retain their developmental potential even after extensive subculturing [98–100]. There are few studies using MPCs to enhance ligament healing. In the tendon, however, it has been shown that implantation of MPCs into an injured tendon significantly improves its mechanical properties [101, 102].

In our research center, cell therapy has been used for enhancement of MCL healing [103]. It was found that nucleated cells, which contain MPCs, survived 3 and 7 days after implantation at the healing site. Furthermore, the transplanted nucleated cells were found to have migrated to the noninjured part of the midsubstance of the ligament at 7 days. These results are encouraging because the survival of implanted nucleated cells and their potential for migration are essential for enhancement of ligament healing process.

Scaffolding Materials

Scaffolding materials, such as synthetic polymers including polyglycolic acid (PGA), polylactic acid (PLA), polylactic-co-glycolic acid (PLGA), calcium phosphate ceramics, alginate, and hyaluronate [104], provide a mechanical support and a biological base for tissue development. An ideal scaffolding material should serve as a template that can guide cells to regenerate new tissues with similar structure as intact tissues. One remarkable advantage of the synthetic polymers is that their fabrication is relatively easy and reproducible. Therefore, they have been used for ligament replacement [105] and, after seeding with fibroblasts, used as a scaffolding material to repair or replace injured ligaments [106].

In recent years, biological scaffolding materials such as small intestinal submucosa (SIS) obtained from porcine small intestines, have received much attention. SIS is mainly composed of collagen (90% of dry weight) and contains a small amount of growth factors such as FGF and TGF-β [107]. In ligaments and tendons, SIS serves as a matrix to hold cells and nutrients necessary for healing, while also providing a load bearing collagenous structure to be remodeled [108].

Mechanical Factors

It has been well recognized that mechanical stimuli can induce change in the composition, mechanical properties, and structure of connective tissues. In healing ligaments, mechanical loading has been shown to affect the organization of collagen fibers and alignment of fibroblasts. [109]. *In vitro,* mechanical stretching of collagen gels induce the alignment of collagen fibrils along the stretching direction [110], further confirming that mechanical loading can influence the organization of collagenous matrices. Recent studies in our research center showed that when grown in microgrooved surfaces instead of smooth culture surfaces, tendon fibro-blasts become elongated and aligned in the microgrooves. Furthermore, when cyclically stretched, the cells in the microgrooves remained aligned in the stretching direction (Fig. 2.8). It is known that tendon and ligament fibroblasts are aligned with collagen fibrils *in vivo.* Therefore, if fibroblasts are implanted in a microgrooved substrate, they can organize similar to cells in the intact ligament and produce a better organized collagen matrix.

FIGURE 2.8. Human patellar tendon fibroblasts were grown in silicone microgrooves. The cells became elongated in shape and aligned in microgrooves with cyclic stretching (*arrows*). The shape and alignment of the tendon fibroblasts look similar to the cells *in vivo*.

Future Directions

As this chapter has outlined, normal ligaments have a wide array of mechanical behaviors stemming from a complex combination of tissue structure and biochemical constituents. Yet, all of these components are harmoniously intertwined to meet the functional demands on each ligament. Injury to a nonhealing ligament, such as the ACL typically requires reconstruction. For healing ligaments, normal mechanical behavior is not regained. Thus, functional tissue engineering of ligaments after injury will ideally have to restore the complicated network of structure and biochemistry to return the ligament's normal biomechanical behavior.

As functional tissue engineering of healing ligaments is still in its infancy, studies have addressed several questions that have lead to more questions regarding the complicate mechanisms of the healing process. Of primary importance at this early stage is the necessity to fully understand the function of each individual component in order to determine optimal strategies to enhance ligament healing. Using antisense gene therapy, Dr. Cy Frank and co-workers successfully demonstrated that decorin, a proteoglycan, plays an important role in decreasing collagen fibril diameters after ligament injury [89, 94]. As other proteins are also upregulated after injury, specifically decreasing the concentrations of these molecules using this technique may be the key to understanding their function. In order to perform such studies, our research center is currently sequencing the genes of collagen types III and V in various animal models. Using super-

computers, we are determining which oligonucleotides may target these genes such that the roles of collagen types III and V can be elucidated during the healing process.

Further, natural occurring biological scaffolds, such as SIS, offer a natural tissue architecture that cells can infiltrate and remodel with minimal degradation products. Early results from our research center show that SIS, when used as a soft-tissue scaffold, increases the tangent modulus and tensile strength of a healing MCL of the rabbit after gap injury by 50% compared with nontreated controls [111]. By seeding fibroblasts on SIS *in vitro* and stimulating the construct with mechanical stretching, our research center is improving the alignment of collagen within the scaffold. Using this tool, future investigations can begin to determine how tissue structure (i.e., collagen alignment) influences the biological and biochemical response of fibroblasts during the healing process of ligaments. Eventually, the basis for future treatments may use a combination of approaches to treat ligament injuries that includes seeding cells on a scaffold that is conditioned with the right combination of mechanical stimuli, growth factors, and gene therapy. As these experiments will require a combination of techniques that simultaneously address the structure, biochemistry, and biology of a ligament in order to enhance healing, a significant effort must be made to encourage collaboration among investigators from different disciplines. By combining appropriate engineering mechanics with other basic sciences, improved outcomes in the process of ligament healing can be achieved.

Acknowledgments

The authors acknowledge the financial support provided by the National Institute of Health Grant AR41820, and the Biomedical Pilot Initiative Grant from the Rockefeller Brothers Fund (#00-53).

References

1. Bruesch, M. and P. Holzach. 1993. Epidemiologie, Behandlung und Verlaufsbeobachtung frischer ligamentarer Kniebinnenverletzungen im alpinen Skisport. *Zeitschrift fur Unfallchirurgie und Versicherungsmedizin* Suppl 1:144–155.
2. Watkins, J. and P. Peabody. 1996. Sports injuries in children and adolescents treated at a sports injury clinic. *J. Sports Med. Phys. Fit.* 36:43–48.
3. Bijur, P.E., et al. 1995. Sports and recreation injuries in US children and adolescents. *Arch. Pediatr. Adolesc. Med.* 149:1009–1016.
4. Simon, S.R. 1994. Kinesiology. In: Orthopaedic Basic Science. S.R. Simon, Amer Academy of Orthopaedic Surgeons.
5. Allen, C.R., et al. 2000. Importance of the medial meniscus in the anterior cruciate ligament-deficient knee. *J. Orthop. Res.* 18:109–115.
6. Indelicato, P.A. 1983. Non-operative treatment of complete tears of the medial collateral ligament of the knee. *J. Bone Joint Surg. Am.* 65:323–329.
7. Woo, S.L-Y., et al. 1997. Medial collateral knee ligament healing. Combined medial collateral and anterior cruciate ligament injuries studied in rabbits. *Acta Orthop. Scand.* 68:142–148.
8. Weiss, J.A., et al. 1991. Evaluation of a new injury model to study medial collateral ligament healing: primary repair versus nonoperative treatment. *J. Orthop. Res.* 9:516–528.
9. Chimich, D., et al. 1991. The effects of initial end contact on medial collateral ligament healing: a morphological and biomechanical study in a rabbit model. *J. Orthop. Res.* 9:37–47.
10. Niyibizi, C., et al. 1995. Collagens in an adult bovine medial collateral ligament: immunofluorescence localization by confocal microscopy reveals that type XIV collagen predominates at the ligament-bone junction. Matrix Biol. 14:743–751.
11. Frank, C., D. McDonald, and N. Shrive. 1997. Collagen fibril diameters in the rabbit medial collateral ligament scar: a longer term assessment. *Connect. Tissue Res.* 36:261–269.

12. Woo, S.L-Y., et al. 1991. Tensile properties of the human femur-anterior cruciate ligament-tibia complex. The effects of specimen age and orientation. *Am. J. Sports Med.* 19:217–225.
13. Kastelic, J. and E. Baer. 1980. Deformation in tendon collagen. *Symp. Soc. Exp. Biol.* 34:397–435.
14. Woo, S.L-Y., et al. 1987. The biomechanical and morphological changes in the medial collateral ligament of the rabbit after immobilization and remobilization. *J. Bone Joint Surg. Am.* 69:1200–1211.
15. Yagi, M., et al., The biomechanical analysis of an anatomical ACL reconstruciton. [Monograph Citation], Accepted paper. Journal Amer J. of Sports Med As of Sept 11th 2002 status still Accepted.
16. Girgis, F.G., J.L. Marshall, and A. Monajem. 1975. The cruciate ligaments of the knee joint. Anatomical, functional and experimental analysis. *Clin. Orthop.* 106:216–231.
17. Jackson, D.W., et al. 1987. Freeze dried anterior cruciate ligament allografts. Preliminary studies in a goat model [published erratum appears in *Am. J. Sports Med.* 1987 Sep—Oct;15(5):482]. *Am. J. Sports Med.* 15:295–303.
18. Kennedy, J.C., H.W. Weinberg, and A.S. Wilson. 1974. The anatomy and function of the anterior cruciate ligament. As determined by clinical and morphological studies. *J. Bone Joint Surg. Am.* 56:223–235.
19. Feagin, J.A., Jr. and K.L. Lambert. 1985. Mechanism of injury and pathology of anterior cruciate ligament injuries. *Orthop. Clin. N. Am.* 16:41–45.
20. Woo, S.L-Y., et al. 2000. Anatomy, biology, and biomechanics of tendon and ligament. In: *Orthopaedic Basic Science Biology and Biomechanics of the Musculoskeletal System.* Buekwalter J.A., Einhorn T.A., and Simon S.R., (Eds). American Academy of Orthopaedic Surgeons, Rosemont, IL, pp. 581–616.
21. Linsenmayer, T.F., et al. 1993. Type V collagen: molecular structure and fibrillar organization of the chicken alpha 1(V) NH2-terminal domain, a putative regulator of corneal fibrillogenesis. *J. Cell Biol.* 121:1181–1189.
22. Birk, D.E. and R. Mayne. 1997. Localization of collagen types I, III and V during tendon development. Changes in collagen types I and III are correlated with changes in fibril diameter. *Eur. J. Cell Biol.* 72:352–361.
23. Lee, T.Q. and S.L-Y. Woo. 1988. A new method for determining cross-sectional shape and area of soft tissues. *J. Biomech. Eng.* 110:110–114.

24. Woo, S.L-Y., et al. 1990. The use of a laser microm-eter system to determine the cross-sectional shape and area of ligaments: a comparative study with two existing methods. *J. Biomech. Eng.* 112:426–431.

25. Puso, M.A. and J.A. Weiss. 1998. Finite element implementation of anisotropic quasi-linear vis-coelasticity using a discrete spectrum approxima-tion. *J. Biomech. Eng.* 120:62–70.

26. Woo, S.L-Y., et al. 1990. The effects of transec-tion of the anterior cruciate ligament on healing of the medial collateral ligament. A biomechanical study of the knee in dogs. *J. Bone Joint Surg.-Am.* 72:382–392.

27. Scheffler, S.U., et al. 2001. Structure and function of the healing medial collateral ligament in a goat model. *Ann. Biomed. Eng.* 29:173–180.

28. Quapp, K.M. and J.A. Weiss. 1998. Material characterization of human medial collateral liga-ment. *J. Biomech. Eng.* 120:757–763.

29. Fuss, F.K. 1989. Anatomy of the cruciate ligaments and their function in extension and flexion of the human knee joint. *Am. J. Anat.* 184:165–176.

30. Harner, C.D., et al. 1995. The human posterior cruciate ligament complex: an interdisciplinary study. Ligament morphology and biomechanical evaluation. *Am. J. Sports Med.* 23:736–745.

31. Sakane, M., et al. 1997. In situ forces in the ante-rior cruciate ligament and its bundles in response to anterior tibial loads. *J. Orthop. Res.* 15:285–293.

32. Woo, S.L-Y., et al. 1992. A comparative evalua-tion of the mechanical properties of the rabbit me-dial collateral and anterior cruciate ligaments. *J. Biomech.* 25:377–386.

33. Butler, D.L., et al. 1992. Location-dependent var-iations in the material properties of the anterior cruciate ligament. *J. Biomech.* 25:511–518.

34. Race, A. and A.A. Amis. 1994. The mechanical properties of the two bundles of the human poste-rior cruciate ligament. *J. Biomech.* 27:13–24.

35. Bigliani, L.U., et al. 1992. Tensile properties of the inferior glenohumeral ligament. *J. Orthop. Res.* 10:187–197.

36. Pfaeffle, H.J., et al. 1996. Tensile properties of the interosseous membrane of the human forearm. *J. Orthop. Res.* 14:842–845.

37. Chazal, J., et al. 1985. Biomechanical properties of spinal ligaments and a histological study of the supraspinal ligament in traction. *J. Biomech.* 18:167–176.

38. Neumann, P., et al. 1992. Mechanical properties of the human lumbar anterior longitudinal liga-ment. *J. Biomech.* 25:1185–1194.

39. Pintar, F.A., et al. 1992. Biomechanical properties of human lumbar spine ligaments. *J. Biomech.* 25:1351–1356.

40. Kwan, M.K., T.H. Lin, and S.L-Y. Woo. 1993. On the viscoelastic properties of the anteromedial bun-dle of the anterior cruciate ligament. *J. Biomech.* 26:447–452.

41. Johnson, G.A., et al. 1994. Tensile and viscoelas-tic properties of human patellar tendon. *J. Orthop. Res.* 12:796–803.

42. Woo, S.L-Y., M.A. Gomez, and W.H. Akeson. 1981. The time and history-dependent viscoelastic properties of the canine medical collateral liga-ment. *J. Biomech. Eng.* 103:293–298.

43. Woo, S.L-Y., et al. 1980. Quasi-linear viscoelastic properties of normal articular cartilage. *J. Biomech. Eng.* 102:85–90.

44. Fung, Y. 1972. Stress-strain-history relations of soft tissues in simple elongation. In: *Biomechanics: Its Foundations and Objectives.* Y. Fung, Ed. Prentice-Hall, Englewood Cliffs, New Jersey, pp. 181–208.

45. Lin, H.C., M. Kwan, and S.L-Y. Woo. 1987. On the stress relaxation properties of anterior cruciate ligament (ACL). *Amer. Soc. Mech. Eng.* BED3:5–6

46. Lyon, R.M., et al. 1988. Stress relaxation of the anterior cruciate ligament (ACL) and the patellar tendon. *Trans Ortho. Res. Soc.* 13:81, Atlanta, GA.

47. Johnson, G.A., et al. 1996. A single integral finite strain viscoelastic model of ligaments and tendons. *J. Biomech. Eng.* 118:221–226.

48. Woo, S.L-Y. 1982. Mechanical properties of ten-dons and ligaments. I. Quasi-static and nonlinear viscoelastic properties. Biorheology 19:385–396.

49. Woo, S.L-Y., et al. 1982. Mechanical properties of tendons and ligaments. II. The relationships of im-mobilization and exercise on tissue remodeling. *Biorheology* 19:397–408.

50. Noyes, F.R. 1977. Functional properties of knee ligaments and alterations induced by immobiliza-tion: a correlative biomechanical and histological study in primates. *Clin. Orthop.* (123):210–242.

51. Larsen, N.P., M.R. Forwood, and A.W. Parker. 1987. Immobilization and retraining of cruciate ligaments in the rat. *Acta Orthop. Scand.* 58: 260–264.

52. Newton, P.O., et al. 1990. Ultrastructural changes in knee ligaments following immobilization. *Matrix* 10:314–319.

53. Woo, S.L-Y., et al. 1979. Effects of immobiliza-tion and exercise on the strength characteristics of bone-medial collateral ligament-bone complex. *ASME Biomech. Symp.* 32:67–70.

54. Viidik, A. 1968. Elasticity and tensile strength of the anterior cruciate ligament in rabbits as influenced by training. *Acta Physiolo. Scand.* 74:372–380.

55. Laros, G.S., C.M. Tipton, and R.R. Cooper. 1971. Influence of physical activity on ligament insertions in the knees of dogs. *J. Bone Joint Surg. Am.* 53:275–286.

56. Woo, S.L-Y., et al. 1981. The effects of exercise on the biomechanical and biochemical properties of swine digital flexor tendons. *J. Biomech. Eng.* 103:51–56.

57. Noyes, F.R. and E.S. Grood. 1976. The strength of the anterior cruciate ligament in humans and Rhesus monkeys. *J. Bone Joint Surg. Am.* 58:1074–1082.

58. Woo, S.L-Y., et al. 1990. The effects of strain rate on the properties of the medial collateral ligament in skeletally immature and mature rabbits: a biomechanical and histological study. *J. Orthop. Res.* 8:712–721.

59. Woo, S.L-Y., et al. 1986. Tensile properties of the medial collateral ligament as a function of age. *J. Orthop. Res.* 4:133–141.

60. Woo, S.L-Y., K.J. Ohland, and P.J. McMahon. 1992. Biology, healing and repair of ligaments. In: *Biology and Bimechanics of the Traumetized Synovial Joint: The Knee as a Model.* G.A.H. Finerman and F.R. Noyes, Eds. American Academy of Orthoaedic Suregons, Chicago, IL, pp 241–273.

61. Grood, E.S. and W.J. Suntay. 1983. A joint coordinate system for the clinical description of three-dimensional motions: application to the knee. *J. Biomech. Eng.* 105:136–144.

62. Holden, J.P., et al. 1994. *In vivo* forces in the anterior cruciate ligament: direct measurements during walking and trotting in a quadruped. *J. Biomech.* 27:517–526.

63. Korvick, D.L., et al. 1996. The use of an implantable force transducer to measure patellar tendon forces in goats. *J. Biomech.* 29:557–561.

64. Malaviya, P., et al. 1998. *In vivo* tendon forces correlate with activity level and remain bounded: evidence in a rabbit flexor tendon model. *J. Biomech.* 31:1043–1049.

65. Rudy, T.W., et al. 1996. A combined robotic/universal force sensor approach to determine in situ forces of knee ligaments. *J. Biomech.* 29:1357–1360.

66. Carlin, G.J., et al. 1996. In-situ forces in the human posterior cruciate ligament in response to posterior tibial loading. *Ann. Biomed. Eng.* 24:193–197.

67. Fox, R.J., et al. 1998. Determination of the in situ forces in the human posterior cruciate ligament using robotic technology. A cadaveric study. *Am. J. Sports Med.* 26:395–401.

68. Li, G., et al. 1999. The importance of quadriceps and hamstring muscle loading on knee kinematics and in-situ forces in the ACL. *J. Biomech.* 32:395–400.

69. Kanamori, A., et al. 2000. The forces in the anterior cruciate ligament and knee kinematics during a simulated pivot shift test: A human cadaveric study using robotic technology. *Arthroscopy (Online)* 16:633–639.

70. Allen, C.R., et al. 1999. Injury and reconstruction of the anterior cruciate ligament and knee osteoarthritis. *Osteoarthritis Cartilage* 7:110–121.

71. Vogrin, T.M., et al. 2000. Effects of sectioning the posterolateral structures on knee kinematics and in situ forces in the posterior cruciate ligament. *Knee Surg. Sports Traumatol. Arthrosc.* 8:93–98.

72. Ishibashi, Y., et al. 1997. The effect of anterior cruciate ligament graft fixation site at the tibia on knee stability: evaluation using a robotic testing system. *Arthroscopy* 13:177–182.

73. Hoher, J., et al. 1998. In situ forces in the posterolateral structures of the knee under posterior tibial loading in the intact and posterior cruciate ligament-deficient knee. *J. Orthop. Res.* 16:675–681.

74. Frank, C., et al. 1983. Medial collateral ligament healing. A multidisciplinary assessment in rabbits. *Am. J. Sports Med.* 11:379–389.

75. Woo, S.L-Y., et al. 1987. Treatment of the medial collateral ligament injury. II: Structure and function of canine knees in response to differing treatment regimens. *Am. J. Sports Med.* 15:22–29.

76. Oakes, B. 1982. Acute soft tissue injuries: Nature and management. *Aust. Fam. Physician* 10(suppl): 3–16.

77. Ohland, K.J., et al. 1991. Healing of combined injuries of the rabbit medial collateral ligament and its insertions: A long term study on the effects of conservative vs. surgical treatment. In: *The Winter Annual Meeting of the American Society of Mechanical Engineers.* Atlanta, GA.

78. Loitz-Ramage, B.J., C.B. Frank, and N.G. Shrive. 1997. Injury size affects long-term strength of the rabbit medial collateral ligament. *Clin. Orthop.* (337):272–280.

79. Yamaji, T., et al. 1996. Medial collateral ligament healing one year after a concurrent medial collateral ligament and anterior cruciate ligament injury: an interdisciplinary study in rabbits. *J. Orthop. Res.* 14:223–227.

80. Woo, S.L-Y., et al. Tissue engineering of ligament and tendon healing. *Clin. Orthop.* (367 Suppl): S312–S323.

81. Sciore, P., R. Boykiw, and D.A. Hart. 1998. Semi-quantitative reverse transcription-polymerase chain reaction analysis of mRNA for growth factors and growth factor receptors from normal and healing rabbit medial collateral ligament tissue. *J. Orthop. Res.* 16:429–437.

82. Panossian, V., et al. 1997. Fibroblast growth factor and epidermal growth factor receptors in ligament healing. *Clin. Orthop.* 342:173–180.

83. Pierce, G.F., et al. 1989. Platelet-derived growth factor and transforming growth factor-β enhance tissue repair activities by unique mechanisms. *J Cell Biol.* 109:429–440.

84. Duffy, F.J., Jr., et al. 1995. Growth factors and canine flexor tendon healing: initial studies in uninjured and repair models. *J. Hand Surg. (Am)* 20:645–649.

85. Steenfos, H.H. 1994. Growth factors and wound healing. *Scand. J. Plast. Reconstr. Hand Surg.* 28:95–105.

86. Schmidt, C.C., et al. 1995. Effect of growth factors on the proliferation of fibroblasts from the medial collateral and anterior cruciate ligaments. *J. Orthop. Res.* 13:184–90.

87. Marui, T., et al. 1997. The effect of growth factors on matrix synthesis by ligament fibroblasts. *J. Orthop. Res.* 15:18–23.

88. Hildebrand, K.A., et al. 1998. The Effects of PDGF-BB on healing of the rabbit medial collateral ligament: An *in vivo* study. *Am. J. Sports Med.* 26(9):549–54

89. Nakamura, N., et al. 1998. Early biological effect of *in vivo* gene transfer of platelet-derived growth factor (PDGF)-B into healing patellar ligament. *Gene Ther.* 5:1165–1170.

90. Natsu-ume, T., et al. 1997. Temporal and spatial expression of transforming growth factor-beta in the healing patellar ligament of the rat. *J. Orthop. Res.* 15:837–843.

91. Lou, J., et al. 1997. *In vivo* gene transfer and over-expression of focal adhesion kinase (pp 125 FAK) mediated by recombinant adenovirus-induced tendon adhesion formation and epitenon cell change. *J. Orthop. Res.* 15:911–918.

92. Xiao, X., et al. 2000. Full functional rescue of a complete muscle (TA) in dystrophic hamsters by adeno-associated virus vector-directed gene therapy. *J. Virol.* 74:1436–1442.

93. Xiao, X., et al. 1997. Gene transfer by adeno-associated virus vectors into the central nervous system. *Exp. Neurol.* 144:113–124.

94. Nakamura, N., et al. 1998. A comparison of *in vivo* gene delivery methods for antisense therapy in ligament healing. *Gene Ther.* 5:1455–1461.

95. Neame, P.J., et al. 2000. Independent modulation of collagen fibrillogenesis by decorin and lumican. *Cell. Mol. Life Sci.* 57:859–863.

96. Kuc, I.M. and P.G. Scott. 1997. Increased diameters of collagen fibrils precipitated *in vitro* in the presence of decorin from various connective tissues. *Connect. Tissue Res.* 36:287–296.

97. Lazarus, H.M., et al. 1995. Ex vivo expansion and subsequent infusion of human bone marrow-derived stromal progenitor cells (mesenchymal progenitor cells): implication for therapeutic use. Bone Marrow Transplant 16:557–564.

98. Goshima, J., V.M. Goldberg, and A.I. Caplan. 1991. Osteogenic potential of culture-expanded rat marrow cells as assayed *in vivo* with porous calcium phosphate ceramic. Biomaterials 12:253–238.

99. Haynesworth, S.E., et al. 1992. Characterization of cells with osteogenic potential from human marrow. *Bone* 13:81–88.

100. Bruder, S.P., N. Jaiswal, and S.E. Haynesworth. 1997. Growth kinetics, self-renewal, and the osteogenic potential of purified human mesenchymal stem cells during extensive subcultivation and following cryopreservation. *J. Cell Biochem.* 64:278–294.

101. Young, R.P., et al. 1998. Use of mesenchymal stem cells in a collagen matrix for Achilles tendon repair. *J. Orthop. Res.* 16:406–413.

102. Awad, H.A., et al. 1999. Autologous mesenchymal stem cell-mediated repair of tendon. *Tissue Eng.* 5:267–277.

103. Watanabe, N., et al. 1998. New method of distinguishing between intrisin cells in situ and extrinsic cells supplied by the autogeneic transplantation, employing transgenic rats. In: *Orthopaedic Research Society.* New Orleans, Louisiana 23:1035

104. Kim, B.S., et al. 1998. Optimizing seeding and culture methods to engineer smooth muscle tissue on biodegradable polymer matrices. *Biotechnol. Bioeng.* 57:46–54.

105. Ambrosio, L., et al. 1998. Viscoelastic behavior of composite ligament prostheses. *J. Biomed. Mater. Res.* 42:6–12.

106. Dunn, M.G., et al. 1995. Development of fibroblast-seeded ligament analogs for ACL reconstruction. *J. Biomed. Mater. Res.* 29:1363–1371.

107. Badylak, S., et al. 1999. Naturally occurring extracellular matrix as a scaffold for musculoskeletal repair. *Clin. Orthop.* (367 Suppl):S333–S343.

108. Badylak, S.F., et al. 1995. The use of xenogeneic small intestinal submucosa as a biomaterial for Achilles tendon repair in a dog model. *J. Biomed. Mater. Res.* 29:977–985.

109. Gomez, M.A., et al. 1991. The effects of increased tension on healing medical collateral ligaments. *Am. J. Sports Med.* 19:347–354.

110. Huang, D., et al. 1993. Mechanisms and dynamics of mechanical strengthening in ligament-equivalent fibroblast-populated collagen matrices. *Ann. Biomed. Eng.* 21:289–305.

111. Musahl, V., et al. 2002. Functional tissue engineering of ligament healing. In: 28th Annual Meeting, American Orthopaedic Society for Sports Medicine, Orlando, FL.

3

Native Properties of Cardiovascular Tissues: Guidelines for Functional Tissue Engineering

Jay D. Humphrey

Introduction

Twenty years ago it was thought that the most distinguishing characteristics of the biomechanics of soft tissues were their complex mechanical properties: they often exhibit nonlinear, anisotropic, heterogeneous, nearly incompressible, viscoelastic behaviors over finite strains. Indeed, these properties endow soft tissues with unique structural capabilities that continue to be extremely challenging to quantify or mimic with synthetic materials. Over the last 20 years, however, we have discovered an even more important property of soft tissues, their homeostatic tendency to adapt in response to changes in their environment. Thus, to understand well the native properties of a soft tissue, we must not only quantify their structure and function at a given time, we must also quantify how their structure and function changes in response to altered stimuli. With regard to tissue replacements, therefore, beyond the need that the construct survives its introduction into the host, the most important characteristic is likely its ability to adapt as needed to maintain near optimal function in an often-changing environment. Whereas synthetic materials cannot meet this design criterion, tissue engineered replacements may—herein lies their greatest potential.

To motivate further the need to understand the constitutive behavior of native tissue, consider the following:

". . . one of the best ways to study tissue engineering is to investigate the changes that can occur in normal organs when the stress and strain fields are disturbed from the normal homeostatic condition." (Fung, 1995)

In this chapter, we shall consider the characteristic structure and biomechanical behaviors of two representative cardiovascular tissues, arteries and the heart wall. We shall see that a layered structural motif appears to be conserved over multiple length scales: parenchymal and supportive tissue is bounded by mechanically protective and biologically functional membranes. It would seem prudent, therefore, to engineer replacement tissues having similar structure. By considering chronologically the advancements in our understanding of the biomechanics of cardiovascular tissue, we shall see that the more complexity that we add to our constitutive relations, the more uniform the computed stress field becomes. That is, simplicity appears to arise from complexity. To engender an equally favorable mechanical environment for all cells across the wall of a tissue-engineered vessel or heart may thus require a "functionally graded" design. Although there are many examples of the adaptation of native tissues to changes in their mechanical environment, there is much to learn about the associated biomechanics. Hence, there is a clear need for biomechanicists to support tissue engineers via the development of constitutive relations for na-

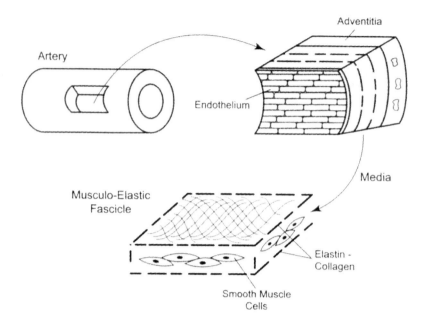

FIGURE 3.1. Schema of the layered make-up of the arterial wall, which consists of a parenchymal tissue (media) bounded by structurally protective and biologically functional layers (intima and adventitia). Also shown is a "musculo-elastic fascicle," that is the repeating layer that constitutes the parenchymal layer.

tive tissues that include biological growth and re-modeling.

Arteries

Structure

Figure 3.1 shows that arteries consist of three layers: the intima, media, and adventitia. One can think of this layered structure simply as a central parenchymal tissue bounded by two protective and biologically functional "membranes." That is, the intima protects the thrombogenic media from direct contact with the blood while allowing selective communication between the two (e.g., the transport of oxygen, hormones, etc.); the adventitia separates the media from the surrounding perivascular tissue and acts as a stiff sheath that protects the smooth muscle from overdistension. The function of the media is supported further by the production of a multitude of biomolecules by the intima (endothelium), including growth factors, vasoactive molecules, adhesion molecules, etc., as well as by the sympathetic nerves

and metabolic support of the vasa vasorum that are within the adventitia. As noted by Clark and Glagov (1985), the structure of the media in elastic arteries can also be described as multiple layers of parenchymal tissue bounded by protective and functional "membranes." In particular, they referred to each of the repeating layers of smooth muscle cells bounded above and below by elastin and collagen as musculo-elastic fascicles. Likewise, in muscular arteries, one sees smooth muscle cells that are invested by collagenous sheaths, which are both protective and functional. Indeed, one can even think of an individual cell as consisting of a parenchymal and supportive material (cytoplasm) surrounded by a protective and functional membrane (e.g., the cell membrane and glycocalyx). Hence, it appears that this simple layered structural motif is conserved across several length scales from the cell to the organ.

Constitutive Behavior and Stress Distributions

In discussing native properties, it is useful to review historically our step-wise increases in un-

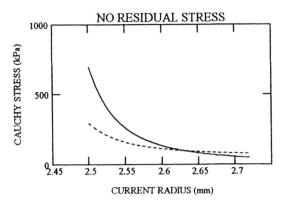

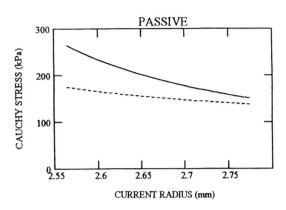

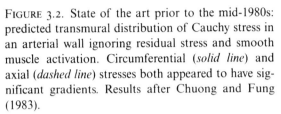

FIGURE 3.2. State of the art prior to the mid-1980s: predicted transmural distribution of Cauchy stress in an arterial wall ignoring residual stress and smooth muscle activation. Circumferential (*solid line*) and axial (*dashed line*) stresses both appeared to have significant gradients. Results after Chuong and Fung (1983).

FIGURE 3.3. Predicted transmural distribution of stresses in a passive artery based on the residual stress formulation of Chuong and Fung (1986). When compared with Figure 3.2, note that incorporation of residual stresses dramatically decreases the predicted gradients in stress.

derstanding. Whereas Roy (1880) understood qualitatively many of the complexities of large artery structure and properties, it was the use of finite elasticity in the late 1960s that provided the first significant advance in modeling. Yet, based on isotropic, and later anisotropic, finite elasticity, early computations showed monotonic transmural changes in stress and strain, with tremendous gradients in the stress (e.g., Vaishnav et al., 1973). For an example, see Figure 3.2. Given these early observations, it was suggested (incorrectly) that the sub-intima is susceptible to atherosclerosis because of the high circumferential stresses in this region. Later, von Maltzahn et al. (1981) emphasized the importance of accounting for the structural heterogeneity of the arterial wall; they assumed that the wall consists of two structurally important homogeneous layers, the media and a (stiffer) adventitia. Computations of transmural stresses reveal that including this heterogeneity decreases slightly the transmural gradient in the computed stresses (see Humphrey, 2002). One of the most important discoveries with regard to wall mechanics, however, was that the arterial wall is residually stressed (residual stresses are those stresses that exist independent of external loads). In particular, Chuong and Fung (1986)

showed that because of the strong nonlinearities, even small residual stresses (3–5 kPa) result in tremendous reductions in the computed transmural gradient in the Cauchy stress (cf. Figs. 3.2 and 3.3)—the mathematical formulation of this boundary value problem is in the Appendix. Thought to arise because of nonuniform growth and remodeling, the existence of residual stress now seems teleologically reasonable for it tends to homogenize the stress field. Although many experimental studies over the years have shown the importance of smooth muscle contraction on the mechanical behavior of arteries, detailed quantification has not kept pace. There is one notable exception, however. Rachev and Hayashi (1999) showed that including a simple model of basal tone further reduces the computed transmural gradient in the stress (cf. Figs. 3.3 and 3.4). And most recently, Taber and Humphrey (2001) suggested that the arterial wall may well exhibit a continuous distribution in stiffness. If so, this too could also help homogenize the transmural distribution of stress (cf. Figs. 3.3 and 3.5).

Although our understanding of the complex, multiaxial behavior of contracting arteries is yet incomplete, with each increasing level of complexity that we include in the stress-strain rela-

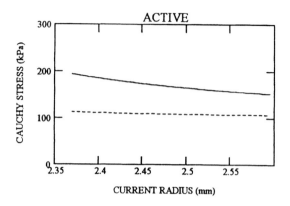

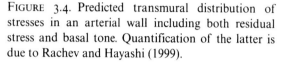

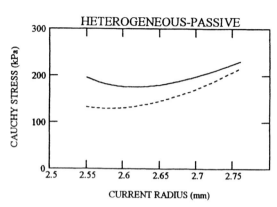

FIGURE 3.4. Predicted transmural distribution of stresses in an arterial wall including both residual stress and basal tone. Quantification of the latter is due to Rachev and Hayashi (1999).

FIGURE 3.5. Predicted transmural distribution of stress in an arterial wall assuming a continuous (quadratic) change in material stiffness from the inner to the outer wall. After Taber and Humphrey (2001).

tion (material heterogeneity, residual stress, muscle activation, etc.), we find that the gradient in the computed stress field tends to become less and less. Hence, ignoring the native structural complexity when engineering a replacement vessel could lead to a stress field that is very different than that in the normal arterial wall, which we must assume is optimal or nearly so.

Growth and Remodeling

There are many examples of vascular adaptations in response to sustained changes in the mechanical environment. These include changes that are induced by local sustained alterations in flow (e.g., in arterio-venous fistulas or due to placental flow during pregnancy), by a microgravity environment, by altered flows and wall stresses due to the process of atherosclerosis, by sustained exercise or the lack thereof, by using veins as arterial grafts, by compliance mismatch at an anastomosis created by the implanting of a synthetic vascular graft, by hypertension, etc. See, for example, Langille (1993).

Although each vessel has unique structure and thus unique functions, growth and remodeling appear to occur via similar pathways even though the gross manifestation often differs (e.g., hypertrophy of smooth muscle versus altered deposition of matrix). In particular, it appears that the two most important features of

growth and remodeling are the local rates of turnover of the individual constituents and the configurations in which this turnover occurs (Humphrey and Rajagopal, 2002). That is, fundamental to the alteration of structure and function is the relative production and removal of material in specific (stressed) configurations. Of course, cells are the effectors of such change, and balances or imbalances in production and removal occur via a few fundamental cellular processes: mitosis, apoptosis, hypertrophy, migration, differentiation, etc. Because these few fundamental processes lead to very different outcomes, depending on the specific stimulus and specific vessel, there is a need to compare and synthesize the associated data.

Of particular relevance to tissue engineering is a comparison of the normal development of an artery with its response to hypertension. Briefly, arteries form in stages. First is the formation of a thin tube consisting largely of endothelial cells as well as the basement membrane on which they reside. Smooth muscle cells are then laid down to form an outer layer for this "endothelial tube." These smooth muscle cells replicate and lay down extracellular matrix in layers (elastin, collagen, etc.), thus resulting in a repeated process that produces the aforementioned layered structural motif. Finally, the primarily collagenous adventitia forms as a protective outer layer (see Stenmark and Mecham, 1997, for more details).

Clark and Glagov (1985) noted in the aorta that there appears to be a direct relationship between the number of musculo-elastic fascicles (and hence the overall wall thickness h) and the inner radius a of the vessel. For example, a normal, mature mouse aorta has five such fascicles and radius of 0.6 mm, whereas a comparable pig aorta has 72 fascicles and a radius of 11.5 mm. Whereas the mean wall tension T ($= Pa$ where P is the distension pressure and a the current inner radius) is much greater in the pig aorta ($T \sim 200$ N/m) than the mouse aorta ($T \sim 8$ N/m), the mean tension per musculo-elastic fascicle is nearly constant ($\sim 2 \pm 0.4$ N/m) across species. A corollary, therefore, is that the thickness of the aortic wall normally increases during development via the addition of additional fascicles, and thus by ~ 0.05 mm for each 0.5 mm increase in the luminal radius. Mathematically, this implies that normal wall thickness appears to be governed largely by the circumferential stress field, or on average, the mean circumferential stress $\sigma = Pa/h$. The luminal radius is thought, of course, to be governed largely by the volumetric flow rate Q that the vessel must accommodate due to the (growing) demands of distal tissue for O_2, etc. More specifically, the radius appears to change such that the mean wall shear stress τ_w is maintained nearly constant, where $\tau_w = 4\mu Q/\pi a^3$ and μ is the viscosity (in large arteries the normal value of τ_w is about 1.5 Pa).

In hypertension, luminal pressure P tends to increase without a significant change in volumetric flow Q or viscosity μ. Hence, due in part to endothelial production of vasoactive molecules such as nitric oxide, endothelin-1, angiotensin-II, thromboxane, etc., the luminal radius a appears to seek to remain the same (to keep τ_w nearly constant) despite the competing effects of the increased pressure tending to distend the wall elastically (i.e., increase a) and the (mild) myogenic reflex tending to constrict the wall (i.e., decrease a). Hence, the only way the wall can maintain/restore σ to normotensive levels is to increase the wall thickness h. This is, of course, the most conspicuous change in the aorta in hypertension [the interested reader should consult Rachev et al. (1998) for a theoretical look at adaptation in response to hypertension]. It is interesting that this adaptive response is accomplished via increases in the thickness of the existing musculo-elastic fascicles (see Matsumoto and Hayashi, 1996) rather than via an increase in their number as would be the case in normal development; moreover, it appears that the increases in fascicle thickness occur first in the inner wall and then in the outer wall. Although the vessel is able to achieve a structure and thus function that is optimized somewhat in each case, we can assume that normal development results in better optimization. If so, this might suggest that a tissue-engineered aorta should have the same number of musculo-elastic fascicles of the same thickness as the normal aorta not fewer layers of greater thickness. This issue raises an additional question—in many cases, vessel grafts are needed for implantation in patients having various diseases, not in "normal" individuals, and thus the host tissues will reflect some underlying adaptation. Thus, should a tissue-engineered aorta for implantation into a person with a history of hypertension be designed with additional musculo-elastic fascicles of the normal thickness or with a normal number but thicker musculo-elastic fascicles? This and likely many similar situations have not been addressed directly, but they clearly ought to be.

The Heart Wall

Structure

Like arteries, the heart consists of three primary layers: the endocardium, myocardium, and epicardium (Fig. 3.6). That is, the parenchymal and supportive tissue (cardiomyocytes and extracellular matrix) are delimited by protective and functional membranes (epicardium and endocardium). Similar to the mechanical role of the adventitia, it is thought that the epicardium (visceral pericardium) and the (parietal) pericardium serve to protect the heart from acute overdistension; in addition, these membranes also form the pericardial space and produce pericardial fluid, which allows the heart to move relatively freely within a constrained region within the thorax. In contrast to the arterial wall, the transmural orientations of muscle fibers within the myocardium are more varied

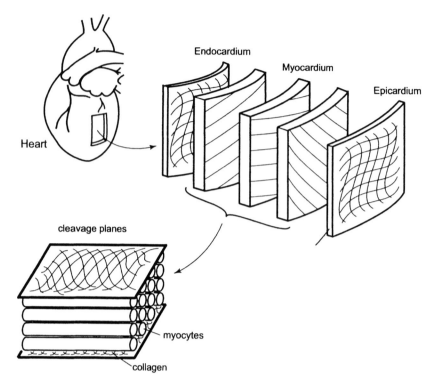

FIGURE 3.6. Schema of the layered make-up of the heart wall, which consists of parenchymal tissue (myocardium) bounded by structurally protective and biologically functional layers (endocardium and epicardium). Also shown is the "collagen-muscle-collagen" unit that repeats across the wall, each separated by so-called cleavage planes according to Hunter et al. (1997).

(Fig. 3.6). A detailed discussion can be found in Hunter et al. (1997), but in general note that the muscle fibers tend to change in orientation from about −60° in the subepicardial region to near 0° in the midwall region to near 60° in the subendocardial region (all relative to the circumferential direction). This transmural splay of muscle fibers gives rise to the large twisting action that is associated with the ejection of blood from the heart during systole. Moreover, note that the muscle fibers tend to be organized into layers between thin sheets of collagen, and of course each myocyte consists of a central parenchyma protected by the cell membrane—this again shows that a layered structural motif is conserved across various length scales. Hunter et al. (1997) refer to the demarcation between the layers of "collagen-muscle fiber-collagen" units as cleavage planes for they tend to separate naturally upon dissection or preparation for microscopy.

Constitutive Relations and Stress Analysis

Just as in arterial mechanics, a major step forward in cardiac mechanics was the use of finite elasticity in the late 1960s. Again, however, the initial thick-walled, isotropic models predicted large transmural gradients in wall stress. In the late 1970s and early 1980s, it was recognized that one must account for the anisotropic behavior of the myocardium and indeed the transmural changes in muscle fiber direction that give rise to the torsional motion of the heart. Incorporation of torsion decreased the computed gradients in wall stress (Arts et al., 1979). Paralleling work in arterial mechanics, the late 1980s and early 1990s saw the incorporation of residual stress in calculations of wall stress. Again, adding this "complexity" simplified the computed transmural wall stress significantly (see Costa et al., 1996). The mid-1990s revealed more information on material heterogeneity, including transmural differences in myocardial properties as well as the distinctly different behaviors of the endocardial/epicardial membranes and the myocardium (Kang et al., 1996; Novak et al., 1994). In hindsight, the former should have been anticipated for smooth transitions in properties across the myocardial-endocardial and myocardial-epicardial boundaries minimizes the potential for jumps in stress. Unfortunately, no detailed analyses of wall stress in the heart have included these transmural variations in myocardial properties or the epicardial and endocardial membranes. If similar to arterial mechanics, including these complexities will likely predict even greater homogeneity in the stress fields. Finally, as in arterial mechanics, quantification of contractile behavior in a multiaxial manner has yet to be completed (see Zahalak et al., 1999), but it is again expected that the combination of active plus passive properties will serve to minimize the computed transmural gradients in stress.

Growth and Remodeling

Although cardiac myocytes do not turnover, there is continual turnover of constitutents within the heart wall. This includes turnover of proteins within the myocytes as well as significant turnover of the extracellular matrix, which consists largely of collagen. It is for this reason, therefore, that the heart can hypertrophy in response to increased load (which is favorable in exercise but less so in hypertension, the intrinsic differences of which need to be understood), it can form collagenous scars following local myocardial infarction, or it can dilate considerably in congestive heart failure or cardiomyopathy (Weber, 1989). Again, there is much to learn from the different adaptations that occur in structure and function in the mature organ and to contrast these to the dramatic changes that occur during development (see Taber, 1995).

Discussion

A long-standing tenet in biology is that structure is intimately related to function—this is certainly supported from the perspective of mechanics. It is interesting that within the cardiovascular system, there appears to be a conservation of a simple structural motif over multiple length scales: parenchymal and supportive tissue are delimited by protective and functional membranes. The concept of layering is certainly not foreign to the engineering of materials, such as composites, and it would seem that this concept should be employed when engineering replacement cardiovascular tissue.

From the perspective of mechanics, arteries, veins, and the heart can be described as thick-walled pressure vessels. It is well known that simply because of the pressure traction conditions and the geometry, such structures are expected to have significant gradients in wall stress. It is also not foreign to engineers that it can be advantageous to minimize such gradients, as, for example, in the design of cannons, and that this can be accomplished via the introduction of residual stresses. Although our understanding of residual stresses in native tissue is yet incomplete, given the geometric and material nonlinearities in cardiovascular mechanics even small residual stresses can provide tremendous reductions in the otherwise large transmural gradient in wall stress. Hence, residual stresses should be built into tissue-engineered cardiovascular constructs. Residual stresses likely result from nonuniform growth and remodeling, which is to say via the differential control of the rates of production and removal of constituents at different radial locations within the wall. For example, the residual stress-related opening angle in arteries first increases with the abrupt initiation of hypertension in coarctation animal models but then decreases toward baseline (Liu and Fung, 1989). This is consistent with an initially greater growth in the inner wall, where stresses would be expected to increase the most due to the hypertension, but a subsequent greater growth in the outer wall. The latter suggests that growth and remodeling occurs in the media and adventitia at different rates and to different extents, which appear to be controlled by the transmural differences in stress. Residual stresses may be able to be built into tissue engineered constructs if they are "grown" under the appropriate time-varying loads and boundary conditions. This again emphasizes the importance of the aforementioned quote from Fung—we need to understand better the mechanics of normal developmental biology as well as the mechanics of adaptive growth and remodeling in mature tissue.

It appears that nature does not rely solely on residual stresses to homogenize the transmural stress field, however. Clever regional variations in material symmetry (e.g., transmural splay of muscle fiber directions in the heart) and material stiffness (e.g., transmural differences in volume fractions or cross-linking of constituents in the heart and arteries) contribute significantly to the homogenization of the stress field. Using such regional variations to control stress fields is not new for engineers either, a good example being the recent focus on "functionally graded composites." It appears that we would be well advised to quantify and then mimic regional variations in properties when engineering a replacement cardiovascular tissue. Unfortunately, despite some longstanding evidence that regional variations likely exist in myocardial (Buccino et al., 1969) and vascular medial (Feldman and Glagov, 1971) composition, there is a pressing need for histologists and biomechanicists to give more attention to its quantification.

As revealed by this brief review, perhaps the least well known and yet most important regulator of multiaxial wall stress (and thus cellular function) in the cardiovascular system is the contribution of muscle activation. In contrast to the classical idea that muscle generates active force only in the direction of the muscle fiber, recent data show that this is not the case (e.g., Strumpf et al., 1993). There is pressing need, therefore, for an increased understanding of the multiaxial stresses that result from muscle activation. Indeed, when we talk of *functional* cardiovascular tissue, what we really mean is primarily its ability to contract or relax in an optimal way under changing conditions. Hence, we must also know how the supporting tissue in the parenchyma (i.e., extracellular matrix) and the delimiting membranes, including effects due to residual stress and regional variations in properties, ensure that the musculature will be optimized. There is clearly much more to learn in this regard.

In closing, it is acknowledged that many of the above arguments suggest that the complex

structure and properties of cardiovascular tissue appear to homogenize the (Cauchy) stress field and that certain adaptations appear to occur such that stresses (e.g., τ_w and σ in arteries) are maintained at or restored to near normal levels. Moreover, it has been suggested that this is reasonable teleologically for it suggests that, regardless of their location within the wall of an artery or the heart, like cells would "prefer" to operate under like mechanical conditions. Related to this observation is an issue that has raised considerable debate—is it stress, strain, strain-rate, strain-energy, etc. that the cell truly senses and thus responds to? At first glance the current presentation might appear to favor stress in contrast to copious arguments for strain as the primary stimulus for mechanotransduction in cells. Actually, this is not the intent—I submit that it is highly unlikely that cells sense directly either of these continuum-based, mathematically defined quantities. Rather, cells likely sense much more primitive quantities, forces or conformational changes, at the molecular or atomic level. This is not to say, however, that stress and strain are not convenient metrics for establishing empirical correlations between mechanical stimuli and cellular response—they are. We must simply remember that in this context, mechanotransduction falls within the domain of constitutive theory wherein one seeks reliable predictors of responses for given stimuli under particular conditions. Indeed, different constitutive relations are often used to describe different behaviors of the same material under different conditions. In some cases, therefore, stress may provide a better or a more convenient correlation whereas in other cases strain may be preferred; as long as the correlation provides predictive capability, it is not necessary that it be unique or that it reflect directly the underlying biophysical mechanism. That said, it must be acknowledged that the apparent preference for stress herein is related, in part, to the semi-inverse approach that is adopted in analytical formulations of the boundary value problems of interest (e.g., finite inflation, extension, and torsion of a thick-walled tube). Strain is prescribed *a priori* to vary monotonically through the wall whereas the associated stresses are calculated from the equations of motion, constitutive relations, and boundary conditions. As shown above, these calculated stresses depend strongly on the presence of a prescribed residual strain field, material symmetries, material heterogeneities, etc. It is because such formulations appear to capture much of the associated mechanics that (Cauchy) stress appears to be a useful metric for predicting many adaptations in cardiovascular tissue. This is an important issue, but one that must be kept in perspective: constitutive relations typically need only enable reliable predictions of responses under conditions of interest, they need not model precisely the underlying mechanisms. The interested reader is referred to Humphrey (2001) for an additional discussion of this issue. Of course, we must remember that information on fundamental mechanisms must remain as our ultimate goal even though tissues can be engineered based on a phenomenological understanding.

Closure

In summary, it appears that there is a common, layered structural motif in the healthy cardiovascular system that should be mimicked in tissue engineering. This layering appears to admit the existence of residual stresses and a transmural complexity in material properties—in particular, heterogeneities and regionally varying material symmetries that suggest a functionally graded make-up—that together result in a simplified, perhaps homogenized stress field. Such a stress field may provide an optimal environment for cellular function under normal conditions. These specific complexities in properties should be quantified and mimicked as well.

Normal adaptations occur in response to deviations in the applied loads, displacement boundary conditions, and pathophysiological state that perturb the otherwise favorable (near optimal) environment for cell function. Whereas such adaptations seek to restore the optimal conditions, they often do not. There is a need, therefore, to identify the specific constraints that cause a tissue to achieve a suboptimal rather than an optimal adaptation, repair, or healing response. Such constraints may include intrinsic differences in cellular function in the adult versus the developing organism (e.g., with respect to the production of elastin) as well as perturbed conditions within the host tissue (local scarring, etc.) that result from the initial injury or disease. Some of the latter constraints will be removed medically or surgically when the replacement is implanted, but others may remain. Hence, there is a need to identify the specific parameters that endow a tissue with its structural integrity and biological functionality and then to engineer the replacement within the context of a *constrained optimization problem* with respect to these parameters. For example, does an aorta normally respond to hypertension by thickening the existing musculoelastic fascicles because there is an intrinsic constraint against the addition of fascicles having a normal thickness? In order to answer this and similar questions, there will remain a need for increased interactions between biologists, biomechanicists, clinicians, mathematicians, and tissue engineers to pose and solve the important challenge of engineering tissue replacements. To engineer truly functional cardiovascular tissues, for example, we must move from the perspective that functional means an ability to pass a simple burst test to the perspective that functional means that the tissue engineered construct can grow and remodel in response to local changes in environment so as to maintain near normal functionality.

Acknowledgments

This work was supported, in part, by grants from the NIH (HL-54957 to JDH), NSF (BES-0084644 to JDH), and Duke University (through NIH Award HL-58856 to M. Friedman). The schematic drawings were done by Mr. William Rogers.

Appendix

Consider the finite extension and inflation of a straight section of a cylindrical artery. Albeit more complex in general, let the deformation be described by a simple mapping of the form,

$$r = r(R), \quad \theta = \frac{\pi}{\Theta_o}\Theta, \quad z = \Lambda\lambda Z \quad \rightarrow \quad \text{(A1)}$$

$$\mathbf{F} = diag\left[\frac{\partial r}{\partial R}, \frac{\pi r}{\Theta_o R}, \lambda\Lambda\right]$$

where (r,θ,z) and (R,Θ,Z) locate a material particle in the current and original configuration, respectively, Θ_0 and Λ are the residual stress related "opening angle" and axial stretch, and λ is a load induced axial stretch (note: Θ_0 was introduced by Y.C. Fung as a simple metric of the underlying residual strain; it is obtained by introducing a radial cut in an excised vascular ring). Rachev and Hayashi (1999) suggest that the active-passive behavior can be described via the following relation for the Cauchy (true) stress **t**,

$$\mathbf{t} = -p\mathbf{I} + \mathbf{F}\cdot\frac{\partial W}{\partial \mathbf{E}}\cdot\mathbf{F}^T + T(\lambda_\theta)\mathbf{n}\otimes\mathbf{n} \quad \text{(A2)}$$

where p is a Lagrange multiplier that enforces incompressibility (if desired), **F** is the deformation gradient, **E** $(=0.5(\mathbf{F}^T\cdot\mathbf{F}-\mathbf{I}))$ is the Green strain tensor, W is the pseudostrain-energy function for the passive response, $T(\lambda_\theta)$ is the active stress, with λ_θ the circumferential stretch, and **n** is the direction of the smooth muscle fibers (typically taken to be circumferential). Numerous specific forms of W have been proposed, but the most popular is the Fung-exponential, which can be written as,

$$W = c\left(e^Q - 1\right) \quad \text{(A.3)}_1$$

where,

$$\begin{aligned}
Q = &\; c_1 E_{RR}^2 + c_2 E_{\Theta\Theta}^2 + c_3 E_{ZZ}^2 + \\
&\; 2c_4 E_{RR}E_{\Theta\Theta} + 2c_5 E_{\Theta\Theta}E_{ZZ} + \\
&\; 2c_6 E_{ZZ}E_{RR} + c_7(E_{R\Theta}^2 + E_{\Theta R}^2) + \\
&\; c_8(E_{\Theta Z}^2 + E_{Z\Theta}^2) + c_9(E_{ZR}^2 + E_{RZ}^2)
\end{aligned} \quad \text{(A3)}_2$$

and c and c_i ($i = 1,2,\ldots,9$ for orthotropy) are material parameters. One form for the active stress consistent with the work of Rachev and Hayashi is,

$$T(\lambda_\theta) = A(Ca^{++})\lambda_\theta\left(1 - \frac{\lambda_m - \lambda_\theta}{\lambda_m - \lambda_o}\right) \quad \text{(A4)}$$

where A is a function of the strength of the contraction (~50 kPa for basal tone) and λ_m and λ_0 are the circumferential stretches at which the active force generation is maximum and minimum, respectively.

Independent of the specific forms of W and T, the primary equation of motion both *in vivo* and *in vitro* (in straight segments) is the radial equation. For quasi-static loading, it is

$$\frac{\partial t_{rr}}{\partial r} + \frac{1}{r}\left(t_{rr} - t_{\theta\theta}\right) = 0 \quad \rightarrow$$
$$p(r) = P + t_{rr}^e - \int_{r_i}^r \left(t_{\theta\theta} - t_{rr}\right)\frac{dr}{r} \quad \text{(A5)}$$

where $\mathbf{t}^e$ is the so-called extra stress, which is to say that part of equation (A2) other than the reaction part $(-p\mathbf{I})$. Equation A5 thus allows one to solve for the Lagrange multiplier at each radial location in the deformed configuration, which in turn allows the stress **t** to be computed at each radial location in the wall. See Figures 3.2–3.5. (Note: These equations represent the current state of the art, but they are yet incomplete. The residual strain field appears to be much more complicated than that described by equation A1 via the parameters Θ_0 and Λ; the constitutive equations A3–A4 clearly need refinement, particularly with regard to the active stress component. Recognition of needs provides guidance for future work, but again it is emphasized that the understanding we have gained through these types of relations has been significant.)

References

Arts, T., Reneman, R.S., and Veenstra, P.C. 1979. A model of the mechanics of the left ventricle. *Ann. Biomed. Eng.* 7:299–318.

Buccino, R.A., Harris, E., Spann, J.F. and Sonneblick, E.H. 1969. Response of myocardial connective tissue to development of experimental hypertrophy. *Am. J. Physiol.* 216:425–428.

Chuong, C.J. and Fung, Y.C. 1983. Three-dimensional stress distribution in arteries. *ASME J. Biomech. Eng.* 105:268–274.

Chuong, C.J. and Fung, Y.C. 1986. On residual stress in arteries. *ASME J. Biomech. Eng.* 108:189–192.

Clark, J.M. and Glagov, S. 1985. Transmural organization of the arterial media. *Arteriosclerosis* 5:19–34.

Costa, K.D., Hunter, P.J., Wayne, J.S., Waldman, L.K., Guccione, J.M. and McCulloch, A.D. (1996) A three-dimensional finite element method for large elastic deformations of ventricular myocardium: II-prolate spheroidal coordinates. *J. Biomech. Eng.* 118:464–472.

Davies, P.F. 1995. Flow-mediated endothelial mechanotransduction. *Physiol. Rev.* 75:519–560.

Feldman, S.A. and Glagov, S. 1971. Transmural collagen and elastin gradients in human aortas: Reversal with age. *Atherosclosis* 13:385–394.

Fung, Y.C. 1995. Stress, strain, growth, and remodeling of living organisms. ZAMP 46:S469–482.

Humphrey, J.D. 2001. Stress, strain, and mechanotransduction in cells. *ASME J. Biomech. Eng.* 123:638–641.

Humphrey, J.D. 2002. *Cardiovascular Solid Mechanics: Cells, Tissues and Organs.* Springer-Verlag, New York.

Humphrey, J.D. and Rajagopal, K.R. 2002. A constrained mixture model for growth and remodeling in soft tissues. *Math. Model Meth. Appl. Sci.* 12:407–430.

Hunter, P.J., Nash, M.P. and Sands, G.B. 1997. Computational electromechanics of the heart. In: *Computational Biology of the Heart.* A.V. Panfilov and A.V. Holden, Eds. Wiley & Sons, U.K.

Kang, T., Humphrey, J.D. and Yin, F.C.P. 1996. Comparison of the biaxial mechanical properties of excised endocardium and epicardium. *Am. J. Physiol.* 270:H2169–H2176.

Langille, B.L. 1993. Remodeling of developing and mature arteries: endothelium, smooth muscle, and matrix. *J. Cardiovasc. Pharmacol.* 21:S11–S17.

Liu, S.Q. and Fung, Y.C. 1989. Relationship between hypertension, hypertrophy, and opening angle of zero-stress state of arteries following aortic coarctation. *ASME J. Biomech. Eng.* 111:325–335.

Matsumoto, T. and Hayashi, K. 1996. Stress and strain distribution in hypertensive and normotensive rat aorta considering residual strain. *ASME J. Biomech. Eng.* 118:62–73.

Nerem, R.M., Braddon, L.G., and Seliktar, D. 1998. Tissue engineering and the cardiovascular system. In: *Frontiers in Tissue Engineering.* C.W. Patrick, A.G. Mikos, L.V. McIntire, Eds. Pergamon, U.K.

Novak, V.P., Yin, F.C.P. and Humphrey, J.D. 1994. Regional mechanical properties of passive myocardium. *J. Biomech.* 27:403–412.

Rachev, A. and Hayashi, K. 1999. Theoretical study of the effects of vascular smooth muscle contraction on strain and stress distributions in arteries. *Ann. Biomed. Eng.* 27:459–468.

Rachev, A., Stergiopulos, N. and Meister, J-J. 1998. A model for geometric and mechanical adaptation of arteries to sustained hypertension. *ASME J. Biomech. Eng.* 120:9–17.

Roy, C.S. 1880. The elastic properties of the arterial wall. *Phil. Trans. R. Soc. Lond. B* 99:1–31.

Stenmark, K.R. and Mecham, R.P. 1997. Cellular and molecular mechanisms of pulmonary vascular remodeling. *Annu. Rev. Physiol.* 59:89–144.

Strumpf, R.K., Humphrey, J.D. and Yin, F.C.P. 1993. Biaxial mechanical properties of passive and tetanized canine diaphragm. *Am. J. Physiol.* 265:H469–475.

Taber, L.A. 1995. Biomechanics of growth, remodeling, and morphogenesis. *Appl. Mech. Rev.* 48:487–545.

Taber, L.A. and Humphrey, J.D. 2001. Stress-modulated growth, residual stress, and vascular heterogeneity. *ASME J. Biomech. Eng.* 123:528–535.

Vaishnav, R.N., Young, J.T. and Patel, D.J. 1973. Distribution of stresses and strain-energy density through the wall thickness in a canine aortic segment. *Circ. Res.* 32:577–583.

von Maltzahn, W.W., Besdo, D. and Wiemer, W. 1981. Elastic properties of arteries: A nonlinear two-layer cylindrical model. *J. Biomech.* 14:389–397.

Weber, K.T. 1989. Cardiac interstitium in health and disease. The fibrillar collagen network. *J. Am. Coll. Cardiol.* 13:1637.

Zahalak, G.I., de Laborderie, V. and Guccione, J.M. 1999. The effects of cross-fiber deformation on axial fiber stress in myocardium. *ASME J. Biomech. Eng.* 12:376–385.

4

Functional Properties of Native Articular Cartilage

Gerard A. Ateshian and Clark T. Hung

Introduction

Articular cartilage is the bearing material of diarthrodial joints. Its primary mechanical function is to transmit large loads across the articular surfaces of joints with minimal friction and wear; under normal conditions, cartilage can maintain this function for seven to eight decades. From an engineering perspective, the mechanical behavior of cartilage is considered to be remarkable, unmatched by any traditional engineering bearing material. However, despite several decades of sophisticated biomechanical studies of cartilage, an accurate understanding of cartilage mechanics remains elusive due to its remarkable versatility and complexity. With each new testing modality, cartilage exhibits yet another level of complexity, often raising more questions than providing answers.

A proper understanding of the unique mechanical properties of articular cartilage is essential in several respects. Firstly, from a basic science perspective, a knowledge of the normal mechanical function of cartilage can provide an explanation as to how cartilage can withstand the relatively harsh environment of diarthrodial joints; it is notable that, despite considerable work in the area of cartilage mechanics, the complete state of stress and strain within contacting articular cartilage layers has not been reliably determined, and consequently the physio-logic loading environment of chondrocytes has not been well characterized. Secondly, proper knowledge of cartilage properties can lead to a better understanding of the pathomechanical processes that might lead to cartilage degeneration and osteoarthritis through mechanical pathways (Fig. 4.1). This may serve to potentially avoid or retard such degenerative processes through various clinical modalities that properly recognize the favorable and unfavorable loading conditions of cartilage. Finally, knowledge of the loading environment of articular cartilage (such as extracellular matrix stresses and strains and interstitial fluid hydrostatic pressure) is essential for the development of strategies to enhance tissue healing and tissue engineering of cartilage substitutes; furthermore, the functional properties of such healed or engineered tissues can be correctly compared with those of normal cartilage.

Studies have demonstrated that the mechanical response of articular cartilage may vary as a function of duration and rate of loading or deformation, that is, cartilage exhibits viscoelasticity. Furthermore, it has been shown that the tensile stiffness of cartilage differs when testing the tissue parallel and perpendicular to the split line directions, that is, it exhibits anisotropy. It has also been established that the stiffness of cartilage in compression may be one to two orders of magnitude smaller than in tension, that is, it ex-

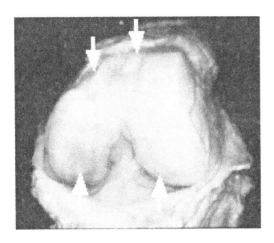

FIGURE 4.1. Distal femoral articular surface from a human knee joint. Arrows indicate osteoarthritic defects.

hibits tension-compression nonlinearity. Various studies have also confirmed that these measured properties may vary from the superficial to the deep zone of cartilage, that is, the tissue exhibits depth-dependent inhomogeneity. These complex properties have been recognized in the literature, though they have neither been completely characterized, nor are they all well understood.

Bioengineers have recognized that an essential step toward a better understanding of articular cartilage mechanics is the development of appropriate constitutive models for this tissue, which relate the state of stress to the state of strain, and fluid and solute fluxes to electrochemical gradients. These mathematical models, when validated through experiments, can provide valuable insight into the functioning of this tissue. Nevertheless, to date, no single constitutive model of articular cartilage has been able to describe its mechanical response to all the testing conditions described in the literature.

Basic Articular Cartilage Structure and Composition

The organic matrix of cartilage is composed of a dense network of fine collagen (type II) fibrils enmeshed in a concentrated solution of proteoglycans. The collagen content of cartilaginous

tissue ranges from 10% to 30% by wet weight and the proteoglycan content from 3% to 10% by wet weight (Muir, 1980); the remaining 60%–87% is water, inorganic salts, and small amounts of other matrix proteins, glycoproteins, and lipids (Linn and Sokoloff, 1965; Maroudas, 1979). Chondrocytes, the sparsely distributed cells in articular cartilage, account for less than 10% of the tissue's volume (Muir, 1980; Stockwell, 1979). Collagen fibrils and proteoglycans are the structural components supporting the internal mechanical stresses that result from loads being applied to the joint cartilage. Moreover, these structural components, together with water, determine the biomechanical behavior of this tissue (e.g., Kempson et al., 1976; Mow et al., 1980; Zhu et al., 1993). The collagen in articular cartilage is inhomogeneously distributed, giving the tissue a layered character. Three separate structural zones have been identified. In the superficial tangential zone (STZ), which represents 10–20% of the total thickness, there are sheets of fine, densely packed fibers randomly woven in planes parallel to the articular surface (Redler and Zimny, 1970; Weiss et al., 1968). In the middle zone (40–60% of the total thickness), there are greater distances between the randomly oriented and homogeneously dispersed fibers. Below this, in the deep zone (about 30% of the total thickness), the fibers come together forming larger, radially oriented fiber bundles. These bundles then cross the tidemark (Bullough and Jagannath, 1983), the interface between articular cartilage and the calcified cartilage beneath it. This anisotropic fiber orientation is mirrored by the inhomogeneous zonal variations in the collagen content, which is highest at the surface and then remains relatively constant throughout the deeper zones (Lipshitz et al., 1976). Proteoglycans are a large protein-polysaccharide molecule composed of a protein core to which one or more glycosaminoglycans are attached (Buckwalter et al., 1985; Hardingham amd Fosang, 1992; Heinegard and Oldberg, 1989; Muir, 1980). Proteoglycan aggregation promotes immobilization of the proteoglycan within the fine collagen meshwork adding structural rigidity to the extracellular matrix (Muir, 1980). Furthermore, cartilage proteoglycans are inhomogeneously distributed throughout the matrix with their concentration generally being highest in the middle zone and lowest in the superficial and deep zones (Maroudas et al., 1969; Venn, 1978). Water, the most abundant component of articular cartilage, is most concentrated near the articular surface (~80%) and decreases in a near linear fashion with increasing depth to a concentration of approximately 65% in the deep zone (Lipshitz et al., 1976). This fluid contains many free mobile cations (e.g., Na^+, K^+ and Ca^{2+}) that greatly influence the mechanical and physicochemical behaviors of cartilage. When loaded, about 70% of the water may be moved (Maroudas, 1979; Torzilli et al., 1982). As will be described below, variations in the structure and composition of cartilage through the depth are closely related to the inhomogeneous and anisotropic mechanical properties of the tissue.

Physiologic Loading Environment of Articular Cartilage

The loading environment of diarthrodial joints is generally well understood. Various classical studies have reported the magnitude of physiologic loads acting across lower and upper extremity joints (Cooney and Chao, 1977; Paul, 1967; Poppen and Walker, 1978; Rydell, 1965). In general, it has been found that the peak magnitudes of such loads are a multiple factor of body weight (BW) in the lower extremities, for example, 2.5 to 4.9 BW in the hip during walking (Armstrong et al., 1979; Paul, 1967; Rydell, 1965) and 3.4 BW in the knee (Paul, 1967), or comparable to body weight in the upper extremities, for example, 0.9 BW in the glenohumeral joint during abduction (Poppen and Walker, 1978). Under normal circumstances, the loading duration of diarthrodial joints is generally cyclical and/or intermittent (Dillman, 1975; Paul, 1967), even for seemingly static activities such as standing or sitting, which involve intermittent shifting of the body weight to relieve loading of the joints; similarly, upper extremity activities rarely involve sustained static loading for durations in excess of a few minutes, though sustained dynamic (cyclical) loading may occur over a half-hour or more. Joint loads result in

contact stresses at the articular surfaces, which have been extensively measured in the literature using various techniques (Ahmed, 1983; Ahmed and Burke, 1983a, 1983b, Brown and Shaw, 1983, 1984; Fukubayashi and Kurosawa, 1980; Huberti and Hayes, 1984, 1988; Kurosawa et al, 1980; Manouel et al., 1992; Singerman et al., 1987; Stormont et al., 1985; Tencer et al., 1988). Typically, it has been found that activities of daily living produce mean contact stresses on the order of 2 MPa, while moderately strenuous activities result in mean contact stresses up to 6 MPa (Ahmed and Burke, 1983a, 1983b; Brown and Shaw, 1983, 1984; Huberti and Hayes, 1984, 1988; Tencer et al., 1988); it has been estimated that the largest magnitude of mean contact stresses that may occur under nontraumatic conditions is about 12 MPa (Matthews et al., 1977), although *in vivo* measurements using instrumented endoprostheses (which relate indirectly to cartilage-on-cartilage contact) have reported contact stresses as high as 18 MPa (Hodge et al., 1989). While in situ cartilage contact stresses have been extensively investigated, there is less information about in situ cartilage deformation. One radiographic cadaver study of hip joints reported a reduction of cartilage thickness by 20% or less in normal intact joints under physiologic loading of 5 BW (Armstrong et al., 1979); another similar radiographic study has been reported on porcine joints (Wayne et al., 1998). Ultrasound measurements of cartilage thickness in a cadaver hip experiment similarly demonstrated changes in cartilage thickness on the order of 10% or less, under 1.2 BW (Macirowski et al., 1994). *In vivo* magnetic resonance imaging (MRI) measurements of cartilage volumetric changes in the knees of human volunteers, prior and subsequent to strenuous activities, has demonstrated a reduction of 6% in cartilage volume (Eckstein et al., 1998). Theoretical contact analyses of biphasic cartilage layers under rolling or sliding motion have demonstrated that in a congruent joint, the cartilage layer thickness decreases by 6% under a contact load of 1 BW (Ateshian and Wang, 1995).

Knowledge of the contact stresses at the articular surface is however insufficient to fully determine the state of stress inside cartilage. Recent joint contact and fluid pressure measurement studies (Ateshian and Wang, 1995; Ateshian et al., 1994; Donzelli and Spilker, 1998; Kelkar and Ateshian, 1999; Macirowski et al., 1994; Oloyede and Broom, 1991; Soltz and Ateshian, 1998, 2000a; Van Der Voet et al., 1993) (see below) have confirmed the hypothesis that the interstitial water of articular cartilage pressurizes considerably when the joint is loaded (Lai and Mow, 1980; Linn and Sokoloff, 1965; McCutchen, 1962; Zarek and Edwards, 1963), contributing significantly to supporting the load transmitted across the articular layers. Various analyses have suggested that the hydrostatic fluid pressure that develops in the interstitial water may contribute up to 90% or more of the contact stress measured at the articular surface (Ateshian and Wang, 1995; Ateshian et al., 1994; Kelkar and Ateshian, 1999; Macirowski et al, 1994; Oloyede and Broom, 1993). Thus, if a mean contact stress of 6.0 MPa is produced at the articular surfaces under physiologic loading, the cartilage interstitial fluid would pressurize to a mean value of approximately 5.4 MPa (depending on the joint congruence, cartilage properties, and loading rate). This pressurization occurs because the interstitial water attempts to squeeze out of the loaded region, but is impeded by the extremely low permeability of the collagen matrix. The significant contribution of interstitial fluid pressurization to the load support explains the findings that in situ cartilage deformation is generally of moderate magnitude (~20% of the thickness or less). While the interstitial fluid pressure has been shown to subside after several hours under purely static loading in a controlled *in vitro* laboratory environment (Grodzinsky et al., 1978; McCutchen, 1962; Mow and Lai, 1980; Mow et al., 1989; Oloyede and Broom, 1991; Soltz and Ateshian, 1998), recent experimental studies (Soltz and Ateshian, 2000a) have suggested that pressure subsidence and tissue consolidation are not likely to occur under *in vivo* physiologic conditions; therefore cartilage deformation caused by physiologic joint loading is always accompanied by significant interstitial fluid pressurization, that is, cartilage deformation and interstitial fluid hydrostatic pressurization are synchronous and inseparable mechanisms *in vivo*. Furthermore, the normal physiologic loading environment can

never be truly static; it is intermittent or cyclical, with loading duration ranging from fractions of a second (e.g., during gait; Dillman, 1975; Paul, 1967) to a few minutes. These findings suggest that the normal loading environment of chondrocytes involves a combination of intermittent cyclical hydrostatic fluid pressurization and moderate matrix deformation.

Cartilage Mechanics

Overview

Investigations of cartilage mechanics have covered several important topics. One of the fundamental elements of tissue mechanics is to formulate a constitutive model for the material being analyzed; such a model provides a mathematical relation between the stress and strain in the material, for example, which needs to be validated from experiments. Different constitutive models may require a different set of material properties to relate the stress to the strain, thus it is not possible to report experimental measurements of cartilage material properties without first proposing a constitutive model for which those properties are obtained. A validated model is one which can successfully predict the experimental response of cartilage under a wide variety of loading conditions, using the same material properties. Since the 1940s, several different constitutive models have been proposed or adopted for articular cartilage, typically spanning from the classical theory of elasticity (Atkinson et al., 1998a; Eberhardt et al., 1990, 1991; Hirsch, 1944; Kelly and O'Connor, 1996a, 1996b; Mente and Lewis, 1994; Sokoloff, 1966), to viscoelasticity (Hayes and Bodine, 1978; Hayes and Mockros, 1971; Simon et al., 1984; Woo et al., 1980; Zhu et al., 1993), to microstructural theories of composite materials (Bursac et al., 2000; Schwartz et al., 1994; Wren and Carter, 1998), porous media theories (Armstrong and Mow, 1982, 1983; Armstrong et al., 1984; Ateshian and Wang, 1995; Ateshian et al., 1994, 1997; Cohen et al, 1994, 1998; Holmes and Mow, 1990; Holmes et al., 1985; Hou et al., 1989; Kwan et al., 1984, 1990; Lai and Mow, 1980; Lai et al., 1981; Lee et al., 1981; Mow et al., 1980, 1989, 1992, 1998; Setton et al., 1993, 1997; Soltz et al., 1998, 2000a, 2000b; Soulhat et al., 1999; Spilker et al., 1990, 1992; Suh et al., 1995; Torzilli and Mow, 1976; van der Voet et al., 1993; Wayne et al., 1991a, 1991b; Wu et al., 1996, 1997, 1998;), physical chemistry theories (Grushko et al., 1989; Katz et al., 1986; Maroudas, 1970, 1972, 1976, 1979; Maroudas and Bannon, 1981; Maroudas et al., 1968, 1985, 1991; Mizrahi et al., 1986; Schneiderman et al., 1986; Tobias et al., 1992; Wachtel and Maroudas, 1998), and combinations thereof (Berkenblit et al., 1994; Bonassar et al., 1994, 1995a, 1995b; Buschmann and Grodzinsky, 1995; Eisenberg and Grodzinsky, 1985, 1987; Frank and Grodzinsky, 1987a, 1987b; Grodzinsky et al., 1978, 1981; Gu et al., 1993, 1997, 1998; Kim et al., 1994, 1995; Lai et al., 1991; Lee et al., 1981; Sachs and Grodzinsky, 1995; Sah et al., 1989; Setton et al., 1995a, 1998). Material properties for these various models have been determined experimentally using various testing configurations; these include, most commonly, in situ indentation of a cartilage layer with a flat- or spherical-tip, impervious, or free-draining porous indenter (Hale et al., 1993; Hayes et al., 1972; Hirsch, 1944; Hori and Mockros, 1976; Kempson et al., 1971; Lyyra et al., 1995; Mak et al., 1987; Mow et al., 1989; Shin et al., 1997; Spilker et al., 1992); uniaxial tensile testing of cartilage strips (Akizuki et al., 1986, 1987; Bader et al., 1981; Grodzinsky et al., 1981; Kempson, 1982, 1991; Kempson et al., 1973; Roth and Mow, 1980; Schmidt et al., 1990; Woo et al., 1976, 1979, 1980); confined compression of cartilage cylindrical plugs in a chamber with impermeable side wall, using a free-draining porous filter (Armstrong and Mow, 1982, 1983; Ateshian et al., 1997; Bursac et al., 2000; Grodzinsky et al., 1978; Holmes and Mow, 1990; Holmes et al., 1985; Khalsa and Eisenberg, 1997; Lai et al., 1981; Lee et al., 1981; Mow et al., 1980; Schinagl et al., 1996, 1997; Setton et al., 1993; Soltz and Ateshian, 1998, 2000a); unconfined compression of cartilage cylindrical plugs between impermeable platens (Armstrong et al., 1984; Brown and Singerman, 1986; Bursac et al., 1999; Cohen et al., 1998; Guilak et al., 1995; Jurvelin et al., 1997; Lanir, 1987; Mizrahi et al., 1986; Soltz and Ateshian, 2000b; Soulhat

et al., 1999; Spilker et al., 1990); shearing of rectangular blocks and torsional shearing of cylindrical plugs (Hayes and Bodine, 1978; Hayes and Mockros, 1971; Setton et al., 1995b; Woo et al., 1987; Zhu et al., 1993, 1994); and other related testing configurations. Other experimental studies have focused on the failure mechanisms of articular cartilage during impact loading or under standard failure tests (Atkinson et al., 1995, 1998a,b; Chin-Purcell and Lewis, 1996; Haut, 1989; Haut et al., 1995; Kelly and O'Connor, 1996b; Li et al., 1995; Newberry et al., 1998; Oegema et al., 1993; Vener et al., 1992; Wren and Carter, 1998), to better understand the various injury mechanisms that may occur in diarthrodial joints. Theoretical and finite element analyses have also been employed to investigate the state of stress inside cartilage, either under a standard testing configuration or under configurations representative of diarthrodial joint contact (Ateshian and Wang, 1995; Ateshian et al., 1994; Donzelli and Spilker, 1998; Donzelli et al., 1999; Eberhardt et al., 1990, 1991; Hayes et al., 1972; Kelkar and Ateshian, 1999; Levenston et al., 1996; Spilker et al., 1990, 1992; Suh and Bai, 1998; van der Voet et al., 1993; Wayne et al., 1991a, 1991b; Wu et al., 1998). The majority of the above studies have been motivated by the need to better understand the function of normal articular cartilage and the mechanical pathways by which the tissue might degenerate (Armstrong and Mow, 1982; Berkenblit et al., 1994; Grushko et al., 1989; Guilak et al., 1994; Oegema et al., 1993; Radin et al., 1991; Setton et al., 1995b, 1997; Vener et al., 1992), as well as to analyze structure-composition-function relationships in the tissue (Bonassar et al., 1995; Brocklehurst et al., 1984; Linn and Sokoloff, 1965; Lipshitz et al., 1975; Maroudas and Venn, 1977; Setton et al., 1997; Venn, 1978; Venn and Maroudas, 1977). From these various studies, it has become evident that cartilage exhibits many complexities, such as viscoelasticity, anisotropy, inhomogeneity, tension-compression nonlinearity, and mechano-electrochemical transduction.

Cartilage Permeability

The collagen-proteoglycan matrix of articular cartilage is porous and permeable to the interstitial fluid. The velocity of this fluid relative to the solid phase is regulated by the gradient in water chemical potential and the electrochemical potential of ions in the fluid, with the dominant effect generally attributed to the water pressure gradient. The most commonly used constitutive law that describes how hydrostatic pressure gradients regulate the flow of interstitial fluid in cartilage is Darcy's law, which linearly relates the relative fluid flux to the pressure gradient with a proportionality constant (a material property) known as the tissue hydraulic permeability. Cartilage permeability has been measured either through direct permeation experiments (Gu et al., 1995; Mansour and Mow, 1976; Maroudas et al., 1968), or indirectly from measuring the transient, flow-dependent response of cartilage under creep, stress-relaxation, or dynamic loading (Ateshian et al., 1997; Athanasiou et al, 1991; Frank and Grodzinsky, 1987; Lai et al., 1981; Lee et al., 1981; Mow et al., 1980). It has been observed that permeability is sensitive to the amount of cartilage deformation, or strain (Mansour and Mow, 1976), thus some studies have extended the constitutive relation for fluid flux by employing a permeability function describing that relation (Holmes and Mow, 1990; Lai and Mow, 1980). Most of the studies of cartilage permeability have reported consistent measurements of permeability, which ranges from 8×10^{-15} m^4/N·s for normal cartilage under small strains, down to the order of 1×10^{-16} m^4/N·s for cartilage under 50% compression. The low permeability of articular cartilage considerably retards fluid flow within the tissue upon loading; it is one of the factors contributing to the large time constant of cartilage, that is, the characteristic time for the tissue to reach equilibrium conditions under a constant load, which is on the order of hundreds to thousands of seconds, typically well in excess of physiologic loading durations.

Articular Cartilage Viscoelasticity

Numerous experiments have confirmed that the response of cartilage is time-dependent, exhibiting creep and stress-relaxation, and varies with the rate of loading or deformation; thus, cartilage exhibits viscoelasticity. There may be potentially two sources for the viscoelasticity of cartilage: 1) "intrinsic" viscoelasticity of the solid collagen-proteoglycan matrix, which arises from internal friction of the solid component molecules (viscoelastic solid); and 2) frictional drag between fluid and solid components, as fluid permeates through the solid matrix. Some of the early studies investigating cartilage viscoelasticity proposed models that attribute the viscoelasticity primarily to the solid phase (Hayes and Bodine, 1978; Hayes and Mockros, 1971; Woo et al., 1976, 1980). Hayes and Bodine (1978) argued that by testing cartilage under small shear deformations, no change in volume would occur in the tested samples, as predicated by the theory of infinitesimal deformations (e.g., Lai et al., 1993); this powerful argument signifies that there should be no flow of the interstitial water into or out of the solid matrix during shear deformations because of the principle of conservation of mass and the near-incompressibility of water and collagen. Thus, dynamic shear testing can be used to assess the intrinsic viscoelasticy of the solid phase of cartilage alone, and indeed significant viscoelastic effects were reported by Hayes and co-workers (1971, 1978) as well as by other investigators who used that very same argument to interpret shear testing results (Setton et al., 1995b; Zhu et al., 1993, 1994). From the studies of Zhu et al. (1993), the dynamic shear modulus of immature bovine knee cartilage was found to increase on average from 0.19 MPa at a loading frequency of 0.01 Hz, to 0.38 MPa at 10 Hz, under an axial compression of 3%. At 16% axial compression, the dynamic shear modulus increased on average from 1.0 MPa at 0.01 Hz to 1.8 MPa at 10 Hz. The phase shift angle, which is a measure of viscoelastic energy dissipation under dynamic loading, ranged from 9° to 15° on average (a purely elastic material having 0° phase shift). The uniaxial tensile response of cartilage has also been analyzed with the quasilinear viscoelasticity (QLV) theory of Fung (Fung, 1981; Woo et al., 1980), however, only limited data has been presented in the literature on the magnitude of the material constants of this theory for articular cartilage. The study of Woo et al. (1980) provides one representative set for bovine humeral head articular cartilage, $c = 2.0$, $\tau_1 = 0.006$ s, $\tau_2 = 9.4$ s, where [$1/\tau_2$, $1/\tau_1$] represents the frequency range over which most of the viscoelastic energy dissipation occurs under dynamic loading, whereas $(1 + c \ln \tau_2/\tau_1)$ is the ratio of instantaneous to equilibrium modulus under sudden loading.

Conversely, many studies suggest that cartilage viscoelasticity results primarily from the flow of interstitial fluid within the solid matrix (i.e., frictional drag between the solid and fluid) (Frank and Grodzinsky, 1987a; Lai and Mow, 1980; McCutchen, 1962; Mow et al., 1980, 1998; Oloyede and Broom, 1991, 1993; Zarek and Edward, 1963; Torzilli and Mow, 1976). Most of these studies employed various forms of porous media theories, most notably mixture theory (Bowen, 1976; Kenyon, 1976; Mow et al., 1980) and poroelasticity (Biot, 1962), where the solid matrix is assumed to be elastic, porous-permeable, and filled with fluid. In these models, the frictional drag between the solid and fluid phase is typically assumed to be proportional to the relative velocity of the two phases, with the proportionality constant inversely related to the tissue permeability (Lai and Mow, 1980) (in agreement with Darcy's law for flow through porous media). The biphasic mixture theory of Mow et al. (1980) was the first of these models to demonstrate good agreement between theory and experiment in confined compression of cartilage cylindrical plugs, predicting a tissue permeability that compared favorably with direct permeation experiments (Mansour and Mow, 1976). From the confined compression studies of Armstrong and Mow (1982) on human patellar cartilage, the confined compression equilibrium modulus (the "aggregate modulus," often denoted by H_A) was found to range from 0.3 to 1.6 MPa. For immature bovine cartilage, Soltz and Ateshian (1998) reported a confined compression aggregate modulus ranging on average from 0.55 to 0.97 MPa, and permeability k ranging from 2.9 to 6.2×10^{-16} m^4/N·s. An example of experimental response under confined compression creep is given in Figure 4.2a; under a constant applied load, the cartilage deformation slowly increases to an equilibrium value, exhibiting its characteristic viscoelastic response. A simultaneous measurement of the interstitial fluid pressure at the specimen face opposite the porous loading indenter demonstrates that the

fluid immediately pressurizes upon loading but that this pressure eventually subsides as the tissue reaches its equilibrium deformation (Figure 4.2b) (Soltz and Ateshian, 1998). For this confined compression configuration, the measured fluid pressure equals the aplied total stress at the instant of loading, emphasizing the importance of fluid load support in articular cartilage. Furthermore, very good agreement can be observed between this experimental response and the corresponding theoretical analysis from the biphasic theory of Mow et al. (1980).

Nevertheless, some studies showed that agreement between theory and experiment often depended on the testing configuration (Armstrong et al., 1984; Athanasiou et al., 1991; Brown and Singerman, 1986; Bursac et al., 1999; Buschmann et al., 1998; Cohen et al., 1998; Mow et al., 1989; Setton et al., 1993). Not surprisingly, some authors have explored the possibility that both intrinsic solid matrix viscoelasticity and diffusive drag play equally important roles in the viscoelasticity of articular cartilage. Mak (1986) developed a poroviscoelastic biphasic theory that combines the biphasic model of Mow et al. (1980) with the QLV theory of Fung (1981) for the solid phase. This model was also adopted by Setton et al. (1993), who demonstrated improved theoretical curvefits of confined compression experiments (with $H_A = 0.54$ MPa, $k = 5.0 \times 10^{-14}$ m^4/N·s, $c = 0.39$, $\tau_1 = 0.06$ s and $\tau_2 = 8.0$ on average, for intact bovine knee cartilage), and by Suh and Bai (1997, 1998), who also showed a good curvefit of indentation results (with $H_A = 0.81$ MPa, $k = 0.8 \times 10^{-15}$ m^4/N·s, $c = 2.3$, $\tau_1 = 0.001$ s and $\tau_2 = 30$ s for a representative sample of bovine knee cartilage).

Finally on this topic, some studies have suggested that since the characteristic time constant of the viscoelastic response of cartilage is on the order of hundreds or thousands of seconds, while physiologic loading durations are typically on the order of seconds or less, it may be reasonable to model cartilage as an elastic material, since rapid loading and unloading does not exhibit substantial energy dissipation in the material. However, it should be noted that the dynamic modulus of cartilage in compression is highly dependent upon interstitial fluid pressurization; thus, even though interstitial fluid flow

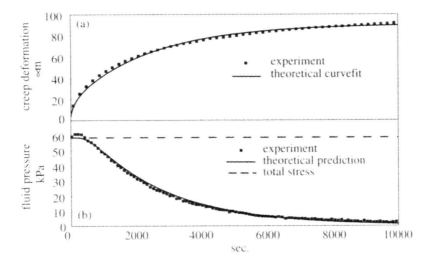

FIGURE 4.2. (*a*) Creep deformation response of articular cartilage from a bovine joint, showing experimental measurements and theoretical curvefit using the linear isotropic biphasic theory of Mow et al. (1980). (*b*) Corresponding interstitial fluid pressure response on surface opposite to the porous loading indenter, showing experimental measurements and theoretical prediction from the biphasic theory. The total applied stress is also indicated on the figure, confirming that the interstitial fluid pressure matches the applied stress immediately upon loading, before eventually subsiding to zero (Soltz and Ateshian, 1998).

may not be significant during physiological loading, the interstitial fluid pressurization contributes considerably to the cartilage load support. The dynamic confined compression studies of Lee et al. (1981), Frank and Grodzinsky (1987a, 1987b), and Soltz and Ateshian (2000a), and the dynamic unconfined compression experiments of Kim et al. (1995) and Buschmann et al. (1999) illustrate the frequency-dependent variation in cartilage dynamic modulus.

Articular Cartilage Anisotropy

The anisotropic response of articular cartilage has been investigated primarily from uniaxial tension tests (Huang et al., 1999; Kempson et al., 1968, 1973; Roth and Mow, 1980; Woo et al., 1976, 1979) though in unconfined compression as well (Mizrahi et al., 1986). It has been shown that the equilibrium tensile modulus of articular cartilage is significantly greater for specimens harvested parallel to the split line directions than for those perpendicular to the split lines. For example, Huang et al. (1999) reported an average tensile equilibrium modulus of human humeral

head cartilage of 7.8 MPa, at small strains, in the superficial zone when measured parallel to the local split line direction, and 5.9 MPa when measured perpendicular to the split line direction, under the same conditions; this disparity was exacerbated at higher strains, with equilibrium moduli averaging 43 MPa parallel to the split lines, and 26 MPa perpendicular to the split lines. The differences in tensile properties along these two directions is a manifestation of tissue anisotropy. Since the tensile properties of cartilage have been found to differ along these two mutually perpendicular directions, there is a very high likelihood that cartilage is at best orthotropic, with its three planes of symmetry defined in situ by the split line direction in a plane tangent to the surface (1-direction), the perpendicular to the split line direction in the same tangent plane (2-direction), and the direction normal to this plane (3-direction), i.e., the "radial" direction of the cartilage layer (Fig. 4.3). A further complication is that the orientation of split lines varies from site to site on the articular layer. Because the standard protocol for testing cartilage in tension has generally required speci-

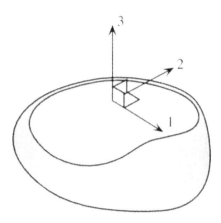

FIGURE 4.3. An orthogonal coordinate system is defined at each surface point based on the local split line direction. 1-direction: parallel to split line; 2-direction: perpendicular to split line; 3-direction: normal to the articular surface.

mens several millimeters in length, it is not surprising that there have been few results reported in the literature on the tensile properties of cartilage along the radial 3-direction, with the exception of the osmotic swelling studies of Basser et al. (1998), who report a tensile modulus of 4 MPa for human femoral head cartilage, and Narmoneva et al. (1999), who report an average tensile modulus of 26 MPa for cartilage from the knees of mature mongrel dogs, along this radial direction.

There exist further literature data in support of cartilage anisotropy in tension; several studies (Chang et al., 1999; Elliott et al., 1999; Huang et al., 1999; Woo et al., 1979) have demonstrated that Poisson's ratio (the ratio of lateral to axial strain) for cartilage in uniaxial tension is in excess of 0.5, which is permissible only for anisotropic materials (e.g., Lai et al., 1993). For human humeral head cartilage, Huang et al. (1999) reported an average Poisson ratio value of $v_{12} = 1.3$ for uniaxial tension along the 1-direction and measurement of the contraction along the 2-direction, and a similar average of $v_{21} = 1.3$ for the converse configuration, in the superficial zone of cartilage; in the middle zone, these values reduced to $v_{12} = 1.2$ and $v_{21} = 1.0$. For human patellar cartilage, Chang et al. (1999) reported average values of $v_{12} = 0.9$ and $v_{13} = 1.8$

in the surface-to-middle zone, and $v_{12} = 0.5$ and $v_{13} = 0.7$ in the deep zone, whereas Elliott et al. (1999) reported $v_{12} = 2.2$ in the surface zone and $v_{12} = 0.6$ in the middle zone. All these measurements suggest that cartilage decreases in volume when subjected to tension; coupled with our understanding of the cartilage response in compression, this means that the cartilage interstitial fluid pressure is always compressive, and fluid exudes from the tissue, whether the tissue is loaded in compression or in tension. This would not be true of an isotropic porous material, where the interstitial fluid pressure and fluid flow direction would reverse signs under tensile versus compressive loading.

Interestingly, in contrast to tensile measurements, there are only two preliminary studies that have investigated the anisotropy of cartilage in compression; Jurvelin et al. (1996) have reported measurements of the compressive modulus of cartilage in unconfined compression, on cylindrical specimens whose axis was oriented either along the radial 3-direction (which is the most commonly tested direction in compression studies) or along a direction parallel to the cartilage surface (though no relation to the split line direction was reported). They found the compressive modulus of human knee cartilage in the 3-direction to be smaller (0.58 MPa versus 0.85 MPa), suggesting that cartilage is also anisotropic in compression. However, in a study by Soltz et al. (1999) where immature bovine carpometacarpal joint cartilage cubes were tested in unconfined compression along the 1-, 2-, or 3-directions, no significant differences were found in the compressive moduli among the three directions (with mean values ranging from 0.41 MPa along the 1-direction to 0.47 MPa in the 3-direction). From compression and indentation experiments, Poisson's ratio at equilibrium has been found to be small in human and bovine articular cartilage, whether measured indirectly from theoretical curvefits of experimental data (Athanasiou et al., 1991; Froimson et al., 1997; Khalsa and Eisenberg, 1997; Mak et al., 1987; Mow et al., 1989) or directly using optical methods (Jurvelin et al., 1997; Wang et al., 2000; Wong, 1999; Wong et al., 1998). Optical measurements from unconfined compression have yielded an average equilibrium Poisson ratio of 0.18 in bovine humeral head cartilage (Jurvelin et al., 1997) and 0.06 in bovine carpometacarpal joint cartilage (Wang et al., 2000). Furthermore, measurements of the instantaneous Poisson ratio (immediately upon loading) in unconfined compression yielded an average value of 0.47 (Wong, 1999), approaching the theoretical value of 0.5 predicted for a biphasic material with intrinsically incompressible solid and fluid phases (Armstrong et al., 1984), and in agreement with the findings of cartilage incompressibility reported by Bachrach et al. (1997). While all of these results in compression suggest that cartilage does not exhibit the same anisotropy in compression as it does in tension, it does not necessarily follow that it is isotropic in compression. Based on literature findings to date, it can be hypothesized that cartilage exhibits material symmetry no higher than orthotropy in tension and no lower than cubic symmetry in compression.

In addition to tensile and compressive studies, it is also theoretically possible to investigate cartilage anisotropy in shear, e.g., by testing cartilage cylindrical specimens harvested with their axis along the 1-, 2-, or 3-direction, using a torsion apparatus. Such experiments have not yet been reported for articular cartilage but are available for meniscal tissue where statistically significant but small differences in the three shear moduli have been found (Zhu et al., 1994). Finally, it is of interest to note that the vast majority of theoretical models of cartilage employed in the literature, whether elastic, viscoelastic or biphasic, have assumed that the material is isotropic. Notable exceptions are select studies (Bursac et al., 1999; Cohen et al., 1993, 1998; Lanir, 1987; Mow and Mansour, 1977) that assumed transverse isotropy.

Inhomogeneity of Articular Cartilage Properties

As indicated above, it is well established that the composition and structure of cartilage vary through the depth of the tissue. Not surprisingly, it has been observed that the mechanical properties of cartilage also vary through the depth. Indeed, the tensile stiffness of normal cartilage in the STZ has been reported to be significantly

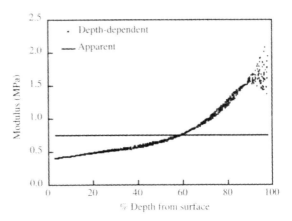

FIGURE 4.4. Depth-dependent unconfined compression modulus of articular cartilage from immature bovine carpometacarpal joint cartilage (Wang et al., 2000).

greater than in the deep zone (Akizuki et al., 1986, 1987; Kempson et al., 1968, 1976; Roth and Mow, 1980; Woo et al., 1976), by up to a factor of 20. Akizuki et al. (1986) found that the cartilage equilibrium tensile modulus measured along the split line direction on the lateral patellar groove of human knee joints decreased from 6.2 MPa in the superficial zone to 0.93 MPa in the middle-to-deep zone, in a high-weight-bearing region, and from 21 MPa down to 1.0 MPa in a low-weight-bearing region. In a series of recent studies (Guilak et al., 1995; Schinagl et al., 1996, 1997; Shin et al., 1997; Wang et al., 2000), significant inhomogeneity of cartilage has also been reported in compression. For example, Shinagl et al. (1996, 1997) used a fluorescence microscopy technique for measuring cartilage strain with chondrocytes as fiducial markers during confined compression; they reported that the confined compression modulus of bovine knee joint cartilage increased from 0.08 MPa in the superficial zone to 2.1 MPa in the deep zone, on average. Similar trends were reported by Wang et al. (2000) in their study of immature bovine carpometacarpal joint cartilage, with representative results presented in Figure 4.4 Less dramatic but qualitatively similar results were reported by Guilak et al. (1995) who also tracked the motion of chondrocytes, in unconfined compression, using confocal microscopy.

As for the case of anisotropy, most theoretical analyses of articular cartilage have assumed homogeneous properties through the depth of the tissue. Some of the exceptions are the studies by Mow and Mansour (1977), who modeled a STZ with properties distinct from the rest of the tissue; Mow et al. (1980), who modeled the inhomogeneity in water content in a biphasic analysis; and Chen et al. (1998) and Wang et al. (2001) who incorporated depth-dependent compressive stiffness in confined compression. Sun et al. (1998) have also examined the influence of depth-dependent fixed charge density on cartilage mechano-electrochemical behaviors. From these most recent studies, it appears that incorporating tissue inhomogeneity into the modeling of articular cartilage improves the agreement between theory and experiment, particularly with regard to the transient response to stress-relaxation or creep.

Tension-Compression Nonlinearity of Articular Cartilage

The studies reported above on the tensile and compressive properties of articular cartilage confirm that the cartilage stiffness can be very different in tension and compression, with the tensile stiffness up to two orders of magnitude greater than the compressive stiffness (e.g., Akizuki et al., 1986, versus Armstrong and Mow, 1982). Indeed, though the equilibrium elastic properties of cartilage have been most commonly reported for the 3-direction in compression and the 1- and 2-directions for tension, there is sufficient preliminary data for measurements of compressive properties along the 1- and 2-directions (Jurvelin et al., 1996; Soltz et al., 1999), and tensile properties along the 3-direction (Basser et al., 1998; Narmoneva et al., 1999), to confirm that the difference between tensile and compressive moduli exists along all three directions. Nevertheless, very few studies have reported on the tensile and compressive properties of cartilage from the same tissue source. Recently, Huang et al. (1999) provided results for the uniaxial tensile properties of human glenohumeral joint cartilage along the 1- and 2-directions, and confined compression modulus along the 3-direction, using samples

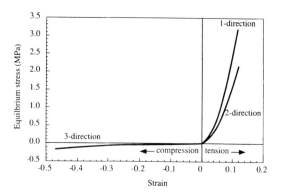

FIGURE 4.5. Typical stress-strain response of human humeral head articular cartilage in uniaxial tension along the 1- and 2-directions, and in confined compression along the 3-direction; this tissue sample is from the surface zone (Huang et al., 1999).

harvested from adjacent sites. At 0% strain, they found the equilibrium tensile modulus on the humeral head to average 7.8 MPa along the split lines and 5.9 MPa perpendicular to the split lines, in the superficial zone, while the corresponding aggregate modulus in compression averaged 0.10 MPa; at 16% strain, the corresponding tensile properties were 43 MPa and 26 MPa, respectively, and the compressive modulus was 0.12 MPa (Fig. 4.5).

This material response is characteristic of fibrous tissues, that which can generally resist tension better than compression. This tension-compression nonlinearity of the stress-strain response is sometimes referred to as a "bimodular" response because, in the range of small strains, the tissue can be thought of having two moduli of elasticity, one in tension and the other in compression. Interestingly, this disparity in moduli had not been a focus of investigation until recently; interest in this topic came about indirectly, as various authors investigated ways to improve the agreement between experimental results and theoretical predictions in the unconfined compression of cartilage between rigid impermable platens, since unconfined compression produces axial compressive stresses simultaneously with radial and circumferential tensile stresses. The theoretical isotropic biphasic solution for unconfined compression with frictionless platens (Armstrong et al., 1984) could not be found to agree with corresponding experimental results (Armstrong et al., 1984; Brown and Singerman, 1986). Subsequent analyses that assumed adhesive platens provided only moderate improvement in the prediction of experimental results (Kim et al., 1995; Spilker et al., 1990). Therefore, following the work of Lanir (1987), Cohen et al. (1993, 1998) proposed that a transversely isotropic biphasic model of cartilage could successfully bridge the gap between theory and experiment, provided that the stiffness in the 3-direction matched the compressive modulus of cartilage, while the stiffness along the 1- and 2-directions matched the tensile modulus. The material properties reported by these authors were in good agreement with the known disparity in tensile and compressive moduli. However, there are limitations to using a transversely isotropic model to describe a bimodular material (Bursac et al., 1999). Circumventing such limitations, Soulhat et al. (1999) proposed an isotropic biphasic model to represent the ground matrix of cartilage, reinforced by tension-only cable-like elements to represent the collagen fibrils. They demonstrated good agreement between theory and experiment in unconfined compression stress-relaxation.

Soltz and Ateshian (2000b) developed an alternative model of cartilage tension-compression nonlinearity that employs the recently developed octantwise orthotropic, conewise linear elasticity (CLE) model of Curnier et al. (1995) to represent the bimodular solid matrix of a biphasic material; they demonstrated that this model can successfully curvefit experimental measurements in unconfined compression, confined compression, and torsional shear of cartilage cylindrical plugs oriented along the 3-direction, while independently predicting experimental measurements of interstitial fluid pressure. The material properties obtained from their study on bovine humeral head cartilage averaged 0.64 MPa for the equilibrium confined compression modulus along the 3-direction, 12.7 MPa for the equilibrium tensile modulus averaged along the 1- and 2-directions, 0.03 for the equilibrium Poisson ratio in unconfined compression along the 3-direction, and 0.17 MPa for the equilibrium shear modulus for torsion about the 3-direction. Additionally, these authors noted that the bimodular response of cartilage served to enhance the ability of the interstitial fluid to pressurize; whereas an isotropic linear model of cartilage would predict that the load supported by fluid pressurization could not exceed 33% of the total applied load (Armstrong et al., 1984), this new bimodular model could predict a fluid load support asymptotically approaching 100% as the ratio of tensile to compressive moduli increases. As a result of this effect, the theoretical analysis suggested that the dynamic modulus of cartilage in unconfined compression would equal approximately one-half of its equilibrium modulus in tension, at physiologic loading frequencies. To date, it has been a challenge to predict the experimentally measured dynamic modulus in unconfined compression, which has been shown to reach as high as 12–20 MPa for bovine calf femoropatellar groove cartilage (Buschmann et al., 1999; Kim et al., 1995), from theoretical analyses. The lack of agreement between theory and experiment is generally an indication that our understanding of cartilage mechanics is incomplete. Incorporation of tension-compression nonlinearity in the modeling of cartilage can thus improve the agreement between theory and experiments (Buschmann et al., 1999; Soltz and Ateshian, 2000b), as can the incorporation of intrinsic solid matrix viscoelasticity (DiSilvestro et al., 1999; Huang et al., 2001), and these topics remain important ongoing research questions.

Mechano-Electrochemical Response of Articular Cartilage

The proteoglycans in cartilage are fixed to its collagen matrix; their glycosaminoglycan chains have negatively charged groups that thus confer a fixed charge density to the tissue (Eisenberg and Grodzinsky, 1985, 1987; Hardingham and Fosang, 1992; Lai et al., 1991; Maroudas, 1979). Electrolytes present in the interstitial water of articular cartilage provide counter-ions to maintain overall tissue electroneutrality; relative to its external bathing environment, however, cartilage contains a higher concentration of ions, which gives rise to a differential fluid pressure, known as the Donnan osmotic pressure, in order to maintain balance of chemical potential (Maroudas, 1979). This osmotic pressure, which has been measured at approximately 0.17 MPa in a 0.15 M

NaCl bathing solution, contributes to the compressive stiffness of cartilage. (More strictly, it is the rate of change in osmotic pressure with changing strain that contributes to the compressive modulus of cartilage, not the absolute value of the osmotic pressure.) Dynamic and transient loading of cartilage also induces streaming and diffusion potentials or currents, due to the flow of ions through the charged-hydrated solid matrix (Frank and Grodzinsky, 1987a, 1987b; Grodzinsky et al., 1978, 1981; Gu et al., 1993; Kim et al., 1995; Lai et al., 2000). A few models have been proposed in the literature to describe the various phenomena arising from the presence of electrolytes and fixed charges within the tissue, including the classical physical chemistry laws (Katchalsky and Curran, 1965). Other theories employ porous media models that subsume both the Donnan theory and the purely mechanical analyses described in the previous sections (Frank and Grodzinsky, 1987a, 1987b; Gu et al., 1993, 1997, 1998; Lai et al., 1991, 2000; Mow et al., 1998). Some investigators have systematically measured the electrochemical response of cartilage in conjunction with its mechanical properties, to provide a more complete characterization of the tissue response (Bonassar et al., 1994, 1995; Eisenberg and Grodzinsky, 1985, 1987; Frank et al., 1987; Grodzinsky et al., 1978, 1981; Kim et al., 1995; Lee et al., 1981; Sachs and Grodzinsky, 1995). These studies firmly establish the need to consider electrical and chemical responses, in addition to mechanical response, when attempting to fully characterize the complex mechanics of articular cartilage. However, there exists sufficient theoretical and experimental data to suggest that, under normal physiological conditions where the molarity and pH of the bathing environment of cartilage do not vary significantly, it is possible to investigate mechanical responses independently of electrical and chemical effects.

Cartilage Lubrication

Diarthrodial joints in humans and animals, such as the knee, hip, or shoulder, operate at relatively high loads and low sliding velocities under normal physiologic conditions. Articular cartilage,

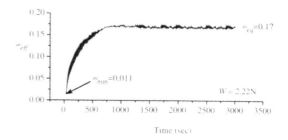

FIGURE 4.6. Time-dependent friction coefficient of bovine articular cartilage under a constant applied load W.

which is its primary bearing material, exhibits remarkable frictional properties that cannot be explained by traditional lubrication theories, despite the presence of synovial fluid in the joint that had led to early speculations of a fluid film lubrication mechanism (Dintenfass, 1963; Dowson, 1967; MacConaill, 1932; Tanner, 1966). Some studies of the biotribology of diarthrodial joints have instead hypothesized that pressurization of the cartilage interstitial fluid may contribute predominantly to reducing the friction coefficient at the contact interface of its articular layers (Ateshian, 1997; Forster et al., 1996; Malcom, 1976; McCutchen, 1962). Experimental measurements of the frictional response of cartilage have demonstrated that it can exhibit a very low friction coefficient upon loading (e.g., in the range of 0.002 to 0.02), which may be favorable for maintaining the high wear resistance of the tissue (Lipshitz et al., 1975); but if the load is maintained constant for durations that exceed normal physiologic conditions, this friction coefficient can become quite elevated (Fig. 4.6). Clearly, the time-dependent increase of the frictional coefficient under a constant load parallels the creep deformation of the tissue (Fig. 4.2a) and follows an opposite trend to the fluid load support (Fig. 4.2b) (McCutchen, 1962). Using the framework of mixture theory for modeling cartilage (Mow et al., 1980), a theoretical formulation describing the dependence of the cartilage friction coefficient on the interstitial fluid pressurization was proposed and verified in an experimental configuration in a recent study (Ateshian et al., 1998). This model also predicates a smaller friction coefficient with increased cartilage water content, which is consistent with the finding that water content is greatest near the cartilage surface. A low friction coefficient is therefore understood to be an essential functional property of articular cartilage, and this functional response is closely dependent on the interstitial fluid pressurization and the water content.

Considerations For Cartilage Tissue Engineering

From this review of the literature, it is apparent that articular cartilage exhibits many levels of complexity, even when focusing on its mechanical properties alone. Clearly, these complexities are matched, if not exceeded, by the biochemistry and biology of articular cartilage, which are not addressed here, although it is firmly established that cartilage mechanics and biology interact through the process of mechanotransduction. It is also evident that a considerable number of studies have successfully tackled various elements of cartilage mechanics, bringing great advances to our current understanding, though a final comprehensive framework for understanding the mechanical response of cartilage has not yet emerged.

The extracellular matrix, intersitial fluid, and ions serve as a signal transducer in articular cartilage, converting joint or explant loading to various extracellular signals (e.g., deformation, pressure, and electrokinetic phenomena) that may in turn influence chondrocyte biosynthetic or catabolic activities (Mow and Wang, 1999). A greater understanding of cartilage mechanics may help to further elucidate the role that changes in the physical environment of the chondrocyte play in the etiology of joint-related pathologies, such as osteoarthritis. In addition, this knowledge may also serve as a guide in the design of articular cartilage-specific tissue engineering bioreactors. As an example, Mauck et al. (2000) have demonstrated a ~5-fold increase, compared to free-swelling control, in the equilibrium confined compressive modulus of chondrocyte-seeded hy-

drogel constructs subjected to applied physiologic deformational loading (daily 10% cyclic deformation in unconfined compression) that reached ~0.1 MPa (approximately one-quarter the native tissue) in 28 days. With a corresponding increase in GAG and collagen content, this study demonstrates for the first time that applied physiologic deformational loading is able to enhance the development of functional articular cartilage *in vitro.*

Given this review of the literature on cartilage mechanics, which properties should be deemed essential for cartilage tissue engineering? Clearly, an engineered cartilage construct must be able to sustain physiological loads in order to remain viable within the joint over the long term. Certainly, for smaller cartilage defects, it is reasonable to assume that surrounding normal tissue may shield the engineered construct from excessive loading. Nevertheless, for larger defects (Fig. 4.1), it is more likely that the construct will be supporting contact stresses within the normal physiological levels ranging from 2 to 6 MPa or higher. Native articular cartilage is able to sustain such stresses because its dynamic modulus in compression can range from 12 to 20 MPa, or even higher (under greater compressive strains). Hence, the dynamic modulus of cartilage in unconfined compression is possibly the most important functional property that would need to be reproduced by engineered cartilage tissue constructs, because it determines whether the tissue can sustain its physiological loading environment. Indeed, a normal dynamic modulus would maintain the physiological tissue strain on the order of 20%, which may be essential for maintaining chondrocyte viability. However, since the dynamic modulus in unconfined compression is dependent on the equilibrium moduli in tension and compression, the tissue permeability, and the intrinsic solid matrix viscoelasticity, it becomes evident that functional properties of engineered tissue constructs would have to match those of native cartilage in all of these aspects in order to reproduce the same function. To date, most biomechanical analyses of engineered cartilage have focused on the determination of the equilibrium-confined compression modulus and the hydraulic permeability. Clearly these are valuable measures of the functionality of the tissue engineered constructs; however, as progress is made in matching these properties with those of native cartilage, it becomes necessary to also investigate the tensile response of the engineered tissues. This can be achieved either by performing uniaxial tensile measurements of the engineered constructs, or dynamic unconfined compression tests. Furthermore, as the mechanics of cartilage are becoming better understood, it is becoming apparent that a high ratio of tensile to compressive properties of the solid matrix serves to enhance interstitial fluid pressurization, and thus the dynamic modulus of the tissue. The detailed measurements of depth-dependent properties that have been reported in the literature indicate that the greatest ratio occurs at the articular surface, where the tensile modulus is greatest and compressive modulus is smallest. This suggests that the greatest interstitial fluid load support occurs at the articular surface, which may be essential to maintain a low friction coefficient and possibly low wear; in the middle and deep zones, this functionality is not required and hence the ratio of tensile to compressive properties is not as high. These findings suggest that an ideal engineered tissue construct may also need to exhibit depth-dependent properties that reproduce those of the native articular cartilage.

Whether an engineered cartilage construct must behave identically to the native tissue to remain viable is a question that can only be answered from experimental trials. However, it is reasonable to assume that a construct that mimics cartilage more closely has a greater potential for success. This review of the literature attempts to describe our current understanding of cartilage mechanics and the various levels of complexity in the functional properties of this tissue. If native properties do need to be reproduced in cartilage tissue engineering, as we believe they do, it is essential to have a basic understanding of the tissue function and its normal native mechanical properties.

Acknowledgments

The authors are grateful for the support of the National Institutes of Health (NIAMS AR43628, AR46532, AR46568).

References

Ahmed, A.M. 1983. A pressure distribution transducer for in-vitro static measurements in synovial joints. *J. Biomech. Eng.* 105:309–314.

Ahmed, A.M., and Burke, D.L. 1983a. In-vitro measurement of static pressure distribution in synovial joints—part I: Tibial surface of the knee. *J. Biomech. Eng.* 105:216–225.

Ahmed, A.M., Burke, D.L., and Yu, A. 1983b. In-vitro measurement of static pressure distribution in synovial joints—part II: Retropatellar surface. *J. Biomech. Eng.* 105:226–236.

Akizuki, S., Mow, V.C., Muller, F., Pita, J.C., Howell, D.S., and Manicourt, D.H. 1986. Tensile properties of human knee joint cartilage: I. Influence of ionic conditions, weight bearing, and fibrillation on the tensile modulus. *J. Orthop. Res.* 4:379–392.

Akizuki, S., Mow, V.C., Muller, F., Pita, J.C., and Howell, D.S. 1987. Tensile properties of human knee joint cartilage. II. Correlations between weight bearing and tissue pathology and the kinetics of swelling. *J. Orthop. Res.* 5:173–186.

Armstrong, C.G., and Mow, V.C. 1982. Variations in the intrinsic mechanical properties of human articular cartilage with age, degeneration, and water content. *J. Bone Joint Surg.* 64A:88–94.

Armstrong, C.G., and Mow, V.C. 1983. The mechanical properties of articular cartilage. *Bull. Hosp. Joint Dis. Orthop. Inst.* 43:109–117.

Armstrong, C.G., Bahrani, A.S., and Gardner, D.L. 1979. *In vitro* measurement of articular cartilage deformations in the intact human hip joint under load. *J. Bone Joint Surg.* 61A:744–755.

Armstrong, C.G., Lai, W.M., and Mow, V.C. 1984. An analysis of the unconfined compression of articular cartilage. *J. Biomech. Eng.* 106:165–173.

Ateshian, G.A. 1997. A theoretical formulation for boundary friction in articular cartilage. *J. Biomech. Eng.* 119:81–86.

Ateshian, G.A., and Wang, H. 1995. A theoretical solution for the frictionless rolling contact of cylindrical biphasic articular cartilage layers. *J. Biomech.* 28:1341–1355.

Ateshian, G.A., Lai, W.M., Zhu, W.B., and Mow, V.C. 1994. An asymptotic solution for the contact of two biphasic cartilage layers. *J. Biomech.* 27:1347–1360.

Ateshian, G.A., Warden, W.H., Kim, J.J., Grelsamer, R.P., and Mow, V.C. 1997. Finite deformation biphasic material properties of bovine articular cartilage from confined compression experiments. *J. Biomech.* 30:1157–1164.

Ateshian, G.A., Wang, H., and Lai, W.M. 1998. The role of interstitial fluid pressurization and surface porosities on the boundary friction of articular cartilage. *J. Tribology* 120:241–251.

Athanasiou, K.A., Rosenwasser, M.P., Buckwalter, J.A., Malinin, T.I., and Mow, V.C. 1991. Interspecies comparison of in situ intrinsic mechanical properties of distal femoral cartilage. *J. Orthop. Res.* 9:330–340.

Atkinson, P.J., and Haut, R.C. 1995. Subfracture insult to the human cadaver patellofemoral joint produces occult injury. *J. Orthop. Res.* 13:936–944.

Atkinson, T.S., Haut, R.C., and Altiero, N.J. 1998a. An investigation of biphasic failure criteria for impact-induced fissuring of articular cartilage. *J. Biomed. Eng.* 120:536.

Atkinson, T.S., Haut, R.C., and Altiero, N.J. 1998b. Impact-induced fissuring of articular cartilage: An investigation of failure criteria. *J. Biomed. Eng.* 120:181–188.

Bachrach, N.M., Mow, V.C., and Guilak, F. 1998. Incompressibility of the solid matrix of articular cartilage under high hydrostatic pressures. *J. Biomech.* 31:445–451.

Bader, D.L., Kempson, G.E., Barrett, A.J., and Webb, W. 1981. The effects of leucocyte elastase on the mechanical properties of adult human articular cartilage in tension. *Biochim. Biophys. Acta* 677:103-108.

Basser, P.J., Schneiderman, R., Bank, R.A., Wachtel, E., and Maroudas, A. 1998. Mechanical properties of the collagen network in human articular cartilage as measured by osmotic stress technique. *Arch. Biochem. Biophys.* 351:207–219.

Berkenblit, S.I., Frank, E.H., Salant, E.P., and Grodzinsky, A.J. 1994. Nondestructive detection of cartilage degeneration using electromechanical surface spectroscopy. *J. Biomech. Eng.* 116:384–392.

Biot, M.A. 1962. Mechanics of deformation and acoustic propagation in porous media. *J. Appl. Phys.* 33:1482–1498.

Bonassar, L.J., Paguio, C.G., Frank, E.H., Jeffries, K.A., Moore, V.L., Lark, M.W., Caldwell, C.G., Hagmann, W.K., and Grodzinsky, A.J. 1994. Effects of matrix metalloproteinases on cartilage biophysical properties *in vitro* and *in vivo*. *Ann. NY Acad. Sci.* 732:439–443.

Bonassar, L.J., Frank, E.H., Murray, J.C., Paguio, C.G., Moore, V.L., Lark, M.W., Sandy, J.D., Wu, J.J., Eyre, D.R., and Grodzinsky, A.J. 1995a. Changes in cartilage composition and physical properties due to stromelysin degradation. *Arthritis Rheum.* 38:173–183.

Bonassar, L.J., Jeffries, K.A., Paguio, C.G., and Grodzinsky, A.J. 1995b. Cartilage degradation and associated changes in biochemical and electromechanical properties. *Acta Orthop. Scand. Suppl.* 266:38–44.

Bowen, R. 1976. Theory of mixtures. In: *Continuum Physics*, ed. A.E. Eringen, Vol. 3, pp. 1–127. New York: Academic Press.

Brocklehurst, R., Bayliss, M.T., Maroudas, A., Coysh, H.L., Freeman, M.A., Revell, P.A., and Ali, S.Y. 1984. The composition of normal and osteoarthritic articular cartilage from human knee joints. With special reference to unicompartmental replacement and osteotomy of the knee. *J. Bone Joint Surg.* 66A:95–106.

Brown, T.D., and Shaw, D.T. 1983. *In vitro* contact stress distributions in the natural human hip. *J. Biomech.* 16:373–384.

Brown, T.D., and Shaw, D.T. 1984. *In vitro* contact stress distribution on the femoral condyles. *J. Orthop. Res.* 2:190–199.

Brown, T.D., and Singerman, R.J. 1986. Experimental determination of the linear biphasic constitutive coefficients of human fetal proximal femoral chondroepiphysis. *J. Biomech.* 19:597–605.

Buckwalter, J.A., Kuettner, K.E., and Thonar, E.J. 1985. Age-related changes in articular cartilage proteoglycans: electron microscopic studies. *J. Orthop. Res.* 3:251.

Bullough, P.G., and Jagannath, A. 1983. The morphology of the calcification front in articular cartilage. *J. Bone Joint Surg.* 65B:72.

Bursac, P.M., Obitz, T.W., Eisenberg, S.R., and Stamenovic, D. 1999. Confined and unconfined stress relaxation of cartilage: appropriateness of a transversely isotropic analysis. *J. Biomech.* 32:1125–1130.

Bursac, P., McGrath, C.V., Eisenberg, S.R., and Stamenovic, D. 2000. A microstructural model of elastostatic properties of articular cartilage in confined compression. *J. Biomech. Eng.* 122:347–353.

Buschmann, M.D., and Grodzinsky, A.J. 1995. A molecular model of proteoglycan-associated electrostatic forces in cartilage mechanics. *J. Biomech. Eng.* 117:170.

Buschmann, M.D., Soulhat, J., Shirazi-Adl, A., Jurvelin, J.S., and Hunziker, E.B. 1998. Confined compression of articular cartilage: linearity in ramp and sinusoidal tests and the importance of interdigitation and incomplete confinement. *J. Biomech.* 31:171–178.

Buschmann, M.D., Kim, Y.J., Wong, M., Frank, E., Hunziker, E.B., and Grodzinsky, A.J. 1999. Stimulation of aggrecan synthesis in cartilage explants by cyclic loading is localized to regions of high interstitial fluid flow. *Arch. Biochem. Biophys.* 366.

Chang, D.G., Lottman, L.M., Chen, A.C., Schinagl, R.M., Albrecht, D.R., Pedowitz, R.A., Brossmann, J., Frank, L.R., and Sah, R.L. 1999. The depth-dependent, multi-axial properties of aged human patellar cartilage in tension. *Trans. Orthop. Res. Soc.* 24:655.

Chen, A.C., Schinagl, R.M., and Sah, R.L. 1998. Inhomogeneous and strain-dependent electromechanical properties of full thickness articular cartilage. *Trans. Orthop. Res. Soc.* 23:225.

Chin-Purcell, M.V., and Lewis, J.L. 1996. Fracture of articular cartilage. *J. Biomech. Eng.* 118:545–556.

Cohen, B., Gardner, T.R., and Ateshian, G.A. 1993. The influence of transverse isotropy on cartilage indentation behavior—A study of the human humeral head. *Trans. Orthop. Res. Soc.* 18:185.

Cohen, B., Chorney, G.S., Phillips, D.P., Dick, H.M., and Mow, V.C. 1994. Compressive stress-relaxation behavior of bovine growth plate may be described by the nonlinear biphasic theory. *J. Orthop. Res.* 12:804–813.

Cohen, B., Lai, W.M., and Mow, V.C. 1998. A transversely isotropic biphasic model for unconfined compression of growth plate and chondroepiphysis. *J. Biomech. Eng.* 120:491–496.

Cooney, W.P., and Chao, E.Y.S. 1977. Biomechanical analysis of static forces in the thumb during hand functions. *J. Bone Joint Surg.* 59-A:27–36.

Curnier, A., He, Q.C., and Zysset, P. 1995. Conewise linear elastic materials. *J. Elasticity* 37:12419.

Dillman, C.J. 1975. Kinematic analyses of running. In *Exercise and Sport Sciences Review,* ed. J.H. Wilmore, Vol. 3, pp. 193–218. New York: Academic Press.

Dintenfass, L. 1963. Lubrication in synovial joints. *Nature* 197:496–497.

DiSilvestro, M.R., Zhu, Q., and Suh, J-K. DiSilvestro, M.R., Zhu, Q., and Suh, J.-K. 1999. Biphasic poroviscoelastic theory predicts the strain rate dependent viscoelastic behavior of articular cartilage. *Proc. 1999 Bioeng. Conf. ASME* BED-42:105–106.

Donzelli, P.S., and Spilker, R.L. 1998. A contact finite element formulation for biological soft hydrated tissues. *Comp. Meth. Appl. Mech. Eng.* 153:62–79.

Donzelli, P.S., Spilker, R.L., Ateshian, G.A., and Mow, V.C. 1999. A finite element investigation of contact between transversely isotropic layers of biphasic cartilage. *J. Biomech.* 32:1037–1047.

Dowson, D. 1967. Modes of lubrication in human joints. *Proc. Inst. Mech. Eng. [H]* 181:45–54.

Eberhardt, A.W., Keer, L.M., Lewis, J.L., and Vithoontien, V. 1990. An analytical model of joint contact. *J. Biomech. Eng.* 112:407–413.

Eberhardt, A.W., Lewis, J.L., and Keer, L.M. 1991. Normal contact of elastic spheres with two elastic layers as a model of joint articulation. *J. Biomech. Eng.* 113:410–417.

Eckstein, F., Tieschky, M., and Faber, S. 1998. *In vivo* quantification of patellar cartilage volume and thickness changes after strenuous dynamic physical activity—a magnetic resonance imaging study. *Trans. Orthop. Res. Soc.* 23:486.

Eisenberg, S.R., and Grodzinsky, A.J. 1985. Swelling of articular cartilage and other connective tissues: electromechanochemical forces. *J. Orthop. Res.* 3:148–159.

Eisenberg, S.R., and Grodzinsky, A.J. 1987. The kinetics of chemically induced nonequilibrium swelling of articular cartilage and corneal stroma. *J. Biomech. Eng.* 109:79–89.

Elliott, D.M., Kydd, S.R., Perry, C.H., and Setton, L.A. 1999. Direct measurement of the Poisson's ratio of human articular cartilage in tension. *Trans. Orthop. Res. Soc.* 24:649.

Forster, H., and Fisher, J. 1996. The influence of loading time and lubricant on the friction of articular cartilage. *Proc. Inst. Mech. Eng. [H]* 210:109–119.

Frank, E.H., and Grodzinsky, A.J. 1987a. Cartilage electromechanics—I. Electrokinetic transduction and the effects of electrolyte pH and ionic strength. *J. Biomech.* 20:615–627.

Frank, E.H., and Grodzinsky, A.J. 1987b. Cartilage electromechanics—II. A continuum model of cartilage electrokinetics and correlation with experiments. *J. Biomech.* 20:629–639.

Frank, E.H., Grodzinsky, A.J., Koob, T.J., and Eyre, D.R. 1987. Streaming potentials: a sensitive index of enzymatic degradation in articular cartilage. *J. Orthop. Res.* 5:497–508.

Froimson, M.I., Ratcliffe, A., Gardner, T.R., and Mow, V.C. 1997. Differences in patellofemoral joint cartilage material properties and their significance to the etiology of cartilage surface fibrillation. *Osteoarthritis Cartilage* 5:377–386.

Fukubayashi, T., and Kurosawa, H. 1980. The contact area and pressure distribution pattern of the knee. *Acta Orthop. Scand.* 51:871–879.

Fung, Y.C. 1981. *Biomechanics: Mechanical Properties of Living Tissues.* New York: Springer-Verlag.

Grodzinsky, A.J., Lipshitz, H., and Glimcher, M.J. 1978. Electromechanical properties of articular cartilage during compression and stress relaxation. *Nature* 275:448–450.

Grodzinsky, A.J., Roth, V., Myers, E., Grossman, W.D., and Mow, V.C. 1981. The significance of electromechanical and osmotic forces in the nonequilibrium swelling behavior of articular cartilage in tension. *J. Biomech. Eng.* 103:221–231.

Grushko, G., Schneiderman, R., and Maroudas, A. 1989. Some biochemical and biophysical parameters for the study of the pathogenesis of osteoarthritis: a comparison between the processes of ageing and degeneration in human hip cartilage. *Connect. Tissue Res.* 19:149–176.

Gu, W.Y., Lai, W.M., and Mow, V.C. 1993. Transport of fluid and ions through a porous-permeable charged-hydrated tissue, and streaming potential data on normal bovine articular cartilage. *J. Biomech.* 26:709–723.

Gu, W.Y., Rabin, J., Lai, W.M., and Mow, V.C. 1995. Measurement of streaming potential of bovine articular and nasal cartilage in 1-D permeation experiment. *ASME Adv. Bioeng.* BED31:49–50.

Gu, W.Y., Lai, W.M., and Mow, V.C. 1997. A triphasic analysis of negative osmotic flows through charged hydrated soft tissues. *J. Biomech.* 30:71–78.

Gu, W.Y., Lai, W.M., and Mow, V.C. 1998. A mixture theory for charged hydrated soft tissues containing multi-electrolytes: passive transport and swelling behaviors. *J. Biomech. Eng.* 102:169.

Guilak, F., Ratcliffe, A., Lane, N., Rosenwasser, M.P., and Mow, V.C. 1994. Mechanical and biochemical changes in the superficial zone of articular cartilage in canine experimental osteoarthritis. *J. Orthop. Res.* 12:474–484.

Guilak, F., Ratcliffe, A., and Mow, V.C. 1995. Chondrocyte deformation and local tissue strain in articular cartilage: a confocal microscopy study. *J. Orthop. Res.* 13:410–421.

Hale, J.E., Rudert, M.J., and Brown, T.D. 1993. Indentation assessment of biphasic mechanical property deficits in size-dependent osteochondral defect repair. *J. Biomech.* 26:1319–1325.

Hardingham, T.E., and Fosang, A. 1992. Proteoglycans: many forms and many functions. *FASEB J.* 6:861–870.

Haut, R.C. 1989. Contact pressures in the patellofemoral joint during impact loading on the human flexed knee. *J. Orthop. Res.* 7:272–280.

Haut, R.C., Ide, T.M., and DeCamp, C.E. 1995. Mechanical responses of the rabbit patello-femoral joint to blunt impact. *J. Biomech. Eng.* 117:402–408.

Hayes, W.C., and Bodine, A.J. 1978. Flow-independent viscoelastic properties of articular cartilage matrix. *J. Biomech.* 11:407–419.

Hayes, W.C., and Mockros, L.F. 1971. Viscoelastic properties of human articular cartilage. *J. Appl. Physiol.* 31:562–8.

Hayes, W.C., Keer, L.M., Herrmann, G., and Mockros, L.F. 1972. A mathematical analysis for indentation tests of articular cartilage. *J. Biomech.* 5:541–551.

Heinegard, D., and Oldberg, A. 1989. Structure and biology of cartilage and bone noncollagenous macromolecules. *FASEB J.* 3:2042–2051.

Hirsch, C. 1944. The pathogenesis of chondromalacia of the patella. *Acta Chir. Scand.* 83(Suppl):1–106.

Hodge, W.A., Carlson, K.L., Fijan, R.S., Burgess, R.G., Riley, P.O., Harris, W.H., and Mann, R.W. 1989. Contact pressures from an instrumented hip endoprosthesis. *J. Bone Joint Surg.* 71A:1378–1386.

Holmes, M.H., and Mow, V.C. 1990. The nonlinear characteristics of soft gels and hydrated connective tissues in ultrafiltration. *J. Biomech.* 23:1145–1156.

Holmes, M.H., Lai, W.M., and Mow, V.C. 1985. Singular perturbation analysis of the nonlinear, flow-dependent compressive stress relaxation behavior of articular cartilage. *J. Biomech. Eng.* 107:206–218.

Hori, R.Y., and Mockros, L.F. 1976. Indentation tests of human articular cartilage. *J. Biomech.* 9:259–268.

Hou, J.S., Holmes, M.H., Lai, W.M., and Mow, V.C. 1989. Boundary conditions at the cartilage-synovial fluid interface for joint lubrication and theoretical verifications. *J. Biomech. Eng.* 111:78–87.

Huang, C.Y., Stankiewicz, A., Ateshian, G.A., Flatow, E.L., Bigliani, L.U., and Mow, V.C. 1999. Anisotropy, inhomogeneity, and tension-compression nonlinearity of human glenohumeral cartilage in finite deformation. *Trans. Orthop. Res. Soc.* 24:95.

Huang, C-Y., Mow, V.C., and Ateshian, G.A. 2001. The role of flow-independent viscoelasticity in the biphasic tensile and compressive responses of articular cartilage. *J. Biomech. Eng.* 123:410–417.

Huberti, H.H., and Hayes, W.C. 1984. Patellofemoral contact pressures. *J. Bone Joint Surg.* 66A:715–724.

Huberti, H.H., and Hayes, W.C. 1988. Contact pressures in chondromalacia patellae and the effects of capsular reconstructive procedures. *J. Orthop. Res.* 6:499–508.

Jurvelin, J.S., Buschmann, M.D., and Hunziker, E.B. 1996. Mechanical anisotropy of human knee articular cartilage in compression. *Trans. Orthop. Res. Soc.* 21:7.

Jurvelin, J., Buschmann, M., and Hunziker, E. 1997. Optical and mechanical determination of Poisson's ratio of adult bovine humeral articular cartilage. *J. Biomech.* 30:235–241.

Katchalsky, A., and Curran, P.F. 1965. *Nonequilibrium Thermodynamics in Biophysics.* Cambridge: Harvard University Press.

Katz, E.P., Wachtel, E.J., and Maroudas, A. 1986. Extrafibrillar proteoglycans osmotically regulate the molecular packing of collagen in cartilage. *Biochim. Biophys. Acta* 882:136–139.

Kelkar, R., and Ateshian, G.A. 1999. Contact creep of biphasic cartilage layers: Identical layers. *J. Appl. Mech.* 66:137–145.

Kelly, P.A., and O'Connor, J.J. 1996a. Transmission of rapidly applied loads through articular cartilage. Part 1: Uncracked cartilage. *Proc. Inst. Mech. Eng.* 210:27–37.

Kelly, P.A., and O'Connor, J.J. 1996b. Transmission of rapidly applied loads through articular cartilage. Part 2: Cracked cartilage. *Proc. Inst. Mech. Eng.* 210:39–49.

Kempson, G.E. 1982. Relationship between the tensile properties of articular cartilage from the human knee and age. *Ann. Rheum. Dis.* 41:508–511.

Kempson, G.E. 1991. Age-related changes in the tensile properties of human articular cartilage: a comparative study between the femoral head of the hip joint and the talus of the ankle joint. *Biochim. Biophys. Acta* 1075:223–230.

Kempson, G.E., Freeman, M.A., and Swanson, S.A. 1968. Tensile properties of articular cartilage. *Nature* 220:1127–1128.

Kempson, G.E., Freeman, M.A., and Swanson, S.A. 1971. The determination of a creep modulus for articular cartilage from indentation tests of the human femoral head. *J. Biomech.* 4:239–250.

Kempson, G.E., Muir, H., Pollard, C., and Tuke, M. 1973. The tensile properties of the cartilage of human femoral condyles related to the content of collagen and glycosaminoglycans. *Biochim. Biophys. Acta* 297:456–472.

Kempson, G.E., Take, M.A., Dingle, J.T., Barrett, A.J., and Horsefield, D.H. 1976. The effects of proteolytic enzymes on the mechanical properties of adult human articular cartilage. *Biochim. Biophys. Acta* 428:741.

Kenyon, D.E. 1976. The theory of an incompressible solid-fluid mixture. *Arch. Rat. Mech. Anal.* 62:131–147.

Khalsa, P.S., and Eisenberg, S.R. 1997. Compressive behavior of articular cartilage is not completely explained by proteoglycan osmotic pressure. *J. Biomech.* 30:589–594.

Kim, Y.J., Sah, R.L., Grodzinsky, A.J., Plaas, A.H., and Sandy, J.D. 1994. Mechanical regulation of cartilage biosynthetic behavior: physical stimuli. *Arch. Biochem. Biophys.* 311:35441.

Kim, Y.J., Bonassar, L.J., and Grodzinsky, A.J. 1995. The role of cartilage streaming potential, fluid flow and pressure in the stimulation of chondrocyte biosynthesis during dynamic compression. *J. Biomech.* 28:1055–1066.

Kurosawa, H., Fukubayashi, T., and Nakajima, H. 1980. Load-bearing mode of the knee joint, physi-cal behavior of the knee joint with or without menisci. *Clin. Orthop.* 149:283–290.

Kwan, M.K., Lai, W.M., and Mow, V.C. 1984. Fundamentals of fluid transport through cartilage in compression. *Ann. Biomed. Eng.* 12:537–558.

Kwan, M.K., Lai, W.M., and Mow, V.C. 1990. A finite deformation theory for cartilage and other soft hydrated connective tissues—I. Equilibrium results. *J. Biomech.* 23:145–155.

Lai, W.M., and Mow, V.C. 1980. Drag-induced compression of articular cartilage during a permeation experiment. *Biorheology* 17:111–123.

Lai, W.M., Mow, V.C., and Roth, V. 1981. Effects of nonlinear strain-dependent permeability and rate of compression on the stress behavior of articular cartilage. *J. Biomech. Eng.* 103:61–66.

Lai, W.M., Hou, J.S., and Mow, V.C. 1991. A triphasic theory for the swelling and deformation behaviors of articular cartilage. *J. Biomech. Eng.* 113:245–258.

Lai, W.M., Rubin, D., and Krempl, E. 1993. *Introduction to Continuum Mechanics,* 3rd ed. New York: Pergamon Press.

Lai, W.M., Mow, V.C., Sun, D.D., and Ateshian, G.A. 2000. On the electric potentials inside a charged soft hydrated biological tissue: Streaming potential vs. diffusion potential. *J. Biomech. Eng.* 122:336–346.

Lanir, Y. 1987. Biorheology and fluid flux in swelling tissues. II. Analysis of unconfined compressive response of transversely isotropic cartilage disc. *Biorheology* 24:189–205.

Lee, R.C., Frank, E.H., Grodzinsky, A.J., and Roylance, D.K. 1981. Oscillatory compressional behavior of articular cartilage and its associated electromechanical properties. *J. Biomech. Eng.* 103:280–292.

Levenston, M.E., Frank, E.H., and Grodzinsky, A.J. 1996. Variationally derived 3-field lagrange multiplier and augmented lagrangian poroelastic finite elements for soft tissue. *ASME Adv. Bioeng.* BED-33:145–146.

Li, X., Haut, R.C., and Altiero, N.J. 1995. An analytical model to study blunt impact response of the rabbit P-F joint. *J. Biomech. Eng.* 117:485–491.

Linn, F.C., and Sokoloff, L. 1965. Movement and composition of interstitial fluid of cartilage. *Arthritis Rheum.* 8:481.

Lipshitz, H., Etheredge, R., and Glimcher, M.J. 1975. *In vitro* wear of articular cartilage. *J. Bone Joint Surg.* 57A:527.

Lipshitz, H., Etheredge, R., and Glimcher, M.J. 1976. Changes in the hexosamine content and swelling ratio of articular cartilage as functions of depth from the surface. *J. Bone Joint Surg.* 58A:1149.

Lyyra, T., Jurvelin, J., Pitkanen, P., Vaatainen, U., and Kiviranta, I. 1995. Indentation instrument for the measurement of cartilage stiffness under arthroscopic control. *Med. Eng. Phys.* 17:395–399.

MacConaill, M.A. 1932. The function of intra-articular fibrocartilages, with special references to the knee and inferior radio-ulnar joints. *J. Anat.* 66:210–227.

Macirowski, T., Tepic, S., and Mann, R.W. 1994. Cartilage stresses in the human hip joint. *J. Biomech. Eng.* 116:35720.

Mak, A.F. 1986. The apparent viscoelastic behavior of articular cartilage—the contributions from the intrinsic matrix viscoelasticity and interstitial fluid flows. *J. Biomech. Eng.* 108:123–130.

Mak, A.F., Lai, W.M., and Mow, V.C. 1987. Biphasic indentation of articular cartilage—I. Theoretical analysis. *J. Biomech.* 20:703–714.

Malcom, L.L. 1976. *An experimental investigation of the frictional and deformational responses of articular cartilage interfaces to static and dynamic loading.* Ph.D. Thesis. San Diego: University of California, San Diego.

Manouel, M., Pearlman, H.S., Belakhlef, A., and Brown, T.D. 1992. A miniature piezoelectric polymer transducer for *in vitro* measurement of the dynamic contact stress distribution. *J. Biomech.* 25:627–635.

Mansour, J.M., and Mow, V.C. 1976. The permeability of articular cartilage under compressive strain and at high pressures. *J. Bone Joint Surg.* 58A:509–516.

Maroudas, A. 1970. Distribution and diffusion of solutes in articular cartilage. *Biophys. J.* 10:365–379.

Maroudas, A. 1972. Physical chemistry and the structure of cartilage. *J. Physiol. (Lond.)* 223:21–22.

Maroudas, A.I. 1976. Balance between swelling pressure and collagen tension in normal and degenerate cartilage. *Nature* 260:808–809.

Maroudas, A. 1979. Physicochemical properties of articular cartilage. In *Adult Articular Cartilage,* 2nd ed. ed. M.A.R. Freeman, pp. 215–290. Tunbridge Wells, England: Pitman Medical.

Maroudas, A., and Bannon, C. 1981. Measurement of swelling pressure in cartilage and comparison with the osmotic pressure of constituent proteoglycans. *Biorheology* 18:619–632.

Maroudas, A., and Bullough, P. 1968. Permeability of articular cartilage. *Nature* 219:1260–1261.

Maroudas, A., and Venn, M. 1977. Chemical composition and swelling of normal and osteoarthrotic femoral head cartilage. II. Swelling. *Ann. Rheum. Dis.* 36:399–406.

Maroudas, A., Bullough, P., Swanson, S.A., and Freeman, M.A. 1968. The permeability of articular cartilage. *J. Bone Joint Surg.* 50B:166–177.

Maroudas, A., Muir, H., and Wingham, J. 1969. The correlation of fixed negative charge with glycosaminoglycan content of human articular cartilage. *Biochim. Biophys. Acta* 177:492–500.

Maroudas, A., Ziv, I., Weisman, N., and Venn, M. 1985. Studies of hydration and swelling pressure in normal and osteoarthritic cartilage. *Biorheology* 22:159–169.

Maroudas, A., Wachtel, E., Grushko, G., Katz, E.P., and Weinberg, P. 1991. The effect of osmotic and mechanical pressures on water partitioning in articular cartilage. *Biochim. Biophys. Acta* 1073:285–294.

Matthews, L.S., Sonstegard, D.A., and Hanke, J.A. 1977. Load bearing characteristics of the patellofemoral joint. *Acta Orthop. Scand.* 48:511–516.

Mauck, R.L., Soltz, M.A., Wang, C.C.B., Wong, D.D., Chao, P.H.G., Valhmu, W.B., Hung, C.T., and Ateshian, G.A. 2000. Functional tissue engineering of articular cartilage through dynamic loading of chondrocyte-seeded agarose gels. *J. Biomech. Eng.* 122:252–260.

McCutchen, C.W. 1962. The frictional properties of animal joints. *Wear* 5:1–17.

Mente, P.L., and Lewis, J.L. 1994. Elastic modulus of calcified cartilage is an order of magnitude less than that of subchondral bone. *J. Orthop. Res.* 12:637–647.

Mizrahi, J., Maroudas, A., Lanir, Y., Ziv, I., and Webber, T.J. 1986. The "instantaneous" deformation of cartilage: effects of collagen fiber orientation and osmotic stress. *Biorheology* 23:311–330.

Mow, V.C., and Lai, W.M. 1980. Recent developments in synovial joint biomechanics. *SIAM Rev.* 22:275–317.

Mow, V.C., and Mansour, J.M. 1977. The nonlinear interaction between cartilage deformation and interstitial fluid flow. *J. Biomech.* 10:31–39.

Mow, V.C., and Wang, C.C. 1999. Some bioengineering considerations for tissue engineering of articular cartilage. *Clin. Orthop.* 367(Suppl):204–223.

Mow, V.C., Kuei, S.C., Lai, W.M., and Armstrong, C.G. 1980. Biphasic creep and stress relaxation of articular cartilage in compression: Theory and experiments.. *J. Biomech. Eng.* 102:73.

Mow, V.C., Gibbs, M.C., Lai, W.M., Zhu, W.B., and Athanasiou, K.A. 1989. Biphasic indentation of articular cartilage—II. A numerical algorithm and an experimental study. *J. Biomech.* 22:853–861.

Mow, V.C., Ratcliffe, A., and Poole, A.R. 1992. Cartilage and diarthrodial joints as paradigms for hierarchical materials and structures. *Biomaterials* 13:67–97.

Mow, V.C., Ateshian, G.A., Lai, W.M., and Gu, W.Y. 1998. Effects of fixed charges on the stress-relaxation behavior of hydrated soft tissues in a confined compression problem. *Int. J. Solids Struct.* 35:4945–4962.

Muir, H. 1980. The chemistry of the ground substance of joint cartilage. In: *The Joints and Synovial Fluid,* Vol. II. ed. L Sokoloff, pp. 27–94. New York: Academic Press.

Narmoneva, D.A., Wang, J.Y., and Setton, L.A. 1999. Nonuniform swelling-induced residual strains in articular cartilage. *J. Biomech.* 32:401–408.

Newberry, W.N., Mackenzie, C.D., and Haut, R.C. 1998. Blunt impact causes changes in bone and cartilage in a regularly exercised animal model. *J. Orthop. Res.* 16:348–354.

Oegema, T.R. Jr., Lewis, J.L., and Thompson, R.C. Jr. 1993. Role of acute trauma in development of osteoarthritis. *Agents Actions* 40:220–223.

Oloyede, A., and Broom, N.D. 1991. Is classical consolidation theory applicable to articular cartilage deformation. *Clin Biomech.* 6:206–212.

Oloyede, A., and Broom, N. 1993. Stress-sharing between the fluid and solid components of articular cartilage under varying rates of compression. *Connect. Tissue Res.* Res30:127–141.

Paul, J.P. 1967. Forces transmitted by joints in the human body. *Proc. Inst. Mech. Eng.* 181(3J):8.

Poppen, N.K., and Walker, P.S. 1978. Forces at the glenohumeral joint in abduction. *Clin. Orthop.* 135:165–170.

Radin, E.L., Burr, D.B., Caterson, B., Fyhrie, D., Brown, T.D., and Boyd, R.D. 1991. Mechanical determinants of osteoarthrosis. *Semin. Arthritis Rheum.* 21:35784.

Redler, I., and Zimny, M.L. 1970. Scanning electron microscopy of normal and abnormal articular cartilage and synovium. *J. Bone Joint Surg.* 52A:1395.

Roth, V., and Mow, V.C. 1980. The intrinsic tensile behavior of the matrix of bovine articular cartilage and its variation with age. *J. Bone Joint Surg.* 62A:1102.

Rydell, N. 1965. Forces in the hip joint: Part (II) Intravital measurements. In: *Biomechanics and Related Bio-Engineering Topics,* ed. R.M. Kenedi, pp. 351–357. Oxford: Pergamon Press.

Sachs, J.R., and Grodzinsky, A.J. 1995. Electromechanical spectroscopy of cartilage using a surface probe with applied mechanical displacement. *J. Biomech.* 28:963–976.

Sah, R.L., Kim, Y.J., Doong, J.Y., Grodzinsky, A.J., Plaas, A.H., and Sandy, J.D. 1989. Biosynthetic response of cartilage explants to dynamic compression. *J. Orthop. Res.* 7:619–636.

Schinagl, R.M., Ting, M.K., Price, J.H., and Sah, R.L. 1996. Video microscopy to quantitate the inhomogeneous equilibrium strain within articular cartilage during confined compression. *Ann. Biomed. Eng.* 24:500–512.

Schinagl, R.M., Gurskis, D., Chen, A.C., and Sah, R.L. 1997. Depth-dependent confined compression modulus of full-thickness bovine articular cartilage. *J. Orthop. Res.* 15:499–506.

Schmidt, M.B., Mow, V.C., Chun, L.E., and Eyre, D.R. 1990. Effects of proteoglycan extraction on the tensile behavior of articular cartilage. *J. Orthop. Res.* 8:353–363.

Schneiderman, R., Keret, D., and Maroudas, A. 1986. Effects of mechanical and osmotic pressure on the rate of glycosaminoglycan synthesis in the human adult femoral head cartilage: an *in vitro* study. *J. Orthop. Res.* 4:393–408.

Schwartz, M.H., Leo, P.H., and Lewis, J.L. 1994. A microstructural model for the elastic response of articular cartilage. *J. Biomech.* 27:865–73.

Setton, L.A., Zhu, W., and Mow, V.C. 1993. The biphasic poroviscoelastic behavior of articular cartilage: role of the surface zone in governing the compressive behavior. *J. Biomech.* 26:581–592.

Setton, L.A., Mow, V.C., Muller, F.J., Pita, J.C., and Howell, D.S. 1994. Mechanical properties of canine articular cartilage are significantly altered following transection of the anterior cruciate ligament. *J. Orthop. Res.* 12:451–463.

Setton, L.A., Gu, W.Y., Lai, W.M., and Mow, V.C. 1995a. Predictions of the swelling induced pre-stress in articular cartilage. In *Mechanics of Porous Media,* ed. A.P.S. Selvadurai, pp. 299–322. Kluwer Academic Pubs.

Setton, L.A., Mow, V.C., and Howell, D.S. 1995b. Mechanical behavior of articular cartilage in shear is altered by transection of the anterior cruciate ligament. *J. Orthop. Res.* 13:473–482.

Setton, L.A., Mow, V.C., Muller, F.J., Pita, J.C., and Howell, D.S. 1997. Mechanical behavior and biochemical composition of canine knee cartilage following periods of joint disuse and disuse with remobilization. *Osteoarthritis Cartilage* 5:35445.

Setton, L.A., Tohyama, H., and Mow, V.C. 1998. Swelling and curling behavior of articular cartilage. *J. Biomed. Eng.* 120:355–361.

Shin, D., Lin, J.H., and Athanasiou, K. 1997. Microindentation of the individual layers of articular cartilage. *ASME Adv. Bioeng.* BED-36:155–156.

Simon, B.R., Coats, R.S., and Woo, S.L. 1984. Relaxation and creep quasilinear viscoelastic models for normal articular cartilage. *J. Biomech. Eng.* 106:159–164.

Singerman, R.J., Pedersen, D.R., and Brown, T.D. 1987. Quantitation of pressure-sensitive film using digital image scanning. *Exp. Mech.* March:99–105.

Sokoloff, L. 1966. Elasticity of aging cartilage. *Fed. Proc.* 25:1089–1095.

Soltz, M.A., and Ateshian, G.A. 1998. Experimental verification and theoretical prediction of cartilage interstitial fluid pressurization at an impermeable contact interface in confined compression. *J. Biomech.* 31:927–934.

Soltz, M.A., and Ateshian, G.A. 2000a. Interstitial fluid pressurization during confined compression cyclical loading of articular cartilage. *Ann. Biomed. Eng.* 28:150–159.

Soltz, M.A., and Ateshian, G.A. 2000b. A conewise linear elasticity mixture model for the analysis of tension-compression nonlinearity in articular cartilage. *J. Biomech. Eng.* 122:576–586.

Soltz, M.A., Palma, C., Barsoumian, S., Wang, C.C.B., Hung, C.T., and Ateshian, G.A. 1999. Multi-axial loading of bovine articular cartilage in unconfined compression. *Trans. Orthop. Res. Soc.* 24:888.

Soulhat, J., Buschmann, M.D., and Shirazi-Adl, A. 1999. A fibril-network reinforced model of cartilage in unconfined compression. *J. Biomech. Eng.* 121:340–347.

Spilker, R.L., Suh, J.K., and Mow, V.C. 1990. Effects of friction on the unconfined compressive response of articular cartilage: a finite element analysis. *J. Biomech. Eng.* 112:138–146.

Spilker, R.L., Suh, J.K., and Mow, V.C. 1992. A finite element analysis of the indentation stress-relaxation response of linear biphasic articular cartilage. *J. Biomech. Eng.* 114:191–201.

Stockwell, R.S. 1979. *Biology of Cartilage Cells.* Cambridge: Cambridge University Press.

Stormont, T.J., An, K.N., Morrey, B.F., and Chao, E.Y. 1985. Elbow joint contact study: Comparison of techniques. *J. Biomech.* 18:329–336.

Suh, J.K., and Bai, S. 1997. Biphasic poroviscoelastic behavior of articular cartilage in creep indentation test. *Trans. Orthop. Res. Soc.* 22:823.

Suh, J.K., and Bai, S. 1998. Finite element formulation of biphasic poroviscoelastic model for articular cartilage. *J. Biomed. Eng.* 120:195–201.

Suh, J.K., Li, Z., and Woo, S.L. 1995. Dynamic behavior of a biphasic cartilage model under cyclic compressive loading. *J. Biomech.* 28:357–364.

Sun, D.N., Gu, W.Y., Guo, X.E., Lai, W.M., and Mow, V.C. 1998. The influence of inhomogeneous fixed charge density of cartilage mechano-electrochemical behaviors. *Trans. Orthop. Res. Soc.* 23:484.

Tanner, R.I. 1966. An alternative mechanism for the lubrication of synovial joints. *Phys. Med. Biol.* 11:119–127.

Tencer, A.F., Viegas, S.F., Cantrell, J., Chang, M., Clegg, P., Hicks, C., O'Meara, C., and Williamson, J.B. 1988. Pressure distribution in the wrist joint. *J. Orthop. Res.* 6:509–517.

Tobias, D., Ziv, I., and Maroudas, A. 1992. Human facet cartilage: swelling and some physico-chemical characteristics as a function of age. Part 1: Swelling of human facet joint cartilage. *Spine* 17:694–700.

Torzilli, P.A., and Mow, V.C. 1976. On the fundamental fluid transport mechanisms through normal and pathological articular cartilage during function—II. The analysis, solution and conclusions. *J. Biomech.* 9:587–606.

Torzilli, P.A., Rose, D.E., and Dethemers, S.A. 1982. Equilibrium water partition in articular cartilage. *Biorheology* 19:519.

Van Der Voet, A, Shrive, N., and Schachar, N. 1993. Numerical modelling of articular cartilage in synovial joints—Poroelasticity and boundary conditions. In: *Recent Advances in Computer Methods in Biomechanics & Biomedical Engineering,* ed. J. Middleton, G. Pande, and K. Williams. United Kingdom: Books & Journals International Ltd.

Vener, M.J., Thompson, R.C. Jr., Lewis, J.L., and Oegema, T.R. Jr. 1992. Subchondral damage after acute transarticular loading: an *in vitro* model of joint injury. *J. Orthop. Res.* 10:759–765.

Venn, M.F. 1978. Variation of chemical composition with age in human femoral head cartilage. *Ann. Rheum. Dis.* 37:168.

Venn, M., and Maroudas, A. 1977. Chemical composition and swelling of normal and osteoarthrotic femoral head cartilage. I. Chemical composition. *Ann. Rheum. Dis.* 36:121–129.

Wachtel, E., and Maroudas, A. 1998. The effects of pH and ionic strength on intrafibrillar hydration in articular cartilage. *Biochim. Biophys. Acta* 1381:37–48.

Wang, C.C.B., Soltz, M.A., Mauck, R.L., Valhmu, W.B., Ateshian, G.A., and Hung, C.T. 2000. Comparison of equilibrium axial strain distribution in articular cartilage explants and cell-seeded alginate disks under unconfined compression. *Trans. Orthop. Res. Soc.* 25:131.

Wang, C.C.B., Hung, C.T., and Mow, V.C. 2001. An analysis of the effects of depth dependent aggregate modulus on articular cartilage stress-relaxation behavior in compression. *J. Biomech.* 34:75–84.

Wayne, J.S., Woo, S.L.Y., and Kwan, M.K. 1991a. Application of the u-p finite element method to the study of articular cartilage. *J. Biomech. Eng.* 113:397–403.

Wayne, J.S., Woo, S.L.Y., and Kwan, M.K. 1991b. Finite element analyses of repaired articular surfaces. *Proc. Inst. Mech. Eng.* 205:155–162.

Wayne, J.S., Brodrick, C.W., and Mukherjee, N. 1998. Measurement of cartilage thickness in the articulated knee. *Ann. Biomed. Eng.* 26:96–102.

Weiss, C., Rosenberg, L., and Helfet, A.J. 1968. An ultrastructural study of normal young adult human articular cartilage. *J. Bone Joint Surg.* 50A:663.

Wong, M. 1999. The incompressibility of adult articular cartilage is insensitive to PH. *Trans. Orthop. Res. Soc.* 24:651.

Wong, M., Jurvelin, J., Ponticiello, M., Tammi, M., Kovanen, V., and Hunziker, E.B. 1998. Simultaneous determination of Poisson's ratio and elastic modulus of mature and immature cartilage. *Trans. Orthop. Res. Soc.* 23:489.

Woo, S.L.Y., Akeson, W.H., and Jemmott, G.F. 1976. Measurements of nonhomogeneous, directional mechanical properties of articular cartilage in tension. *J. Biomech.* 9:785–791.

Woo, S.L.Y., Lubock, P., Gomez, M.A., Jemmott, G.F., Kuei, S.C., and Akeson, W.H. 1979. Large deformation nonhomogeneous and directional properties of articular cartilage in uniaxial tension. *J. Biomech.* 12:437–446.

Woo, S.L.Y., Simon, B.R., Kuei, S.C., and Akeson, W.H. 1980. Quasi-linear viscoelastic properties of normal articular cartilage. *J. Biomech. Eng.* 102:85–90.

Woo, S.L.Y., Kwan, M.K., Lee, T.Q., Field, F.P., Kleiner, J.B., and Coutts, R.D. 1987. Perichondrial autograft for articular cartilage. Shear modulus of neocartilage studied in rabbits. *Acta Orthop. Scand.* 58:510–515.

Wren, T.A.L., and Carter, D.R. 1998. A microstructural model for the tensile constitutive and failure behavior of soft skeletal connective tissues. *J. Biomech. Eng.* 120:55–61.

Wu, J.Z., Herzog, W., and Ronsky, J. 1996. Modeling axi-symmetrical joint contact with biphasic cartilage layers—an asymptotic solution. *J. Biomech.* 29:1263–1281.

Wu, J.Z., Herzog, W., and Epstein, M. 1997. An improved solution for the contact of two biphasic cartilage layers. *J. Biomech.* 30:371–375.

Wu, J.Z., Herzog, W., and Epstein, M. 1998. Evaluation of the finite element software ABAQUS for biomechanical modelling of biphasic tissues. *J. Biomech.* 31:165–169.

Zarek, J.M., and Edward, J. 1963. The stress-structure relationship in articular cartilage. *Med. Electron. Biol. Eng.* 1:497–507.

Zhu, W., Chern, K.Y., and Mow, V.C. 1994. Anisotropic viscoelastic shear properties of bovine meniscus. *Clin. Orthop.* 306:34–45.

Zhu, W., Mow, V.C., Koob, T.J., and Eyre, D.R. 1993. Viscoelastic shear properties of articular cartilage and the effects of glycosidase treatments. *J. Orthop. Res.* 11:771–781.

5

Excitability and Contractility of Skeletal Muscle: Measurements and Interpretations

John A. Faulkner and Robert G. Dennis

"To move is all mankind can do and for such the sole executant is *muscle.*"
Sir Charles Sherrington, Croonian Lecture, 1937–1938.

Introduction

Because muscle is the sole executant for movement, the ultimate goal of tissue engineering of skeletal muscle is to restore normal movement through the restoration of control values for structure, function, and structure-function relationships. The purpose of this chapter is to describe the current state of our knowledge of the hierarchical nature of the structure and of the function of skeletal muscles and the critical structure-function relationships, the technical challenges that exist, and the future perspectives and directions for research in these areas. Developing and aging muscle, muscle diseases such as muscular dystrophy, and tissue-engineered muscle constructs each provide examples of variations in excitability and contractility. The contractions of skeletal muscles provide stability and power for all movements of every species, but this chapter will restrict itself to mammalian skeletal muscle contractions and movement. The rationale is that the basic science of tissue engineering is designed to translate biotechnology into new diagnostic and treatment modalities that are of clinical significance. With skeletal muscle, as with many other tissues, the basic biologic experiments are inevitably performed on mice, rats, cats, rabbits, dogs, or, even on rarer

occasions, primates, with the subsequent translation to humans (Weindruch and Masoro, 1991). For tissue engineering of skeletal muscle, one of the major advantages of the initial research on small mammals is that whole skeletal muscles, either control or experimental, can be isolated. Excitability and contractility can be measured *in vitro* or *in situ* and subsequently the whole muscle can be excised and the structure of the whole muscle, as well as parts of the muscle, may be studied.

The structure, function, and structure-function relationships of skeletal muscle may be studied at the level of the single cross-bridge (Vale and Milligan, 2000), single permeabilized (Macpherson et al., 1997) or intact (Lannergren and Westerblad, 1987) fiber, motor unit (Burke et al., 1973), whole muscle, or groups of muscles, such as the dorsi-flexors or plantar-flexors of foot (Miller et al., 1998). Upon activation of fibers, at any one of these levels of organization, the fibers will attempt to "contract," but whether the muscle actually shortens, stays at a fixed length, or is stretched depends on the relative magnitude of the force developed by the muscle and the load on the muscle. The three types of contraction that result from the interaction between force and load are 1) a shortening, when the force is greater; 2) a fixed length, when force

Types of Contractions

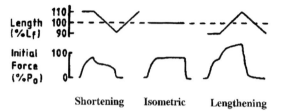

FIGURE 5.1. Representative recordings of isovelocity displacements of the muscle length (*upper trace*) and the force developed during shortening of the muscle during the contraction (shortening contraction), fixed end contraction (isometric contraction), and stretching of the muscle during the contraction (lengthening contraction). Changes in muscle length are expressed relative to optimum fiber length (L_f), and forces are given relative to maximum isometric force (P_o). The figure is modified from McCully and Faulkner (1985).

and load are equal, or the load is immovable; and 3) a forced-stretch, when the load is greater (Fig. 5.1). The three different types of contraction occur during co-contractions of agonist and antagonist muscles. The contractions provide stability and controlled power for the wide diversity of movements skeletal muscles of mammals make during daily activities. When activated muscles are stretched, either repeatedly through small strains, or through single large strains >30%, muscle fibers are injured (McCully and Faulkner, 1985). The magnitude of the injury may be evaluated either by direct measures of morphological damage or indirectly by the deficit in maximum force in the absence of fatigue (Faulkner et al., 1993). For a complete evaluation of the functionality of an experimental skeletal muscle, maximum force, power, and force deficit after a protocol of forced-stretches of a maximally-activated muscle must be measured, normalized, and compared with control values.

Current Status

Very close to a half century has passed since Andrew Huxley and Rolf Niedergerke (1954) and Hugh Huxley and Jean Hanson (1954) independently outlined the sliding filament theory for the contraction of muscle. The subsequent 15 years of research on the structural and functional properties of single fibers culminated in the concept of the actin filaments sliding past the myosin filaments through the movement of the head of the myosin molecule, termed a "cross-bridge" (H. Huxley, 1969). Muscle biomechanists had to wait almost a quarter of a century for the clarification of the mechanism responsible for the movement of the cross-bridge. The breakthrough was based on a combination of "the molecular structures of individual proteins and low-resolution electron density maps of the complex derived by cryo-electron microscopy and image analysis" (Rayment et al., 1993a, 1993b). The current status of measurements of the structure and function of skeletal muscle and of muscle fibers is at an all time high. The advent of a wide range of *in vitro* motility assays (Howard, 1997) coupled with sophisticated numerical and computational models permit assessments of movement from the molecular level (Vale and Milligan, 2000) to the whole organism, including robotics (Dickinson et al., 2000).

Skeletal Muscle Structure

A critical element in the structure of skeletal muscle is the hierarchical nature of muscles (Fig. 5.2). The hierarchical structure begins with the structure of a single myosin head and the myosin binding site on the actin molecule (Rayment et al., 1993b). The hierarchical structure of muscle fibers continues with the organized relationship between the thin actin and thick myosin filaments and z-lines that give rise to the repeating pattern of sarcomeres. The sarcomeres are arranged serially within myofilaments and with a parallel alignment of myofilaments and z-lines within fibers. The fibers are then organized in parallel with a specific architectural design for a given muscle. The specificity of the architectural design provides for the requirements of a muscle for shortening, force, velocity of shortening, power output, and power input (Gans, 1982). The architectural designs vary from a simple parallel alignment, through pennate and bipennate, to multipennate (Gans, 1982). The fiber

Skeletal Muscle Hierarchical Structure

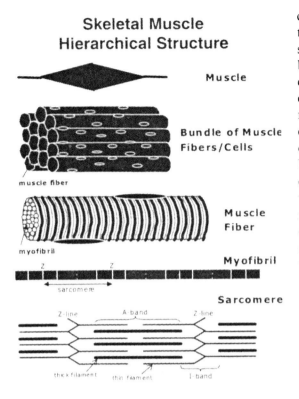

Muscle

Bundle of Muscle Fibers/Cells

muscle fiber

Muscle Fiber

myofibril

Myofibril

sarcomere

Sarcomere

FIGURE 5.2. Levels of anatomical organization within a skeletal muscle. (From Bloom, W., and Fawcett, D.W., 1968. *A Textbook of Histology,* 9th ed., Saunders, Philadelphia, with permission.)

length is dependent not only on the distance between the origin and the insertion, but also the angle of pennation and the mass of the muscle (Gans, 1982; Weeks, 1989).

The approximately 600 skeletal muscles constitute about 40% of the mass of the mammalian organism. For humans, small muscles in the hands, feet, and craniofacial region contain hundreds of fibers, whereas those in the large muscles in the limbs and trunk are composed of hundreds of thousands of fibers. The connective tissue surrounding muscles and muscle fibers is structured into three separate but interconnected sheaths (Sanes, 1994). The epimysium surrounds the total muscle with coarse collagen fibers extending inward to form a perimysium around bundles, or fasciles of fibers. Lastly, each individual fiber is surrounded by an endomysium. The innermost layer of the endomysium, the basement membrane, lies immediately adja-

cent to the plasma membrane of the fiber, termed the sarcolemma. The sarcolemma is a semipermeable membrane that acts as a selective barrier between the extracellular matrix and the cytoplasm of the muscle fiber. Within the sarcolemma are the cytosolic proteins: contractile, regulatory, metabolic, and structural; many nuclei; and cytosolic organelles, such as mitochondria (Lehninger, 1973). The mitochondrion have been termed the power-house of the cell, because of its role in the generation of adenosine triphosphate (ATP). ATP provides the energy for the vast majority of cellular functions, including contraction (Mommaerts, 1969). The sarcolemma also serves an important role in signal transduction and in anchoring the muscle fiber to the extracellular matrix (Engel and Franzini-Armstrong, 1994). Satellite cells lie between the basement membrane and the sarcolemma. Satellite cells are quiescent embryonic stem cells that under normal circumstances are destined to be myogenic precursor cells and form muscle fibers (Bischoff, 1994), but satellite cells under carefully controlled conditions differentiate into a variety of other cell types, including adipocytes (Morrison et al., 2000). Conversely, bone marrow–derived cells transplanted into damaged muscle were transformed into myogenic precursor cells and participated in the regeneration of the damaged fibers (Ferrari et al., 1998). Aside from the growing realization of the potential, of not only embryonic stem cells, but a wide variety of stem cells in different tissues, satellite cells continue to function effectively in the repair of muscle tissue even in extremely old mammals (Carlson et al., 2000). Following damage to muscle fibers, satellite cells are activated, divide mitotically, form myoblasts that fuse to form myotubes, and subsequently with innervation differentiate terminally into muscle fibers (Carlson and Faulkner, 1983).

The number of fibers in a specific muscle is highly conserved from birth to maturity (Goldspink, 1983). The enormous increase in the mass of muscles from the neonate to maturity is achieved by hypertrophy of individual fibers through an increase in the number of myofibrils per fiber (Goldspink, 1983) and the addition of sarcomeres at the ends of fibers (Williams and Goldspink, 1971). Similarly, hypertrophy of muscles in response strength conditioning (Goll-nick et al., 1981), or atrophy with bed rest (Saltin et al., 1968), result from increases or decreases, respectively, in muscle fiber cross-sectional areas (CSAs), not a change in the number of fibers. In contrast, the immutable atrophy of muscles of 80-year-old humans results from a decrease of about 50% in the number of fibers (Lexell, 1995). Although many very old humans also show atrophy in muscle fiber CSAs, this phenomenon is reversible with weight training (Grimby, 1995). Part, if not all, of the loss in fiber number is attributable to losses in whole motor units as observed in both rats (Cederna et al., 1999) and humans (Campbell et al., 1973). Losses in motor units and fiber number, similar to those associated with aging, are associated with some debilitating muscle diseases such poliomyocytis and dystrophy (McComas, 1996). The traditional measurements of structure through the morphological techniques of histology, histochemistry, and immunocytochemistry have been aided by quantitative fluorescent and computer systems to measure fiber CSAs and fiber types. The identification of myosin isoform with electrophoretic as opposed to histochemical techniques has been particularly enlightening. Electron (H. Huxley, 1957) and confocal (Lipp et al., 1996; Tsugorska et al., 1995) microscopy have also provided tools to probe the structural and functional elements of single fibers.

Skeletal Muscle Function

Similar to that of structure, the function of skeletal muscles has a hierarchical design of its own. The hierarchical design begins with the function of individual cross-bridges as independent force generators (Gordon et al., 1966), transitions through the sarcomere dynamics of single fibers (Macpherson et al., 1997), four different types of motor units with unique functional properties (Burke et al., 1973), and culminates in the function of whole muscles during locomotion (McMahon, 1984). In addition, groups of muscles may act synergistically, and flexor and extensor muscles and muscle groups may act antagonistically (Dickinson et al., 2000; McComas, 1996).

The most basic function of muscle is the generation of force by activated cross-bridges. A single cross-bridge generates approximately 4 pN of force, although a great deal of uncertainty surrounds this number (A. Huxley, 2000). Within each fiber, cross-bridges on the myosin filament attach strongly to binding sites on the actin filament. The cross-bridge head then undergoes a conformational change that slides the thin actin filament past the thick myosin filament pulling the z-lines toward one another (Fig. 5.2). The understanding of the three-dimensional structural change associated with the interaction of actin and myosin began with observations of asynchronous flight muscles of insects by electron microscopy (Reedy et al., 1965). The long axis of cross-bridges, that were perpendicular to the fiber axis at rest, were at approximately 45° in rigor. The tilting of the cross-bridge head was interpreted widely as the cause of force development and shortening (H. Huxley, 1969; A. Huxley, 2000). Sophisticated techniques, including cryo-electron microscopy and image analysis (Rayment et al., 1993a, 1993b) have indicated that a hinge exists in the bulky motor domain of the myosin head between the catalytic domain that binds to actin and the long, thin tail that includes the light chain binding sites. Three-dimensional reconstructions from electron micrographs support tilting at both sites (Schmitz et al., 1997; Taylor et al., 1999). Controversy still surrounds the number of steps, either one or two (A. Huxley, 2000), and the estimates of the total movement that range from 6 nm (Rayment et al., 1993b) to 10 to 12 nm (Dominguez et al., 1998; Duke, 1999). Biochemical assays and electrophoretic techniques provide precise assessments of metabolic potential and myosin isoforms (Reiser et al., 1985). The conformational change requires energy and initially estimates had the hydrolysis of one ATP for each rotation, or cycle, of the cross-bridge (Woledge et al., 1985). The concept of one ATP per cycle has been challenged by observations that the distance filaments slide past one another are greater than the distance a cross-bridge could stay attached to the same actin binding site (Higuchi and Goldman, 1995). A likely scenario is that cross-bridges detach from one site and reattach immediately to a site on an adjacent actin monomer (Kitamura et al., 1999).

Many insightful experiments on skeletal muscle function have been performed on single "skinned," or intact muscle fibers. Stienen (2000) has provided a short chronicle on the experiments on the "skinned" or permeabilized fibers. The permeabilized fibers have had their membrane either disrupted by a "skinning" solution (Eastwood et al., 1979), or removed mechnically (Natori, 1954). In either case, the "skinning" procedure eliminated the membrane potential and permeabilized fibers, whether chemically or mechanically "skinned," were kept in relaxing solution until activated by various concentrations of calcium that mimicked the cytosolic conditions (Godt and Maughan, 1977). More recently, Posterino and his associates (2000) with major precautions in the maintenance of the fiber have successfully activated mechanically skinned single fibers through the use of voltages more than 10 times the normal field strength. The single permeabilized fiber preparation has already enabled a wide range of studies to be performed on mammalian muscle fibers at a time when only a few laboratories had the capability of dissecting single intact fibers (Allen et al., 1989; Lannergren and Westerblad, 1987). The studies on single permeabilized fibers have included thin filament activation (Metzger and Moss, 1988), aging (Thompson and Brown, 1999), contraction-induced injury (Brooks and Faulkner, 1996), and sarcomere heterogeneity (Macpherson et al., 1997). Despite attempts on many other muscles and species, single intact mammalian skeletal muscle fibers have been obtained only from the flexor digitorum brevis muscle of the mouse (Lannergren and Westerblad, 1987) and published studies have been limited to just two groups, for studies of temperature effects (Lannergren and Westerblad, 1987), fatigue (Lannergren and Westerblad, 1991), and the specific role of calcium in fatigue (Allen et al., 1989; Westerblad and Allen, 1991). Hopefully, this highly desirable, yet technically challenging preparation will be mastered by other investigators and applied to a wider range of investigations. Meanwhile, the electrical activation of permeabilized fibers opens up a whole new vista of possible experiments on the permeabilized fiber preparation (Posterino et al., 2000).

In 1949, A.V. Hill proposed that during a maximum isometric tetanic contraction, stronger regions of the fiber shorten the most and weaker regions shorten the least. Abbot and Aubert (1952) extended the concept of weaker and stronger regions to a forced-stretch of a maximally activated fiber and concluded that the weaker regions would be stretched more than the stronger regions. The laser diffraction technique with the beam deflected along a fiber by a rotating prism has provided direct measurements of the sarcomere dynamics of single permeabilzed rat fiber segments during maximum isometric contractions and during stretches (Macpherson et al., 1997). As proposed by the earlier investigators, during a maximum isometric contraction, some groups of sarcomeres shortened, others remained at the same length, and some were stretched to longer lengths. The regions stretched to longer lengths during the isometric contraction were already on the descending limb of their length-force curve (Fig. 5.3) and when stretched were at considerable risk of being stretched excessively and injured. A subsequent 40% strain of the maximally activated single permeabilized fibers produced a 10% force deficit and a loss of the diffraction pattern in the stretched regions. Studies of sarcomere dynamics in permeabilized fibers have clarified the roles of weak and strong groups of sarcomeres in the development of a force deficit and sarcomere injury (Macpherson et al., 1997). Despite the insights provided by studies of permeabilized fibers, investigations of sarcomere dynamics in intact fibers are critical in the final analysis of this phenomenon because of the importance of the lateral transmission of force (Huijing, 1999; Street, 1983).

Within a skeletal muscle, fibers are organized into motor units (Burke et al., 1973). A motor unit consists of the cell body of the motor nerve, the motor nerve trunk, or axon, all of the branches of the axon going to individual muscle fibers, and the muscle fibers innervated by the single motor nerve. Consequently, a motor unit is the smallest unit of muscle fibers that can be recruited volitionally by an organism, although it is unlikely that a single unit is recruited normally. Most contractions, even when only minimal force or power is required, involve a number of motor units acting synchronously, or asynchronously. The fibers in a motor unit tend to be homogeneously of one type, but different motor

Variation in isometric tetanic tension

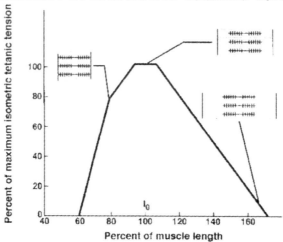

FIGURE 5.3. Variation in active isometric tetanic tension with muscle-fiber length. The light-gray band represents the range of length changes that can occur physiologically in the body while muscles are attached to bones. (From Vander, Sherman & Luciano, McGraw-Hill Company, 2001, with permission.)

units have widely different innervation ratios, number of muscle fibers per motor unit, and widely different functional properties for the fibers in the motor unit. The force developed by a whole muscle during a contraction is a function of the force developed by each motor unit and the number of motor units recruited. Motor units are classified generally into three groups on the basis of the functional properties of the fibers: slow-fatigue resistant, fast-fatigue resistant, and fast-fatigable (Burke et al., 1973). Small mammals, mice and rats, tend to have muscles composed almost exclusively of one type of fiber that, with the exception of the slow soleus muscle, tend to be all fast fibers, either fast-fatigue resistant, or fast-fatigable (Burkholder et al., 1994). Humans, on the other hand, tend to have almost all of their muscles composed of almost equal proportions of slow and fast fibers (Johnson et al., 1973). As humans age, the motor units that are lost are fast motor units and consequently the fibers that are lost are predominantly fast fibers (Larsson, 1983). Consequently, the skeletal muscles of old people are not only atrophied (Lexell, 1995) and weak with a low force production per CSA (Brooks and Faulkner, 1988), but also contract more slowly

and are less powerful (Brooks and Faulkner, 1991).

The current status of measurement of excitability and contractility is at an all time high due to developments and improvements in both the apparatus available to make measurements and the methods used to make the measurements. Improved apparatus is available commercially, as in the servo-motors developed initially by Cambridge Instruments (Brooks and Faulkner, 1988) and more recently by Aurora Instruments (Lynch et al., 2000), and in custom-made minaturized force transducers (Dennis and Kosnik, 2000; Dennis et al., 2001). The minaturized force transducers have been designed specifically for measurements of tissue engineered muscle constructs (Dennis and Kosnik, 2000). In addition, a wide variety of *in vitro* motility assays have been designed and constructed (Block, 1996; Howard, 1998; Vale and Milligan, 2000). Innovative design and construction of customized minaturized force transducers, laser tweezers, and servo-motors have made possible measurements of excitability and contractility of tissue engineered muscle constructs (Dennis and Kosnik, 2000) and single myosin heads and actin filaments (Howard, 1997). The muscle constructs, with a greatly diminished maximum capability for force development, only approximately 1% that of skeletal muscles in adult mammals (Dennis and Kosnik, 2000), pose an additional challenge in the measurement of functional capacity. The commercially available force transducers and servo-motors of variable size and design permit measurements of excitability and contractility of whole muscles of young and old mice and rats (Brooks and Faulkner, 1988, 1991; Carlson and Faulkner, 1996), motor units of small mammals (Cederna et al., 1999; Larsson, 1983; Kadhiresan et al., 1996), and single permeabilized (Brooks and Faulkner, 1994), and intact (Lannergren and Westerblad, 1987) fibers from all mammals, including humans (Larsson and Moss, 1993).

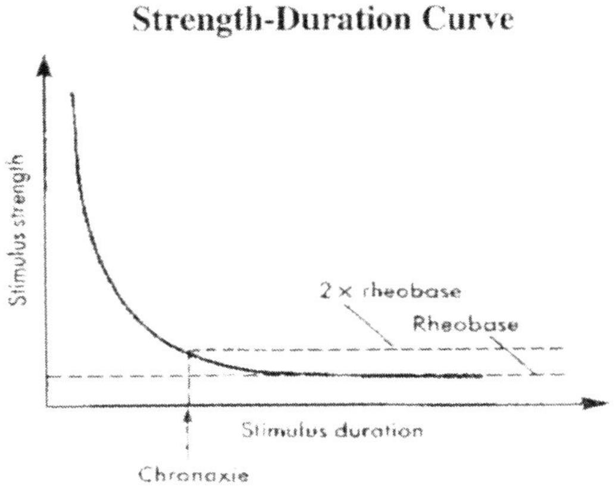

Excitability

Excitable cells have a membrane potential that can be depolarized by a suitable stimulus that then transmits an action potential throughout the cell without loss in strength. Muscle—cardiac, skeletal and smooth—and nerves are excitable tissues. A skeletal muscle is composed of fibers each with a membrane potential of approximately 70 mV. Normally, activation of fibers is initiated by action potentials from the motor cortex that depolarize the anterior horn cells of single motor units that activate each of the muscle fibers within the motor units. Muscle fibers, or all of the fibers within a whole muscle, either *in situ* or *in vitro*, may also be activated by an electrical stimulus applied to the motor nerve, or directly to the muscle. In contrast, motor units may only be activated by stimulation of the appropriate single axons either teased out (Kadheresan et al., 1996), or impaled (Burke et al., 1973) at the level of the ventral roots. The threshold values for the stimulus for depolarization has characteristics of amplitude (rheobase) and duration (chronaxie). The rheobase and chronaxie, taken collectively, define the excitability of the tissue (Fig. 5.4). Mature skeletal muscles are highly excitable, but embryonic, neonatal, and engineered muscle constructs are much less excitable and more variable in their excitability. Despite the importance of the excitability to valid assessments of the function of

Strength-Duration Curve

FIGURE 5.4. The strength-duration curve. The ordinate shows stimulus strength, and the abscissa shows the minimal time a stimulus of that strength must be applied to produce an action potential. (From Berne and Levy, *Physiology, Human Physiology—The Mechanisms of Body Function,* 3rd ed., Mosby, 1993, with permission.)

an excitable tissue, currently, when making measurements of contractility, few investigators make measurements of excitability (Dennis and Kosnik, 2000). Because the excitability of skeletal muscles of young, adult, and even old mammals does not vary, apparently, the assumption is made that the excitability of all muscle tissue is constant. Such an assumption is erroneous, since the excitability of skeletal muscles in adult mammals are one to two orders of magnitude more excitable than tissue engineered muscle constructs (Dennis et al., 2001).

Contractility

Contractility encompasses the totality of the contractile response of a submaximally or maximally activated muscle fiber, a motor unit, or whole muscle. Most evaluations of contractility are restricted to the responses of maximally activated muscles, because the range of submaximum values is essentially limitless as well as being difficult to standardize. Comparisons of maximum values may be made among different muscle types, mammals of different species or ages, or specific muscles exposed to different experimental interventions. The data on maxi-

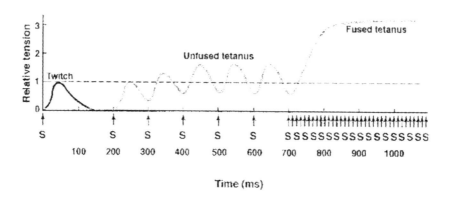

FIGURE 5.5. Isometric contractions produced by multiple stimuli at 10 stimuli per second (unfused tetanus) and 100 stimuli per second (fused tetanus), as compared with a single twitch. (From Vander, Sherman & Luciano, McGraw-Hill Company, 2001, with permission.)

mally activated muscles, motor units, or fibers make the comparison more definitive and have the added value of defining the maximum value for a given measure of contractility. Frequently, the characterizations of skeletal muscles after experimental interventions are limited to morphological, biochemical, or genetic assessments that do not assure a comparable level of function (Partridge, 1993).

Even when function is measured, the structure-function comparisons have frequently involved only measurements of maximum isometric tetanic force (Fig. 5.5) measured at optimum length for force development (Fig. 5.3). Although of considerable significance and interpretive value, such a limited comparison may lead to erroneous conclusions. Skeletal muscles, motor units, and fibers may not differ in one measure of contractility, such as maximum force, yet be quite different in measures of power output, power input, and resistance to injury. The development of force across joints by agonists and antagonists is vitally important for posture and to provide a stable platform for movements of the limbs and trunk. All bodily movements require the development of power output, the product of force and

velocity (Jones et al., 1984). Furthermore, most activities require controlled movements that result from co-contractions of agonist and antagonist muscles or groups of muscles operating alternatively to extend and flex joints (Dickinson et al., 2000). With high-velocity movements of a limb, or of the total body, antagonistic muscles or groups of muscles are required to decelerate the limb or body, and act as a brake on the velocity of the movement before full extension or flexion is reached. A failure to decelerate a limb or the total body can result in severe soft or hard tissue injury (Radin, 1986).

Contraction-Induced Injury

The phenomenon of contraction-induced injury was first described by Hough (1902), but not investigated thoroughly as to the underlying factors responsible for the injury until the early eighties (Friden et al., 1983: Newham et al., 1983). In 1985, McCully and Faulkner demonstrated unequivocally through experiments on *in situ* extensor digitorum longus (EDL) muscles of young mice that maximally activated muscles exposed to 225 forced stretches of 20% strain from optimum length for force development (L_o) were injured severely, whereas muscles allowed to shorten or held at fixed length showed transient fatigue, but no evidence of injury. Similarly, non-stimulated muscles stretched by the same protocol as maximally activated muscles were not injured. Subsequently, the phenomenon of contraction-induced injury was investigated during single stretches of maximally activated whole muscles (Brooks et al., 1995) and single permeabilized fibers (Brooks and Faulkner, 1996; Macpherson et al., 1996) that permitted accurate measurement of the force deficit in the absence of fatigue. For muscles and single fibers from young and adult mice and rats, the initial injury was identified as a purely mechanical event. The force deficit had a coefficient of correlation of 0.80 with the product of the strain and the average force during the stretch, or the work done to stretch the muscle fibers. The susceptibility of single muscle fibers or of whole muscles to be injured by their own contractions provides a powerful tool to assess the status of experimental compared with control muscles (Brooks, 1998: Deconinck et al., 1996).

Structure-Function Relationships

The structure-function relationship of a whole skeletal muscle is dependent on each and every one of the complex hierarchical steps in structure from single cross-bridges, through sarcomeres, myofibrils, and muscle fibers, to the architectual design of the whole muscle (Gans, 1982). The force generated by a single fiber is a function of the number of force generating cross-bridges per half sarcomere and the number of half sarcomeres in parallel. In a maximally activated muscle fiber, measures of stiffness indicates that approximately 40% of the cross-bridges in a half sarcomere are in their driving stroke and generating force and any one point in time (Linari et al., 1998). The best indirect estimate of the number of cross-bridges per half sarcomere is the true cross-sectional area of a muscle fiber or of all of the muscle fibers in a muscle (Close, 1965). Close (1965) measured values of 235 kN/m^2 for soleus and EDL muscles of rats, a value confirmed in later measures of muscles from both mice (Brooks and Faulkner,

1988) and rats (Carlson and Faulkner, 1989). Lannergren and Westerblad (1987) obtained a value of 350 kN/m^2 for single intact fibers of EDB of the mouse. The lower values for the whole muscles are attributed to the angle of pennation (Gans, 1982, Weeks, 1989) and the compliance of the whole muscle preparation (Griffiths, 1991). The velocity of unloaded shortening is a function of the fiber length and the myosin isoform, with fast EDL muscles having a velocity about four-fold greater than slow soleus muscles in any given species (Ranatunga, 1984). Smaller mammals have faster velocities of unloaded shortening than larger mammals with the relationship between velocity and body mass fairly linear (Close, 1967). Similarly, fast-fatigable, Type 2B, fibers have higher velocities than fast fatigue-resistant, Type 2A, fibers and slow, Type 1, fibers have the slowest velocities. The relationship is dependent on the type of myosin isoform present and the more recently discovered Type 2X, or 2D, intermediate between the Type 2A and the 2B isoforms. Neonatal and embryonic isoforms have even slower velocities than Type 1 (Close, 1965). The power output, a function of the force and velocity, depends on a complex interaction amongst fiber length, total fiber CSA, and isoforms of structural, regulatory and contractile proteins, particularly myosin. Since the product of fiber length and total fiber CSA, corrected for muscle density equals the muscle mass, power output is normalized by muscle mass (watts/kg). The hierarchical structure of a whole muscle is highly complex with a wide variety of interactions possible among the performance characteristics of the individual motor units (Burke et al., 1973) and the architectural design of the muscle (Gans, 1982; Weeks, 1989). The breadth and diversity of the structural measurements coupled with sophistication of the devices to measure contractility provide ample opportunities for the formulation of mechanistic hypotheses regarding structure-function relationships.

Technical Challenges

The technical challenge of the measurement of the excitability of tissue engineered muscle constructs has been resolved by the return to the concepts of rheobase and chronaxie to ensure that these constructs are maximally activated despite their low levels of excitability (Dennis and Kosnik, 2000; Dennis et al., 2001). The measurement of the contractility of engineered skeletal muscle constructs has constituted a challenge because of the large baseline forces and low values for maximum isometric tetanic force. Dennis and Kosnik (2000) resolved each of these difficulties through a combination of equipment design and inventive techniques. No attempt has been made to date to measure the power output of muscle constructs or the susceptibility of the constructs to contraction-induced injury. These are challenges that must be met to fully characterize the contractility of tissue-engineered muscle constructs.

Future Perspectives and Directions for Research

The most important area for future research on functional tissue engineering of skeletal muscle is in the provision of optimum envionmental cues to ensure optimum development of the muscle constructs. The embryonic development of skeletal muscle tissue *in utero* and the neonatal development during the early stages of postnatal life, both depend critically on a wide variety of hormonal, chemical, and mechanical cues (Engel and Franzini-Armstrong, 1994). The understanding of this complex array of environmental cues is incomplete even for adult skeletal muscles *in vivo*. The controlled application of these cues for skeletal muscle constructs engineered *in vitro* presents additional technical challenges. Consequently, functional tissue engineers have a long way to go before even the remotest hope exists of optimizing environmental cues. Requirements are for a better understanding of the sequence and timing of the expression and the function of each of the genes related to muscle development and the myogenic regulatory factors (MRFs), that control the expression of these genes (Hauschka, 1994). An MRF is a molecule that is able to convert a nonmuscle cell into a muscle cell. The MRFs, MyoD, myogenin,

Myf-5, and MRF4, are transcriptional factors that control the specification and differentiation of developing muscle (Sabourin and Rudnicki, 2000). The discovery of the MRFs and their binding to E-boxes, DNA control target elements, with well-established capability in the regulation of skeletal muscle genes has clarified the specific biologic roles the MRFs play in embryogenesis, as well as in repair following muscle injury. The MRFs undoubtedly will be found to play a similar role in primary cultures of skeletal muscle.

For individual skeletal muscle fibers within a tissue-engineered muscle construct, the excitability of the fiber and of the muscle construct, depends on the development of an appropriate sarcolemmal membrane with an approximately 70 mV membrane potential. In addition, the sarcolemma must be associated structurally and functionally with the contractile proteins, the cytoskeleton of the muscle fibers, as well as with the T-tubule system and the sarcoplasmic reticulum (Franzini-Armstrong and Fischman, 1994). The early signs are that these conditions are reasonably well met under the current conditions of culturing skeletal muscle constructs (Dennis and Kosnick, 2000; Dennis et al., 2001), although the excitability of the muscle constructs is low.

The importance of the lateral transmission of force through the cell membrane to the extra cellular matrix (ECM), and the lateral coupling between individual muscle fibers (Street, 1983) has been largely overlooked in tissue engineering of skeletal muscle. An important element in the design of synthetic scaffolds for functional engineered skeletal muscle is an understanding of the force transmission from muscle fibers to the connective tissue surrounding them, and ultimately to tendons and the skeletal system (Dennis and Kosnik, 2000; Dennis et al., in press; Vandenburgh et al., 1991). In muscle fibers, the absence of dystrophin and the dystrophin-associated glycoprotein complex that links actin in the cytoskeleton to laminin in the ECM results in an inherently unstable contractile apparatus (Lynch et al., 2000). Within dystrophic muscles, the fibers that lack such elements are highly susceptible to contraction-induced injury (Brooks, 1998; Deconinck et al., 1996). Similarly, single permeabilized fibers from muscles of control mice that have had their lateral transmembrane mechanical interfaces interrupted by chemical "skinning" are equally susceptible to injury when stretched (Lynch et al., 2000). The lateral and longitudinal transmission of force between each developing muscle fiber and the surrounding fibers play a major role in the growth and development of not only the individual muscle fiber but also of the total muscle or muscle construct (Vandenburgh, 1982; Vandenburgh and Kaufman, 1979, Vandenburgh et al., 1989). Mechanical interactions with specific transmembrane structures such as integrins play a major role at each stage in the development and maturation of muscle fibers (Huttenlocher et al., 1998: Sastry and Horowitz, 1996). Although tissue-cultured myotubes can be activated by electrical stimulation (Brevet et al., 1976; Naumann and Pette, 1994), innervation of the muscle construct is the very essence of the interface between the organism and the muscle. Nerves and muscles are highly dependent on one another for normal development, maintenance, and the attainment of control values for physiological properties (Grinnell, 1994, 1995). In particular, muscle tissue will never fully express the adult phenotype in the absence of innervation and the specific trophic (Funakoshi et al., 1995; Li et al., 1995) and mechanical (Chromiak and Vandenburgh, 1992) signals that accrue from nerve-muscle interactions. A more complete understanding of the complex interactions between muscle fibers, motor nerves, and their environment is essential for the design of artificial scaffolds and the development of fully differentiated engineered skeletal muscle constructs.

Acknowledgments

The editorial assistance of Carol Davis, particularly in putting together the reference list, and the critical review of the manuscript by Susan V. Brooks are gratefully acknowledged. The research reported from the Muscle Mechanics Laboratory in the Institute of Gerontology, University of Michigan, was supported by the National Institute on Aging grant AG-06157 and the Contractility Core of the Nathan Shock Center for the Basic Biology of Aging P-30 AG-13283.

References

Abbott, B.C., Aubert, X.M. 1952. The force exerted by active striated muscle during and after changes of length. *J. Physiol.* 117:77–86.

Allen, D.G., Lee, J.A., and Westerblad, H. 1989. Intracellular calcium and tension during fatigue in isolated single muscle fibres from *Xenopus laevis. J. Physiol. (Lond)* 415:433–458.

Bischoff, R. 1994. The satellite cell and muscle regeneration. In: *Myology.* Engel, A.G. and Franzini-Armstrong, C., eds. New York: McGraw-Hill, pp. 97–133.

Block, S.M. 1996. Fifty ways to love your lever: myosin motors. *Cell* 87:151–157.

Brevet, A., Pinto, E., Peacock, J., Stockdale, F.E. 1976. Myosin synthesis increased by electrical stimulation of skeletal muscle cell cultures. *Science* 193:1152–1154.

Brooks, S.V. 1998. Rapid recovery following contraction-induced injury to in situ skeletal muscles in mdx mice. *J. Muscle Res. Cell Motil.* 19:179–187.

Brooks, S.V., Faulkner, J.A. 1988. Contractile properties of skeletal muscles from young, adult and aged mice. *J. Physiol. (Lond)* 404:71–82.

Brooks, S.V., Faulkner, J.A. 1991. Forces and powers of slow and fast skeletal muscles in mice during repeated contractions. *J. Physiol. (Lond)* 436:701–710.

Brooks, S.V., Faulkner, J.A. 1994. Isometric, shortening, and lengthening contractions of muscle fiber segments from adult and old mice. *Am. J. Physiol.* 267:C507–C513.

Brooks, S.V., Faulkner, J.A. 1996. The magnitude of the initial injury induced by stretches of maximally activated muscle fibres of mice and rats increases in old age. *J. Physiol. (Lond)* 497(Pt 2):573–580.

Brooks, S.V., Zerba, E., Faulkner, J.A. 1995. Injury to muscle fibres after single stretches of passive and maximally stimulated muscles in mice. *J. Physiol. (Lond)* 488(Pt 2):459–469.

Burke, R.E., Levine, D.N., Tsairis, P., Zajac, F.E., III 1973. Physiological types and histochemical profiles in motor units of the cat gastrocnemius. *J. Physiol. (Lond)* 234:723–748.

Burkholder, T.J., Fingado, B., Baron, S., Lieber, R.L. 1994. Relationship between muscle fiber types and sizes and muscle architectural properties in the mouse hindlimb. *J. Morphol.* 221:177–190.

Campbell, M.J., McComas, A.J., Petito, F. 1973. Physiological changes in ageing muscles. *J. Neurol. Neurosurg. Psychiatry* 36:174–182.

Carlson, B.M., Dedkov, E., Borisov, A., and Faulkner, J.A. (2001). Skeletal muscle regeneration in very old rats. *J. Gerontol. Biol. Sci.* 56:B224–33.

Carlson, B.M., Faulkner, J.A. 1983. The regeneration of skeletal muscle fibers following injury: a review. *Med. Sci. Sports Exerc.* 15:187–198.

Carlson, B.M., Faulkner, J.A. 1989. Muscle transplantation between young and old rats: age of host determines recovery. *Am. J. Physiol.* 256:C1262–C1266.

Carlson, B.M., Faulkner, J.A. 1996. The regeneration of noninnervated muscle grafts and marcaine-treated muscles in young and old rats. *J. Gerontol. Biol. Sci.* 51:B43–B49.

Cederna, P.S., van der Meulen, J.H., Faulkner, J.A., Kuzon, W.M. 1999. Force deficits in aging skeletal muscle: motor unit properties. *Surgical Forum* L:608–610.

Chromiak, J.A., Vandenburgh, H.H. 1992. Glucocorticoid-induced skeletal muscle atrophy *in vitro* is attenuated by mechanical stimulation. *Am. J Physiol.* 262:C1471–C1477.

Close, R. 1965. The relation between intrinsic speed of shortening and duration of the active state of muscle. *J. Physiol. (Lond)* 180:542–559.

Close, R. 1967. Properties of motor units in fast and slow skeletal muscles of the rat. *J. Physiol.* 193:45–55.

Deconinck, N., Ragot, T., Marechal, G., Perricaudet, M., Gillis, J.M. 1996. Functional protection of dystrophic mouse (mdx) muscles after adenovirus-mediated transfer of a dystrophin minigene. *Proc. Natl. Acad. Sci. U. S. A.* 93:3570–3574.

Dennis, R.G., Kosnik, P.E. 2000. Excitability and isometric contractile properties of mammalian skeletal muscle constructs engineered *in vitro. In Vitro Cell. Dev. Biol. Anim.* 36:327–335.

Dennis, R.G., Kosnik, P.E., Gilbert, M.E., Faulkner, J.A. (2001). Excitability and contractility of skeletal muscle engineered from primary cultures and cell lines. *Am. J. Physiol. Cell. Physiol.* 280:C288–95.

Dickinson, M.H., Farley, C.T., Full, R.J., Koehl, M.A., Kram, R., Lehman, S. 2000. How animals move: an integrative view. *Science* 288:100–106.

Dominguez, R., Freyzon, Y., Trybus, K.M., Cohen, C. 1998. Crystal structure of a vertebrate smooth muscle myosin motor domain and its complex with the essential light chain: visualization of the pre-power stroke state. *Cell* 94:559–571.

Duke, T.A. 1999. Molecular model of muscle contraction. *Proc. Natl. Acad. Sci. U. S. A.* 96:2770–2775.

Eastwood, A.B., Wood, D.S., Bock, K.L., Sorenson, M.M. 1979. Chemically skinned mammalian skeletal muscle. I. The structure of skinned rabbit psoas. *Tissue Cell* 11:553–566.

Engel, A.G., Franzini-Armstrong, C. 1994. *Myology.* New York: McGraw-Hill.

Faulkner, J.A., Brooks, S.V., Opiteck, J.A. 1993. Injury to skeletal muscle fibers during contractions: conditions of occurrence and prevention. *Phys. Ther.* 73:911–921.

Ferrari, G., Cusella-De Angelis, G., Coletta, M., Paolucci, E., Stornaiuolo, A., Cossu, G., Mavilio, F. 1998. Muscle regeneration by bone marrow-derived myogenic progenitors. *Science* 279:1528–1530.

Franzini-Armstrong, C., Fischman, D.A. 1994. Morphogenesis of skeletal muscle fibers. In: *Myology.* Engel, A.G., Franzini-Armstrong, C., eds, New York: McGraw Hill, pp. 74–96.

Friden, J., Seger, J., Sjostrom, M., Ekblom, B. 1983. Adaptive response in human skeletal muscle subjected to prolonged eccentric training. *Int. J. Sports Med.* 4:177–183.

Funakoshi, H., Belluardo, N., Arenas, E., Yamamoto, Y., Casabona, A., Persson, H., Ibanez, C.F. 1995. Muscle-derived neurotrophin-4 as an activity-dependent trophic signal for adult motor neurons. *Science* 268:1495–1499.

Gans, C. 1982. Fiber architecture and muscle function. *Exerc. Sport Sci. Rev.* 10:160–207.

Godt, R.E., Maughan, D.W. 1977. Swelling of skinned muscle fibers of the frog. Experimental observations. *Biophys. J.* 19:103–116.

Goldspink, G. 1983. Alterations in myofibril size and structure during growth, exercise, and changes in environmental temperature. In: *Handbook of Physiology: Skeletal Muscle.* Peachey, L.D., Peachey, R.H., Adrian, R.H., and Geiger, S.R., eds. Bethesda, MD: American Physiology Society, pp. 539–554.

Gollnick, P.D., Timson, B.F., Moore, R.L., and Riedy, M. 1981. Muscular enlargement and number of fibers in skeletal muscles of rats. *J. Appl. Physiol.* 50:936–943.

Gordon, E.E., Kowalski, K., Fritts, M. 1966. Protein changes in quadriceps muscle of rat with repetitive exercises. *Arch. Phys. Med. Rehabil.* 48:297–303.

Griffiths, R.I. 1991. Shortening of muscle fibres during stretch of the active cat medial gastrocnemius muscle: the role of tendon compliance. *J. Physiol. (Lond)* 436:219–236.

Grimby, G. 1995. Muscle performance and structure in the elderly as studied cross-sectionally and longitudinally. *J. Gerontol. Biol. Sci.* 50 Spec No:17–22.

Grinnell, A.D. 1994. Trophic interaction between nerve and muscle. In *Myology.* Engel, A.G., Franzini-Armstrong, C., eds, New York: McGraw-Hill, pp. 303–332.

Grinnell, A.D. 1995. Dynamics of nerve-muscle interaction in developing and mature neuromuscular junctions. *Physiol. Rev.* 75:789–834.

Hauschka, S.D. 1994. Development, anatomy, and cell biology. In: *Myology.* Engel, A.G., Franzini-

Armstrong, C., eds, New York: McGraw-Hill, pp. 3–73.

Higuchi, H., Goldman, Y.E. 1995. Sliding distance per ATP molecule hydrolyzed by myosin heads during isotonic shortening of skinned muscle fibers. *Biophys. J.* 69:1491–1507.

Hough, T. 1902. Ergographic studies in muscular soreness. *Am. J. Physiol.* 7:76–92.

Howard, J. 1997. Molecular motors: structural adaptations to cellular functions. *Nature* 389:561–567.

Howard, J. 1998. How molecular motors work in muscle. *Nature* 391:239–240.

Huijing, P.A. 1999. Muscle as a collagen fiber reinforced composite: a review of force transmission in muscle and whole limb. J. Biomech. 32:329–345.

Huttenlocher, A., Lakonishok, M., Kinder, M., Wu, S., Truong, T., Knudsen, K.A., Horwitz, A.F. 1998. Integrin and cadherin synergy regulates contact inhibition of migration and motile activity. *J. Cell Biol.* 141:515–526.

Huxley, A.F. 2000. Mechanics and models of the myosin motor. *Philos. Trans. R. Soc. Lond. B Biol. Sci.* 355:433–440.

Huxley, A.F., Niedergerke, R. 1954. Interference microscopy of living muscle fibres. *Nature* 173:971–973.

Huxley, H.E. 1957. The double array of filaments in cross-striated muscle. *J. Biophys. Biochem. Cytol.* 3:631–648.

Huxley, H.E. 1969. The mechanism of muscular contraction. *Science* 164:1356–1365.

Huxley, H.E., Hanson, J. 1954. Changes in the cross-striations of muscle during contraction and stretch and their structural interpretation. *Nature* 173:973–976.

Johnson, M.A., Polgar, J., Weightman, D., Appleton, D. 1973. Data on the distribution of fibre types in thirty-six human muscles. An autopsy study. *J. Neurol. Sci.* 18:111–129.

Jones, D.A., Jackson, M.J., McPhail, G., Edwards, R.H. 1984. Experimental mouse muscle damage: the importance of external calcium. *Clin. Sci.* 66:317–322.

Kadhiresan, V.A., Hassett, C.A., Faulkner, J.A. 1996. Properties of single motor units in medial gastrocnemius muscles of adult and old rats. *J. Physiol. (Lond)* 493(Pt 2):543–552.

Kitamura, K., Tokunaga, M., Iwane, A.H., Yanagida, T. 1999. A single myosin head moves along an actin filament with regular steps of 5.3 nanometres. *Nature* 397:129–134.

Lannergren, J., Westerblad, H. 1987. The temperature dependence of isometric contractions of single, intact fibres dissected from a mouse foot muscle. *J. Physiol. (Lond)* 390:285–293.

Lannergren, J., Westerblad, H. 1991. Force decline due to fatigue and intracellular acidification in isolated fibres from mouse skeletal muscle. *J. Physiol. (Lond)* 434:307–322.

Larsson, L. 1983. Histochemical characteristics of human skeletal muscle during aging. *Acta Physiol. Scand.* 117:469–471.

Larsson, L., Moss, R.L. 1993. Maximum velocity of shortening in relation to myosin isoform composition in single fibres from human skeletal muscles. *J. Physiol. (Lond)* 472:595–614.

Lehninger, A.L. 1973. Bioenergetics. Menlo Park, CA: W. A. Benjamin, pp. 2–34.

Lexell, J. 1995. Human aging, muscle mass, and fiber type composition. *J. Gerontol. Biol. Sci.* 50 Spec No:11–16.

Li, L., Wu, W., Lin, L.F., Lei, M., Oppenheim, R.W., Houenou, L.J. 1995. Rescue of adult mouse motoneurons from injury-induced cell death by glial cell line-derived neurotrophic factor. *Proc. Natl. Acad. Sci. U. S. A.* 92:9771–9775.

Linari, M., Dobbie, I., Reconditi, M., Koubassova, N., Irving, M., Piazzesi, G., Lombardi, V. 1998. The stiffness of skeletal muscle in isometric contraction and rigor: the fraction of myosin heads bound to actin. *Biophys. J.* 74:2459–2473.

Lipp, P., Luscher, C., Niggli, E. 1996. Photolysis of caged compounds characterized by ratiometric confocal microscopy: a new approach to homogeneously control and measure the calcium concentration in cardiac myocytes. *Cell Calcium* 19:255–266.

Lynch, G.S., Rafael, J.A., Chamberlain, J.S., Faulkner, J.A. 2000. Contraction-induced injury to single permeabilized muscle fibers from mdx, transgenic mdx, and control mice. *Am. J. Physiol. Cell Physiol.* 279:C1290–C1294.

Macpherson, P.C., Dennis, R.G., Faulkner, J.A. 1997. Sarcomere dynamics and contraction-induced injury to maximally activated single muscle fibres from soleus muscles of rats. *J. Physiol. (Lond)* 500(Pt 2):523–533.

Macpherson, P.C., Schork, M.A., Faulkner, J.A. 1996. Contraction-induced injury to single fiber segments from fast and slow muscles of rats by single stretches. *Am. J. Physiol.* 271:C1438–C1446.

McComas, A.J. 1996. *Skeletal Muscle: Form and Function.* Champaign, IL: Human Kinetics.

McCully, K.K., Faulkner, J.A. 1985. Injury to skeletal muscle fibers of mice following lengthening contractions. *J. Appl. Physiol.* 59:119–126.

McMahon, T.A. 1984. *Muscles, Reflexes, and Locomotion.* Princeton, NJ: Princeton University Press.

Metzger, J.M., Moss, R.L. 1988. Thin filament regulation of shortening velocity in rat skinned skeletal muscle: effects of osmotic compression. *J. Physiol. (Lond)* 398:165–175.

Miller, S.W., Hassett, C.A., Faulkner, J.A. 1998. Recovery of muscle transfers replacing the total plantar flexor muscle group in rats. *J. Appl. Physiol.* 84:1865–1871.

Mommaerts, W.F. 1969. Energetics of muscular contraction. *Physiol. Rev.* 49:427–508.

Morrison, S.J., Csete, M., Groves, A.K., Melega, W., Wold, B., Anderson, D.J. 2000. Culture in reduced levels of oxygen promotes clonogenic sympathoadrenal differentiation by isolated neural crest stem cells. *J. Neurosci.* 20:7370–7376.

Natori, R. 1954. The role of myofibrils, sarcoplasma and sarcolemma in muscle-contraction. *Jikei Med. J.* 1:18–28.

Naumann, K., Pette, D. 1994. Effects of chronic stimulation with different impulse patterns on the expression of myosin isoforms in rat myotube cultures. *Differentiation* 55:203–211.

Newham, D.J., McPhail, G., Mills, K.R., Edwards, R.H. 1983. Ultrastructural changes after concentric and eccentric contractions of human muscle. *J. Neurol. Sci.* 61:109–122.

Partridge, T.E. 1993. *Molecular and Cell Biology of Muscular Dystrophy.* London: Chapman and Hall

Posterino, G.S., Lamb, G.D., Stephenson, D.G. 2000. Twitch and tetanic force responses and longitudinal propagation of action potentials in skinned skeletal muscle fibres of the rat. *J. Physiol.* 527 Pt 1:131–137.

Radin, E.L. 1986. Role of muscles in protecting athletes from injury. *Acta Med. Scand. Suppl.* 711:143–147.

Ranatunga, K.W. 1984. The force-velocity relation of rat. *J. Physiol. (Lond)* 351:517–529.

Rayment, I., Rypniewski, W.R., Schmidt-Base, K., Smith, R., Tomchick, D.R., Benning, M.M., Winkelmann, D.A., Wesenberg, G., Holden, H.M. 1993a. Three-dimensional structure of myosin subfragment-1: a molecular motor. *Science* 261:50–58.

Rayment, I., Holden, H.M., Whittaker, M., Yohn, C.B., Lorenz, M., Holmes, K.C., Milligan, R.A. 1993b. Structure of the actin-myosin complex and its implications for muscle contraction. *Science* 261:58–65.

Reedy, M.K., Holmes, K.C., Tregear, R.T. 1965. Induced changes in orientation of the cross-bridges of glycerinated insect flight muscle. *Nature* 207:1276–1280.

Reiser, P.J., Moss, R.L., Giulian, G.G., Graeser, M.L. 1985. Shortening velocity and myosin heavy chains of developing rabbit muscle fibers. *J. Biol. Chem.* 260:14403–14405.

Sabourin, L.A., Rudnicki, M.A. 2000. The molecular regulation of myogenesis. *Clin. Genet.* 57:16–25.

Saltin, B., Blomqvist, G., Mitchell, J.H., Johnson, R.L., Jr., Wildenthal, K., and Chapman, C.B. 1968. Response to exercise after bed rest and after training. *Circulation* 38:VII1–78.

Sanes, J.R. 1994. The extracellular matrix. In: *Myology.* Engel, A.G. and Franzini-Armstrong, C., eds. New York: McGraw-Hill, pp. 242–260.

Sastry, S.K., Horwitz, A.F. 1996. Adhesion-growth factor interactions during differentiation: an integrated biological response. *Dev. Biol.* 180:455–467.

Schmitz, H., Reedy, M.C., Reedy, M.K., Tregear, R.T., and Taylor, K.A. 1997. Tomographic three-dimensional reconstruction of insect flight muscle partially relaxed by AMPPNP and ethylene glycol. *J. Cell Biol.* 139:695–707.

Stienen, G.J. 2000. Chronicle of skinned muscle fibres. *J. Physiol.* 527 Pt 1:1.

Street, S.F. 1983. Lateral transmission of tension in frog myofibers: a myofibrillar network and transverse cytoskeletal connections are possible transmitters. *J. Cell Physiol.* 114:346–364.

Taylor, K.A., Schmitz, H., Reedy, M.C., Goldman, Y.E., Franzini-Armstrong, C., Sasaki, H., Tregear, R.T., Poole, K., Lucaveche, C., Edwards, R.J., Chen, L.F., Winkler, H., and Reedy, M.K. 1999. Yomographic 3D reconstruction of quick-frozen, Ca2+-activated contrasting insect flight muscle. *Cell* 99:421–431.

Thompson, L.V., Brown, M. 1999. Age-related changes in contractile properties of single skeletal fibers from the soleus muscle. *J. Appl. Physiol.* 86:881–886.

Tsugorka, A., Rios, E., Blatter, L.A. 1995. Imaging elementary events of calcium release in skeletal muscle cells. *Science* 269:1723–1726.

Vale, R.D., Milligan, R.A. 2000. The way things move: looking under the hood of molecular motor proteins. *Science* 288:88–95.

Vandenburgh, H., Kaufman, S. 1979. *In vitro* model for stretch-induced hypertrophy of skeletal muscle. *Science* 203:265–268.

Vandenburgh, H.H. 1982. Dynamic mechanical orientation of skeletal myofibers *in vitro. Dev. Biol.* 93:438–443.

Vandenburgh, H.H., Hatfaludy, S., Karlisch, P., Shansky, J. 1989. Skeletal muscle growth is stimulated by intermittent stretch-relaxation in tissue culture. *Am. J. Physiol.* 256:C674–C682.

Vandenburgh, H.H., Swasdison, S., Karlisch, P. 1991. Computer-aided mechanogenesis of skeletal muscle organs from single cells *in vitro. FASEB J* 5:2860–2867.

Weeks, O.I. 1989. Vertebrate skeletal muscle: power source for locomotion. *Bioscience* 39:791–799.

Weindruch, R., Masoro, E.J. 1991. Concerns about rodent models for aging research [editorial]. *J. Gerontol. Biol. Sci.* 46:B87–B88.

Westerblad, H., Allen, D.G. 1991. Changes of myoplasmic calcium concentration during fatigue in single mouse muscle fibers. *J. Gen. Physiol.* 98: 615–635.

Williams, P.E., Goldspink, G. 1971. Longitudinal growth of striated muscle fibres. *J. Cell Sci.* 9:751–767.

Winkler, H., and Reedy, M.K. 1999. Tomographic 3D reconstruction of quick-frozen, Ca2+-activated contracting insect flight muscle. *Cell* 99: 421–431.

Woledge, R.C., Curtin, N.A., Homsher, E. 1985. *Energetic Aspects of Muscle Contraction.* Orlando, FL: Academic Press.

Part II

Functional Requirements of Engineered Tissues

6

Functional Requirements for the Engineering of a Blood Vessel Substitute

Robert M. Nerem and Jan P. Stegemann

Introduction

There is a critical need for a blood vessel substitute that can be used in coronary artery bypass graft surgery. This is because many patients for a variety of reasons do not have available the native vessels normally used. The tissue engineering of a small-diameter blood vessel substitute thus has long represented a holy grail for those in the cardiovascular area (Conte, 1998).

To engineer such a vascular substitute will require addressing numerous issues, ones that range all the way from cell source to the integration of a blood vessel substitute into the living system. It includes achieving a three-dimensional construct that mimics native tissue in both architecture and function, scaling up fabrication processes to manufacture the substitute in a quantity that will address the widespread patient need, and preserving the product to provide off-the-shelf availability. It is such core, enabling technologies that are being addressed by the Georgia Tech/Emory Center for the Engineering of Living Tissues (GTEC) which was established in 1998 with an Engineering Research Center Award from the National Science Foundation (Nerem et al., 1998).

Although the term *tissue engineering* wasn't established until 1987, the concept of engineering living, cellularized implants goes back a quarter of a century. In fact, some would argue that there were activities prior to the Second World War that were of a tissue engineering nature (Carrel and Lindbergh, 1938). Furthermore, the engineering of functional tissue substitutes has in a sense always been the goal of those engaged in tissue engineering. Even so, frequently, too little attention has been paid to what today is termed *functional tissue engineering* and to the role of mechanics. Thus, the recent emphasis on function, although certainly not new for cardiovascular substitutes, is both timely and important.

There is no more important example of the role of biomechanics in all of tissue engineering than that of a blood vessel substitute. An artery has as its primary purpose the delivery of blood with its nutrients to tissue, and for myocardial tissue, the coronary artery is critical to the perfusion required. Although this delivery of blood may be viewed as a mechanical function, the vascular system cannot be viewed as a passive, piping system. Rather, it is an active system, one in which, if it is to carry out its mechanical function, there are important biological functions. Furthermore, not only are these biological functions critical to the vessel performing its mechanical function, but there are important influences of biomechanics on this functionality and thus on the biology of the vascular system.

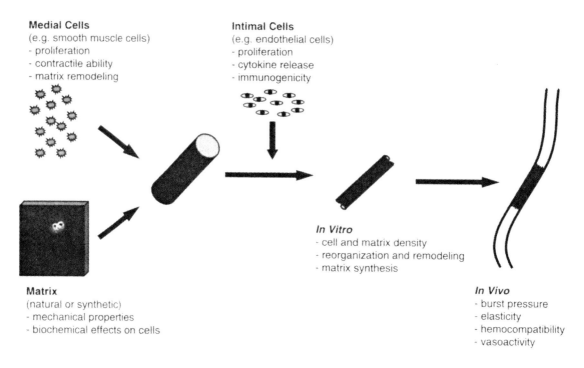

FIGURE 6.1. Functional requirements for the engineering of a blood vessel substitute—the properties and functions of both the cellular and matrix materials must be considered when constructing an engineered blood vessel. *In vitro,* this function can be controlled to improve construct properties, and subsequent *in vivo* behavior must be appropriate to ensure proper integration and function as a living vascular tissue.

In identifying the functional requirements to be engineered into a blood vessel substitute, any potential list certainly must include the following:

- nonthrombogenicity
- sufficient burst pressure
- viscoelastic mechanical properties
- appropriate remodeling responses
- vasoactivity

In each of these there is a role for biomechanics. Furthermore, in spite of some of the important advances in the tissue engineering of a blood vessel substitute, each of the currently available concepts falls short in achieving at least one of these desirable characteristics. These concepts range from acellular approaches to cell-seeded scaffolds and include the use of collagen-based, polymeric, and cell-secreted materials (Huynh et al., 1999; L'Heureux et al., 1998; Niklason et al., 1999; Seliktar et al., 2000). For a description and evaluation of the different approaches now under development, please see the review of Nerem and Selkitar (2001).

Figure 6.1 shows schematically some of the main functional requirements that must be considered when developing a blood vessel substitute. These include inherent properties and functions of both the cellular and matrix components of the engineered tissue. The cells used in the medial layer must be able to populate and remodel the matrix material, while the intimal cells must provide a confluent, hemocompatible surface. The matrix material may function purely as a mechanical support, but ideally would also provide appropriate biochemical and mechanical signals to the cellular component of the tissue. Functional requirements also include the response of the cell-matrix combination to its environment both *in vitro* and *in vivo*. During production of the tissue in culture, it is likely that enhanced cell proliferation, as well as active matrix deposition and reorganization, would be desirable to produce a robust tissue. *In vivo,* however, it is probable that a more quiescent tissue that exhibits only the appropriate mechanical and functional properties would be preferable.

The purpose of this chapter thus is to discuss the functional requirements of a blood vessel substitute in the context of the role of biomechanics, not to provide a review of the state-of-the-art. This will be done by first addressing requirements for the cells to be employed, then considering the requirements for the mechanical properties of a blood vessel construct or substitute, and finally focusing on the requirements associated with the remodeling of an engineered substitute, both *in vitro* and *in vivo*.

Nonthrombogenicity and the Vascular Endothelial Cell

A primary failure mode of small diameter blood vessel substitutes made from synthetic materials is thrombotic closure. Thus, a critical issue for a tissue-engineered blood vessel substitute is to provide for nonthrombogenicity. This suggests having an endothelium, or at least an endothelial-like cell layer. Such an interfacial, cellular monolayer should be adherent, confluent, and biologically nonactivated. In all of these, there is a role of biomechanics.

The vascular endothelial cell (EC) has gone from 30 years ago being a cell thought to be somewhat uninteresting, simply providing a nonthrombogenic inner layer for a blood vessel, to a cell recognized as being a dynamic participant in vascular biology. It literally serves as a signal-transduction interface, linking the chemical and mechanical signals associated with flowing blood to the underlying vessel wall (Nerem et al., 1998). It is now among the most studied mammalian cell types.

As part of achieving this recognition, there have been extensive investigations of the influence on the vascular EC of its mechanical environment, i.e. the stresses and strains which it undergoes. This environment is very complex, being one in which the endothelial cell sees pulsatile pressure, rides on a basement membrane undergoing cyclic stretch, and is exposed to a time-varying viscous shear stress. As a way of understanding the endothelial responses to this complicated environment, most investigators have studied the effects of a specific mechanical component. Even so, there are potential synergies of pressure, cyclic stretch, and shear stress which must be recognized.

What we now know is that different mechanical environments can have very different effects. This is just as different biochemical stimuli can have different agonist effects. As an example of this, let us examine the influence of flow on the expression of one particular, but important molecule. This is vascular cell adhesion molecule-1 (VCAM-1), a molecule important in the recruitment of monocytes, which themselves are a member of the white cell family and have a major role in inflammatory responses. Vascular cell adhesion molecule-1 (VCAM-1) is known to be cytokine inducible; however, this induction can be inhibited by anti-oxidants.

A key issue is how the expression of VCAM-1 may be regulated by its mechanical environment. To consider the role of flow and the associated shear stress in the regulation of VCAM-1, let us first look at the influence of steady shear stress. Based on cell culture studies in which endothelial monolayers are exposed to the sudden onset of a steady flow and thus a steady viscous shear stress, what is observed is that there is no upregulation of VCAM-1 (Varner et al., 1998). In contrast, if one preconditions an endothelial monolayer with steady flow for 24 hours and then exposes it to the cytokine IL-1β, there is a very important effect of shear stress. Here preconditioning with steady flow reduces the normal cytokine-induced upregulation of VCAM-1 by 90%, this as compared with what one finds for a static culture control. In contrast, if one investigates the effect of a purely oscillatory shear stress environment, what one finds is a significant upregulation, one which is almost cytokine-like, but without the presence of a cytokine (Chappell et al., 1998). Interestingly, preconditioning with a purely oscillatory flow produces only a 40% inhibition of the IL-1β, cytokine-induced up-regulation. This is obviously much less of an effect than that observed for steady flow preconditioning.

This underscores what has been known for a number of years. This is that ECs residing in different mechanical environments exhibit different characteristics. Thus, an aortic endothelial cell is not the same as a coronary artery endothelial cell is not the same as a microvascular endothelial cell, for example, one obtained from adipose tissue.

This may be in part due to differences in genetic makeup, but it is also due to differences in environment, that is, the mechanical environment. Whatever the case, the result is that an endothelial cell is not an endothelial cell is not an endothelial cell, something that the senior author of this chapter has been saying for almost two decades.

In creating an engineered blood vessel substitute, all of the above must be taken into account. A key issue is the identification of a source for the endothelial cells to be used. Are these to be autologous, perhaps recruited from surrounding vessels or obtained either from adipose tissue or from circulating endothelial cells? Or is one's strategy to be the employment of allogeneic endothelial cells in which case the engineering of immune acceptance will be critical? Alternatively, at least long term, stem cell technology offers considerable potential. Whatever the case, the functional characteristics including how the cells respond to mechanical forces will be important. Although vascular ECs have been studied extensively in this regard, little is yet known about the effects of mechanical stimuli on circulating endothelial cells and stem cells.

In summary, the vascular EC is the critical element in a normal blood vessel. One of its roles is to provide for nonthrombogenicity, and its ability to do this, as well as to carry out other functions, is very much influenced by its hemodynamic environment, that is, by the influence of biomechanics. Now it may be possible to provide for this nonthrombogenicity in a variety of ways, perhaps even employing some other cell type; however, it should not be forgotten that the vascular endothelial cell also plays other, very different roles. One is that it serves as a signaling interface, and in doing this controls vasoactivity and the maintenance of vascular tone. This leads us to the next requirement for an engineered blood vessel substitute and the role of the vascular smooth muscle cell.

Vasoactivity and the Vascular Smooth Muscle Cell

Vasoactivity represents a biomechanical function that is mediated by cellular processes. The two vascular cell types critical to vasoactivity are the endothelial cell and the smooth muscle cell. The former receives and sends important signals that control vascular tone, and was discussed in the previous section. Here we turn our attention to the vascular smooth muscle cell (SMC), the natural neighbor of the endothelial cell that is responsible for generating the contractile force necessary to change vessel dimensions. For such vasoactivity to be exhibited, the vascular SMC must exist in the appropriate contractile phenotype, characterized by the expression of proteins involved in the rapid shortening of cell dimensions. In addition, these cells must be oriented in a circumferential direction within the blood vessel wall, so that when there is SMC contraction, there is vessel constriction.

The contractile nature of the vascular smooth muscle cell thus is a critical issue; however, when SMC are taken from their native environment and placed in culture and are then passaged, they are observed to rapidly take on what sometimes is called a synthetic phenotype. This shift is shown both schematically and using micrographs in Figure 6.2. It is characterized by a change in cell morphology, and an increase in protein synthetic activity and cell proliferation. It might be that vascular SMCs in a petri dish have no need to exhibit a contractile function, and, if this is the case, then in the artificial environment of the petri dish these cells might go into a highly secretory mode, this in order to remodel their *in vitro* environment. If this is the case, then perhaps such cells, when incorporated into a three-dimensional construct and placed in a mechanical stress environment, might revert to a more contractile phenotype, and there is actually some evidence of this (Kanda et al., 1993; Shirinsky et al., 1995). There is, however, also experimental evidence that SMCs exposed to cyclic mechanical strain can exhibit an increase in matrix synthesis (O'Callaghan and Williams, 2000; Lee et al., 2001) and cell proliferation (Li et al., 1997; Mills et al., 1997), characteristics more associated with the synthetic phenotype.

There may be other approaches as well, as there are many factors associated with the extracellular environment that may influence SMC phenotype. There also may be possibilities that involve the altering of phenotype using genetic

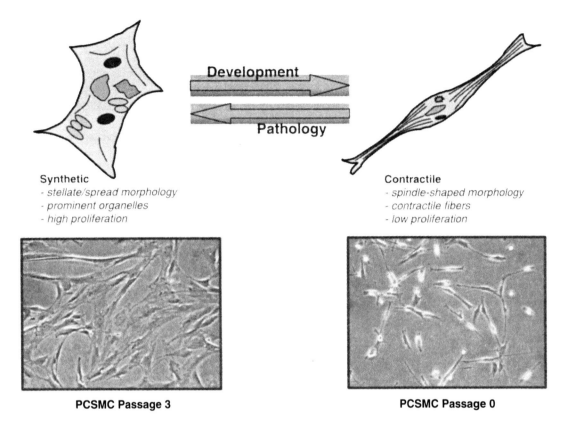

Synthetic
- *stellate/spread morphology*
- *prominent organelles*
- *high proliferation*

Contractile
- *spindle-shaped morphology*
- *contractile fibers*
- *low proliferation*

PCSMC Passage 3

PCSMC Passage 0

FIGURE 6.2. Smooth muscle cell phenotype—cells in the medial layer of the blood vessel wall are generally in what is referred to as the contractile phenotype, however, when these cells are removed from their native environment and placed in culture, they shift toward a more synthetic, proliferative phenotype. This is shown above both schematically and using micrographs of isolated porcine carotid smooth muscle cells (PCSMC).

engineering. This in fact has been demonstrated by the laboratory of Dr. T.M. Lincoln (see Boerth et al., 1997), where an approach was used in which SMCs were transfected so as to enhance the cyclic guanosine monophosphate (GMP)-dependent protein kinase, PKG. The result of this approach is a very enhanced expression of smooth muscle α-actin, an indicator of a more contractile phenotype. GTEC is now collaborating with Lincoln's laboratory in using these genetically engineered SMCs in cell-seeded collagen constructs. The initial data demonstrate that such genetically engineered SMCs exhibit enhanced contractile characteristics.

It should be noted that in many ways genetic engineering should be viewed as an ally to tissue engineering. In addition to being used as a tool to control SMC phenotype, other possible applications include using genetic engineering to en-

hance expression of antithrombogenic substances for endothelial cells or to increase the synthesis of elastin by SMCs so as to create a blood vessel substitute with enhanced mechanical properties. This latter will be discussed further in the next section.

Returning to the subject of vasoactivity, another issue is of course SMC orientation. As noted previously, these cells need to be oriented circumferentially so that, when there is SMC contraction or relaxation, there will be vessel constriction or dilation. Obtaining contractility at the single cell level is therefore not sufficient; there must be a coordinated, directional cellular response in order to achieve contraction at the tissue level. In the next section, it will be noted that there is evidence that this can be induced as part of an *in vitro* remodeling caused by mechanical stimulation.

The discussion in this section has been based on the premise that vasoactivity is a desired functional characteristic. This of course is true if one seeks to engineer a blood vessel substitute that mimics the complete functionality of a native artery. If the goal, however, is first and foremost long-term patency, then is vasoactivity required? Intuitively, the answer would appear to be yes; however, there is evidence to the contrary. An example of this is the long-term, clinical patency that has been achieved by some using endothelialized ePTFE arterial grafts. Such a substitute is rigid and not vasoactive. Zilla and his co-workers have demonstrated excellent nine-year clinical results (see Deutsch et al., 1999), and this suggests that, although vasoactivity may be a desired characteristic, it is not necessary.

Finally, as noted earlier, vascular SMCs are the natural neighbors to the endothelial cells. As such, another important role for SMCs may be to provide through their presence a more natural environment for the endothelial cells. In doing so, if patency is enhanced, then this may be at least as important, if not more important, than providing for vasoactivity.

Mechanical Properties

There are at least two important mechanical properties for a blood vessel substitute. The first and most basic of these is ultimate strength, frequently denoted simply as burst pressure, that is, the pressure required to rupture or burst the vessel. For native arteries this is in excess of 2000 mm Hg. Since arterial pressures range up to several hundred mm Hg, this then implies that native vessels are designed with a safety factor of approximately 10. As far as tissue engineering approaches are concerned, several of these have resulted in burst pressures on the order of 2000 mm Hg (Huynh et al., 1999; L'Heureux at al., 1998; Niklason et al., 1999). It is, however, not clear how much of the natural 10-fold safety factor must be incorporated to obtain a functionally suitable engineered blood vessel.

A second, potentially very important characteristic has been more difficult to achieve. This is to engineer into a construct a viscoelasticity that matches that of native vessels. In this a critical

problem has been the incorporation of elastin into the wall of the construct so as to have elasticity. The various tissue engineering approaches pursued to date have resulted in constructs with a very limited amount of elastin. To address this there are three approaches that have been pursued within GTEC. One of these is the use of elastin mimetic peptides that, as an example, could be woven into a scaffold into which cells could be seeded. The second is the genetic engineering of cells so as to enhance the expression of elastin where such cells could be incorporated into a cellularized construct. Such a transfection could be transient in nature as one may simply wish to accelerate the process of elastin synthesis beyond what is normally taking place. Finally, a third approach is the use in a cell-seeded collagen gel construct as a foundation the elastin network obtained by digesting away the collagen and cells in a xenogeneic animal artery (dog or porcine).

At the beginning of the previous paragraph, the matching of the viscoelastic properties of a blood vessel substitute with those of native vessels was raised as potentially important. Why only potentially when it seems intuitively obvious to anyone in biomechanics that this should be a requirement? The reason again comes from the excellent clinical results achieved by some with endothelialized ePTFE arterial grafts (Deutsch et al., 1999). In spite of the rigidity of such a prosthesis, they work. This suggests that what is intuitively obvious in fact may not be necessary at all, although different anatomical graft placements may have different biomechanical property requirements.

In Vitro versus
In Vivo Remodeling

When a tissue-engineered blood vessel substitute is implanted, there is an integration into the host system that takes place, and associated with this there will be a remodeling. Thus, six months after implantation the characteristics of the blood vessel substitute should not be expected to be the same as those exhibited at the time of surgery.

There also is a remodeling, however, that can be induced prior to implantation as part of the process used to grow the tissue. Thus, the *in vitro* environment can be used to advantage in engineering the remodeling of the tissue substitute. A number of laboratories have developed bioreactor systems for doing this (Niklason et al., 1999; Seliktar et al., 2000). These combine a mechanical environment with a chemical environment, with this combined environment being tailored so as to induce the desired remodeling. At GTEC we have developed a bioreactor which uses a 1-Hz, 10% strain, cyclic distension to remodel SMC-seeded collagen gel, tubular constructs (Seliktar et al., 2000). Doing this over a four-day period results in a 138% increase in ultimate stress for a human SMC-seeded collagen construct.

This enhancement in mechanical properties is the result of a remodeling that involves a variety of changes. These include morphology, where both the collagen fibers and the SMC become oriented circumferentially, and composition, where there is an upregulation in elastin mRNA expression and a down-regulation of collagen. There also are changes in the expression of the matrix metalloproteinases (MMPs) that could play an important role in the changes induced. As an example, MMP-2 is significantly upregulated by this cyclic strain environment (Nerem and Seliktar, 2001).

MMP-2 should be viewed simply as an example, as there are a myriad of biologically active molecules playing a role in the remodeling process. It will be critical, however, to understand these biological mechanisms, the role of biomechanics, and how the biology and the biomechanics integrate with one another in the remodeling which takes place.

Conclusion

From the preceding, one can see that there are important roles for biomechanics in the tissue engineering of a blood vessel substitute. This includes the integration of a blood vessel substitute into an animal model or ultimately as an implant in a human, what can be called integration into the living system. This integration will determine how successful one has been in addressing the critical issues identified here. Once

implanted there will be not just thrombotic events, but a variety of potentially important biologic responses, and it will be essential to understand the influence of biomechanics on these biologic responses and its role in possible strategies for controlling them. This influence of biomechanics, as observed for *in vivo* models, may be far different from what is observed *in vitro,* and it will be important to understand such differences. In this there will need to be a focus on the mechanisms involved and on how they may differ for the *in vivo* situation as compared to the *in vitro* environment.

Although generally thought not to be related in any way to biomechanics, a critical issue will be the immune acceptance of the implant. If a tissue-engineered blood vessel substitute is to be available on a widespread basis and off-the-shelf, nonautologous cell approaches will need to be developed. Although it may be possible to use allogeneic SMC, this is not true for the vascular endothelial cell. Without the engineering of immune acceptance, the use of allogeneic endothelial cells will not be viable. As an antigen-presenting cell, endothelial cells are just too talented. Fortunately, there is progress being made in the development of strategies to overcome the immunological barrier (Larsen et al., 1996).

In addition to the development of a blood vessel substitute that will exhibit long term patency and thus address patient need, there are a number of other excellent applications of tissue engineering. One of these involves the development of model systems for research studies. These can be used not only to study vascular biology, but also to systematically investigate some of the issues identified in this chapter. In order to obtain a physiologically relevant blood vessel model that can be used to study vascular biology *in vitro,* it will be necessary to have a high degree of control over the phenotype and function of the cells that make up the engineered tissue. This control could be used to vary the characteristics of a blood vessel substitute, and observe the effects on function.

As an example of the latter, one could assess the importance of matching the mechanical properties of an implanted blood vessel substitute with those of the native arteries. This could be achieved by creating a family of substitutes, each with a different set of mechanical properties, and employing these in a series of experiments. Thus, using a very systematic approach, one could investigate how differences in mechanical properties result in different *in vivo* outcomes.

There obviously is much yet to be done before a tissue-engineered blood vessel substitute is available for widespread clinical use. Taking all the relevant issues together, we are still a decade away from having a Food and Drug Administration–approved, tissue-engineered blood vessel substitute. Of course, not all of the issues are ones related to biomechanics; however, the biomechanical ones described here are of real significance and thus are deserving of the attention of the entire cardiovascular tissue engineering community.

Acknowledgments

The authors acknowledge the support provided by the National Science Foundation through an Engineering Research Center Award, which funds the Georgia Tech/Emory Center for the Engineering of Living Tissues. The authors also acknowledge the many stimulating discussions with their colleagues at Georgia Tech and Emory University.

References

Boerth, N.J., Dey, N.B., Cornwell, T.L., Lincoln, T.M. 1997. Cyclic GMP-dependent protein kinase regulates vascular smooth muscle cell phenotype. *J. Vasc. Surg.* 34:245–259.

Carrel, A., Lindbergh, C.A. 1938. *The Culture of Organs.* New York:Paul B. Hoeber.

Chappell, D.C., Varner, S.E., Nerem, R.M., Medford, R.M., Alexander, R.W. 1998. Oscillatory shear stress stimulates adhesion molecule expression in cultured human endothelium. *Circ. Res.* 82:532–539.

Conte, M.S. 1998. The ideal small arterial substitute: a search for the holy grail? *FASEB J.* 12:43–45.

Deutsch, M., Meinhart, J., Fishchlein, T., Preiss, P., Zilla, P. 1999. Clinical autologous *in vitro* endothelialization of infrainguinal ePTFE grafts in 100 patients: a 9-year experience. *Surgery* 126:847–855.

Huynh, T., Abraham, G., Murray, J., Brockbank, K., Hagen, P-O., Sullivan, S. 1999. Remodeling of an acellular collagen graft into a physiologically responsive neovessel. *Nat. Biotechnol.* 17:1083–1086.

Kanda, K., Matsuda, T., Oka, T. 1993. Mechanical stress induces cellular orientation and phenotypic modulation of 3-D cultured smooth muscle cells. *ASAIO J.* 39:M686–M690.

Larsen, C.P., Elwood, E.T., Alexander, D.Z., Ritchie, S.C., Hendrix, R., Tucker-Burden, C., Cho, H.R., Aruffo, A., Hollenbaugh, D., Linsley, P.S., Winn, K.J., Pearson, T.C.. 1996. Long-term acceptance of skin and cardiac allografts by blockade of the CD40 and CD28 pathways. *Nature* 381:434–438.

Lee R.T., Yamamoto C., Feng Y., Potter-Perigo S., Briggs W.H., Landschulz K.T., Turi T.G., Thompson J.F., Libby P., Wight T.N. 2001. Mechanical strain induces specific changes in the synthesis and organization of proteoglycans by vascular smooth muscle cells. *J. Biol. Chem.* 276:13847–13851.

L'Heureux, N., Paquet, S., Lacoe, R., Germain, L., Auger, F.A. 1998. A completely biological tissue engineered human blood vessel. *FASEB J.* 12:47–56.

Li Q, Muragaki Y., Ueno H., Ooshima A. 1997. Stretch-induced proliferation of cultured vascular smooth muscle cells and a possible involvement of local renin-angiotensin system and platelet-derived growth factor (PDGF). *Hypertens. Res.* 20:217–223.

Mills I., Cohen C.R., Kamal K., Li G., Shin T., Du W., Sumpio B.E. 1997. Strain activation of bovine aortic smooth muscle cell proliferation and alignment: study of strain dependency and the role of protein kinase A and C signaling pathways. *J. Cell. Physiol.* 170:228–234.

Nerem, R.M., Seliktar, D. 2001. Vascular tissue engineering. *Annu. Rev. Biomed. Eng.* 3:225–243.

Nerem, R.M., Alexander, R.W., Chappell, D.C., Medford, R.M., Varner, S.E., Taylor, W.R. 1998a. The study of the influence of flow on vascular endothelial biology. *Am. J. Med. Sci.* 316:169–175.

Nerem, R.M., Ku, D.N., Sambanis, A. 1998b. Core technologies for the development of tissue engineering. In: *Proceedings of the Second International Symposium of Tissue Engineering for Therapeutic Use (Tissue Engineering for Therapeutic Use 2),* Tokyo, Japan, October 1997. Elsevier, New York, pp. 19–29.

Niklason, L.E., Gao, J., Abbott, W.M., Hirschi, K.K., Houser, S., Marini, R., Langer, R. 1999. Functional arteries grown *in vitro. Science* 284:489–493.

O'Callaghan C.J., Williams B. 2000. Mechanical strain-induced extracellular matrix production by human vascular smooth muscle cells: role of TGF-beta1. *Hypertension* 36:319–324.

Seliktar, D., Black, R.A., Vito, R.P., Nerem, R.M. 2000. Dynamic mechanical conditioning of collagen gel blood vessel constructs induces remodeling *in vitro, Ann. Biomed. Eng.* 28:351–362.

Shirinsky, V.P., Birukov, K.G., Stepanova, O.V., Tkachuk, V.A., Hahn, A.W.A., Resink, T.J. 1995. Mechanical stimulation affects phenotype features of vascular smooth muscle cells. In: *Atherosclerosis X, Proceedings of the Xth International Symposium on Atherosclerosis,* held in Montreal, Canada, Oc-

tober 9–14, 1994. F.P. Woodford, J. Davignon, and A. Sniderman, eds. Elsevier Science B.V., Amsterdam, pp. 822–826.

Varner, S.E., Alexander, R.W., Medford, R.M., Nerem, R.M. 1998. Endothelial VCAM-1 gene regulation by steady and oscillatory shear stress. In: *Atherosclerosis XI, Proceedings of the XIth International Symposium on Atherosclerosis,* held in Paris, France, October 5–9, 1997. B. Jacotot, D., Mathe, and J-C. Fruchart, eds. Elsevier Science, Singapore, pp. 957–961.

7

In Vivo Force and Strain of Tendon, Ligament, and Capsule

Kai-Nan An

Introduction

In vivo force and strain is important when providing design criteria in tissue engineering for tissue repair or regeneration. In addition, such information will be essential for understanding the potential mechanical stimulation in newly generated tissue and interaction with the surrounding host tissue. In this chapter, *in vivo* force and strain measurements in soft tissue will be presented in six groups: tendon tension, ligament deformation, capsular pressure, tendon surface friction, soft tissue stress, and strain. Finally, the related topics for future research will be discussed.

Tendon Tension

Tendon connects and transmits forces generated by muscle to the skeletal system. The magnitude of tendon tension has been estimated by various analytical methods, and also by combined experimental and analytical methods. Direct measurements have also been attempted. In the analytical approach, the model based on the anatomic structure through the biomechanical free-body analysis must be established. The unknown muscle and tendon forces have been obtained by solving a system of equilibrium equations along with the additional constraints of the potential muscle forces and joint static constraints. One of the mathematical problems commonly encountered in this approach is the indeterminate problem—the number of unknowns usually exceeds the number of the available equations. Therefore, the solution is not unique. To resolve this indeterminate problem, the forward solution based on the additional experimental data of muscle activity has been considered. The normalized electromyographic data has been commonly used as the indicator of muscle activity. To further validate the solutions obtained from the analytic model, direct experimental measurements of the muscle and tendon forces have been attempted by numerous investigators. Three typical transducers used for such measurements are the buckle transducer, the intramuscular or intratendinous force transducer, and the pressure transducer.

The buckle transducer has been used extensively for the measurement of the axial tensile force in tendons and ligaments. (An et al., 1990; Barnes and Pinder, 1974; Biewener et al., 1988; Herzog et al., 1993; Komi, 1990; Lewis et al., 1982). In general, the buckle transducer converts the axially directed force into transverse force and bending moment. There are different designs, either in the E shape or S shape, which could register, respectively, the bending moment or shear force. The transducer would cause slight elongation of the tissue to be evaluated, which usually causes more problems for short tissue, such as ligaments, than for longer tissues, such as tendon. In situ calibrations were required during measurements.

Because of space restrictions and potential impingement artifacts, these external buckle transducers cannot be used for force measurements in all tendons and ligaments (Herzog et al., 1996). In these situations, a transducer placed inside the tissue has also been developed that includes the implantable force transducer (Glos et al., 1993; Herzog et al., 1996; Malaviya et al., 1998). The suitability and performance of an internally placed force transducer has been assessed. It was noted that many factors may influence the measurements. A small angular displacement of the transducer within the tissue could result in substantially changed transducer output for given externally applied loads. Also, the transducer output was found to depend on the rate of load application. It was also noted that there is influence of sensor size on the accuracy of *in vivo* ligament and tendon force measurements (Rupert et al., 1998). The measurement can be affected by differences in tendon stress distribution between calibration and experimental conditions. These errors could be minimized by using a larger transducer, which could sample the tension in a large percentage of the tendon fibers. Calibration and application of the arthroscopically implantable force probe (AIFP) for the measurement of patellar tendon graft forces has also been performed (Fleming et al., 1999). The postimplantation approach provided a better estimate of the anterior cruciate ligament (ACL) graft force than the preimplantation technique. However, the errors for the postimplantation approach were still high and suggested that caution should be employed when using implantable force probes for *in vivo* measurement of ACL graft forces.

More recently, the optic fiber method has been used for direct *in vivo* tendon force measurements (Kaufman and Sutherland, 1995; Komi et al., 1996). This method is based on light intensity modulation by mechanical modification of the geometric properties of the optic fiber. A special optic fiber with a plastic covering buffer and a total diameter of either 265 microns or 500 microns was carefully prepared at both ends for receiving and transmitting light. In dynamic loading conditions the optic fiber followed well the response of a strain gauge transducer, which was also attached to the tendon (Komi et al., 1996).

Most commonly studied tendons include patellar tendon (PT) of the knee, flexor tendon of the finger, and the Achilles tendon (AT) of the foot. Fukashiro et al. (1995) measured the *in vivo* AT loading during jumping in humans. A buckle-type transducer was implanted under local anesthesia around the right AT of an adult subject. The peak AT force by the calf muscles was 2233 N in the squat jump, 1895 in the counter movement jump, and 3786 N while hopping. Maganaris and Paul (1999) measured the force and mechanical properties of human tibialis anterior (TA) tendon at the neutral ankle position using percutaneous electrical stimulation, real-time ultrasonography, and magnetic resonance imaging. The tendon force stress and strain at maximum isometric load were 530 N, 25 MPa, and 2.5%, respectively. These results are in agreement with previous reports on *in vitro* testing of isolated tendons, and suggest that under physiological loading the TA tendon operates within the elastic "toe" region.

PT force was measured during activity with an implantable force transducer (IFT) in adult goats (Korvick et al., 1996). PT force and vertical ground reaction force (VGRF) were recorded for standing, walking, and trotting. Following data collection, animals were euthanized and the IFT calibrated *in vitro*. Standing PT force averaged 207 N. Maximum PT force was approximately 800 N for walking and 1000 N for trotting, and occurred at mid-stance. PT force dropped from 200 N at toe-off to 0 N by midswing. For each activity, the PT force increased with increases in VGRF. Maximum *in vivo* PT stress occurred during trotting and measured 29 MPa. This study demonstrates the IFT's usefulness in measuring tendon force directly.

In vivo measurements of flexor tendon force were conducted by Schuind et al. (1992). S-shaped force transducers were developed for measurement of the forces along intact tendons. After calibration, the transducers were applied to the flexor pollicis longus, and the flexor digitorum superficialis (FDS) and profundus (FDP)

TABLE 7.1. Tendon forces involved in various activities

Tension (N)	Tendon	Activity	Reference
207	Goat PT	Standing	Korvick, 1996
801	Goat PT	Walking	Korvick, 1996
999	Goat PT	Trotting	Korvick, 1996
13	Rabbit FDP	Standing	Malaviya, 1998
50	Rabbit FDP	Level hopping	Malaviya, 1998
82	Rabbit FDP	Inclinal hopping	Malaviya, 1998
2600	Human AT	Walking	Komi et al., 1992
9000	Human AT	Running	Komi et al., 1992
4000	Human AT	Running	Komi et al., 1992
480–661	Human AT	Cycling	Komi et al., 1992
1895–2233	Human AT	Squat jump	Fukashiro et al., 1995
3786	Human AT	Hopping	Fukashiro et al., 1995
1–9	Human FDP	Passive motion	Schuind et al., 1992
35	Human FDP	Active motion	Schuind et al., 1992
120	Human FDP	Pinch	Schuind et al., 1992
8–16	Human FDS	Keystroke	Dennerlein et al., 1999

TABLE 7.2. Tendon stress involved in various activities

Stress (MPa)	Tendon	Activity	Reference
23	Goat PT	Walking	Korvick, 1996
29	Goat PT	Trotting	Korvick, 1996
59	Human AT	Walking	Komi et al., 1992
110	Human AT	Running	Komi et al., 1992

tendons of the index finger in patients operated on for treatment of carpal tunnel syndrome. The tendon forces generated during passive and active motion of the wrist and fingers were recorded. For pinch function, the amount of applied load was measured with a special pinch meter. Tendon forces in the range of 1–6 N were measured during passive mobilization of the wrist. Tendon forces up to 9 N were present during passive mobilization of the fingers. Tendon forces up to 35 N were present during active unresisted finger motion. Tendon forces up to 120 N were recorded during tip pinch, with a mean applied pinch force of 35 N. More recently, measurements of finger flexor force in tapping on keyswitch were performed. During the open carpal tunnel release surgery, a tendon-force transducer was inserted on the flexor digitorum superficialis of the long finger (Dennerlein et al., 1999). The average tendon maximum forces during a keystroke ranged from 8.3 to 16.6 N. These tendon tensions were four to seven times larger than the maximum forces observed at the fingertip. Tendon forces estimated from an isometric tendon-force model were only one to two times larger than tip force, significantly less than the observed tendon forces.

The values of tendon forces and tendon stress measured and reported in the literature are summarized in Tables 7.1 and 7.2, respectively. The stress in AT measured (Komi et al., 1992) are quite high, and likely approaching failure stresses for these tissues.

Another way of evaluating tendon tension is to compare the value with the potential failure strength of the tendon (Table 7.3). Investigators (Biewener et al., 1988; Korvick et al., 1996) have shown that peak knee and ankle tendon forces approach one-quarter to one-third of ultimate or failure force values. Malaviya et al. (1998) used the rabbit flexor digitorum profundus tendon model to test the hypotheses that peak *in vivo* forces increase with increasing activity but do not exceed one-third of their ultimate or fail-

TABLE 7.3. Ratio of tendon force and failure strength

Ratio (%)	Tissue	Reference
28.6	Rabbit flexor digitorum	Malaviya et al., 1998
32	Goat PT	Butler et al., 1986, 1989
9	Goat ACL	Jackson et al., 1993
25	Rat ankle extensor	Biewener et al., 1988
10–20	Human finger flexor	An et al., 1995
25–30	Human rotator cuff	Itoi et al., 1995
100	Human AT	Komi et al., 1992

ure values. The FDP tendon was instrumented in three animals, and each rabbit was subjected to an experimental design involving three activity levels. Peak tensile forces and rates of rise and fall in tendon force increased significantly with increasing activity. For the most vigorous activity, the tensile force was on average within 30% of the ultimate strength. The ratios of tendon tension to failure strength of rotator cuff tendon and human finger flexors have also been obtained by Itoi et al. (1995) and An et al. (1995), respectively. The ratio for AT is extremely high during running (Komi et al., 1992). Table 7.3.

Ligament Deformation

Numerous studies have demonstrated the capacity of mechanical strains to modulate cell behavior through several different signaling pathways. Understanding the response of ligament fibroblasts to mechanically induced strains may provide useful knowledge for treating ligament injury and improving rehabilitation regimens (Hsieh et al., 2000).

Various transducers have been designed and developed for the measurement of soft tissue deformation and strain. In the past, the commonly used transducer was the liquid metal strain gauge. The performance and characteristics of this strain gauge have been evaluated by both static and dynamic bench testing (Brown et al., 1986). Statically, the devices were found to have outputs closely proportional to engineering strains, up to strain levels of 40%. While individual gauge factors varied appreciably, each of the gauges studied showed excellent reproducibility of behavior.

Dynamically, the response to sinusoidal strain inputs was frequency-independent up to 50 Hz, and there was no detectable phase shift.

A strain transducer was also developed which employs a magnetic field sensing device to detect linear displacement (Arms et al., 1983). The transducer has been used to evaluate strains in medial collateral ligaments (MCL) of human autopsy specimens, minimally influencing their physiologic behavior. A strain "map" of the MCL as a function of knee flexion (full extension to 120°) both with and without abduction force was obtained. Their investigation revealed consistent differences in the strain patterns between proximal, middle, and distal segments of the anterior and posterior borders of the MCL. Howe et al. (1990) first described a new arthroscopic technique to study the anterior cruciate ligament *in vivo*. A Hall effect strain transducer (HEST) was inserted arthroscopically into the anterior medial band (AMB) of the ACL. The strain was calculated from HEST displacement data. Arthroscopic implantation of the Differential Variable Reluctance Transducer (DVRT) has also been used to assess the strain of the ACL while subjects are under local anesthesia. Movement of the knee from a flexed to an extended position, either passively or through contraction of the leg muscles, produces an increase in ACL strain values (Beynnon and Fleming, 1998). Isolated contraction of the dominant quadriceps with the knee between 50° flexion and extension creates substantial increases in strain. In contrast, isolated contraction of the hamstrings at any knee position does not increase strain.

Beynnon et al. (1995) also measured the strain behavior of the ACL during rehabilitation activities *in vivo* in patients with normal anterior cruciate ligaments instrumented with the Hall effect transducer. At 10° and 20° of flexion, ligament strain values for active extension of the knee with a weight of 45 N applied to a subject's lower leg were significantly greater than active motion without the weight. Isometric quadriceps muscle contraction at 15° and 30° also produced a significant increase in ligament strain, while at 60° and 90° of knee flexion there was no change in ligament strain relative to relaxed muscle condition. Simultaneous contraction of the quadriceps and hamstrings muscles at 15° produced a significant increase in ligament strain compared with the relaxed state, but did not strain the ligament at 30°, 60°, and 90° of flexion. Isometric contraction of hamstring muscles did not produce change in ligament strain at any flexion angle.

Stationary bicycling is commonly prescribed after anterior cruciate ligament injury or reconstruction. Fleming et al. (1998) measured ligament strain in eight patients who were candidates for arthroscopic meniscectomy under local anesthesia. Six different riding conditions were evaluated at three power levels (75, 125, and 175 W), each of which was performed at two cadences (60 and 90 rpm). The peak ligament strain values ranged from 1.2% for the 175-W, 90-rpm condition, to 2.1% for the 125-W, 60-rpm condition. No significant differences were found in peak strain values due to changes in power level or cadence. The mean peak strain value was 1.7%, a value that is relatively low compared with other rehabilitation activities previously tested.

Stair climbing is a closed kinetic chain exercise that is also thought to be useful for knee rehabilitation following ACL reconstruction while protecting the graft from excessive strain. Fleming et al. (1999) measured the strain produced in the AMB of the normal ACL during stair climbing *in vivo*. AMB strain was measured *in vivo* using the DVRT. Two different climbing cadences were evaluated; 80 and 112 steps per minute. Strain values increased as the knee was moved from a flexed to an extended position. The mean peak AMB strain values for the 80 and 112 steps per minute conditions were 2.69% (±2.89%) and 2.76% (±2.68%), respectively. These values were not significantly different.

Compared with other rehabilitation activities previously tested in the same manner, the AMB strain values produced during stair climbing were highly variable across subjects.

Functional knee braces are widely used to protect injured or reconstructed anterior cruciate ligaments. Beynnon et al. (1992) studied seven functional braces, representative of both the typical custom-fit and off-the-shelf designs. The braces were tested on subjects who had a normal anterior cruciate ligament and were scheduled for arthroscopic meniscectomy or exploration of the knee under local anesthesia. After the operative procedure, a HEST was applied to the anterior cruciate ligament. Under low anterior shear loads, two braces provided some protective strain-shielding effect compared with no brace, but this strain-shielding effect did not occur at the higher anterior shear loads expected during the high-stress activities common to athletic events.

Strain within the MCL was measured in human knee specimens to determine both the single and combined external loads most likely to cause injury (Hull et al., 1996). Using a load application system that allowed six degrees of freedom, both single loads of anterior/posterior force, medial/lateral force, varus/valgus torque, and internal/external axial torque and all pairs of these loads were applied. Liquid mercury strain gages were used to measure strain at four sites in the MCL. Two of the sites were the anterior fibers superior and inferior to the joint line, and the other two were posterior to the two anterior sites. A factorial analysis revealed a significant interaction between the site experiencing the greatest strain and flexion angle. Most strain measurements have been performed in the ACL and MCL (Table 7.4).

Capsule Pressure

The loading and deformation on the joint capsule has impact on soft tissue remodeling, which may be directly and indirectly responsible for the joint contracture or joint laxity after surgery or during inmobilization. The capsule loading can be associated with the intraarticular pressure (IAP) measurement.

TABLE 7.4. Ligament elongation and strain

Strain (%)	Ligament	Activity	Reference
4.4	ACL	Isometric Q	Beynnon et al., 1995
3.8	ACL	Active flex/ext W	Beynnon et al., 1995
2.8	ACL	Active flex/ext noW	Beynnon et al., 1995
3.6	ACL	Squatting without weight	Beynnon et al., 1997
4.0	ACL	Squatting with weight	Beynnon et al., 1997
4.7	ACL	Anterior shear load	Bern et al., 1992 (in vitro)
1.2–2.1	ACL	Bicycling	Fleming et al., 1998
4	MCL (knee)	Passive motion	Arms et al., 1983
7.7	MCL (knee)	Combined loading	Hull et al., 1996

The effects of joint position and continuous passive motion (CPM) on IAP and joint compliance were studied in 40 knee joints of 20 rabbits (O'Driscoll et al., 1983). Increments of 0.5 ml normal saline were injected into the knee joint and the intraarticular pressures were recorded with the knees flexed from 40° to 160° and during CPM. Increasing degrees of flexion of the knee produced significantly higher intraarticular pressures. A sinusoidal oscillation in pressure was observed during CPM. The optimum volume of effusion required to simulate the pressures associated with a hemarthrosis was found to be 2 ml. Viscoelastic stretching of the capsule could be demonstrated prior to synovial rupture, which occurred with a mean of 4 ml of injected saline.

Later, IAP was continuously monitored during continuous passive motion of five normal and 11 abnormal human knees using a new fiberoptic, transducer-tipped Camino catheter (Pedowitz et al., 1989). IAP varied in a consistent pattern in the normal knees, with subatmospheric pressures recorded at intermediate angles of joint flexion. A similar pattern was recorded in the abnormal knees without cruciate ligament pathology, whereas considerable variability was noted in the knees with cruciate ligament abnormality. IAP was lower in extension to flexion than in the flexion to extension portion of the CPM cycle, providing evidence of intraarticular fluid flow during portions of the CPM cycle. Capsular viscoelastic changes and/or synovial fluid volume changes were observed during CPM. It was concluded that the therapeutic mechanism of continuous passive motion might be related to cyclic variation of the IAP.

The compliance, capacity, and the position of minimum intraarticular pressure were also measured in human elbow specimens (O'Driscoll et al., 1990). The capacity of the joint capsule was 23 ml. The IAP was the lowest at 80° of flexion. Capsular rupture occurred at relatively low IAP of 80 mm Hg. The "resting position" of 80° of flexion minimizes capsular tension and therefore might contribute to the development of joint contracture associated with prolonged immobilization in this position. This would be consistent with the observation that patients with posttraumatic elbow stiffness have an average arc of flexion of 60–90°.

Tendon Surface Friction

Tendon not only transmits force from the muscle to the skeletal system, but also provides the displacement of the bony structure due to muscle shortening. When excursion of the flexor tendon through the pulley system takes place, frictional forces are encountered on the tendon surface. This frictional force on tendon surface has been hypothesized to be a potential mechanism related to the development of tendinosis. In addition, such frictional force has been considered as one of the important factors in the surgical repair of the tendon laceration.

A unique device has been developed for documenting the frictional force or gliding resistance between tendon and the surrounding tissue (Uchiyama et al., 1997). The gliding ability of the flexor digitorum profundus tendon and of the palmaris longus (PL) tendon through the A2 pulley was compared. The average gliding resis-

TABLE 7.5. Surface friction of tendon under 5 N tension

Friction (N)	Tendon	Reference
0.2	FDP/A2 pulley	Uchiyama et al., 1997
0.3–0.7	PL/A2 pulley	Uchiyama et al., 1997
0.3–0.4	BIC/grove	Heers et al., 2000
0.5–1.2	PTT	Uchiyama et al., 2000

tance at the interface between the palmaris longus tendon and the A2 pulley was found to be much greater than that between the flexor digitorum profundus tendon and the A2 pulley under similar loading conditions. The findings demonstrated that the intrasynovial and extrasynovial tendon have dramatically different gliding characteristics. More recently, gliding resistance of the biceps (BIC) tendon through the biceps groove at the shoulder and that of the posterior tibial tendon (PTT) throughout bony tunnels at the ankle have been measured as well (Heers et al., 2000; Uchiyama et al., 2000). With 5 N tension applied to the tendon, the surface friction and gliding resistance were measured (Table 7.5).

Soft Tissue Stress and Strain

In addition to the axial tensile force, the tendon or ligament encountered compressive force when wrapping around a bony structure or pulley system. Such compressive force, in combination with the tensile force, makes the stress and strain fields at that region in the tendon relatively complex. Such complex stress and strain fields have been associated with the modeling and remodeling of the soft tissue. In one study, the location of the major fibrocartilaginous area in the flexor tendon at the metacarpophalangeal joint correlated well with the region predicted by biomechanical modeling to be under greatest compressive loads during standing and claw movement (Okuda et al., 1987). Comparative biochemical analysis showed an elevated water content, a fivefold higher hexuronic acid content, and a larger hydroxylysine/hydroxyproline ratio in this region relative to that for more tendinous areas. The major glycosaminoglycan component of fibro-

cartilaginous areas was chondroitin sulfate, whereas in other areas, dermatan sulfate and hyaluronic acid dominated. Cell density and DNA analyses indicated a slightly higher cellularity for fibrocartilaginous areas and the region of vinculum insertion. These data document the existence of discrete areas of specialization within the flexor tendon that appear to be an adaptation to nutritional and mechanical factors.

The hypothesis that eliminating *in vivo* compression to the wrap-around, fibrocartilage-rich zone of the flexor digitorum profundus tendon results in rapid depletion of fibrocartilage and changes in its mechanical properties, microstructure, extracellular matrix composition, and cellularity was tested again recently (Malaviya et al., 2000). The right flexor digitorum profundus tendons of 2.5- to 3-year-old rabbits were translocated anteriorly to eliminate *in vivo* compression and shear to the fibrocartilage zone and, at 4 weeks after surgery, were compared with tendons that had sham surgery and with untreated tendons. The translocated tissue showed a significant increase in equilibrium strain under a compressive creep. The thickness and area of the fibrocartilage zone also decreased significantly. The tightly woven basket weave-like mesh of collagen fibers in the zone appeared more loosely organized, suggesting matrix reorganization due to translocation. With use of this unique *in vivo* model, this research clearly elucidates how changing tissue function by removing compressive forces rapidly alters tissue form.

Finite element analysis has been used to determine whether the regions of increased development of cartilaginous matrix in tendons that wrap around bones correspond to regions in which tendon cells are subjected to higher pressures, and whether the maintenance and rearrangement of fibrous extracellular matrix in these tendons is associated with regions of stretching and distortion of cells (Giori et al., 1993). It was found that regions of cartilaginous matrix and fibrous matrix formation and turnover correlate well with patterns of hydrostatic compressive stress and distortional strain in the tendon. More recently, a simplified two-dimensional finite element model was used to investigate the stress environment in the supraspinatus tendon (Luo et al., 1998). The extrafibrillar matrix and collagen fibers were modeled with fiber-reinforced composite elements. The results demonstrated that subacromial impingement generates high stress concentrations in and around the critical zone. Such high stress could initiate a tear; tears that result from stress point to an extrinsic mechanism. It was also found that high stress and potential tears caused by impingement may occur on the bursal side, the articular side, or within the tendon.

Future Development

To better document the load environment in the tendon, ligament and other soft tissue, more direct *in vivo* measurements should be performed. Noninvasive techniques based on image analysis could potentially provide the necessary tools for such measurement.

A technique for quantifying two-dimensional intratendinous rotator cuff strain using magnetic resonance image (MRI) has been developed (Bey and Soslowsky, 1999). MRI images of the shoulder specimen were taken under two testing conditions. The first testing condition simulated an "unstrained" supraspinatus tendon, while the second testing condition simulated a "strained" supraspinatus tendon. MR images were collected on a 1.5 T scanner using a two-dimensional, spin-echo technique. Tissue deformation was quantified from these two images using texture correlation. This approach has been used previously to study the detailed strain distribution within the trabecular bones. This technique uses a pattern-matching algorithm to quantify tissue deformation from the "unstrained" and "strained" images. No physical markers are placed on or within the tissue as used in the traditional methods. Instead, this approach utilizes patterns naturally inherent to the tissue, treating small rectangular regions of the image as discrete, unique markers.

Recently, in an *in vivo* study, Sheehan and Drace (2000) measured the strain in the human patella noninvasively. The *in vivo* three-dimensional velocity profiles for the patellar tendon, femur, and tibia were measured noninvasively in 18 healthy knees during a low-load extensor task using cine phase contrast MRI. These data were used to calculate patellar tendon elongation and strain. Average maximum strains of 6.6% were found for a low load extension task at relatively small knee angles. A newly developed technology, the magnetic resonance elastography (MRE), provides great potential for noninvasive *in vivo* investigation (Muthupillai et al, 1995). The MRE images the response of tissue to acoustic shear waves to determine the shear modulus as well as the tension in the muscle.

Furthermore, for situations where the direct measurement is not possible, analytic calculations need to be considered. Analytical calculation could also provide more detailed and refined stress and strain fields. The model could also provide the information on stress and strain fields at the cellular level where the direct cellular response would take place. Information on the loading environment within the soft tissue associated with postsurgical rehabilitation would be essential for treatment using tissue engineering approaches. Additional study for such results would be warranted.

Refrences

An K.N., Berglund L., Cooney W.P., Chao E.Y.S., Kovacevic N. 1990. Direct *in vivo* tendon force measurement system. *J. Biomech.* 23:1269–1271.

An K.N., Cooney W.P., Morrey B.F. 1995. The relationship between upper limb load posture and tissue loads at the elbow. In: *Repetitive Motion Disorders of the Upper Extremity.* Gordon S.L., Blair S.J., Fine L.J., eds. American Academy of Orthopaedic Surgeons, Rosemont, ILL.

Arms S., Boyle J., Johnson R., Pope M. 1983. Strain measurement in the medial collateral ligament of the human knee: an autopsy study. *J. Biomech.* 16:491–496.

Barnes G.R., Pinder D.N. 1974. *In vivo* tendon tension and bone strain measurement and correlation. *J. Biomech* 7:35–42.

Bey M.J., Soslowsky L.J. 1999. A technique for quantifying two-dimensional intratendinous rotator cuff strain. BED Vol. 42 1999 Bioengineering Conference, ASME, New York, pp. 141–142.

Beynnon B.D., Fleming B.C. 1998. Anterior cruciate ligament strain *in vivo*: a review of previous work. *J. Biomech* 31:519–525.

Beynnon B.D., Pope M.H., Wertheimer C.M., Johnson R.J., Fleming B.C., Nichols C.E., Howe J.G. 1992. The effect of functional knee-braces on strain on the anterior cruciate ligament *in vivo*. *J. Bone. Joint Surg. (Am)* 74:1298–1312.

Beynnon B.D., Fleming B.C., Johnson R.J., Nichols C.E., Renstrom P.A., Pope M.H. 1995. Anterior cruciate ligament strain behavior during rehabilitation exercises *in vivo*. *Am. J. Sports Med.* 23:24–34.

Biewener A.A., Blickhan R., Perry A.K., Heglund N.C., Taylor C.R. 1988. Muscle forces during locomotion in kangaroo rats: force platform and tendon buckle measurements compared. *J. Exp. Biol.* 137:191–205.

Brown T.D., Sigal L., Njus G.O., Njus N.M., Singerman R.J., Brand R.A. 1986. Dynamic performance characteristics of the liquid metal strain gage. *J. Biomech.* 19:165–173.

Dennerlein J.T., Diao E., Mote C.D. Jr, Rempel D.M. 1999. *In vivo* finger flexor tendon force while tapping on a keyswitch. *J. Orthop. Res.* 17:178–184.

Fleming B.C., Beynnon B.D., Renstrom P.A., Peura G.D., Nichols C.E., Johnson R.J. 1998. The strain behavior of the anterior cruciate ligament during bicycling. An *in vivo* study. *Am. J. Sports Med.* 26:109–118.

Fleming B.C., Beynnon B.D., Renstrom P.A., Johnson R.J., Nichols C.E., Peura G.D., Uh B.S. 1999. The strain behavior of the anterior cruciate ligament during stair climbing: an *in vivo* study. *Arthroscopy* 15:185–191.

Fleming B.C., Good L., Peura G.D., Beynnon B.D. 1999. Calibration and application of an intra-articular force transducer for the measurement of patellar tendon graft forces: an in situ evaluation. *J. Biomech. Eng.* 121:393–398.

Fukashiro S., Komi P.V., Jarvinen M., Miyashita M. 1995. *In vivo* Achilles tendon loading during jumping in humans. *Eur. J. Appl. Physiol.* 71:453–458.

Giori N.J., Beaupre G.S., Carter D.R. 1993. Cellular shape and pressure may mediate mechanical control of tissue composition in tendons. *J. Orthop. Res.* 11:581–591.

Glos D.L., Butler D.L., Grood E.S., Levy M.L. 1993. *In vitro* evaluation of an implantable force transducer (IFT) as a patellar tendon model. *J. Biomech. Eng.* 115:335–343.

Heers G., O'Driscoll S., Halder A., Zobitz M., Berglund L.B., Mura N., An K.N. 2000 Gliding properties of the long head of the biceps brachii. Jounal of Orthopaedic Research (In Press)

Herzog W., Leonard T.R., Guimaraes A.C. 1993. Forces in gastrocnemius, soleus, and plantaris tendons of the freely moving cat. *J. Biomech.* 26:945–953.

Herzog W., Archambault J.M., Leonard T.R., Nguyen H.K. 1996. Evaluation of the implantable force transducer for chronic tendon-force recordings. *J. Biomech.* 29:103–109.

Howe J.G., Wertheimer C., Johnson R.J., Nichols C.E., Pope M.H., Beynnon B. 1990. Arthroscopic strain gauge measurement of the normal anterior cruciate ligament. *Arthroscopy* 6:198–204.

Hsieh A.H., Tsai C.M., Ma Q.J., Lin T., Banes A.J., Villarreal F.J., Akeson W.H., Sung K.L. 2000. Time-dependent increases in type-III collagen gene expression in medical collateral ligament fibroblasts under cyclic strains. *J. Orthop. Res.* 18: 220–227.

Hull M.L., Berns G.S., Varma H., Patterson H.A. 1996. Strain in the medial collateral ligament of the human knee under single and combined loads. *J. Biomech.* 29:199–206.

Itoi E., Berglund L.J., Grabowski J.J., Schultz F.M., Growney E.S., Morrey B.F., An K.N. 1995. Tensile properties of the supraspinatus tendon. *J. Orthop. Res.* 13:578–584.

Kaufman K.R., Sutherland D.H. 1995. Dynamic intramuscular pressure measurement during gait. *Oper. Tech. Sports Med.* 3:250–255.

Komi P.V. 1990. Relevance of *in vivo* force measurements to human biomechanics. *J. Biomech.* 23 (Suppl 1):23–34.

Komi P.V., Fukashiro S., Jarvinen M. 1992. Biomechanical loading of Achilles tendon during normal-locomotion. *Clin. Sports Med.* 11:521–531.

Komi P.V., Belli A., Huttunen V., Bonnefoy R., Geyssant A., Lacour J.R. 1996. Optic fibre as a transducer of tendomuscular forces. *Eur. J. Appl. Phys.* 72:278–280.

Korvick D.L., Cummings J.F., Grood E.S., Holden J.P., Feder S.M., Butler D.L. 1996. The use of an implantable force transducer to measure patellar tendon forces in goats. *J. Biomech.* 29:557–561.

Lewis J.L., Lew W.D., Schmidt J. 1982. A note on the application and evaluation of the buckle transducer for knee ligament force measurement. *J. Biomech. Eng.* 104:125–128.

Luo Z.P., Hsu H.C., Grabowski J.J., Morrey B.F., An K.N. 1998. Mechanical environment associated with rotator cuff tears. *J. Shoulder Elbow Surg.* 7:616–520.

Maganaris C.N., Paul J.P. 1999. *In vivo* human tendon mechanical properties. *J. Phys.* 521:307–313.

Malaviya P., Butler D.L., Korvick D.L., Proch F.S. 1998. *In vivo* tendon forces correlate with activity level and remain bounded: evidence in a rabbit flexor tendon model. *J. Biomech.* 31:1043–1049.

Malaviya P., Butler D.L., Boivin G.P., Smith F.N., Barry F.P., Murphy J.M., Vogel K.G. 2000. An *in vivo* model for load-modulated remodeling in the rabbit flexor tendon. *J. Orthop. Res.* 18:116–125.

Muthupillai R., Lomas D.J., Rossman P.J., Greenleaf J.F., Manduca A., Ehman R.L. 1995. Magnetic resonance elastography by direct visualization of propagating acoustic strain waves. *Science* 269:1854–1857.

O'Driscoll S.W., Kumar A., Salter R.B. 1983. The effect of the volume of effusion, joint position and continuous passive motion on intraarticular pressure in the rabbit knee. *J. Rheumatol* 10:360–363.

O'Driscoll S.W., Morrey B.F., An K.N. 1990. Intraarticular pressure and capacity of the elbow. *Arthroscopy* 6:100–103.

Okuda Y., Gorski J.P., An K.N., Amadio P.C. 1987. Biochemical, histological, and biomechanical analyses of canine tendon. *J. Orthop. Res.* 5:60–68.

Pedowitz R.A., Gershuni D.H., Crenshaw A.G., Petras S.L., Danzig L.A., Hargens A.R. 1989. Intraarticular pressure during continuous passive motion of the human knee. *J. Orthop. Res.* 7:530–537.

Rupert M., Grood E., Byczkowski T., Levy M. 1998. Influence of sensor size on the accuracy of in-vivo ligament and tendon force measurements. *J. Biomech. Eng.* 120:764–769.

Schuind F., Garcia-Elias M., Cooney W.P. 3d, An K.N. 1992. Flexor tendon forces: *in vivo* measurements. *J. Hand Surg. (Am)* 17:291–298.

Sheehan F.T., Drace J.E. 2000. Human patellar tendon strain. A noninvasive, *in vivo* study. *Clin. Orthop.* (370):201–207.

Uchiyama E., Fujii T., Kitaoka H., Luo Z., Momose T., An K.N. 2000. Effects of ankle position and foot deformity on friction about the posterior tibial tendon. P191, 46th Annual Meeting, Orthopedic Research Society, Orlando, Florida, March 12–15, 2000.

Uchiyama S., Amadio P.C., Coert J.H., Berglund L.J., An K.N. 1997. Gliding resistance of extrasynovial and intrasynovial tendons through the A2 pulley. *J. Bone Joint Surg. (Am)* 79:219–224.

8

Requirements for Biological Replacement of the Articular Cartilage at the Knee Joint

Thomas Andriacchi

Introduction

Due to the functional demands on the knee, the articular cartilage at this joint is particularly vulnerable to traumatic damage as well as frequent degenerative damage with aging (Mauer, 1979). Damage to articular cartilage occurs in the form of focal lesions as well as generalized structural and biological changes resulting in thinning and loss of mechanical integrity. The loss of mechanical integrity of articular cartilage causes disability, pain, and loss of function. In general, the natural repair of damaged cartilage does not produce functional hyaline cartilage. Treatment for osteoarthritis or cartilage damage has been primarily symptomatic ranging from the use analgesics and antiinflammatory to surgical intervention with total knee replacement. New therapeutic interventions such as cartilage transplantation (Brittberg et al., 1994, 1997), osteochondral allo- or autografting (Bobic, 1996, Stevenson et al., 1989), osteotomies (Nizard, 1998; Peterson et al., 2000), and tissue-engineered cartilage replacement (Grande et al., 1999; Mow and Wang, 1999) are emerging for repair, regeneration or replacement of native articular cartilage (Buckwalter, 1999). The success or failure of each of these therapeutic interventions is dependent on clinical, biological, and mechanical factors.

The same functional demands that place the knee at risk for traumatic damage and frequent degenerative changes must be considered in the design and evaluation of any therapeutic intervention. In the context of functional tissue engineering, mechanical factors are of particular importance. While defining the mechanical and structural properties of native articular cartilage is an important consideration for establishing design requirements for tissue-engineered cartilage replacement, these mechanical properties must be considered in the context of true *in vivo* loads. Tissue-engineered devices can be distinguished from traditional artificial devices by the role of living cells. Cells are present in a tissue-engineered device to maintain structural integrity of the device. Ultimately, the behavior of the cell will be critical to the long-term viability of any tissue-engineered cartilage replacement. While the cells occupy only 10% of the tissue by volume, their metabolic activity is fundamental to the synthesis, assembly, and maintenance of the extracellular matrix (Smith et al., 1996). The functional loading on the joint is particularly important, since *in vivo* loading influences cell metabolic activity.

There are a number of important studies describing the material and structural characteristics of articular cartilage (Mow et al., 1992). The *in vitro* properties of cartilage have been reported over a range of loading modes including compression, shear, tension, and impact. The *in vivo* mechanical, electrical, and biological requirements must be considered together. At present, there is a paucity of information on the *in vivo* interactions between the viability of the

Macro ⟶ Micro

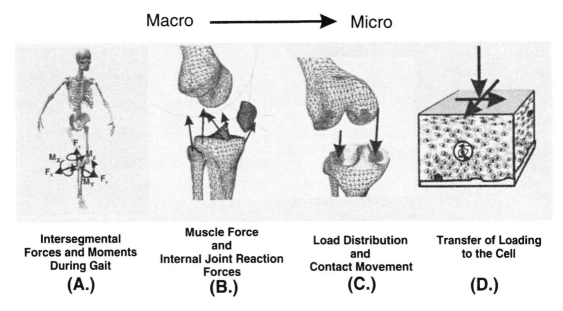

Intersegmental Forces and Moments During Gait (A.)   **Muscle Force and Internal Joint Reaction Forces (B.)**   **Load Distribution and Contact Movement (C.)**   **Transfer of Loading to the Cell (D.)**

FIGURE 8.1. An illustration of the steps involved obtained macro level intersegmental forces and moments during gait (*A*), to determine internal muscle force distribution during gait (*B*), and to determine the load distribution across the articular surface (*C*). Ultimately, the transfer of these true *in vivo* loads to tissue are critical to the maintenance of viable tissue engineered cartilage replacement (*D*).

cells and the physical environment (Smith et al., 2000). Clearly, the problem of defining design requirements is complicated by interdependence of these two areas. In particular, information is needed on the translation of *in vivo* loads at the macro level to deformation of the cell at the micro level (Fig. 8.1).

The purpose of this paper is to examine the relevance and need for considering dynamic *in vivo* loading in the establishing design criteria for tissue-engineered replacement of articular cartilage at the knee. The steps involved for translating macro-scale *in vivo* measurements to micro-level tissue loads will be discussed. Previous work on the treatment of knee osteoarthritis will be used to illustrate that dynamic loading during walking can have a substantial influence on the treatment outcome.

Current State of Art *in Vivo* Functional Loading

An analysis of the *in vivo* mechanical loading on the knee is an important step toward establish-

ing design criteria. It is useful to examine the methods available to obtain measurements of *in vivo* loading. Directly measuring joint loads is not feasible on a large scale in humans due to the invasive nature of the method. However, gait analysis can be used to calculate the external joint loading parameters that can be used to estimate the internal joint loads. The assumptions and accuracy of these methods have been discussed in detail elsewhere (Andriacchi and Hurwitz, 1997). Using inverse dynamics, intersegmental forces and moments (Fig. 8.1*a*) are approximated by modeling the body as a system of rigid segments and measuring the three-dimensional position of the limb segments and external ground reaction force (Andriacchi et al., 1997). The calculation of the muscle forces needed to balance the external joint moments is an indeterminate problem due to the redundancy of muscles (Fig. 8.1*a*). In general, two approaches have been used to solve the indeterminate problem. The first is a "reduction" method which groups muscles into functional units reducing the indeterminate problem to a determinate one (Morrison, 1970; Paul, 1971). The second approach uses optimization methods that assume the force distribution among the muscles is done in such a manner that some physical parameter modeled by an objective function is optimized (Crowninshield and Brand, 1981; Pandy and Anderson, 2000; Seireg and Arvitar, 1975).

Direct measurements of *in vivo* forces at the hip have provided useful information that can be used as a basis to evaluate the accuracy of analytical models for predicting joint forces. Hodge et al. (1989) collected contact measurements from a Moore type endo-prosthesis and measured maximum pressures of 5.5 MPa at the hip during gait. Recent studies (Davy et al., 1988) used telemetrized devices to measure these components of force at the hip. The most recent studies (Bergman et al., 1993, 1995) have the longest postoperative follow-up of any of the instrumented devices and show peak forces as high as 4.3 times body weight (BW). The consistency of the characteristics of the loads at the hip reported from the various studies is also quite important. The forces predicted with statically determinate (Paul, 1971) models compare favorably to more complex optimization models as well as measurements of forces obtained directly from *in vivo* devices. It is likely that the variations seen between the different studies are due to variations in walking speed, or differences between the gait of normal subjects versus patients following total hip replacement. In addition, the transducer studies are often reported in the early postoperative period when the gait of patients is not likely comparable to normal. The results demonstrate that the indirect analytical models provide a reasonable approximation of joint contact forces at the hip.

While the number of studies reporting direct force measurements at the knee are not as extensive as the hip joint, the analytical prediction of forces at the knee are also quite reproducible among various reports. The early work of Morrison (1970) described the forces at the tibiofemoral joint. The general characteristics of the forces at the knee joint show three peaks during stance phase. Morrison reported maximum forces of approximately 4 times BW, while Seireg and Arvikar (1975), using an optimization model found a similar pattern of loads, but with maximum forces of approximately 7 times BW. Taylor et al. (1998) using a telemetrized proximal femur replacement

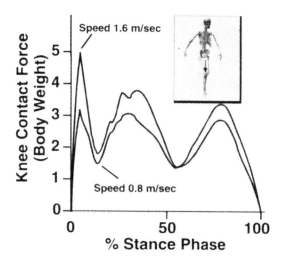

FIGURE 8.2. The resultant contact force at the knee joint during walking was during stance phase. The maximum peak was dependent on walking speed.

measured joint contact forces very similar to those predicted from the analytical model results illustrated in Figure 8.2. Schipplein and Andriacchi (1991) applied a statically determinate model, which accounted for the changing lever arm of the extensor mechanism with flexion and reported forces ranging from 3 to 5 times BW depending on walking speed (Fig. 8.2).

An analysis of the joint forces at the knee during walking provides useful information for the design and evaluation of tissue-engineered cartilage replacement. The magnitude and cyclic nature of the compressive force on the tibiofemoral joint are important consideration for the viability of articular cartilage (Fig. 8.2). There are three cycles of loading peaks occur during stance phase (approximately 0.6 seconds) of gait. The loads during swing phase are quite small relative to stance phase. Therefore, the loading cycle during walking can be characterized by three peaks cycles followed by a steady state low load for each step during gait. The magnitude and frequency content of the loading may be important for the maintenance of cell viability (Smith et al., 2000).

It should also be noted that the peak forces are strongly dependent on walking speed. The influence of walking speed on joint contact force should be considered for the design of tissue-engineered cartilage replacement. The contact

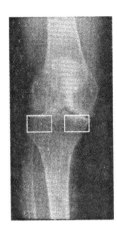
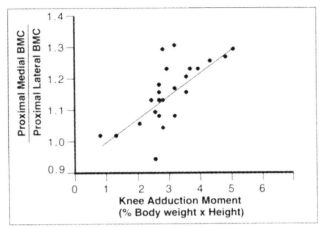

FIGURE 8.3. The ratio of the lateral to medial bone mineral content (BMC) was related to the peak magnitude of the abduction moment during gait. It has been shown that the abduction moment is directly related to the medial-lateral distribution of load across the surface of the tibia.

force at the knee can nearly double for a patient following successful treatment, since increased walking speed is common following relief of symptoms.

The relevance of the *in vivo* joint force measurement to tissue engineered cartilage replacement can be further examined by reviewing several studies where clinical outcome and physical tissue response can be related to forces at the knee during gait. There is indirect evidence that the external intersegmental joint loading measured during gait produces a physical tissue response. The validity and relevance of these methods has been evaluated by testing for relationships between tissue response and predicted loading during gait. For example, it has been shown that the knee adduction moment (Fig. 8.3) during gait influences the distribution of bone density in the proximal tibia (Hurwitz et al., 1998). In that study, the external knee adduction moment during gait was correlated with the distribution of medial and lateral bone mineral content. The best single predictor of the medial-to-lateral ratio of proximal BMC was the adduction moment. The adduction moment has been shown to be the primary intersegmental load component that influences the load distribution between the medial and lateral plateaus (Schipplein and Andriacchi, 1991). The higher the adduction moment, the greater the load on the me-

dial plateau relative to that of the lateral plateau and the higher the bone mineral content in the proximal tibia under the medial plateau (Fig. 8.2) as compared with that under the lateral plateau. This result is important since it demonstrates that there is a biological response to dynamic loads measured during walking.

Treatment outcome for medial compartment osteoarthritis was also related to the magnitude of the adduction moment during gait. This finding is particularly relevant to any attempt to identify criteria for tissue-engineered replacement of articular cartilage. The relationship between the external dynamics loads during walking and the internal load distribution at the knee is consistent with studies that demonstrated that individual variations in the preoperative adduction moment during gait influenced clinical outcome of high tibial osteotomy (HTO) for treatment of varus gonarthrosis (Prodromos et al., 1985; Wang et al., 1990). The objective of HTO is to realign the tibia and femur such that some of the load on the medial arthritic compartment is transferred to the more normal lateral compartment. Surgical candidates with a lower preoperative adduction moment during gait had a better long-term radiographic and clinical outcome then candidates with a higher preoperative adduction moment during gait. The adduction moment during walking has also been shown to influence the breakdown of native articular cartilage (Sharma et al.,1998).

These studies suggest that dynamic loading at the knee can influence the progression of degenerative joint disease. The extrapolation of these results suggest that even if the tissue-engineered replacement exactly replicates the native tissue, it is possible that the disease will continue to advance if the underling mechanical factors are not addressed. There are several important implications from these studies that can be applied to the eventual clinical use of tissue-engineered cartilage replacement. First, the dynamic loads during gait play an important role in the viability of native articular cartilage as well as the outcome of treatment for degenerative joint disease. Second, the loading distribution across the knee joint has a substantial gradient. Thus, design requirements for tissue-engineered cartilage replacement might be different for different regions of the articular surface of the knee. Clearly, medial and lateral articular cartilage at the knee experience different loading. Third, clinical and functional characteristics of the patient may have an important role in the outcome of knee cartilage replacement. Individual variations in function during gait as well as alignment that lead to the breakdown of the native articular cartilage may be an important consideration in establishing design requirements as well as the patient selection criteria for cartilage replacement.

Future Perspectives and Direction for Future Research

Is the magnitude and frequency content of loading at the knee joint sufficient to establish design criteria for tissue-engineered cartilage replacement? Again, if one looks at factors leading to breakdown of the native articular cartilage, instability associated with ligament damage has been related (Buckland-Wright et al., 2000) to an increased incidence of breakdown of the native articular cartilage. Thus, joint kinematics are likely another important consideration in the establishment of appropriate design criteria for tissue engineered cartilage replacement. Identifying the salient features of knee kinematics during

ing *in vivo* activities represents a unique challenge in that the kinematics and dynamics at the knee are quite complex. The knee must transmit large forces while allowing the mobility to permit normal function. Typically, the femoral rollback during knee flexion is a result of a combination of rolling, gliding, and rotation of the condyles over the tibial plateau (Kapandji, 1970), whereby rolling motion is predominant early in flexion (0–20°) as described by Andriacchi et al., (1986) and Nisell (1985).

Again, it is useful to take an historical perspective when evaluating the salient features of knee kinematics. The most common method for replacing articular cartilage at the knee is total knee replacement. The successful evolution of total knee replacement has been made possible in part due to a better understanding of the gait dynamics (Andriacchi and Hurwitz, 1997). Recently, the impact of joint kinematics on factors leading to wear of the articular surface in total knee replacement has been demonstrated (Wimmer and Andriacchi, 1997). Articular contact surface kinematics can produce substantial variations in wear patterns for engineering materials. In particular, surface sliding can produce substantial wear in contrast to rolling under similar load conditions. In addition, the initiation and breaking (acceleration/deceleration) of the rolling motion of the femur on the tibia produce tangential surface loads not present during free rolling. Johnson (1985) defined such conditions as tractive rolling. In a metal-polyethylene knee articulation, the tractive forces generated during tractive rolling (no sliding) can be substantial since the static coefficient of friction is approximately twice the dynamic coefficient of friction which would be applicable during pure sliding.

The coefficient of friction of any articular replacement should be considered among the design parameters. While the coefficient of friction is low for native articular cartilage, friction will increase as the cartilage surface breaks down. Thus, both the repair surface and counter surface should be considered, since tractive forces produce stress potentially damaging to the bearing surface of articular cartilage. It was demonstrated that the coefficient of friction, gait characteristic, muscle activity, or patellofemoral mechanics can influence the generation of tractive

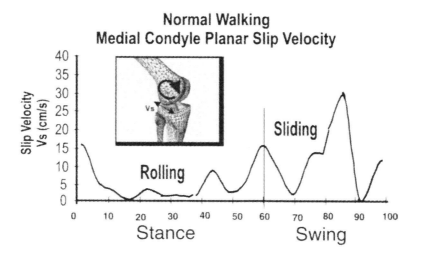

FIGURE 8.4. An illustration of this spatial distribution of the normal force and the anterior posterior surface traction force on the tibial surface during walking.

forces during gait (Wimmer and Andriacchi, 1997). That study also reported the movement of the compressive and tangential force along the articular surface (Fig. 8.4). During the stance phase of gait, the knee oscillates between the initiation and rolling (tractive rolling) and breaking between flexion of 0 and 20°. The initiation of pure rolling and the deceleration of rolling generate forces tangential to the surface. In the anterior region of the tibial plateau, tangential force reached a peak producing a posterior pull on the tibial surface. A second peak occurred in the posterior region as the femoral condyles rolled backward with a relative angular acceleration as high as 50 rad/s². In contrast to sliding, the direction of the tangential force during tractive rolling was not necessarily in the direction of relative motion between first and second body. A reversal of the tractive force occurred at the posterior end of the contact region. The knee rolled forward with knee extension following mid-stance flexion and tangential force did not change its posterior-anterior direction till the end of stance. These forces can rise to a maximum dependent upon the static coefficient of friction (here twice the dynamic coefficient of friction) and, thus, can be greater than those produced during pure sliding motions that are dependent upon the dynamic coefficient of friction.

The study described by Wimmer and Andriacchi (1997) indicates that under certain conditions there can be a substantial tractive force tangential to the surface of the tibia. The transfer of shear stress to the articular surface and ultimately to the cell can influence metabolic activity (Smith et al., 1996) and the structural viability of the extracellular matrix. In addition, this study indicates that there was substantial movement of the contact force along the articular surface. The tractive forces and contact movement along the joint surface are also an important consideration when using tissue flaps for articular cartilage repair. Periosteal flaps are often used to cover a cell suspension placed into a void. Driesang and Hunziker (2000) recently reported a high rate of delamination of periosteal flaps.

The limitation of the work described by Wimmer and Andriacchi (1997) was that the contact anterior-posterior contact movement was derived from *in vitro* studies. The true *in vivo* contact motion will be an important consideration for establishing improved design criteria. It has been difficult to quantify tibiofemoral contact motion under normal *in vivo* conditions. Skin movement relative to the underlying bone is a primary factor limiting the resolution of detailed joint movement using skin-based system (Reinschmidt et al., 1997). In most cases, only large motions such as flexion extension have acceptable error limits with skin-based marker systems. Skeletal movement can be measured using alternative approaches to a skin-based marker

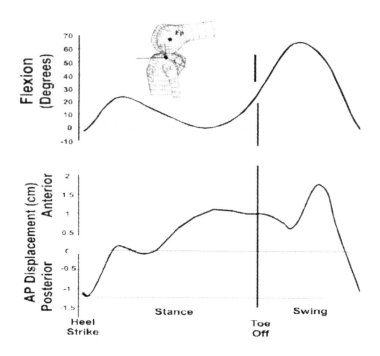

FIGURE 8.5. The anterior posterior motion of the femoral reference point (Fp) relative to the angle of knee flexion during the gait cycle.

system. These approaches include stereoradiography, bone pins (Banks and Hodge, 1996; LaFortune et al., 1992) external fixation devices (Cappozzo et al., 1997), or single-plane fluoroscopic techniques (Banks and Hodge, 1996). All of these methods are invasive or expose the test subject to radiation. Therefore, the widespread applicability of these methods is limited. A newly developed method described as a point cluster technique (PCT) permits direct *in vivo* measurement of the anterior-posterior motion as well as the complete six-degree of freedom motion of the femur on the tibia while performing activities of daily living (Andriacchi et al., 1998, Alexander and Andriacchi, 2001). In that study, the anterior-posterior movement of the femur relative to the tibia was measured during normal gait. The maximum range of the anterior-posterior (AP) displacement of the femur on the tibia during the gait cycle was quantified by the displacement of point Fp (midpoint of transepicondylar axis) as illustrated in Figure 8.5. The pattern of AP movement of the femoral point (Fp) relative to the tibia had several characteristics common to all subjects. At heel strike, the

femur is located posteriorly on the tibia. At midstance the femur moves forward as the knee flexes and continues to move forward as the knee extends to terminal extension. The maximum anterior position typically occurs during swing.

In a later study, Johnson (1999) applied the point cluster technique to an analysis of articular surface motion for patients following total knee replacement. The relative slip velocity was used to characterize the motion of femoral condyles relative to the tibia on the lateral and medial plateaus. The interfacial slip velocity (v_s) was proportional to both the linear displacement rate of change of the femoral component relative to the tibial component and the angular velocity of the femoral component relative to the tibia component. The magnitude of the interfacial slip velocity provides quantification of the rolling versus sliding behavior of the tibiofemoral joint when relative motion occurs. For pure rolling, the interfacial slip velocity will be minimized ($v_s = 0$). These results demonstrated substantial variations in the slip velocity. Thus, regions of potential sliding and rolling will vary over the articular surface of the tibia.

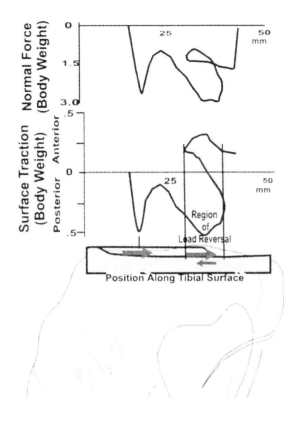

FIGURE 8.6. The slip velocity between the medial femoral condyles and tibial articular surface for a patient with a total knee replacement demonstrated a method for determining regions of relative rolling reversed sliding.

The magnitude and characteristics of the tibiofemoral sliding motion during walking is likely an important consideration in the design criteria for cartilage replacement. Joint instability associated with anterior cruciate ligament injury has been related to degenerative changes of the articular cartilage at the knee joint. Thus, methods for determining *in vivo* dynamic stability will be needed for future development of appropriate design criteria.

Conclusions

The ultimate goals of treatment for an arthritic joint are the elimination of pain and restoration of function. Historically, total joint replacement has been the primary method for the treatment of severely degenerated joints. These devices re-

place damaged cartilage using metal and plastic components as bearing surfaces for the joint. The ultimate design and success of a tissue-engineered cartilage replacement has a number of parallels with the design of a total joint replacement. The goal of restoration of function to patients with arthritic joints requires a quantitative measure of normal function over a range of activities. Experience with conventional joint replacement has demonstrated that the magnitude and nature of the loading and motion at the knee joint cannot be ignored in the design process. Mechanical implant failure and implant wear can be related to a lack of appropriate information on the complex kinematics of the joint.

Tissue-engineered cartilage replacement must sustain the same types of functional loads seen during normal function if it is to meet its ultimate goal of restoration of function. The problem of dealing with cartilage replacement is more complex than total joint replacement, since the unknowns associated with the mechanical properties of the tissue-engineered construct are not well defined. One of the potential advantages of a biological replacement is its ability to adapt to its mechanical environment. However, this condition suggests a need to precondition the extracellular construct derived from the chondrocytes to the *in vivo* physical environment. This type of problem is not present when one considers a total joint replacement made of metal or plastic, since the material properties of these components are known. In addition, these devices are typically used to replace an entire joint surface. Thus, regional variations in load or stress on the joint are not an important consideration for total joint replacement. However, regional variations in loading on articular cartilage will be an important factor for tissue-engineered cartilage replacement.

The knee joint sustains high loads over a relatively large range of motion. These high loads translate to relatively high local stresses on the surface of the joint. Even during a simple activity like walking on level ground, the loading as well as the kinematics of the knee joint are complex. The following conclusions can be drawn from an analysis of dynamic *in vivo* loading:

1. The magnitude and cyclic nature of the compressive force on the tibiofemoral joint are important consideration for the viability of ar-

ticular cartilage. There are three cycles of loading peaks that occur during stance phase (approximately 0.6 seconds) of gait. The magnitude and cyclic nature of the compressive force on the tibiofemoral joint are important consideration for the viability of articular cartilage

2. The contact force at that knee can nearly double for a patient following successful treatment, since increased walking speed is common following relief of symptoms. Individual variations in function during gait as well as alignment that lead to the breakdown of the native articular cartilage may be an important consideration in establishing design requirements as well as the patient selection criteria for cartilage replacement. The clinical and functional characteristics of the patient may have an important role in the outcome of knee cartilage replacement.

3. The medial and lateral articular cartilage at the knee experience different loading. The loading distribution across the knee joint has a substantial gradient. Thus, design requirements for tissue engineered cartilage replacement might be different for different regions of the articular surface of the knee (Andriacchi et al., 2000).

4. Substantial tractive forces can be generated tangential to the surface of the tibia depending on the condition of the articular surface. The transfer of shear stress to the articular surface and ultimately to the cell can influence metabolic activity and thus the structural integrity of the extracellular matrix. More information is needed on the specific characteristic of *in vivo* tibiofemoral contact mechanics during walking and other activities of daily living. In addition, methods to measure *in vivo* variation in cartilage properties will be needed in the treatment planning for tissue engineered cartilage replacement.

5. Regions of potential sliding and rolling will vary over the articular surface of the tibia. The magnitude and characteristics of the tibiofemoral sliding motion during walking is likely an important consideration in the design criteria for cartilage replacement. Methods for determining *in vivo* dynamic stability will be needed for future development of appropriate design criteria.

Clearly, for tissue-engineered cartilage replacement to successfully sustain normal function, we must gain a better understanding of the salient characteristics of the functional loads on the joints under true *in vivo* conditions.

References

Alexander, E.J., Andriacchi, T.P. 2001. Correcting for deformation in skin based marker systems. *J. Biomech.* 34:355–361.

Andriacchi T.P., Hurwitz D.E. 1997. Gait biomechanics and the evolution of total joint replacement. *Gait Posture* 5:256–264.

Andriacchi T.P., Stanwick T.S., Galante J.O. 1986. Knee biomechanics and total knee replacement. *J. Arthroplasty* 1:211–219.

Andriacchi T.P., Natarajan R.N., Hurwitz D.E. 1997. Musculoskeletal dynamics: locomotion and clinical applications. In: *Basic Orthopaedic Biomechanics*, 2nd Ed. Mow V.C., Hayes W.C., eds.: New York, Lippincott-Raven, pp. 37–68.

Andriacchi T.P., Alexander E.J., Toney M.K., Dyrby C.O., Sum J. 1998. A point cluster method for *in vivo* motion analysis: applied to a study of knee kinematics. *J. Biomech. Eng.* 120:743–749.

Andriacchi T.P., Alexander E.J., Lang P.L., Hurwitz D.E. 2000. Methods for evaluating the progression of osteoarthritis. *J. Rehabil. Res. Dev.* 37:163–170.

Banks, S.A., Hodge, W.A. 1996. Accurate measurement of three-dimensional knee replacement kinematics using single-plane fluoroscopy. *IEEE Trans. Biomed. Eng.* 43:638–649.

Bergmann G., Graichen F., Rohlmann A. 1993. Hip joint loading during walking and running measured in two patients. *J. Biomech.* 26:969–990.

Bergmann G., Graichen F., Rohlmann A. 1995. Is staircase walking a risk for the fixation of hip implants? *J. Biomech.* 28:535–553.

Bobic V. 1996. Arthoscopic osteochondral autograft transplantation in anterior cruciate ligament reconstruction: a preliminary clinical study. *Knee Surg. Sports Traumatol. Arthrosc.* 3:262–264.

Brittberg M., Lindahl A., Nilsson A., Ohlsson C., Isaksson O., Peterson L. 1994. Treatment of deep cartilage defects in the knee with autologous chondrocyte transplantation. *N. Engl. J. Med.* 331:889–895.

Brittberg M., Lindahl A., Homminga G., Nilsson A., Isaksson O., Peterson L. 1997. A critical analysis of cartilage repair. *Acta Orthop. Scand.* 68:186–191.

Buckland-Wright J.C., Lynch J.A., Dave B. 2000. Early radiographic features in patients with anterior cruciate ligament rupture. *Ann. Rheum. Dis.* 59:641–646.

Buckwalter, J.A. 1999. Evaluating methods for restoring cartilaginous articular surfaces. *Clin. Orthop.* 367S:224–238.

Cappozzo A., Cappello A., Della Croce U., Pensalfini F. 1997. Surface marker cluster design criteria for 3-d bone movement reconstruction. *IEEE Trans. Biomed. Eng.* 44:1165–1174.

Crowninshield R.D., Brand R.A. 1981. A physiologically based criterion of muscle force prediction in locomotion. *J. Biomech.* 14:793–801.

Davy D.T., Kotzar G.M., Brown R.H., Heiple K.G., Goldberg V.M., Heiple K.G., Berilla J., Burstein A.H. 1988. Telemetric force measurements across the hip after total hip arthroplasty. *J. Bone Joint Surg.* 70A:45–50.

Driesang, I.M.K., Hunziker, E.B. 2000. Delamination rates of tissue flaps used in articular cartilage repair. *J. Orthop. Res.* 18:909–911.

Grande D.A., Brietbart A.S., Mason J. Paulino C., Lasar J., Schwartz R.E., 1999. Cartilage tissue engineering: current limitations and solutions. *Clin. Orthop.* 367S:176–175.

Hodge W.A., Carlson K.L., Fijan S.M., Burgess R.G., Riley P.O., Harris W.H., Mann R.W. 1989. Contact pressures from an instrumented hip endoprosthesis. *J. Bone Joint Surg.* 71A:1378–1385.

Hurwitz D.E., Sumner, D.R., Andriacchi, T.P., Sugar, D.A. 1998. Dynamic knee loads during gait predict proximal tibial bone distribution. *J Biomech.* 31:423–430.

Johnson K.L. 1985. *Contact Mechanics.* Cambridge University Press, Cambridge, U.K.

Johnson T.S. 1999. *In-vivo* contact kinematics of the knee joint: advancing the point-cluster technique. Ph.D. thesis, Department of Mechanical Engineering, University of Minnesota.

Kapandji I.A. 1970. The knee. In: *The Physiology of the Joints,* 2nd ed., Vol. 2 Churchill Livingstone, Edinburgh, UK, pp. 72–106.

LaFortune M.A., Cavanagh P.R., Sommer H.J., Kalenak A. 1992. Three dimensional kinematics of the human knee during walking. *J. Biomech.* 25:347–357.

Maurer K. National Center for Health Statistics. 1979. Basic data on arthritis: knee, hip, and sacroiliac joint, in adults aged 25–74 years: United States, 1971–1975. Vital and Health Statistics Series 11-Number 213. DHEW Publ. No. (PHS) 79-1661. Public Health Service. Washington, U.S. Government Printing Office.

Morrison J.B. 1970. The mechanics of the knee joint in relation to normal walking. *J. Biomech.* 3:51–61.

Mow V.C., Wang C.C. 1999 Some bioengineering considerations for tissue engineering of articular cartilage. *Clin. Orthop.* 367S:204–223.

Mow V.C., Ratcliffe A., Poole A.R. 1992. Cartilage and diarthrodial joints as paradigms for hierarchical materials and structures. 13:67–97.

Nisell R. 1985. Mechanics of the knee. *Acta Orthop. Scand.* 56(Suppl 216):4–42.

Nizard R.S. 1998. Role of tibial osteotomy in the treatment of medial femorotibial osteoarthritis. *Rev. Rheum. Engl. Ed.* 65:443–446.

Pandy M.G., Anderson F.C. 2000. Dynamic simulation of human movement using large-scale models of the body. *Phonetica* 57:219–228.

Paul J.P. 1971. Comparison of EMG signals from leg muscles with the corresponding force actions calculated from walk path measurements. *Human Locomotor Eng.* 6–26.

Peterson L., Minas T., Brittberg M., Nilsson A., Sjogren-Jansson, E. Lindahl. 2000. A two-to-9 year outcome after autologous chondrocyte transplantation of the knee. *Clin. Orthop.* (374):212–234.

Prodromos C.C., Andriacchi T.P., Galante J.O. 1985. A relationship between knee joint loads and clinical changes following high tibial osteotomy. *J. Bone Joint Surg.* 67A:1188–1194.

Reinschmidt C., van den Bogert A.J., Nigg B.M., Lundberg A., Murphy N. 1997. Effect of skin movement on the analysis of skeletal knee joint motion during running. *J. Biomech.* 30:729–732.

Schipplein O.D., Andriacchi T.P. 1991. Interaction between active and passive knee stabilizers during level walking. *J. Orthop. Res.* 9:113–119.

Seireg A., Arvikar R.J. 1975. The prediction of muscular load sharing and joint forces in the lower extremities during walking. *J. Biomech.* 8:89–102.

Sharma L., Hurwitz D.E., Thonar E.J.M.A., Sum J.A., Lenz M.E., Dunlop D.D., Schnitzer T.J., Kirwan-Mellis G., Andriacchi T.P. 1998. Knee adduction moment, serum hyaluronic acid level, and disease severity in medial tibiofemoral osteoarthritis. *Arthritis Rheum.* 41:1233–1240.

Smith R.L., Donlon B.S., Gupta M.K., Mohtai M., Das P., Carter D.R., Cooke J., Gibbons G., Hutchinson N., Schurman D.J. 1995. Effects of fluid-induced shear on articular chondrocyte morphology and metabolism *in vitro. J. Orthop. Res.* 13:824–831.

Smith R.L., Rusk S.F., Ellison B.E., Wessells P., Tsuchiya K., Carter D.R., Caler W.E., Sandell L.J., Schurman D.J. 1996. *In vitro* stimulation of articular chondrocyte mRNA and extracellular matrix synthesis by hydrostatic pressure. *J. Orthop. Res.* 14:53–60.

Smith R.L., Lin J., Trindade M.C., Shida J., Kajiyama G., Vu T., Hoffman A.R., van der Meulen M.C., Goodman S.B., Schurman D.J., and Carter D.R. 2000. Time-dependent effects of intermittent hydrostatic pressure on articular chondrocyte type II collagen and aggrecan mRNA expression. *J. Rehabil. Res. Dev.* 37:153–161.

Stevenson S., Dannucci G.A., Sharkey N.A., Pool R.R. 1989. The fate of articular cartilage after transplantation of fresh and cryopreserved tissue-antigen-matched and mismatched osteochondral allografts in dogs. *J. Bone Joint Surg.* 71:1297–1307.

Taylor S.J., Walker P.S., Perry J.S., Cannon S.R., Woledge R. 1998. The forces in the distal femur and the knee during walking and other activities measured by telemetry. *J. Arthroplasty* 13:428–437.

Wang J.-W., Kuo K.N., Andriacchi T.P., Galante J.O. 1990. The influence of walking mechanics and time on the results of proximal tibial osteotomy. *J. Bone Joint Surg.* 72A:905–913.

Wimmer M.A., Andriacchi T.P. 1997. Tractive forces during rolling motion in total knee replacement. *J. Biomech.* 30:131–138.

9

Functional Requirements: Cartilage

Jack L. Lewis

Introduction

Tissue engineering has become a dynamic scientific and commercial enterprise, with extensive research and development ongoing on replacement tissues for a variety of applications. In the early phase of this work, emphasis has been on "proof of concept," in which the goal has been to demonstrate that a tissue approximating the native tissue could be grown. The field has now entered a second phase, in which the specific design requirements of these tissues need to be more clearly defined, methods and parameters for evaluating candidate tissues identified, and specific target values of these parameters defined. However, this leads to the difficult issue of which properties to measure and what these property values should be. There is no consensus on this issue, and little focused discussion. The goal of this chapter is to provide a framework for discussion for identifying these properties, using the functional or mechanical material properties of engineered articular cartilage as the vehicle for discussion.

In addition to mechanical property requirements, a tissue for cartilage replacement must also satisfy a variety of other clinical and social criteria. It must obviously result in improved patient function without pain. It must have a reasonable rehabilitation protocol, be able to be performed surgically within a reasonable time, and

be cost competitive. The standard by which these features will be measured will be the alternative treatment methods for the treated disease or injury, which will vary depending on the specific medical condition. These will be very important issues, probably more important than the specific mechanical material properties. The result of these multiple criteria is that the functional properties of an engineered tissue will likely not need to be as good as the natural tissue, since tissue properties will be only one of several important criteria for choosing the better treatment. A successful engineered cartilage replacement will need to compare favorably with competing treatment methods. Since this will be a complex decision process, it is difficult to know what role the tissue properties will play in the overall comparison and, therefore, what criteria to apply to tissue properties. The strategy taken here is that the desired functional properties of the tissue should be established without regard to the other considerations, so that the mechanical material properties of a candidate engineered tissue can be compared with these desired properties, and this comparison used as one factor in the evaluation process. How close the properties of the engineered tissue need to be to the natural tissue will need to be established in the context of the total decision process. With this qualification, this chapter will focus on functional properties of engineered cartilage tissue, where functional properties are equivalent to the mechanical material properties. The other factors will not be discussed further, but discussion is presented with the understanding that the material properties would not be considered alone in evaluating a candidate engineered cartilage tissue.

Returning to the issue of material properties and their measurement, there will be different uses for such measurements. After *in vitro* growth to a point to be ready to implant, there are the following questions: 1) "What mechanical properties are suitable as initial implant conditions?" This is more related to ability to survive early joint forces, as modified by pain and rehabilitation, and as suitable initial states to encourage continued growth and remodeling *in vivo*. There is the different state when the implant has been in place for a period of time, either in an animal model or in humans, and then the question might be 2) "Which mechanical properties should be measured, and what values should the properties be, in order to predict future properties, that is, predictive of continued successful remodeling and function?" And then finally, at some point after implantation, the question might be 3) "Is the implant functioning properly and which properties and what values would be indicative of successful function?" Questions 1 and 2 relate to predicting future function based on present properties; question 3 relates to evaluating present function based on present properties. There is very little research available that would allow predicting future properties based on present properties, so answers to questions 1 and 2 would have to be extremely speculative. As discussed previously, there are multiple factors involved in evaluating the successful functioning of an implant; mechanical properties of the implant are only one. Answers to question 3 will involve all of these factors, plus the mechanical material properties, so the question cannot be answered on the basis of material properties alone. The strategy of the author, and the point of view of this chapter, is that a design set of mechanical material properties of the implant should be proposed based on available information, to serve as a framework for discussion. Such a framework would hopefully lead to future studies in which predictive properties and necessary current properties

could be identified more objectively. However, the consequence of this strategy is that at this time the proposed variables and parameters will necessarily be only the opinion of the author.

Critical Questions

In order to define design material properties, the following questions need to be addressed:

1. What functional environment must an engineered tissue withstand in order for it to replace the native tissue?
2. What mechanical qualities should the engineered tissue have in order for it to function properly in the required environment?
3. Which mechanical material parameters (constitutive equation parameters, other material parameters) should be measured as indicators of successful function in the required environment?
4. What specific values of the mechanical parameters would be indicative of a successful implant?

An obvious answer to the last two questions is that the properties of the engineered tissue should be identical to the natural tissue. Although a laudable goal, it is likely unattainable in the foreseeable future. Therefore, the less than ideal candidate tissues need to be evaluated relative to alternative tissues and treatments and the questions remain. Also, we still need to decide which of the many properties of natural cartilage that could be measured would be the most appropriate for evaluating a cartilage replacement. Each of the four questions will be addressed in the following sections.

What Functional Environment Must an Engineered Tissue Withstand in Order for it to Replace the Native Tissue?

Cartilage is the soft thin layer at the ends of long bones that contacts during joint articulation. It serves to provide a low-friction surface for articulation of bony surfaces and a soft layer to attenuate static and dynamic contact stresses. An engineered cartilage tissue must perform these functions, without failure, within a specific mechanical environment. The mechanical environment of articulating joints is still poorly known, but there is information available to provide guidelines.

The only joint for which there are good measures of *in vivo* forces is the hip joint. Bergmann et al. (1993) measured the forces on the head of a femoral component of a total hip replacement in two patients. Peak forces (median) for various activities are shown in Table 9.1.

TABLE 9.1. Median peak resultant forces, times body weight (Bergmann et al., 1993)

Activity	Force
Walking (3 km/hr)	3.2
Jogging (7 km/hr)	5.0
Stumbling	5.4

Stress on the head of the femur *in vivo* was measured directly by Krebs et al. (1998), who used an instrumented hemiarthroplasty with surface-mounted pressure transducers, implanted into an 85-year-old patient. Typical peak stresses on the surface of the femoral head measured during walking were approximately 5 MPa.

Joint force magnitude would be expected to vary between joints. However, since cartilage properties are similar between joints, stress levels may be similar between joints, even though force magnitudes vary. The data in Krebs et al. (1998) provides direct pressure or stress data; the data in Table 9.1 provides a measure of relative stress as function of specific activities.

Joint loads can obviously be higher than those referenced, since cartilage is permanently damaged in joint trauma, that is, there is an upper load carrying capability of cartilage. The relationship between resultant joint force and cartilage failure depends on details of contact stresses and the stress criteria for failure for cartilage, neither of which is known with much precision. A reasonable estimate of failure stress for cartilage in the joint is provided by *in vitro* tests of cartilage failure during impact loading. Impact loading is an appropriate loading condition for single load tolerance, because single load failure is most likely to occur during high-speed trauma, not during low-speed high-level loading. Several investigators have performed tests of impact loading of cartilage explants, with failure stress as the reported output (Borrelli et al., 1997; Jeffrey et al., 1995; Newberry et al., 1997; Repo and Finlay, 1977; Thompson et al., 1991; Torzilli et al., 1999). This stress is really an average stress over the loaded surface, defined as the total force divided by the projected contact area. Since the tests are either unconstrained compression or punch tests, the actual failure stress is unknown and not the reported stress. Average failure stresses range from 15 to 30 MPa, at which point cracks are formed in the cartilage explant surface (Repo and Finlay, 1977; Torzilli et al., 1999). Because live cells are essential for continued cartilage function, tissue loading must not result in cell damage or death. Cell death appears to occur during impact loading at about the same level as cartilage fracture, so the 15–30 MPa level is also a reasonable estimate for critical stress level for cell death (Repo and Finlay, 1977; Torzilli et al., 1999). A reasonable upper level for cartilage loading without failure is 15 MPa.

With this data as a basis, load environment of cartilage can be described as periods of repetitive surface stress of the order of 5 MPa, from the gait studies, with occasionally higher single loads up to 15 MPa. These estimates are obviously only approximate and would vary with activity.

A second important feature of the load environment is the time history of loading. The dominant time feature of cartilage loading is the gait cycle (Figure 3 in Krebs et al., 1998; pressure on femoral head). However, details of the complete daily load history for cartilage are unknown. There are surely long periods of minimal load while sitting or sleeping. There are perhaps long periods of constant loading as the average value of cyclic loading during gait. The number of cycles per load session will surely vary widely with individual. A rough estimate of cycle number would be approximately 1700 cycles per mile walked, two miles walked or approximately 3000 cycles per day, and 10^6 cycles per year. The repetitive surface pressure during the gait cycle given by Krebs et al. (1998) can be analyzed by fast Fourier transform to give an approximate frequency content. There is a prominent average value, reflecting the compression-only nature of cartilage loading, and frequencies from 1 to 5 Hz, with smaller higher-frequency contributions. Isolated loads, such as during rising from a chair or stumbling, likely have times of the order or shorter than the gait cycle, with somewhat higher peak forces (Bergmann et al., 1993). Since these would be smaller versions of an impact load, they probably do not need to be dealt with specifically. Impact loads also occur. A typical impact load during an *in vitro* experiment shows rise time of the order of 1 msec and durations of 10–50 msec (Thompson et al., 1991). These load histories can serve as an estimate of functional or design load histories for cartilage.

Another important mechanical feature of the load environment is the relative velocity of the articulating surfaces, since this will influence the lubrication and friction. Relative joint velocities are in the range of 0.06 m/sec (hip in normal walking) to 0.6 m/sec (shoulder in baseball pitching) (Mow and Ateshian, 1997). *In vitro* studies of cartilage-on-cartilage have produced coefficients of friction from 0.003 to 0.06 (Mow and Ateshian, 1997). These very low coefficients of friction result in minimal shear forces in joint motion, implying that the surface stress is essentially normal to the surface. For a normal stress of 5 MPa, shear stresses would be from 0.015 to 0.3 MPa. The joint environment includes minimal friction while sustaining the normal joint loads.

To summarize the functional environment of cartilage, joint forces are essentially normal to the joint surface because of the low friction, with peak stresses during the gait cycle of 5 MPa. Frequency content of the cyclic load are predominantly greater than 1 Hz, with a constant component. Function requires on the order of 3000 cycles per day, and 10^6 cycles per year. Single-impact loads can create surface stresses up to 15 MPa without failure. Relative surface velocities are of the order of 0.1 m/sec, during which the joint loads occur. These loads and motions occur throughout the life of the individual; cartilage must withstand this environment without failure due to permanent deformation, fracture, or wear. This set of mechanical

conditions can be used as design loads for a replacement tissue.

What Mechanical Qualities Should the Engineered Tissue Have in Order for it to Function Properly in the Required Environment?

Given the environment in which an engineered cartilage must function, it next must be established which mechanical qualities would be necessary for it to function in this environment. Since cartilage is attached to bone and stresses within the bone should be maintained at the natural levels, the geometry and stiffness of the implant should be approximately the same as natural cartilage to ensure normal bone stresses due to the contacting surfaces. The tissue should recover its undeformed geometry after load removal, therefore it should have elastic and time-dependent components such that after loading there is no permanent deformation after application of the design loads. The tissue should also have time dependency similar to natural cartilage in order for energy dissipation and damping to be similar. From the estimated design environment, load frequencies are in the range of 1–5 Hz, with a significant steady state value due to average stress being above zero, and impact loads have rise times of 0.5–2 msec, with durations of 5–50 msec. The tissue should allow these load histories without permanent deformation. Natural cartilage appears to temporarily deform after long-term loading, but to recover to the original geometry after a similar period of no load. This may be acceptable in an engineered cartilage tissue if the temporary deformation is minimal and the time dependency appropriate.

The engineered tissue must also resist fracture and wear in the design environment, including single and cyclic loads. This requires specific values of fracture toughness and wear rate under design conditions.

Other less obvious material requirements are the ability to control geometry of the implanted tissue and adherence to adjacent natural tissue. The tissue must not grow beyond the intended bounds, yet it should completely fill the design region. The engineered tissue will usually be applied within remaining natural cartilage or on bone; the new tissue must adhere to the natural cartilage and underlying hard tissue without failing at the interface.

To summarize, the engineered tissue must have the following qualities:

• Geometry control
• Time dependent with elastic component
• Resists impact loading
• Resists permanent deformation and fracture under cyclic loading
• Resists wear under cyclic articulation
• Adheres to adjacent tissue

Which Mechanical Material Parameters (Constitutive Equation Parameters, Other Material Parameters) Should Be Measured as Indicators of Successful Function in the Required Environment?

With the generic mechanical qualities of an engineered cartilage tissue defined, the specific mechanical material properties that would be indicative of a successful implant must be identified. Since success will be a somewhat relative term, and there are many material properties that could be measured, there is no single answer as to which properties would be predictive of success. It seems appropriate to identify several that seem more necessary, and also identify some that are less important, realizing that this is necessarily a subjective division.

Intimately connected with the concept of mechanical material property is the associated continuum theory. The material parameter of several properties will be curve fits to parameters in constitutive equations in specific continuum theories, solved for the boundary conditions simulating particular test conditions. Some parameters, such as coefficient of friction, will not

require an underlying continuum theory. However, generally, a continuum theory will have to be chosen on which to base interpretation of an experimental test.

Also intimately involved with a specific mechanical material parameter is the method to be used to measure it. Methods for measuring identified parameters must also be identified. Methods will be different for *in vitro* and *in vivo*. Each of the desired material qualities will be discussed separately.

Geometry Control

The main geometric variable for cartilage is thickness, the distance from the bone or calcified cartilage to the articulating surface. This will vary with joint and location within a joint surface. Typical values for human joints are 1–3 mm. The engineered tissue should be approximately the same thickness as the natural cartilage at that site and the same height as surrounding remaining natural cartilage. A variety of methods exist to measure cartilage height (Ateshian and Soslowsky, 1997; Cohen et al., 1999), so this will not be discussed further.

Time Dependent with Elastic Component

There are two time-dependent continuum models of cartilage that are in common use: a single-phase viscoelastic model and a two-phase solid-fluid model (Mow et al., 1980). Both would be candidates for an engineered cartilage. Identifying a particular material parameter will depend on which model is chosen to describe the tissue. The two-phase model implies that fluid flow and its resistance within the solid is the source of the time dependency of the material response. The viscoelastic model implies that the time dependency is a more local process and the time dependency is assigned as a point constitutive relationship without identifying the microstructural source of the time dependency. There is no doubt that loaded cartilage can lose water and that water flow occurs in long-duration loading. However, typical time constants for which the biphasic models have been matched with experimental data have been in the hundreds of seconds, much longer than the dominant frequencies in the design loads discussed previously (Woo and Mow, 1987). Loading these models at gait-type load cycles results in a nearly elastic response (Eberhardt et al., 1990). However, there is energy dissipation at gait-type load cycles, so it is implied that either water flow is not significant at those frequencies or that there is another water-flow dependent mechanism at work at the higher frequencies. Also, water flow may be a more prominent feature of the less-organized engineered tissues, even at gait frequencies. The issue is complicated by the lack of experimental techniques to monitor water movement at gait type cycle frequencies. One approach has been to use a combined viscoelastic solid-nonviscous fluid biphasic model (Mak, 1986; Suh and Bai, 1998). The difficulty with these models is that they are inherently complex and difficult to use for involved boundary value problems such as are needed for the test methods for the engineered tissue, particularly for in situ methods, such as indentation tests. At this time, for the sake of simplicity and practicality, it would seem that if fluid flow cannot be documented for the higher frequencies it would be more appropriate to assume a single-phase viscoelastic model. If fluid flow can be documented, the combined viscoelastic-biphasic model would be more appropriate.

A nonlinear viscoelastic material model may be necessary, because tests have shown the rapid response of cartilage to be nonlinear in stress versus strain (Woo and Mow, 1987). The simplest viscoelastic model to incorporate this type of nonlinearity is the Fung quasilinear viscoelasticity model (Woo and Mow, 1987). The model consists of the instantaneous response and a relaxation function. Methods for measuring these in one-dimensional tension and compression are well developed (Woo and Mow, 1987), and will not be discussed further here. The isotropic biphasic model of Mow and co-workers (Mow and Ratcliffe, 1997), can serve as the continuum model for a biphasic model. A well-developed test and theory system for small deformations are available for the linear theory, although the nonlinearities incorporated in the quasilinear viscoelastic model are not incorporated in the Mow model. The solid elastic constants, E and v, and a permeability constant are

provided by the theory and testing. Mak has extended this model to include quasilinear viscoelasticity into the solid part of the model (Mak, 1986).

The elastic component of the viscoelastic and biphasic models provides a measure of the elastic element needed for geometry recovery. The elastic constants provide a measurable parameter for material comparison.

In addition to the question of model phases, there is also the question of anisotropy and differing properties in tension and compression. This is still an active area of research and there is no consensus on the importance of these effects. This is possibly due to the difficulty of measuring the tensile and compressive properties on the same tissue. If these features are important, the model complexity will increase considerably, probably requiring some simplifying assumptions for realistic testing.

A further consideration of an assumed material model is that of large deformations. Cartilage and its replacement tissue are inherently highly compliant. During normal function they undoubtedly deform into a nonlinear strain region. The question is how significant is this feature and whether a large deformation formulation is necessary. Again, this is an area of current research with no consensus on need and method.

Resists Impact Loading

There is no specific material property that characterizes response to impact. The feature of the material response of interest is material fracture. Since none of the impact tests have been analyzed in detail to provide a detailed stress analysis for the impact, it seems appropriate to compare materials on the basis of response in a specific test. Of the several test methods reported, several variables have been considered. Cartilage explants can be removed from the bone and made flat by also removing surface slices, and impacted with a platen of larger diameter than the explant in unconfined compression (Jeffrey et al., 1995). Alternatively, a flat-ended punch can be loaded into the prepared cartilage specimen (Borrelli et al., 1997). The other approach is to leave the cartilage on the bone and load the cartilage with a punch. Since the test of interest is intended for evaluating engineered tissue, both a test of removed explant and an in situ test should be considered. Impact of a removed plane-ended tissue explant with a larger platen and a punch test of in situ tissue would appear to be suitable tests for engineered cartilage. Outcome variables would be tissue surface fracture, which could be evaluated by either India ink staining or histology, although potential cell damage, that can lead to future cell death, and microscopic matrix damage may need to be considered as well.

Resists Permanent Deformation and Fracture under Cyclic Loading

Fatigue of biological materials is complicated by reparative remodelling that occurs. Assuming there is minimal remodelling within 24 hours, a lower bound of cycle number would be 3000 cycles. This number is undoubtedly too low, since remodeling will retard, but not prevent fatigue failure. A more realistic number will have to await further research on this topic. At a minimum, therefore, the tissue should resist at least 3000 cycles of load without permanent deformation or fracture. The minimal cycle number would seem appropriate until a better definition of design cycles can be developed. Both cyclic tensile tests and unconfined compression should be used. Compressive stresses of 5 Mpa at 1 Hz for 3000 cycles could form a minimally acceptable evaluation test. There is little information on tensile stress values *in vivo*. Roberts et al. (1986), measured tensile strength of human cartilage to be up to 14 MPa. Following the ratio of 3:1 of failure to gait stress levels for compressive stress, a cyclic tensile stress of 5 MPa would seem a reasonable first estimate for cyclic tension. Permanent deformation at the end of the test, and visual fracture could be the outcome measures. This raises the issue of what methods to use to evaluate tissue fracture. This could be done at the microstructural fibril level, or the visual level. It would seem most practical to use a visual method, possibly by India ink staining and grading. This has been shown to correlate with histological evaluation, a supposedly more sensitive measure. Further research may indicate

the need to assess cell damage and microstructural matrix damage.

Surface friction could be measured by coefficient of friction for tissue-on-tissue, using any of several standard friction test methods and apparata (Mow and Ateshian, 1997). Wear is a more difficult test. Wear and surface damage could be quantified for the friction tests, although further work needs to be done in this area to produce convincing data to support a specific method.

Adheres to Adjacent Tissue

Quality of the bond to adjacent tissue, cartilage or bone, can be assessed by direct test of the bond, either by simple tension or by a peel test (Ahsan and Sah, 1999).

Given these several possible test parameters, which are the more important? The material parameters suggested include:

- Thickness
- Viscoelastic: elastic parameters, relaxation function
- Biphasic: solid elastic constants, permeability
- Critical impact load
- Permanent deformation and fracture after cyclic loading
- Coefficient of friction
- Wear rates, surface damage
- Adhesion strength to adjacent tissue

Appropriate thickness would appear to be essential, since too little thickness would result in little load through the tissue and too much thickness would result in too much load through the tissue. Elastic properties would be important, to provide appropriate load transfer through contact stresses and to surrounding tissue, and return to original geometry on unloading. The cyclic loading and wear tests would appear to also be critical, since the most common mode of failure of existing engineered tissue or replacements is breakdown of the tissue. The replacement tissue must be able to remain in place without cracking and wearing away. Adhesion to adjacent tissue is important, especially adhesion to underlying bone, if the defect is full thickness. Adhesion to adjacent cartilage is important, although it is not clear if failure of this interface will result in a failed implant. Time dependency properties are also of unknown importance, although the tissue should not permanently deform under cyclic loading, and relaxation should not be too fast to prevent load carriage. Difference in time response will result in different energy dissipation, but the consequences of this are unknown.

The suggested parameters are somewhat of a wishlist of information, and not realistic as measures for all new candidate engineered cartilage. Some priorities must be set and efforts focused on a smaller set of measures. Ranking the properties in order of importance is necessarily speculative, but in the interest of establishing a starting point for discussion, the properties are placed into groups of primary and secondary importance as follows (there is no implied order within a group):

- Primary importance:
  —Thickness
  —Elastic constants
  —Fatigue strength in tension and compression
  —Permanent deformation after cyclic compression
  —Wear
  —Adhesion to bone
- Secondary importance:
  —Single load fracture or tensile strength
  —Time dependency
  —Adhesion to adjacent cartilage
  —Coefficient of friction

What Specific Values of the Mechanical Parameters Would Be Indicative of a Successful Implant?

As discussed earlier, there will be different uses for tissue material properties, either as predictors of future properties or as measures of current properties. Relationships between current and future properties are virtually unknown, so speculating on suitable initial properties is not warranted. For current properties, success will be defined in comparison to competing methods, so there are no absolute specific values indicative of success. There are also few property measurements for engineered cartilage, so for a starting point, the properties of natural cartilage must be used.

Thickness

The engineered tissue should be able to form thicknesses from 1 to 3 mm, the specific value depending on the site of application. The thickness should be stable and not be thicker or thinner than adjacent natural cartilage, if it exists.

Elastic Constants, Biphasic Model

The most common biphasic model used is the KLM model of Mow and co-workers. This is a linear, isotropic theory. The parameters of the model are the solid modulus and Poisson's ratio and the permeability. Typical values of these can be found in several review papers (Mow and Ratcliffe, 1997).

Elastic Constants, Viscoelastic Model

The parameters in a linear, elastic viscoelastic model are the "instantaneous" elastic modulus and Poisson's ratio, and the creep or relaxation function. These will likely differ in tension and compression (Woo and Mow, 1987).

Permanent Deformation after Cyclic Tensile and Compressive Loading

Permanent deformation after loading should be similar to natural cartilage, since natural cartilage will likely be attached to the replacement cartilage. Since the suggested fatigue test is not a commonly done test, there is no data available on what that permanent deformation would be. It should be expected to be small, on the order of 10% strain.

Fracture after Cyclic Tensile and Compressive Loading

Fracture should not be evident after the design loads. Assuming 10^6 cycles/year as a design load, there is the question of how many cycles should be used in an *in vitro* fatigue test, allowing for remodeling that occurs *in vivo?* It was suggested that 3000 cycles would be a minimal number that would be needed before remodeling, but this is obviously inadequate for long-term performance. More research must be done to choose a more realistic number.

Wear

As with fracture, the question is how to evaluate wear. Thickness change, wear debris analysis, and scanning electron microscopy of the surface are all candidate methods. So little has been done on this topic for engineered cartilage it is not useful to speculate. This is an area in need of future research.

Current State of the Art

The functional requirements of a replacement cartilage are known with reasonable accuracy. Material properties indicative of tissue functional qualities can be identified, based on developed methods in engineering materials. Experimental and theoretical methods exist for *in vitro* measurement of several of these variables, in particular geometry and time-dependent prefailure properties. *In vitro* methods to measure presence of fracture and permanent deformation after cyclic loading, and impact loading are available. Methods to quantify failure properties, particularly fatigue fracture strength and permanent deformation, adhesion to adjacent tissue, and wear are much less developed. Consequently, there is little available on design requirements for failure variables. There are very few methods for measuring any of the variables *in vivo.*

The relationship between properties measured at one time point, including preimplantation, and future properties is virtually unknown. Perhaps measurement at different times can be used as an indicator of future performance, but this must still be shown. The use of material property measurements at this time is limited to assessing current properties as measures of potential current function.

Although properties of engineered cartilage will likely not need to be equal to that of natural cartilage to be successful, few properties of engi-

neered cartilage under development have been published, so there is no standard based on engineered tissue to use for evaluating new candidate tissues. Hopefully, this will change with time, but for now the most realistic standard for new tissues is natural cartilage.

Future Needs

Although methods exist to measure geometry and prefailure properties, these need to be refined for application to engineered cartilage, and some consensus among researchers and developers in this area is needed.

Tissue failure is the main property of an engineered tissue that cannot currently be measured. There is a need to establish practical methods to measure fracture strength, fatigue strength in tension and compression, and wear. There is also a need to refine and establish methods for measuring adhesion strength between new and native tissue.

The importance of anisotropy, material nonlinearities, material nonhomogeneity, and large deformations in the theoretical models and experimental measurements needs to be clarified. If these are important, reasonable simplifications and practical test methods need to be identified.

There is a strong need to establish a science of tissue development and the relationship between initial construct properties and final outcome properties. Effect of *in vivo* environment and rehabilitation program need to be included in this science.

Most methods for measuring cartilage material properties are applicable for *in vitro* conditions. There are very few methods that can be used *in vivo,* and fewer that are noninvasive. There is a need to develop methods for measuring *in vivo* mechanical properties of engineered cartilage.

Finally, the relationship between initial and final properties, and final properties and clinical outcome must be established. Successful clinical outcome is the ultimate goal of tissue engineering. Continually comparing measured mechanical material parameters and outcome could provide a sound basis for identifying particular material properties as appropriate in an engineered cartilage, both as initial conditions and as current properties.

References

Ahsan T., Sah R.L. 1999. Biomechanics of integrative cartilage repair. *J. Osteoarthritis Res. Soc. Intl.* 7:29–40.

Ateshian G.A., Soslowsky L.J. 1997. Quantitative anatomy of diarthrodial joint articular layers. In: *Basic Orthopaedic Biomechanics.* V.C. Mow, W.C. Hayes, eds. Philadelphia: Lippincott-Raven Publishers, pp. 253–273.

Bergmann G., Graichen F., Rohlmann A. 1993. Hip joint loading during walking and running, measured in two patients. *J. Biomech.* 26:969–990.

Borrelli Jr., J., Torzilli P.A., Grigiene R., Helfet D.L. 1997. Effect of impact load on articular cartilage: development of an intra-articular fracture model. *J. Orthop. Trauma* 11:319–326.

Cohen Z.A., McCarthy D.M., Kwak D., Legrand P., Fogarasi F., Ciaccio E.J., Ateshian G.A. 1999. Knee cartilage topography, thickness, and contact areas from MRI: in-vitro calibration and in-vivo measurements. *J. Osteoarthritis Res. Soc. Intl.* 7:95–109.

Eberhardt A.W., Keer L.M., Lewis J.L., Vithoontien V. 1990. An analytical model of joint contact. *J. Biomech. Eng.* 112:407–413.

Jeffrey J.E., Gregory D.W., Aspden R.M. 1995. Matrix damage and chondrocyte viability following a single impact load on articular cartilage. *Arch. Biochem. Biophys.* 322:87–96.

Krebs D.E., Robbins C.E., Lavine L., Mann R.W. 1998. Hip biomechanics during gait. *J. Orthop. Sports Phys. Ther.* 28:51–59.

Mak A.F. 1986. The apparent viscoelastic behavior of articular cartilage—the contributions from the intrinsic matrix viscoelasticity and interstitial fluid flows. *J. Biomech. Eng.* 108:123–130.

Mow V.C., Ateshian G.A. 1997. Lubrication and wear of diarthrodial joints. In: *Basic Orthopaedic Biomechanics.* V.C. Mow, W.C. Hayes, eds. Philadelphia: Lippincott-Raven Publishers, pp. 275–315.

Mow V.C., Ratcliffe A. 1997. Structure and function of articular cartilage and meniscus. In: *Basic Orthopaedic Biomechanics.* V.C. Mow, W.C. Hayes, eds. Philadelphia: Lippincott-Raven Publishers, pp. 113–177.

Mow V.C., Kuei S.C., Lai W.M., Armstrong C.G. 1980. Biphasic creep and stress relaxation of articular cartilage: theory and experiments. *J. Biomech. Eng.* 102:73–84.

Newberry W.N., Zukosky D.K., Haut R.C. 1997. Subfracture insult to a knee joint causes alterations in the bone and in the functional stiffness of overlying cartilage. *J. Orthop. Res.* 15:450–455.

Repo R.U., Finlay J.B. 1977. Survival of articular cartilage after controlled impact. *J. Bone Joint Surg.* 59-A 8:1068–1076.

Roberts S., Weightman B., Urban D., Chappell D. 1986. Mechanical and biochemical properties of human articular cartilage in osteoarthritic femoral heads and in autopsy specimens. *J. Bone Joint Surg.* 68-B 2:278–288.

Suh J.K., Bai S. 1998. Finite element formulation of biphasic poroviscoelastic model for articular cartilage. *J. Biomech. Eng.* 120:195–201.

Thompson, Jr., R.C., Oegema, Jr., T.R., Lewis J.L., Wallace L. 1991. Osteoarthrotic changes after acute transarticular load. An animal model. *J. Bone Joint Surg.* 73-A 7:990–1001.

Torzilli P.A., Grigiene R., Borrelli, Jr., J., Helfet D.L. 1999. Effect of impact load on articular cartilage: cell metabolism and viability, and matrix water content. *J. Biomech. Eng.* 121:433–441.

Woo S.L-Y., Mow V.C. 1987. Biomechanical properties of articular cartilage. In: *Handbook of Bioengineering.* R. Skalak, S. Chien, eds. New York: McGraw-Hill, Inc., pp. 4.1–4.44.

Part III

Design Parameters for Tissue Engineering

10

Design Parameters For Functional Tissue Engineering

Arnold I. Caplan

Introduction

Tissue engineering is a field that developed in an attempt to provide cell- or molecular-based tissue therapies for developmental anomalies, defects arising from excised tissue (tumor excision) or for adult-onset deficiencies (diabetes). In the past, tissue engineering was largely an empirical discipline in which a variety of constructs were implanted into a number of sites with subsequent evaluation of repair events and tissue function. I suspect that, in part, the empirical approaches we used initially (Ohgushi et al., 1989; Wakitani et al., 1989, 1994; Young et al., 1998) led to the popularization of the "engineering" name of this new discipline since engineers often use empirical approaches. Also, clearly, material science engineers were important in the development of new, man-made delivery vehicles (Kim et al., 1994; Langer and Vacanti, 1993; Ohgushi et al., 1989). As in all young disciplines, tissue engineering relied on several oversimplifying assumptions to provide logics for the initial technologies that were tested. Probably the most dissatisfying assumption involves introduction of progenitor cells and/or recombinant molecular triggers to a particular site based on the assumption that the tissue site itself will provide all of the subsequent information and cueing necessary to provide functional restoration of that tissue (Ohgushi et al., 1989; Wakitani et al., 1989, 1994;

Young et al., 1998). Although each individual tissue site does provide unique sets of microenvironmental cues, such cueing, in the adult organism, has proven to be inefficient and insufficient to provide totally satisfactory tissue reformation. It would now seem appropriate to take several steps back and to view tissue engineering in a logical sequence and to attempt to provide a variety of data-based design parameters that can bring tissue engineering into the realm of reliable and reproducible restoration logics for a variety of tissues (Bruder and Caplan, 2000; Bruder et al., 1994; Caplan 2000; Caplan and Bruder, 1996; Caplan and Goldberg, 1999; Caplan et al., 1997). Only skeletal tissues will be considered in this chapter.

Functional Tissue Engineering

Functional tissue engineering can be defined as the controlled restoration of tissue morphology and function. In this light, optimal functional tissue engineering provides perfectly regenerated and functionally integrated tissue. Such regeneration of skeletal tissues must come on a background of tissue repair mechanisms that have been evolutionarily selected. These repair mechanisms function to seal off the injured or surgically intruded-upon tissue and then to very rapidly and imperfectly fix that tissue (Caplan,

1990). Such repair mechanisms provide "functional" repairs and allow the animals to limp away from danger so that the slower and more regenerative remodeling process can be activated to provide the ultimate functional fix. Thus, to optimally regenerate a tissue, there is need to inhibit the intrinsic repair response.

An obvious example of a natural repair response that has all of the characteristics found in almost every tissue can be readily observed when skin is cut or damaged (Mast et al., 1992; McCallion and Ferguson, 1996). The events that occur subsequent to this tissue injury involve sealing off of the injured tissue, the flooding of that tissue with a serum filtrate and digestive cells and enzymes to rid the tissue of bacterial and fungal contaminants, and then the subsequent invasion of that tissue by fibrous-producing cells to quickly join the cut ends of the tissue to provide continuity to protect the animal from outside intrusions of microbes and other dangerous organisms and substances. Last, vasculature once again enters the site to nurture the new tissue mass and to initiate the remolding events. This rapid repair response brings about a scar, which in skin is not a functional problem but could be catastrophic in skeletal tissues since scars represent mechanical fault lines, and tissue breakage and damage are likely to occur at these fault lines under normal mechanical stress.

There are several distinctive differences between repair versus regeneration of tissues:

- Repair is rapid, regeneration is slow.
- Repair fabricates fibrous joining tissue, regeneration fabricates normal and site-specific tissue.
- Repair fabricates mechanically compromised tissue, regeneration fabricates mechanically identical tissue.
- Repair provides imperfect integration with the surrounding normal tissues while regeneration provides complete molecular and mechanical integration.

Embryonic Tissue Formation

When tissue naturally regenerates [usually in small focal defects in very young animals (Pineda

et al., 1992)], the observed regenerative events are comparable in some specific cases to exactly those same events observed in embryonic tissue formation (Caplan, 1997; Zwilling, 1968). This suggests that many of the "design parameters" for successful regeneration of skeletal tissue can be deduced from a careful analysis of the components of embryonic tissue formation or from detailed analysis of regenerating tissue systems in lower species. This premise leads to the working hypothesis that successful and functional tissue engineering will involve recapitulation of selected embryonic events.

Tissue Fabricated by Specialized Cells

All tissue arises from the biosynthetic efforts of specialized or differentiated cells. These differentiated cells arise from progenitors whose numbers and availability is site and age specific. All skeletal tissues arise by the active synthesis and secretion of extracellular matrix (ECM) molecules that are organized in a tissue-specific manner. For example, bone- and tendon-producing cells both fabricate large amounts of type I collagen. Clearly, the type I collagen of bone organizes calcium phosphate crystals while type I collagen of tendon forms large diameter fibers that are organized in parallel and function in a connected and concerted manner in the direction of load. Each of the skeletal tissues is fabricated by cells that originate from a common progenitor cell that we call a mesenchymal stem cell (MSC) (Caplan, 1991, 1994, 2000). Thus, in the embryo and the adult, a sufficient number of receptive MSCs must be present in order to coordinately stimulate the fabrication of a differentiated tissue, for example, bone versus tendon.

The ECM of Undifferentiated Mesencyhme

The microenvironment seen by MSCs will determine when and how it will differentiate. Both the complex composition of the ECM and the presence of specific growth factors or cytokines is necessary to bring cells into and along the differentiation pathway. Additional signals are required to allow cells to progress through that differentiation pathway and, finally, other factors will modulate the final composition and skeletal properties of the tissues so formed. Embryonic mesenchyme, prior to the differentiation of specific skeletal tissues, is composed of an ECM-containing type I collagen, fibronectin, hylauronan (HA), and certain proteoglycans like versican and decorin (Fernandez et al., 1991; Toole, 1981). These ECM components serve a variety of functions amongst which they, importantly, provide a milieu for MSC expansion without differentiation. In addition, these mesenchymal matrices establish the residence cells in a receptive framework to be induced into a specific differentiation pathway. All of these ECM components can bind specific growth factors and cytokines; all such bioactive factors can be selectively released upon injury or can be sequestered and never reach the cell during certain physiological states. For tissue engineering logics, it may be that some of these natural polymers prevalent in embryonic mesenchyme ECM should serve as highly specific vehicles for cell or bioactive factors delivery in regenerative, functional tissue engineering of defective tissues.

HA Is Chondroinductive and Antiangiogenic

In avian embryonic limb development, the central mass of cells in the limb differentiates into a cartilage rod whose shape is similar to that of future bones. Zwilling (1968) suggested that there were two separate yet overlapping phases of embryonic limb development: the first "morphogenetic phase" involves signals that establish a complex pattern of morphological development (hand versus feet) and a second phase of "cytodifferentiation" in which the specific molecules of cartilage are biosynthetically generated and organized in the site-specific ECM. During the morphogenetic phase, Toole and colleagues (Linsenmayer et al., 1973; Toole, 1981, 1997) have shown that HA is high, and, prior to the active cytodifferentiation into cartilage, HAase appears and increases to function to eliminate HA from the ECM. The transition from a HA-rich to a chondroitin sulfate-rich to a heparan sulfate-rich ECM is indicative of embryonic skeletal tissue development (Caplan, 1986).

Based on these observations, we chemically bound HA to tissue culture dishes and showed that a specific size class (200,000–400,000 daltons) was optimal for cartilage formation from undifferentiated embryonic stage-24 limb bud mesenchymal progenitor cells (Caplan, 1970; Caplan et al., 1968; Kujawa and Caplan, 1986; Kujawa et al., 1986). This chondroinductive potential of HA *in vivo* may also be involved with another very important aspect of HA's complex interaction with certain cells. Feinberg and Beebee (1983) have shown in the same age of embryonic chick limb buds that this same molecular weight of HA is antiangiogenic: introduction of HA into limb mesenchyme causes vasculature to be excluded from the central region and/or to regress from that HA-rich region.

These chondroinductive and antiangiogenic capabilities are, in addition to and separate from the well-established correlation of high concentrations of HA around mesenchymal cells, capable of high levels of proliferation (Knudson and Knudson, 1991; Sherman et al., 1994). Again, Toole and others (Toole, 1990) clearly established that HA coats are established around the cells exhibiting high levels of cell division. We now know that the presence of specific HA receptors, such as CD44 and RHAMM (Aruffo et al., 1990; Hall et al., 1994; Lesley et al., 1993; Yang et al., 1994), are responsible for translating this extracellular activity into intracellular signals. Furthermore, high molecular weight HA is a natural component of cartilage matrix and is responsible for tethering aggrecan, the major water structuring proteoglycan in cartilage, to the matrix in which there are high levels of fluid movement (Caplan, 1984). This unique noncovalent relationship between aggrecan and HA is highly specific and can be disrupted by oligomers of HA and/or can be stabilized by additional HA binding proteins such as link protein.

Thus, depending on the developmental state of the cells, the implantation site, and other parameters, HA may be an informational molecule and, thus, maybe a useful delivery vehicle for tissue engineering the regeneration of cartilage and, through endochondral bone formation, the regeneration of bone.

Differentiated Skeletal Cells Fabricate Specific and Abundant ECMs

Skeletal tissues like cartilage, bone, and tendon are characterized by a low cellularity and a high amount of tissue bulk. Thus, the embryonic mesenchyme must transform from a relatively low ECM to cell ratio (high cellularity) to a very high ECM to cell ratio. These cells must fabricate highly specific and voluminous ECMs. As noted above, although both bone and tendon cells fabricate a large amount of type I collagen, this major ECM material is organized quite differently in tendon, which does not mineralize, compared with bone, which organizes mineral into dense arrays. Thus, any delivery vehicle for tissue engineering must have as a design component large spaces or pores to accommodate the voluminous ECMs produced by differentiated skeletal cells.

Integrated Host with *Neo*-Tissue

Although regenerating a sector of skeletal tissue itself is a considerable task, this positive event may not be functional unless the *neo*-tissue is morphologically, molecularly, and mechanically fully integrated into the host's surrounding tissue. This integration is no small task since the molecular and cellular mechanisms controlling integration may be quite different from those involved in the fabrication of the *neo*-tissue. For example, we found that a brief exposure of the digestive enzyme trypsin of the cut edge of host cartilage of a full thickness defect in the medial femoral condyle of a rabbit was necessary to facilitate the integration of *neo*-cartilage with host cartilage (Mochizuk et al., 1995). On a molecular level, we have shown that trypsin clipped the core protein of aggrecan and, thus, released it from the tissue (Caplan et al., 1998). Trypsin does not affect the collagen scaffold of the cartilage and, thus, the absence of aggrecan produces

space for the molecular bridging of host with *neo*-cartilage. Likewise, we would suggest that HA oligomers, greater than 6 sugars, are capable of unloading aggrecan from cartilage matrix (Tammi et al, 1998). Such oligomers are released from the HA delivery vehicles as the HA breaks down. Thus, molecular mechanisms to facilitate *neo*-tissue integration into hosts must be designed into tissue engineering strategies.

HA Oligomers Are Angiogenic

By designing delivery vehicles composed of HA, it is expected is that adult marrow-derived MSCs will sense their microenvironments as "embryonic" and, like embryonic mesenchymal progenitors, they will differentiate into chondrocytes. If this HA microenvironment is stable, the chondrocytes should be stable. Moreover, the HA is antiangiogenic, thus, the HA microenvironment of the pores of a delivery vehicle should initially exclude vasculature (Caplan, 2000). The chondrocytes themselves secrete antiangiogenic factors (Moses et al., 1990; Sorgente et al., 1975), which should also function to reinforce the exclusion of blood vessels. This leads to a complication of using such design parameters, with the most onerous being that center of implanted material may not be accessible to nutrients unless the cartilaginous volume is subjected to high-pressure fluid flow in an *in vitro* bioreactor or mechanical *in vivo* deformation-reformation. Thus, as the porous structure of the delivery vehicle rapidly fills up with cartilage ECM, cells at center of the implant become susceptible to nutrient deprivation. In this regard, 1.0–1.5 mm seems to be the limiting distance for access to vasculature to insure nutrition of cells within an implant or forming tissue.

All of the HA delivery vehicles break down as determined by their chemical modifications and physical format. For example, we have used sponges made of pure HA with simple ester crosslinks (ACP™, Fidia Advanced Biopolymers, srl, Italy) that breaks down *in vivo* in 7 to 10 days (Campoccia et al., 1998). If the uronic acid carboxyl moiety is 100% substituted with a benzyl ester (HYAFF®-11), the benzyl group is first slowly hydrolyzed off with the subsequent HA-backbone then breaking down identically to ACP, with the entire process taking 8 to 10 weeks. The final breakdown products of these HA delivery vehicles are the same, HA oligomers of various sizes.

Such HA oligomers themselves have important biological properties. Importantly, HA oligomers have been shown to be highly angiogenic (Wester et al., 1975). In *in vivo* sites, either heterotopic or orthotopic, the breakdown of the HA delivery vehicle into oligomers triggers the continued differentiation of chondrocytes into their hypertrophic compartment and their subsequent cell death. This sequence is followed by rapid vascular invasion and replacement of the hypertrophic cartilage with newly fabricated woven bone that is highly vascularized. In this case, we believe that the angiogenic capabilities of HA oligomers plays a determinative role in these events. In addition, as stated above, the HA oligomers are suspected of denuding the host's cartilage of aggrecan at the cut edge of the defect. Such matrix unloading may account for the integration of *neo* with host tissue.

These unique properties of HA point to important new design parameters for functional tissue engineering, including material:

- that initially houses or attracts cells
- that induces these cells into the appropriate phenotypic compartment
- that provides space for ECM accumulation
- that provides mechanical stability to fill the entire defect
- whose breakdown products are Informational
- whose breakdown *Triggers* the final developmental cascade

In the case of cartilage, the breakdown of HA matrices has a profound effect on the formation of subchondral bone and the integration of new with old tissue.

An Example of HA-Mediated Design Parameters

My colleagues have developed a systematic series of assays to evaluate delivery vehicles for skeletal tissues (Solchaga et al.). The first step is to insure that MSCs, as progenitor cells for skeletal tissues, bind to the biological delivery vehicles. The simplest assay is to use labeled MSCs and to expose them to the test vehicle in the presence and absence of fibronectin. The MSCs bind to fibronectin and the assay directly compares precoating with fibronectin with no such coating since fibronectin coating is assumed to be optimal for MSC attachment. The second assay is to load cells into/onto the test material and to implant the composite into a subcutaneous location on the back of an immunocompromised rodent or syngeneic or donor animal. We have established the "gold standard" for this assay as a 3-mm (on a side) porous calcium phosphate ceramic cube because of our earlier studies using this material for bone repair (Dennis and Caplan, 1993; Dennis et al., 1992; Ohgushi et al, 1989). By side-by-side comparison of the test material with the ceramic, the *in vivo* differentiation of MSCs can be assessed. Lastly, the test material is placed empty, filled with autologous bone marrow or culture-expanded autologous marrow-derived MSCs in a full-thickness osteochondral defect in the medial condyl of a young adult rabbit femur (Laurent and Fraser, 1992; Solchaga et al, 1999, 2000).

Such assays have been performed for HA-based delivery vehicles ACP and HYAFF-11 and the observations made from this assay indicate that:

- HA, as a porous sponge or fleece, is chondroinductive and anti-angiogenic.
- Fibronectin-coated HA matrices are more effective at binding MSCs, especially for the intrinsically hydrophobic HYAFF-11.
- Empty HA sponges attract young host MSCs from the nearby marrow depots that then differentiate into chondrocytes as a consequence of binding to HA material.
- The intrinsic breakdown characteristics of the HA matrices trigger the conversion of included hypertrophic cartilage into endochondral bone.
- The breakdown of HA into oligomers enhances vascular penetration and separately facilitates host and *neo*-cartilage integration presumably by unloading aggrecan from the

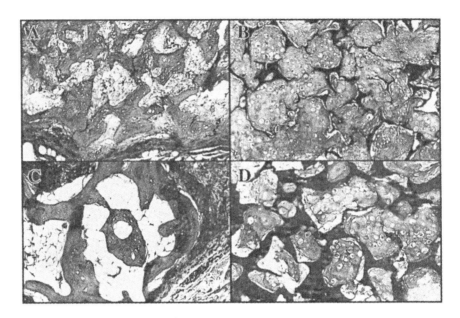

FIGURE 10.1. Light microscopy of the composites after subcutaneous implantation in immunocompromised rodent hosts with HYAFF-11p75HE sponge (*A, C*) or HYAFF-11 sponge (*B, D*), 3 weeks (*A, B*) and 6 weeks (*C, D*) after implantation; fixed, embedded, sectioned and stained with toluidine blue.

cut edge of the cartilage to facilitate molecular integration.

• There appears to be no difference in the cartilage morphology of a full-thickness defect filled with fast-dissolving ACP or slow HYAFF-11 at 12 weeks after HA implants have been inserted.

A new material has been fabricated by Fidia Advanced Biopolymers in which the HA backbone of HYAFF-11 is only 70–80% substituted with benzyl groups instead of the normal 100% (Solchaga et al.,). When placed in a subcutaneous pouch of a nude mouse, this material called HYAFF-11-1175HE (*p75* for short), can be recovered at 3 or 6 weeks postimplantation, fixed, embedded in paraffin, and sectioned and stained with toluidine blue (Fig. 10.1). For comparative purposes, ACP (unsubstituted but ester-cross-linked HA) sponges empty or filled with MSCs cannot be recovered since it dissolves too quickly in this subcutaneous site. Figure 10.1 provides a side-by-side (same host) comparison of *p75* (Fig. 10.1, *A* and *C*) and HYAFF-11 (Fig. 10.1, *B* and *D*) loaded with young adult rabbit marrow-derived MSCs. Most, if not all, of the HA sponge

is absent in the *p75* specimen while the HYAFF-11 specimen is mostly intact. There is an outer shell of endochondrally derived bone with fatty marrow inside in the *p75* in contrast to the uniform distribution of cartilage tissue in the HYAFF-11. We conclude that the breakdown susceptibility of the HA matrix directly controls the differentiation of the MSCs preloaded into them. In this case, ACP, *p75*, and HYAFF-11 exhibit increasingly slower breakdown times (7–10 days, 2–3 weeks, 8–10 weeks, respectively) (Caplan, 2000).

These breakdown time differences are reflected in the kinetics observed in critical sized full-thickness, osteochondral defects in the medial condyl of young adult rabbits (Fig. 10.2). At 4 weeks following implantation into the knees of the same rabbit (Fig. 10.2, *A* and *B*), the HYAFF-11 is mostly intact and filled with chondro-tissue (Fig. 10.2*B*) while the *p75* has broken down and, in most of the defect except for the top sector, the cartilage that formed earlier is replaced by endochondral bone. By 12 weeks, the HYAFF-11 and *p75* in the knees of the same rabbit (Fig. 10.2, *C* and *D*) both have broken down and the top of the defect is covered by well-integrated *neo*-cartilage on top of subchondral bone. It is quite clear that

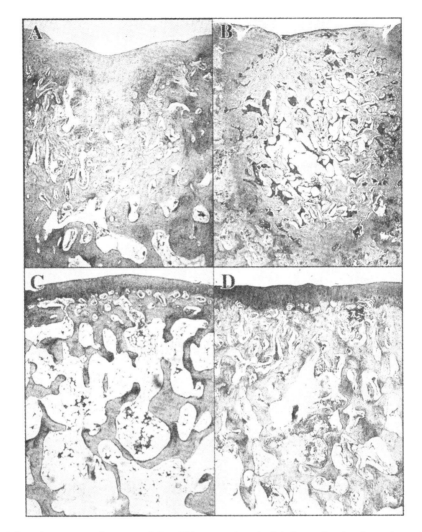

FIGURE 10.2. Light microscopy of young adult rabbit osteochondral defects after implantation with HYAFF-11p75HE sponge (*A, C*) or HYAFF-11 sponge (*B, D*), 4 weeks (*A, B*) and 12 weeks (*C, D*) after implantation; fixed, embedded, sectioned and stained with toluidine blue.

p75 is similar to ACP implants but it exhibits breakdown characteristics somewhat slower than ACP (Caplan, 2000, Solchaga et al., in press). Nonetheless, by 12 weeks, ACP, *p75,* and HYAFF-11 all exhibit the same morphologies. We would infer that the HA-based sponges and the resultant HA oligomers released on complete breakdown are the common drivers in these complex regenerative events. These observations provide excellent examples of key design parameters for generating functional tissue using newly delineated principles of tissue engineering.

Principles of Functional Tissue Engineering

Following are some principles of functional tissue engineering:

- The implant material must establish an embryonic-like microenvironment.
- The implant material must be porous to:
  —accommodate regenerative progenitor or differentiated cells or bioactive factors that can attract such cells

—provide space for abundant and specific ECM biosynthesis and assembly

—provide connected structure to mechanically align or connect the new ECM

—provide channels for nutrient transport to internal cells

- The implant should fill the entire defect and inhibit the natural repair sequence to insure regenerative tissue formation.
- The breakdown of the implant material should be rapid and release breakdown products that trigger or facilitate the late stages of the regenerative cascade.
- The implant material or its breakdown products should facilitate or directly contribute to the integration of *neo-* with host tissue.
- Each tissue and, indeed, each tissue site requires implant materials with distinctive and different design parameters.

Conclusion

We have passed through the early stages of the development of tissue engineering as a new discipline. We have made progress and have moved cautiously and we have made our mistakes. From both our successes and failures, we now are poised to design new regenerative tissue formation strategies to provide for functional tissue repair. The perfection of this new science will be functional tissue engineering.

Acknowledgments

I am particularly indebted to my colleagues at Case Western Reserve University and University Hospitals of Cleveland for their contribution to the research that forms the basis for this treatise. I especially thank Luis Solchaga, Ph.D., for Figures 10.1 and 10.2 and Victor M. Goldberg, M.D., for his consistent encouragement and support. The studies reported here were supported, in part, by funds from the National Institutes of Health.

References

Aruffo A., Stamenkovic I., Melnick M., Underhill C.B., Seed, B. 1990. CD44 is the principal cell surface receptor for hyaluronate. *Cell* 61:1303–1313.

Bruder S.P., Caplan A.I. 2000. Bone regeneration through cellular engineering. In: *Principles in Tissue Engineering,* 2nd Ed. R. Lanza, R. Langer, J. Vancanti, eds. Springer, New York, pp. 683–696.

Bruder S.P., Fink D.J., Caplan A.I. 1994. Mesenchymal stem cells in bone development, bone repair, and skeletal regeneration. *J. Cell Biochem.* 56:283–294.

Campoccia D., Doherty P., Radice M., Brun P., Abatangelo G., Williams D.F. 1998. Semisynthetic resorable materials from hyaluronan esterification. *Biomaterials* 19:2101–2127.

Caplan A.I. 1970. Effects of the nicotinamide-sensitive teratogen 3-acetylpyridine on chick limb cells in culture. *Exp. Cell Res.* 62:341–355.

Caplan A.I. 1977. Muscle, cartilage and bone development and differentiation from chick limb mesenchymal cells. In: *Vertebrate Limb and Somite Morphogenesis.* D.A. Ede, J.R. Hinchliffe, M. Balls eds. Cambridge, U.K. Cambridge University Press, pp. 199–213.

Caplan A.I. 1984. Cartilage. *Sci. Am.* 251:84–94.

Caplan A.I. 1986. The extracellular matrix is instructive. *Bioessays* 5:129–132.

Caplan A.I. 1990. Cartilage begets bone *versus* endochondral myelopoiesis. *Clin. Orthop. Rel. Res.* 261:257–267.

Caplan A.I. 1991. Mesenchymal stem cells. *J. Orthop. Res.* 9:641–650.

Caplan A.I. 1994. The mesengenic process. *Clin. Plastic Surg.* 21: 429–435.

Caplan A.I. 2000. Tissue engineering designs for the future: new logics, old molecules. *Tissue Eng.* 6:1–8.

Caplan A.I., Bruder S.P. 1996. Cell and molecular engineering of bone regeneration. In: *Principles of Tissue Engineering,* R.P. Lanza, W.L. Chick, R. Langer, eds., Springer, New York, pp. 599–618.

Caplan A.I., Elyaderani M., Mochizuki Y., Wakitani S., Goldberg V.M. 1997. The principles of cartilage repair/regeneration. *Clin. Orthop. Rel. Res.* 342: 254–269.

Caplan A.I., Fink D.I., Bruder S.P., Young R.G., Butler D.L. 1998. The regeneration of skeletal tissues using mesenchymal stem cells. In: *Frontiers in Tissue Engineering.* C.W. Patrick Jr., A.G. Mikos, L.V. McIntire, eds. New York: Elsevier Science, pp. 471–480.

Caplan A.I., Goldberg V.M. 1999. The principles of tissue engineered regeneration of skeletal tissues. *Clin. Orthop. Rel. Res.* 367S:12–16.

Caplan A.I., Zwilling E., Kaplan N.O. 1968. 3-Acetylpyridine: effects *in vitro* related to teratogenic activity in chick embryos. *Science* 160:1009–1010.

Dennis J.E., Caplan A.I. 1993. Porous ceramic vehicles for rat-marrow-derived (*Rattus norvegicus*) osteogenic cell delivery: effects of pre-treatment with fibronectin or laminin. *J. Oral Implant.* 19:106–115.

Dennis J.E., Haynesworth S.E., Young R.G., Caplan A.I. 1992. Osteogenesis in marrow-derived mesenchymal cell porous ceramic composites transplanted subcutaneously: effect of fibronectin and laminin on cell retention and rate of osteogenic expression. *Cell Transpl.* 1:23–32.

Feinberg R.N., Beebee D.C. 1983. Hyaluronate in vasculogenesis. *Science* 220:1177–1179.

Fernandez M.S., Dennis J.E., Drushel R.F., Carrino D.A., Kimata K., Yamagata M., Caplan A.I. 1991. The dynamics of compartmentalization of embryonic muscle by extracellular matrix molecules. *Dev. Biol.* 147:46–61.

Hall C.L., Wang C., Lange L.A., Turley E.A. 1994. Hyaluronan and the hyaluronan receptor RHAMM promote focal adhesion turnover and transient tyrosine kinase activity *J. Cell Biol.* 126:575–588.

Kim W.S., Vacanti J.P., Cima L., et al. 1994. Cartilage engineered in predetermined shapes employing cell transplantation on synthetic biodegradable polymers. *Plast. Reconstr. Surg.* 94:233–240.

Knudson W., Knudson C.B. 1991. Assembly of a chondrocyte-like pericellular matrix on non-chondrogenic cells: role of the cell surface hyaluronan receptors in the assembly of a pericellular matrix. *J. Cell Sci.* 99:227–235.

Kujawa M.J., Caplan A.I. 1986. Hyaluronic acid bonded to cell culture surfaces stimulates chondrogenesis in stage 24 limb mesenchyme cell cultures. *Dev. Biol.* 114:504–518.

Kujawa M.J., Carrino D.A., Caplan A.I. 1986. Substrate-bonded hyaluronic acid exhibits a size-dependent stimulation of chondrogenic differentiation of stage 24 limb mesenchymal cells in culture. *Dev. Biol.* 114:519–528.

Langer R., Vacanti J.P. 1993. Tissue engineering. *Science* 260:920–932.

Laurent T.C., Fraser J.R.E. 1992. Hyaluronan. *FASEB J.* 6:2397–2404.

Lesley J., Hyman R., Kincade P.W. 1993. CD44 and its interaction with extracellular matrix. *Adv. Immunol.* 54:271–335.

Linsenmayer T.F., Toole B.P., Trelstad R.L. 1973. Temporal and spatial transitions in collagen types during embryonic limb development. *Dev. Biol.* 35:232–239.

Mast B.A., Diegelmann R.F, Krummel T.M, Cohen, I.K. 1992. Scarless wound healing in the mammalian fetus. *Surg. Gyn. Obst.* 174:441–51.

McCallion R.L., Ferguson M.W.J. 1996. Fetal wound healing and the development of antiscarring therapies for adult wound healing. In: *The Molecular and Cellular Biology of Wound Repair.* Clark RAF, ed., New York: Plenium Press, pp. 561–600.

Mochizuk Y., Elyaderani M., Young R.G., Goldberg V.M., Caplan A.I. 1995. Mesenchymal stem cell implantation into a pretreated full-thickness articular cartilage defect. *Trans. Orthop. Res. Soc.* 20:367.

Moses M.A., Sudhalter J., Langer R. 1990. Identification of an inhibitor of neovascularization. *Science* 248:1408–1410.

Ohgushi H., Goldberg V.M., Caplan A.I. 1989. Heterotopic osteogenesis in porous ceramics induced by marrow cells. *J. Orthp. Res.* 7:568–578.

Ohgushi H., Goldberg V.M., Caplan A.I. 1989. Repair of bone defects with marrow cells and porous ceramic. *Acta Orthop. Scand.* 60:334–339.

Pineda S., Pollack A., Stevenson S., Goldberg V.M., Caplan A.I. 1992. A semi-quantitative scale for histologic grading of articular cartilage repair. *Acta Anat.* 143:335–340.

Sherman L., Sleeman J., Herrlich P., et al. 1994. Hyaluronate receptors: key players in growth, differentiation, migration and tumor progression. *Curr. Opin. Cell Biol.* 6:726–733.

Solchaga L., Goldberg V.M., Caplan A.I. 2000. Hyaluronic acid based biomaterials in tissue engineered cartilage repair. In: *New Frontiers in Medical Sciences: Redefining Hyaluronan.* G. Abantangelo, P.H. Weigel, eds. Excepta Medica, Elsevier Science, pp. 233–246.

Solchaga L., Yoo J.U., Lundberg M., Dennis J.E., Huibregtse B.A., Goldberg V.M., Caplan A.I. 2000. Hyaluronic acid-based polymers in the treatment of osteochondral defects. *J. Orthop. Res.,* 18: 773–780.

Solchaga L., Yoo J.U., Lundberg M., Goldberg V.M., Caplan A.I. 1999. Augmentation of the repair of osteochondral defects by autologous bone marrow in a HA-based delivery vehicle. *Trans. Orthop. Res. Soc.* 24:801.

Solchaga L.A., Goldberg V.M., Caplan A.I. 2002. Hyaluronan and tissue engineering. In: *Hyaluroran.* J.F. Kennedy, G.O. Phillips, V. Hascall, E.A. Balas, eds. Woodhead, Cambridge, UK.

Sorgente N., Kuettner K., Soble L.W., Einstein R. 1975. The resistance of certain tissues to invasion II evidence for extractable factors in cartilage which inhibit invasion by vascularized mesenchyme. *Lab. Invest.* 32:217–222.

Tammi R., MacCallum D., Hascall V.C., Pienimaki J. 1998. Hyaluronan bound to CD44 on keratinocytes is displayed by hyaluronan decasaccharides and not hexasaccharides. *J. Biol. Chem.* 44: 28878–28888.

Toole B.P. 1981. Glycosaminoglycans in morphogenesis. In: *Cell Biology of Extracellular Matrix.* E.D. Hay, ed. New York: Plenum Press, pp 259–94.

Toole B.P. 1990. Hyaluronan and its binding proteins, the hyaladherins. *Curr. Opin. Cell Biol.* 2:839–844.

Toole B.P. 1997. Hyaluronan in morphogenesis. *J. Intern Med.* 242:35–40.

Wakitani S., Goto T., Pineda S.J., Young R.G., Mansour J.M., Goldberg V.M., Caplan, A.I. 1994. Mesenchymal cell-based repair of large full-thickness defects of articular cartilage and underlying bone. *J. Bone Joint Surg.* 76:579–592.

Wakitani S., Kimura T., Hirooka A., et al. 1989. Repair of rabbit articular surfaces with allograft chondrocytes embedded in collagen gel. *J Bone Joint Surg.* 71:74–80.

Wester D.C., Hampston I., Arnold F., Kumar S. 1975. Angiogenesis induced by degradation products of hyaluronan. *Science* 228:1324–1326.

Yang B., Yang B.L., Savani R.C., Turley E.A. 1994. Identification of a common hyaluronan binding motif in the hyaluronana binding proteins RHAMM. CD44 and link protein. *EMBO J.* 13:286–296.

Young R.G., Butler D.L., Weber W., Gordon S.L., Fink D.J., Caplan A.I. 1998. The use of mesenchymal stem cells in Achilles tendon Repair. *J. Orthop. Res.* 16:406–413.

Zwilling E. 1968. Morphogenetic phases in development. *Dev. Biol. Suppl.* 2:184–207.

11

Tissue Engineering a Heart: Critical Issues

Michael V. Sefton and Robert Akins

Introduction

The LIFE Initiative, a multi-institutional affiliation of researchers, was created to use tissue engineering and regeneration to produce an endless supply of human vital organs for transplantation. With an unlimited supply of vital organs, replacing a damaged or failed organ could become not substantially different from any other surgical procedure. Once a surgeon deemed a transplant necessary, an organ of the appropriate type and size could be supplied from a local source, much like any other medical device, as soon as an operating room could be scheduled. There would be no wait for a suitably matched donor organ. Of more significance, a large number of patients may benefit from the implantation of smaller structures—heart muscle, valves, vascular grafts—so that the spin-off benefits of a tissue engineering effort may become even more important than the whole organ.

The need to create an unlimited supply of vital organs and component tissues is easy to recognize. Most compelling is patient need for a dependable source of organs and tissues (Table 11.1). Unfortunately, there are insufficient native tissues to address these needs. Using the heart as an example, approximately 20,000–40,000 patients in the United States could benefit from a transplant (AHA, 2000), but only 2,100 heart transplants are performed each year due to the

lack of organ donors. The development of an unlimited supply of cardiac implant devices will be an enormous undertaking, but one with immense benefit.

The cells of the heart are arranged in precise architectures, and the correct organization and functional coordination of these cells are requisite for efficient contractile activity. The disruption or absence of appropriate organization in the heart can be seriously debilitating and life threatening. In addition to congenital defects, many people born with structurally normal hearts suffer cardiac malfunctions at some point during their lives due to disease, infection, or poor coronary circulation (Gillum, 1994; McGovern et al., 1996). According to U.S. government statistics, approximately 33% of all deaths are related to some type of heart disease.

Experience has shown that the heart has a limited ability to repair itself *in vivo,* whether in response to the underdevelopment of structures or following injury (Butler, 1989; McMahon and Ratliff, 1990; Wicken and Shorey, 1970). Partially because of this limited ability, current treatments are restricted to medical and surgical approaches that address the sequelae of the primary defect, and current approaches do not restore lost structure or function. Medical and surgical treatments for severe cardiac problems have become highly advanced, but when they fail, organ transplantation often remains the

TABLE 11.1. The need for transplants

Type of transplant	Number of transplants performed (1999)	Patients waiting for transplant (09/18/2000)	Number of patients removed from the waiting list due to death (1999)
Kidney	12,483	46,363	3,046
Liver	4,698	16,084	1,753
Pancreas	363	960	17
Pancreas islet cell	—	160	—
Kidney-pancreas	946	2,389	169
Intestine	70	131	45
Heart	2,185	4,070	709
Heart-lung	49	221	53
Lung	885	3,584	587
Total*	21,692	71,663	6,125

Source: UNOS National Patient Waiting List for Organ Transplant. Available at www.unos.org. Accessed September 18, 2000.
*note some patents waiting for more than one transplant.

only other option. Unfortunately, there is a profound shortage of donor hearts, and many patients do not receive the necessary life-saving procedures (Stevenson et al., 1994). The possibility that diminished cardiac function may be recovered through the implantation of biosynthetic constructs has great appeal. The ability to re-establish cardiac tissue architecture and function *de novo* from isolated cells may make it possible to specifically tissue engineer constructs for implantation. Such methods would have potential application in a wide variety of debilitating heart diseases.

To date, there have been only limited advances in cardiac tissue engineering. A recent review provides more details (Akins, 2000). Most promising is the work of Shinoka et al. (1995), who have demonstrated the use of a biosynthetic pulmonary valve in sheep, and Sodian et al. (2000a,b), who have recently reported the ability to construct trileaflet valves using biodegradable polyhydroxyalkanoate scaffolds. One of the more fruitful areas of research has been in the direct injection of cells into the myocardium. The lack of regeneration in the heart has been attributed to the insufficiency of cardiomyocyte proliferative activity and to the absence of a stem cell population (Tam et al., 1995): since there are no cells available to replace damaged areas, regeneration cannot occur. Accordingly, a significant amount of research has been performed to iden-

tify alternative cell sources for use in implantation and in situ repair therapies: sketelal muscle and muscle satellite cells, fetal cardiomyocytes, and various stem cell populations.

A few groups have reported success with the organization of cardiotypic structures *in vitro*. In one set of studies, cardiac tissue structure and aspects of cardiac function were re-established *in vitro* in the absence of three-dimensional cues from extracellular scaffolds (Akins et al., 1997, 1999). These results suggest that cardiac cells possess an innate capacity to re-establish complex, three-dimensional, cardiac organization *in vitro*. In other studies, investigators have used three-dimensional meshes or foams as culture scaffolds. Freed and co-workers (Bursac et al., 1999) have developed a model system using a biodegradeable polymer mesh that can be used for electrophysiological studies. Li and co-workers (2000) have grown three-dimensional cardiac grafts and implanted them into host myocardia. Two critical observations from this work were the survival of functional grafts and the apparent vascularization of the implants by the host circulation.

Overall, the field of cardiac tissue engineering is very much in its infancy. Although the results seen to date are exceedingly encouraging, a good deal of work remains to be done in order to develop clinically relevant approaches, let alone move toward a whole heart. Hence, a recent task

TABLE 11.2. Critical issues associated with tissue engineering a heart

Issues associated with three-dimensional constructs	Issues specific to heart
Controlling phenotype	Growing functional human cardiomyocytes
Multiple cell types	Mechanical elasticity and strength (fatigue)
Vascularization (nutrients)	Thrombogenicity (endothelialization)
Immune/inflammatory response	Pacemaker and electrical conduction
Manufacturing, storage, and quality control	Valves and conduits

force (NIH, 1999) has emphasized development of heart components such as a cardiac patch or a valve before graduating to whole-heart engineering. This, too, is the view of the LIFE Initiative.

Critical Issues

Successful engineering of an organ as complex as the heart will require many technological advancements. To start with: what is the most feasible approach to growing a heart? One possible approach uses knowledge of developmental biology to grow a mature heart from an embryonic or fetal organ or by regeneration from undifferentiated stem cells. Another possibility is to fabricate the organ using cells seeded into a matrix or scaffold and soluble signals. Some combination of these two approaches might be envisioned. Alternatively, a tissue-engineered total heart might be grown in sections such as valves, inlet vessels, and chambers and these parts later fused, in some yet unknown method.

There are numerous critical issues (Table 11.2), some of which apply to all large constructs and some which are more specific to the heart. The generic issues reflect the fundamental nature of how an organ is different from a tissue and how a heart (or liver or kidney) is different from a blood vessel.

A tissue is defined as an aggregation of similarly specialized cells united in the performance of a particular function (Dorland, 1994); an organ consists of a number of tissues. Hence, by definition, engineering an organ requires multiple cell types, each of which phenotype must be maintained for long periods of time and in a form that permits complete intercellular communication. Anyone who has tried to co-culture cells *in vitro* knows this is difficult and that the more rapidly growing cell typically outgrows the slower one. On the other hand, more complex and perhaps more physiological three-dimensional culture systems are being explored. In these systems, it appears that maintaining phenotype of mixed cell systems [e.g., hepatocytes and connective tissue (Michalopoulos et al, 1999)] is less of a concern or at least is not a fundamental barrier to progress. The goal appears to be to find the conditions that allow cells to do what they are genetically programmed to do.

More problematic may be the intrinsic nature of large cell-based constructs (tissues or organs) and the corresponding difficulty of supplying cells deep within the construct with nutrients. Diffusion is fine for 100 μm or so and low cell densities can extend this limit. Hence, thin or essentially two-dimensional (e.g., a tube) constructs are feasible without an internal blood/nutrient supply. However, it is hard to combine cells at tissue densities (10^{10} cells/cm^3) into 1 cm or larger slabs of tissue without some sort of prevascularization or its alternative. Three-dimensional manufacturing or micromachining methods are being explored as one approach to this problem (Kaihara et al., 2000; Kim et al., 1998); angiogenic growth factor delivery may be another approach. Finding ways around the physical limitations of Fick's law may be the most crucial barrier to overcome to tissue engineer a heart or any other large organ.

Manufacturing, quality control, and storage issues apply to all tissue engineering constructs as do questions related to the immune and inflammatory response to these constructs (Babensee et al., 1997). While the former gets little attention from the academic community—it is the primary concern of our industrial colleagues—the immune issues have already drawn considerable attention because they are similar

TABLE 11.3. Mechanical attributes of a heart as drawn from a simple biophysical approach*

Power requirements	Heart contracts 72 min^1 × 5 l/min at 100 mm Hg = ~1.3 W (PV work) at rest; with exercise, power requirements are about six times higher
Total oxygen consumption	~10–20 × pressure-volume work
Fatigue life	10 years of life = 3 × 10^8 cycles
Wall tension	3 × 10^4 N/m based on law of Laplace with radii of curvature (left ventricle) of 20–80 mm and thickness ~8 mm
Frank-Starling law	Cardiac output = venous return; stroke volume increases with circulatory filling pressure or myocardium stretch

*Burton (1972)

to the issues faced by the transplantation community. Using autologous cells, whether fully differentiated or some type of progenitor, is an approach that is practical immunologically, but it does preclude the off-the-shelf concept behind an unlimited supply of organs. It also may raise the concern over the underlying disease states that necessitated the organ replacement in the first place. On the other hand, strategies to induce immune tolerance [e.g., by blocking co-stimulatory factors using, say, CTLA4Ig, [Weber et al., 1997)] are under active study but may not yet be enough to circumvent the very sophisticated human immune system. Perhaps nuclear transfer and therapeutic cloning strategies (McLaren, 2000a) may be necessary assuming the various ethical issues (McLaren, 2000b Perry, 2000) can be resolved.

If these issues were not enough, one can easily identify problems specifically related to a heart; a review by one of us (Akins, 2000) highlights some of these. While obtaining a source of human cardiomyocytes or a functioning equivalent cell may not be insurmountable (see above), getting the cells to function as an organ is problematic. The cells need to form the appropriate subcellular structures, intercellular connections, and matrix arrangements required for functional coordination and directed beating. Highly oriented sarcomeric structures and T-tubules within myocytes are hallmarks of the myocardium and are needed for the cyclical events of force generation and relaxation carried out by the heart. The alignment of myocytes in register and the proper formation of the intercalated disks between myocytes are also critical in enabling electrical pulses to be transmitted in the correct direction at normal speeds and in allowing suitable force

transmission between myocytes. The force-transducing ability and elasticity of the extracellular matrix will likely be important; the composition of the matrix, its organization, and the formation of myofiber units may all be critical to the function of a biosynthetic organ. In addition, the heart contains specialized cells that participate in the electrical conduction routes found throughout the heart (e.g., the SA node, the AV node, and Purkinje fibers). These specialized cells are crucial to the coordination of the heart's contractile effort, and including them in the proper places in a biosynthetic substitute may be critical. Obviously, there are clear differences between the rhythmic twitching of cultured cardiac cells en masse and the organized, efficient, regulated beating of the heart; only the latter will generate the force required to pump blood at systolic pressure levels.

Thus, a critical feature of a heart is both its movement and the complexity of the electrical conduction pathways. Neither problem has received much attention in the tissue engineering literature. However, given the variety of electrical conduction-related diseases in a normal myocardium, there is good reason to suspect that simple mimicry of heart muscle may fall short of the goal. Thus, functional tissue engineering in the context of a heart has specific and well-established attributes (Table 11.3). Simply speaking, the heart must pump blood at a mean pressure of roughly 100 mm Hg. Hence, heart muscle must stretch in response to capillary filling pressure and eject a volume of blood that varies with demand. The latter requires a uniform and well-coordinated contraction that generates the required power. There is a long way to go from the small number of cardiomyocyte-like

TABLE 11.4. Potential spin-off benefits of a tissue-engineered heart

Research and development	Cardiovascular repair
Cells for diagnostics	In situ repair of damaged hearts
Animal models for human disease	Cardiac patches
In vitro models for conduction based diseases	Mechanically robust grafts
Novel scaffold processes	Endothelial seeding of vascular grafts
Degradable materials for other applications	"Capillary beds"
Control of host response for other constructs	Pediatric cardiac valves
Tissue-engineered kidney, liver, other organs	

cells beating in a culture dish to the complex architecture of the myocardium and the duplication of its nonlinear viscoelastic properties. The following quote from Y.C. Fung (1993) is worth remembering at this point:

. . . heart muscle in the resting state is an inhomogeneous, anisotropic and incompressible material. Its properties change with temperature and other environmental conditions. It exhibits stress relaxation under maintained stretch and creep under maintained stress. It dissipates energy and has a hysteresis loop in cyclic loading and unloading. Thus heart muscle in the resting state is viscoelastic.

The properties, once stimulated, are much more complex.

These issues and concerns can be summarized in the form of project milestones, which can be expressed in several ways (Akins and Sefton, 2001). These range from having appropriate large animal models (e.g., cows) that can be used to test its functional attributes to defining the strategies, materials, and cell sources to be used to make the components of the heart (or to induce differentiation of the appropriate adult or embryonic stem cells). The latter is already underway and can be done without reference to the whole heart and is the basis for a number of projects working toward a cardiac patch (Akins et al., 1999; McDevitt et al., 2000; Van Luyn et al., 2000). Thus, the cardiac patch can be considered a stepping stone toward a whole heart, or, looking at it from the opposite direction, the patch is a spin-off benefit (Table 11.4) of working towards a heart. The road to a full heart is obviously long and conditional on these issues being resolved. At the same time, the various components—cardiomyocytes, elastomeric scaffolds, immune control strategies—will be valuable both commercially and clinically in their own right. Some may even be applicable in other areas of biology and biotechnology.

LIFE Initiative

Reaching the goal of tissue engineering vital organs is clearly going to be difficult and expensive. We can easily see the effort costing several billion dollars and taking more than a decade. To this end, the LIFE Initiative was formed to bring the vision to fruition. With approximately 60 participants so far, the LIFE Initiative is in the very early days yet, and it is still working out the details of what should happen. The underlying premise is that no single group or institution has all the expertise needed to engineer a heart. Hence, it is necessary to pool our talents across discipline, institution, and national barriers. Hence, we envisage LIFE as a global, multi-institution project, much like the human genome project, although the nature of the project is much different.

Conclusions

Tissue engineering whole organs such as a functional heart is clearly ambitious, if not, as some would put it, ridiculous. But the problem of growing whole organs can be broken down into more manageable components and interim milestones. Furthermore, reaching an interim goal such as a cardiac patch or a strategy for vascularizing thick slabs of tissue has an intrinsic therapeutic value on its own. In fact, these spin-off benefits may be more valuable and may benefit more patients than the whole organ. Thus, the LIFE Initiative has been created to tissue engineer complex three-dimensional, mechanically robust and dynamic organs like the heart. But the underlying goal is to advance the science and art of tissue engineering, to create novel arrangements for collaborative research on a global scale, and to foster a new industry based on the capacity for treating human disease with replacement tissues and organs.

Acknowledgments

R.A. would like to acknowledge the Nemours Foundation and NASA for their generous support of some of the work discussed here. M.S. would like to thank the Canadian Institutes of Health Research and the Natural Sciences and Engineering Research Council for support of his activities in tissue engineering. Both wish to thank the many participants within LIFE who have contributed their views, words and support for this initiative and for the documents from which this chapter is abstracted.

References

AHA 2000. Heart and Stroke Statistical Update. American Heart Association, available at www.americanheart.org/statistics/index.html.

Akins R.E. 2000. Prospects for the use of cell implantation, gene therapy, and tissue engineering in the treatment of myocardial disease and congenital heart defects. In: *Medizinische Regeneration und Tissue Engineering.* K. Sames, ed. EcoMed: Landsberg, Germany.

Akins R., et al. 1997. Neonatal rat heart cells cultured in simulated microgravity. *In Vitro Cell Dev. Biol. Anim.* 33:337–343.

Akins R.E., et al. 1999. Cardiac organogenesis *in vitro:* reestablishment of three-dimensional tissue architecture by dissociated neonatal rat ventricular cells. *Tissue Eng.* 5:103–118.

Akins R., Sefton M.V. 2001. Tissue Engineering a Heart, New Surgery 1:26–32.

Babensee J.E., Anderson J.M., McIntire L.V., Mikos A.G. 1998. Host response to tissue engineered devices. *Advanced Drug Delivery Reviews* 33:111–139.

Burton A.C. 1972. *Biophysical Basis of the Circulation.* Yearbook Medical Publishers, Chicago.

Bursac N., et al. 1999. Cardiac muscle tissue engineering: toward an *in vitro* model for electrophysiological studies. *Am. J. Physiol.* 277(2 Pt 2): H433–H444.

Butler R. 1989. Evidence for a regenerative capacity in adult mammalian cardiac myocytes. *Am. J. Physiol.* 256(3 Pt 2):797–800.

Dorland 1994. Dorland's Illustrated Medical Dictionary, 28th ed. W.B. Saunders, Philadelphia.

Fung Y.C. 1993. *Biomechanics: Mechanical Properties of Living Tissues,* 2nd ed. Springer-Verlag, New York, p433.

Gillum R.F. 1994. Epidemiology of congenital heart disease in the United States. *Am. Heart J.* 127:919–927.

Kaihara S., Borenstein J., Koka R., Lalan S., Ochoa E.R., Ravens M., Pien H., Cunningham B., Vacanti J.P., 2000. Silicon micromachining to tissue engineer branched vascular channels for liver fabrication. *Tissue Eng.* 6:105–117.

Kim S.S., Utsunomiya H., Koski J.A., Wu B.M., Cima M.J., Sohn J., Mukai K., Griffith L.G., Vacanti J.P. 1998. Survival and function of hepatocytes on a novel three-dimensional synthetic biodegradable polymer scaffold with an intrinsic network of channels. *Ann Surg.* 228:8–13.

Li R.K., et al. 2000. Construction of a bioengineered cardiac graft. *J. Thorac. Cardiovasc. Surg.* 119: 368–375.

McDevitt T.C., Angello J.C., Huschka S.D., Whiteney M.L., Reinecke H., Murry C.E., Kyriakides T.K., Bornstein P., Seatena M., Giachelli C.M., Stayton P.S. 2000. Micropatterning of extracellular matrix components to direct cellular and tissue response. Society For Biomaterials. Sixth World Biomaterials Congress Transactions, p. 1232.

McGovern P.G., et al. 1996. Recent trends in acute coronary heart disease—mortality, morbidity, medical care, and risk factors. *N. Engl. J. Med.,* 334:884–890.

McLaren A. 2000a. Cloning: pathways to a pluripotent future. *Science* 288:1775–1780.

McLaren, A. 2000b. Stem cells: golden opportunities with ethical baggage. Science 288:1778.

McMahon J.T., Ratliff N.B. 1990. Regeneration of adult human myocardium after acute heart transplant rejection. *J. Heart Transplant.* 9:554–567.

Michalopoulos G.K., Bowen W.C., Zajac V.F., Beer-Stolz D., Watkins S., Kostrubsky V., Strom S.C. 1999. Morphogenetic events in mixed cultures of rat hepatocytes and nonparenchymal cells maintained in biological matrices in the presence of hepatocyte growth factor and epidermal growth factor. *Hepatology* 29:90–100.

NIH 1999. Working Group on Tissue Genesis and Organogenesis for Heart, Lung and Blood Applications, National Institutes of Health, August 13, 1999. Available at www.nhlbi.nih.gov/meetings/workshops/tissueg1.htm.

Perry D. 2000. Patients' voices: the powerful sound in the stem cell debate. *Science* 287:1423.

Shinoka T., et al. 1995. Tissue engineering heart valves: valve leaflet replacement study in a lamb model. *Ann. Thorac. Surg.* 60(6 Suppl):513–516.

Sodian R., et al. 2000a. Fabrication of a trileaflet heart valve scaffold from a polyhydroxyalkanoate biopolyester for use in tissue engineering. *Tissue Eng.* 6:183–188.

Sodian R., et al. 2000b. Tissue engineering of heart valves: *in vitro* experiences. *Ann. Thorac. Surg.* 70:140–144.

Stevenson L.W., et al. 1994. The impending crisis awaiting cardiac transplantation. Modeling a solution based on selection. *Circulation* 89:450–457.

Tam S.K., et al. 1995. Cardiac myocyte terminal differentiation. Potential for cardiac regeneration. *Ann. N.Y. Acad. Sci.* 752:72–79.

Van Luyn M.J.A., Plantinga J.A., Berling S., de Leig L.F.M.H., van Wachem P.B. 2000. Cardiac tissue engineering of neonatal rat cardiomyocytes in two and three dimensional cultures. Society For Biomaterials. Sixth World Biomaterials Congress Transactions, p. 103.

Weber C.J., Hagler M.K., Chrussochoos J.T., et al 1997. CTLA4-lg prolongs survival of microencapsulated neonatal porcine islet xenografts in diabetic NOD mice. *Cell Transplant.* 6:505–508.

Wilcken D., Shorey C. Eilense 1970. Ultrastructural evidence for regeneration of heart-muscle cells after experimental infarction. *Lancet* 2(7662): 4:21–23.

12

Design Parameters for Engineering Bone Regeneration

Robert E. Guldberg and Angel O. Duty

Introduction

Bone serves multiple critical functions within the body, including hematopoeisis, mineral homeostasis, protection of vital organs, and mechanical support of the musculoskeletal system. Local loss of bone structural integrity resulting from trauma or long-term degeneration is consequently a major cause of patient disability. Fortunately, bone possesses a remarkable inherent capacity for repair and regeneration due to the local presence of osteoprogenitor cells, osteoinductive proteins, and vascularity. Damage to skeletal tissue typically initiates an ordered sequence of cellular migration, proliferation, and differentiation resulting in the formation of a woven mineralized matrix that is subsequently remodeled to lamellar bone. Thus, bone has the unique ability to go beyond repair and fully regenerate its original structure and mechanical properties. However, an estimated 5–10% of the over 6 million fractures that occur in the United States each year require treatment to augment the natural regeneration process and restore bone function (Einhorn, 1999). Moreover, certain challenging clinical situations such as spine fusions and large segmental defects in long bones require exogenously supplied factors in all cases to stimulate osseous union.

The strategies being developed to regenerate bone are numerous and diverse, reflecting the heterogeneity of the tissue and its functional environment. Bone is a complex hierarchical composite material composed of an apatite mineral phase embedded in a primarily type I collagen extracellular matrix. The organization and interaction of these constituents are responsible for the excellent strength, stiffness, and fracture resistance properties of bone. Bone matrix also contains an intricate cellular communication network of osteocytes and is subject to local remodeling by bone-resorbing osteoclasts and bone-forming osteoblasts, serving under normal conditions to maintain skeletal structural integrity. The forces applied to bone *in vivo* range from negligible in certain craniofacial bones to several times body weight in the lower limb long bones and articular joints. In proportion to its local function, the porosity of bone varies from approximately 5% in cortical bone to over 90% in some trabecular bone regions. This high degree of variability in the form and function of bone presents significant challenges for the design of bone regeneration constructs and suggests that no one solution will be optimal for all defect sites or all clinical applications.

The basic elements required for successful bone regeneration include an extracellular matrix scaffold, cells, and osteoinductive factors (Bruder and Fox, 1999). Whether these elements are provided by the host or must be included alone or in some combination within an implantable construct will depend critically on the local biochemical, mechanical, and vascular environments at the defect site. Autogenous bone grafting is currently the clinical standard for osseous reconstruction associated with fracture non-unions, spinal fusions, and bone loss resulting from metabolic diseases, tumor resections, or adverse remodeling adjacent to total joint implants (Khan et al., 2000). However, the use of autograft bone has several disadvantages, including limited available tissue for transplantation, lack of structural integrity to withstand functional loads, and increased patient morbidity at the site of harvest (Cook et al., 1994). Recent advances in bioresorbable materials and discoveries in bone cell and molecular biology have led to the emergence of tissue engineering as an alternative to traditional bone grafting. Through exogenous provision of matrices, cells, and bioactive factors, bone tissue engineering strategies seek to mimic the natural process of bone regeneration.

This chapter will present an overview of design parameters currently under investigation for bone regeneration constructs and identify several remaining issues and technical challenges. The optimal selection and integration of design parameters will require consideration of the functional *in vivo* mechanical environment of bone and the utilization of test beds to quantitatively benchmark construct performance and accurately predict clinical efficacy. The adaptation of bone to mechanical stimuli during repair and regeneration will therefore be discussed in the context of interactions between mechanical factors and the elements of bone regeneration constructs.

Bioresorbable Scaffolds

The transformation from bone grafting to bone tissue engineering began with the introduction of osteoconductive bone graft substitutes or scaffolds and is now evolving to include local delivery of osteogenic cells and osteoinductive factors (Boden, 1999). Osteoconductive scaffolds facilitate invasion of capillaries, attachment of osteoprogenitor cells, and subsequent appositional mineralized matrix formation in large bone defects (Cornell, 1999; Kenley et al., 1993). Used alone, this approach relies on an adequate

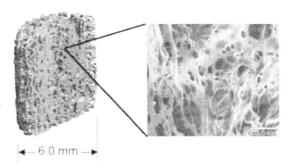

FIGURE 12.1. Microtomography image (*left*) of a porous poly(L-lactide-co-DL lactide 70:30) scaffold created using 15% azodicarbonamide as a pore-forming agent. Stereological analysis of the architecture indicated a porosity of 75%, average strut or plate thickness of 82 μm, average pore size of 255 μm, and degree of anisotropy of 1.5. The compressive modulus and strength of the scaffold were 145.4 MPa and 7.1 MPa, respectively. Viable osteoblast-like MC3T3-E1 cells attached throughout the scaffold microstructure (*right*).

supply of endogenous osteoprogenitor cells to invade the scaffold and differentiate into osteoblasts. The *in vivo* response to osteoconductive scaffolds is characterized by bone formation progressing inward from the margins of the bone defect into the interior of the scaffold. For load-bearing clinical applications, the scaffold must possess mechanical properties that provide adequate initial stability at the defect site and gradual transfer of physiological mechanical stimuli under functional loading to newly formed bone (Davy, 1999). Finally, the construct should resorb in a controlled manner consistent with the rate of new bone formation and release nontoxic degradation products (Yaszemski et al., 1996).

Numerous natural and synthetic scaffold materials have been investigated for bone tissue engineering, including demineralized bone matrix, fibrin, hyaluronic acid, collagen gels or sponges, ceramics such as hydroxyapatite (HA), tricalcium phosphate (TCP), coral, and bioactive glass, and a wide variety of degradable polymers such as polylactides (PLA) (Fig. 12.1) and polyglycolides (PGA) (Cornell, 1999; Ducheyne et al., 1994; Elgendy et al., 1993; Gebhart and Lane, 1991; Hutmacher, 2000; Laurencin et al.,

1996; Mikos et al. 1993; Peter et al., 1998; Sasaki and Watanabe, 1995; Solchaga et al., 1999; Thomson et al, 1999). Some of these materials are manufactured in solid or particulate form for implantation. Others are injected and subsequently polymerized to fill a defect void. Deficiencies related to the efficacy, handling, biocompatibility, bioresorbability, and mechanical properties of currently available scaffolds are driving the rapid development of novel materials and composites designed to effectively regenerate bone.

Scaffolds for bone regeneration are generally porous to facilitate invasion of cells and vascularity from the surrounding tissue bed. The three-dimensional architecture therefore influences the biological response by determining in part the osteoconductivity of the scaffold. For example, a scaffold containing closed voids that does not provide a fully interconnected pore structure will retard revascularization and formation of mineralized matrix throughout the scaffold (Boyan et al., 1999; Bruder and Fox, 1999). In addition to interconnectivity, the porosity, pore size, surface to volume ratio, and anisotropy are potentially important architectural parameters to consider in the design of bone regeneration scaffolds (Whang et al., 1999). The potential importance of the architectural parameters has even led some investigators to explore the use of rapid prototyping technologies to manufacture precisely controlled microstructural scaffolds (Hollister et al., 2000). Although the bioresorbable materials available for use in rapid prototyping systems is currently limited, this approach would allow scaffolds to be created with specified internal architectures and complex external topologies designed to match patient-specific defect geometries.

The optimal porosity or pore size for bone regeneration constructs remains an open question, the answer to which may depend critically on the mechanical and vascular environment at the defect site. Several groups have reported the optimal pore size for bone in growth to be in the range of 150–600 microns (Cornell, 1999; Ishaug et al., 1997). However, Whang et al. (1999) demonstrated substantial bone formation in non-load-bearing defects filled with polymer scaffolds possessing a median pore size less than 50 microns and porosity greater than 90%. These investigators hypothesized that high-porosity scaffolds with small pore sizes provided greater hematoma stabilization in the earliest phases of bone regeneration. This approach appeared to wick in and trap osteoprogenitor cells and bioactive factors resulting in bone formation throughout the scaffold rather than progressive osteoconduction from the defect margins. A possible reason for the uncertainty regarding optimal pore size may relate to interactions with the scaffold mechanical properties and the functional loading environment. The spatial requirement for vascularization and invasion of osteoprogenitor cells should be less than 20 microns based on the dimensions of blood vessels and bone cells (Whang et al., 1999). However, micromotion between the scaffold interface and surrounding tissue may disrupt vascular ingrowth into such small pores. Larger pores, although perhaps not optimal for hematoma stabilization or cell attachment, may allow vascular in growth to occur despite the presence of limited interface micromotion.

The mechanical properties of scaffolds are determined by the scaffold material and its microarchitecture. Highly porous polymer scaffolds do not typically possess sufficient mechanical properties to bear *in vivo* loads (Boyan et al., 1999). Although porous ceramic scaffolds are much stiffer and stronger (Haddock et al., 1999), they are typically too brittle to withstand cyclic *in vivo* loading. If we use trabecular bone as a design goal, typical ranges for compressive strength and modulus are 1–10 MPa and 50–500 MPa, respectively (Keaveny et al., 1999; Mow and Hayes, 1997). Implanting a material that is substantially stiffer than the surrounding bone may cause stress shielding and nonphysiologic remodeling in the surrounding bone tissue (Spector, 1994). Scaffolds with insufficient strength risk plastically deforming under functional loads, leading to collapse of the internal porosity and subsidence of the implant. Given the cyclic nature of *in vivo* loading, monotonic strength and stiffness are not the only mechanical properties that should be considered. Michel et al. (1993) evaluated the compressive fatigue behavior of bovine trabecular bone, reporting the number of cycles to failure to be 20 at 2.1% strain and 400,000 at 0.8% strain. In a study on human vertebral trabecular bone, Keaveny et al. (1999) found the ultimate strain to average 1.28% and that specimens subsequently reloaded beyond the ultimate strain level demonstrated reduced modulus and strength. In addition to monotonic strength and modulus then, the ability to withstand repetitive loading (i.e., fatigue resistance) over time is an important criterion to consider for scaffolds expected to maintain their integrity until bone integration can occur.

Several groups have developed composite scaffolds in an attempt to better match the mechanical properties of normal bone tissue (Du et al., 1999; Kellomaki et al., 2000; Marra et al., 1999; Porter et al., 2000; Thomson et al., 1998; Zhang and Ma, 1999a, 1999b). Marra et al. incorporated 10 μm HA particles into porous biodegradable polymer blends created using a salt leaching technique. The inclusion of HA particles slightly increased the scaffold modulus but did not increase strength. Both strength and modulus were substantially lower than normal trabecular bone. Although the development of novel composite scaffolds is clearly an important future trend, this study demonstrated that the intricate relationship between mineral and organic matrix that lends bone its combination of high strength, stiffness, and fracture toughness cannot be mimicked by simply mixing polymer and ceramic components. The interface strength between the matrix and reinforcing material, particularly when tested in an aqueous environment, as well as the scale of constituent interactions are likely to be critical determinants of composite scaffold mechanical properties.

The degradation profile of scaffolds is also an important design parameter that can significantly influence bone regeneration. If a scaffold resorbs very slowly, as is the case with many hydroxyapatite formulations, it will restrict the bone regeneration response and may fail mechanically due to repetitive *in vivo* loading. In contrast, scaffolds that resorb very quickly may lose their mechanical integrity before the defect site can be stabilized by newly formed bone. The degradation mechanism may also be important since scaffolds may experience hydrolytic, enzymatic, bulk, or surface degradation. Finally, the degradation profile of an implanted scaffold may be significantly altered by the functional *in vivo* environment. Although it is well known that body fluids and elevated temperatures accelerate polymer degradation, cyclic mechanical loading may have an equal or greater impact on the mechanical property degradation rate of bioresorbable polymers (Daniels et al., 2000; Smultz et al., 1991).

Functional modification of biomaterial surfaces is another promising strategy being investigated to improve the efficacy of scaffolds for bone tissue engineering (Garcia et al., 1998; Thomas et al., 1997). The adsorption of extracellular matrix proteins such as fibronectin onto biomaterial surfaces can be used to modulate cell adhesion strength (Garcia et al., 1998). Furthermore, Garcia et al. (1999) have proposed that substrate-dependent changes in the conformation of the absorbed proteins or protein fragments can be utilized to direct integrin receptor binding and thereby control cell proliferation and differentiation. The engineering of "rationally designed" surfaces that control cell function may endow osteoconductive scaffolds with the additional ability to induce osteoblast differentiation and subsequent bone regeneration.

Cell-Based Strategies

Cell-based strategies for engineering bone regeneration involve the implantation of cells with osteogenic potential directly into a defect site. Cellular augmentation may be especially important for difficult clinical cases involving older patients, smokers, patients receiving chemotherapy or radiation, and patients with severely damaged wound beds or metabolic diseases in which the endogenous cellular supply may be diminished (Bruder and Fox, 1999). The choice of cell type is important and different cell-based strategies have been described for bone regeneration. These include transplantation of fresh marrow containing osteoprogenitor cells, delivery of differentiated osteoblasts or chondrocytes, and implantation of purified mesenchymal stem cells (Bruder and Fox, 1999). Cells genetically engineered to express osteoinductive factors have also been developed recently and tested for their ability to regenerate bone. These cells may not themselves be capable of synthesizing bone matrix but instead are expected to recruit endogenous cells and stimulate differentiation to an osteoblastic phenotype through delivery of bioactive factors.

Bone marrow has been used in many instances, either alone or in concert with a scaffold material or osteoinductive factor to augment bone repair. The presence of cell populations capable of osteogenesis in bone marrow has been demonstrated by numerous investigators both *in vivo* and *in vitro* (Ashton et al., 1984; Aubin, 1999; Beresford, 1989). Autologous marrow can be aspirated from the iliac crest of a patient and immediately transplanted to a bone defect site or combined with an appropriate biomaterial scaffold prior to implantation (Connolly et al., 1991). Early studies demonstrated that a combination of bone graft and fresh marrow was more osteogenic than the graft alone (Nade and Burwell, 1977; Salama et al., 1973). More recently, Lane et al. (1999) reported on the relative abilities of cancellous bone graft, bone marrow, and a recombinant osteoinductive protein (rhBMP-2) to heal critical sized segmental defects in rat femora. At 12 weeks, the bone union rate was lowest for the groups receiving bone graft alone (38%) and bone marrow alone (47%), whereas the rhBMP-2 group achieved an 80% union rate. Interestingly, a construct composed of rhBMP-2 and bone marrow yielded a 100% union rate at only 6 weeks, demonstrating that the marrow and rhBMP-2 acted synergistically.

Although the benefits of marrow transplantation are well documented, this approach is limited by the small volume of marrow that can be harvested from a patient and the scarcity of osteoprogenitor cells within marrow. It is estimated that humans have one osteoprogenitor for every 100,000 nucleated bone marrow cells (Bruder and Fox, 1999). Furthermore, there is strong evidence to suggest that osteoprogenitor numbers and their sensitivity to osteogenic stimuli decrease with age (D'Ippolito et al., 1999; Huibregtse et al., 2000; Nishida et al., 1999). As marrow transplantation is dependent on the presence of a sufficient number of osteogenic cells to achieve bone regeneration, an approach that provides a more concentrated cell source would be beneficial.

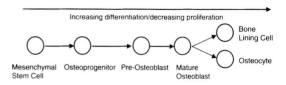

FIGURE 12.2. Differentiation progression from a multipotent mesenchymal stem cell to a committed osteoblast. Cellular proliferation potential decreases with increasing differentiation.

The potential to enhance bone regeneration via implantation of scaffolds seeded with differentiated cells has been tested by several investigators. Differentiated osteoblasts are difficult to isolate and possess limited proliferation capacity (Fig. 12.2). As such, populations of osteoblasts are typically derived from cellular precursors found in rat bone marrow or periosteum using osteogenic cell culture supplements including β-glycerophosphate, dexamethasone, and ascorbic acid (Bruder and Fox, 1999). Although the preparation of implantable constructs involves several steps and is perhaps too lengthy to be clinically practical, differentiated osteoblasts seeded onto ceramic or polymeric scaffolds have been shown to effectively regenerate bone *in vivo* and may provide more rapid bone formation than that achieved using osteoblast precursor cells (Breitbart et al., 1998; Yoshikawa et al., 1996). The well-known sensitivity of osteoblasts to mechanical stimuli suggests that constructs implanted into load-bearing sites may respond differently than those implanted into a mechanically shielded environment. This interaction may have an important impact on the choice of scaffold delivery vehicle, cell seeding density, or other construct parameters.

Other investigators have implanted differentiated chondrocytes into bone defects. The rationale behind these studies is that cartilage typically serves as the natural scaffold for bone-forming cells during the highly regulated process of endochondral ossification. Endochondral ossification is the primary mechanism of bone formation responsible for embryonic skeletal formation, postnatal growth, and fracture repair. Regulatory links exist between hypertrophic differentiation of chondrocytes and several processes essential for bone formation, including cartilage matrix degradation, angiogenesis, osteoblast differentiation, and matrix mineralization. In an attempt to recapitulate endochondral ossification, Vacanti et al. (1995) seeded articular chondrocytes or periosteal cells into PGA scaffolds and implanted the constructs into rat calvarial defects. Both groups formed cartilage matrix but only the periosteal cell constructs continued through the endochondral pathway to form a mineralized matrix. Neoangiogenesis was only observed in the periosteal cell constructs, perhaps explaining the different bone formation responses.

In contrast, Case et al. (2000) demonstrated significant appositional bone formation on tissue-engineered cartilage matrices implanted into bone defects *in vivo* (Fig. 12.3). Rabbit articular chondrocytes were seeded into PGA scaffolds and cultured for 4 weeks prior to implantation into a bone chamber implant model (Fig. 12.6). Intracellular expression of osteopontin within chondrocytes was observed at the junction of cartilaginous and osseous tissue in biopsies taken 3 weeks postimplantation, suggesting hypertrophic differentiation within the constructs. Several recent studies have demonstrated that hypertrophic chondrocytes are functionally coupled to bone formation through the expression of angiogenic factors such as vascular endothelial growth factor (VEGF) (Carlevaro et al., 2000), osteoinductive factors such as bone morphogenetic protein 6 (BMP-6) (Grimsrud et al., 1999), bone-related proteins such as osteopontin (Lian et al., 1993), and core-binding factor 1 (cbfa 1), a transcriptional factor required for normal osteogenesis (Ducy et al., 1997; Inada et al., 1999; Komori et al., 1997). The implications of these findings for tissue engineering applications have not been fully explored. However, if the hypertrophic phenotype could be promoted in tissue-engineered constructs, this approach would provide cell-mediated delivery of a combination of bone-promoting factors within a physiological matrix and thus may represent an integrated strategy to regenerate bone and bone-cartilage composite tissues.

Both differentiated osteoblasts and chondrocytes originate from mesenchymal stem cells (MSCs) found in bone marrow. MSCs have demonstrated the ability to differentiate along

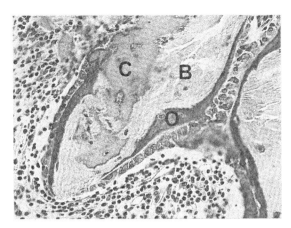

FIGURE 12.3. Goldner's trichrome-stained section showing appositional bone formation (*B*) on implanted tissue-engineered cartilage (*C*) matrix. Unmineralized osteoid (*O*) is shown adjacent to active osteoblasts.

any of several mesenchymal lineages including osteoblastic (Beresford, 1989; Bruder et al., 1997; Haynesworth et al., 1992; Kadiyala et al., 1997), chondrogenic (Kavalkovich et al., 2000; Mackay et al., 1998; Wakitani et al., 1994), myogenic (Rogers et al., 1995), tenogenic (Awad et al., 2000), and adipogenic (Pittenger et al., 1999). Figure 12.2 illustrates the progression from a multipotent MSC to a committed osteoblast. The proliferative capacity of the cells diminishes as they differentiate toward a specific matrix-producing phenotype.

MSC-based approaches to bone regeneration have advanced rapidly in recent years in parallel with an increased understanding of musculoskeletal cell biology. Delivering a source of cells capable of differentiating into bone-forming osteoblasts within an appropriate resorbable scaffold may be a particularly effective strategy for treating large defects or defects in elderly patients for which an adequate supply of osteoprogenitor cells may be the rate-limiting factor for bone regeneration. Ex vivo expanded MSCs provide an accessible and abundant cell source for subsequent implantation into bone defects. MSCs are readily isolated from a simple bone marrow aspirate and can be expanded over many passages without spontaneous differentiation. Recent *in vivo* work has suggested that allo-

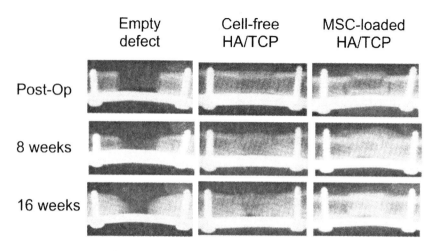

FIGURE 12.4. Mesenchymal stem cell (MSC) mediated bone regeneration in a 21-mm canine femoral gap defect. Radiographic evidence indicated substantial bone regeneration in animals receiving hydroxyapatite (*HA*)/tricalcium phosphate (*TCP*) scaffolds loaded with MSCs at 16 weeks. (Reprinted with permission from Bruder SP. *J. Bone Joint Surg. Am.* 80:985–996, 1998).

genic stem cells may elicit little to no functional immune response, a finding that, if verified, would substantially enhance the clinical benefit of this approach by providing off-the-shelf availability (Peter et al., 2000).

Investigators have recently had success in repairing critical-sized bony defects with MSC-seeded scaffolds implanted *in vivo*. Kadiyala et al. (1997) showed that syngeneic MSCs implanted within a ceramic HA/TCP scaffold into 8-mm rat segmental gap defects provided significantly better bony fill than its controls. Cell-seeded scaffolds produced bone or cartilage in 43% of pores, whereas scaffolds implanted with bone marrow only formed bone in 19% and cell free ceramic implants only produced 10% bony fill (Bruder et al., 1997). A similar study was performed using autologous canine MSCs on ceramic implants to repair 21-mm segmental gap defects in dogs with comparable success (Fig. 12.4) (Bruder et al., 1998). Further work has shown that mesenchymal progenitor cells in a collagen gel matrix injected into a distraction gap will also accelerate bone formation (Richards et al., 1999).

Although several recent studies have demonstrated that MSC-seeded scaffolds significantly increase bone formation *in vivo*, more extensive characterization of MSCs may be a prerequisite for their use in a functional bone scaffold. Studies

have shown that MSCs proliferate into multipotent daughter cells that may be induced to differentiate along respective lineages by such factors as cell density, basal nutrients, spatial organization, growth factors, and cytokines (Pittenger et al., 1999), but the effects of other factors have yet to be determined. Of even more consequence is the absence of characterization studies investigating the interaction between MSCs and their three-dimensional environment. Without such information, many questions remain to be answered. How does the supporting scaffold mediate the behavior of the MSC? What is the ideal delivery vehicle? And, what role does cell density play on the ability of MSCs to stimulate functional bone regeneration? Similarly, how does the local mechanical environment affect cellular differentiation and matrix production?

The optimal use of this technology may require knowledge of how the *in vivo* mechanical environment influences cellular differentiation and mineralized matrix synthesis. Although little is known about the specific effect of mechanical stimulation on mesenchymal stem cells in a three-dimensional architecture, several studies have established that mechanical factors play a role in tissue differentiation during fracture repair (Aro and Chao, 1993; Carter et al., 1988; Goodship and Kenwright, 1985; Goodship et al., 1998) and serve as an adaptive stimulus during normal bone remodeling (Rubin and Lanyon, 1987). Archambault et al. (2000) addressed this issue by casting human MSCs into a bovine collagen I mold and subjecting these constructs to cyclic stretch. The investigators reported that collagen synthesis tripled and the collagen formed had a more organized and crimped orientation parallel to the axis of stretch. This study suggests that MSCs are influenced by their mechanical strain environment, but the implications of these findings for bone tissue engineering are unknown.

Another interesting option might involve predifferentiating MSCs ex vivo into committed osteoprogenitors before reimplantation on a scaffold carrier. MSCs have shown strong osteogenic capacity *in vitro* when cultured in the presence of a standard set of osteogenic supplements, including β-glycerophosphate, L-ascorbic acid-2-phosphate, and the synthetic glucocorticoid dexamethasone. The osteoblastic phenotype is evidenced by a change from a fibroblastic to a cuboidal morphology, expression of alkaline phosphatase, reactivity with anti-osteogenic cell surface monoclonal antibodies, osteocalcin mRNA expression, and mineralized nodule formation (Jaiswal et al., 1997; Kadiyala et al., 1997). This option may be suitable for clinical cases involving a wound bed that has become fibrotic or unstable due to nonunion or other factors. In such instances where the host environment may not be able to provide the factors that normally stimulate osteogenesis, the implantation of committed cells may be beneficial. However, the use of induced MSCs retains the advantages of accessibility and massive expansion that cannot be provided by terminally differentiated cell populations.

Osteoinductive Factors

The incorporation of proteins or genetically engineered cells into a construct that promotes bone regeneration through induction of osteoblast differentiation is a rapidly advancing field and the subject of several current preclinical and clinical trials. In a review by Boden (1999), three strategies for inducing bone regeneration are described: extraction and purification of osteoinductive proteins from animal or human bone, synthesis of recombinant osteoinductive proteins, and gene therapy approaches in which the DNA encoding an osteoinductive protein is delivered to a defect site.

Urist (1965) first demonstrated osteoinduction by implanting demineralized bone matrix into a muscle pouch in rodents, a site that would not normally form bone. Unidentified proteins in the demineralized bone matrix induced MSCs from the surrounding tissue to migrate into the implant, proliferate, differentiate, and synthesize cartilage and bone. This landmark study opened the door for the discovery of an entire family of secretory osteoinductive proteins called bone morphogenetic proteins (BMPs). Cloning and sequencing of specific BMPs has led to the development of recombinant proteins (rhBMPs), several of which also demonstrate the ability to induce cartilage and bone formation *in vivo* (Wang et al., 1990; Wozney and Rosen, 1998; Wozney et al., 1988).

BMP extractions and rhBMPs have been shown to heal critical-sized segmental defects and fuse spines in animal models ranging from rats to nonhuman primates (Boden, 1999). Cook et al. (1994) implanted rhBMP-7 within demineralized bone matrix (DBM) into 2.5-cm canine segmental defects and tested the torsional strength of the defect compared with normal contralateral controls. Defects filled with DBM alone failed to heal, whereas all defects receiving DBM + rhBMP-7 constructs were radiographically bridged by 8 weeks. At 16 weeks, the average torsional strength of healed segmental defects was 72% of normal controls. In a clinical pilot trial, osteoinduction in humans has been reported within interbody spine fusion cages using rhBMP-2 in a collagen sponge delivery scaffold (Boden et al., 2000). Although the sample size was limited, spinal arthrodesis was found to be more consistent in patients receiving the rhBMP-2 construct than in controls treated with bone autograft.

Although currently the least developed of the three approaches, regional gene therapy may ultimately prove to be more effective and less expensive than protein delivery for bone regeneration applications (Boden, 1999). Currently, the state of the art for gene therapy focuses on long-

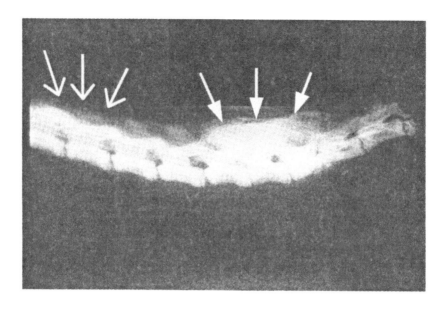

FIGURE 12.5. LIM mineralization protein 1 (LMP-1) induced spine fusion. Lateral radiograph of the thoracolumbar spine of a rat 4 weeks following implantation of devitalized bone matrix loaded with bone marrow cells that were transfected with the cDNA for LMP-1 in either the correct orientation (*solid arrows*) or in a reverse orientation negative control (*open arrows*). Reprinted with permission from Boden SD. *Spine* 23:2486–2492, 1998.)

term and often systemic applications. These are very challenging clinical problems and are typically limited by technological difficulties associated with prolonged gene expression, tissue targeting, transfection efficiency, and immune response to vectors. In contrast, induction of local bone regeneration via transient gene therapy is potentially easier to achieve. Long-term gene expression may not be necessary if the bone regeneration cascade can be initiated with a sufficient nudge by transient expression of an osteoinductive gene. Moreover, local delivery of genes either directly in a scaffold matrix or within transfected cells would address the tissue targeting barrier. Finally, high transfection efficiencies may not be necessary for this application since expression by a few implanted cells may produce enough osteoinductive factors to direct a larger number of host cells near the defect to differentiate into osteoblasts.

The first gene therapy designed for tissue engineering applications involved implanting a gene activated matrix (GAM) composed of a resorbable scaffold combined with plasmid DNA (Fang et al., 1996). A two plasmid GAM encoding BMP-4 and a parathryroid protein fragment

(PTH1-34) yielded increased bone formation within rat segmental defects relative to controls. This technology relies on direct gene transfer to repair cells *in vivo* in order to initiate a local bone formation response. Other investigators have transfected marrow and periosteal cells to overexpress osteoinductive genes *ex vivo* for implantation *in vivo*. DBM constructs containing cells infected with adenovirus expressing BMP-2 have been shown to induce ectopic bone formation and heal segmental defects in rats as well as fuse spines in rabbits (Lieberman et al., 1998, 1999; Riew et al., 1998). Breitbart et al. (1999) demonstrated similar success healing critical size cranial defects using periosteal cells retrovirally transduced with BMP-7 and seeded on PGA scaffolds. Finally, Boden et al. (1998a,b) recently identified a novel osteoinductive protein, LIM mineralization protein 1 (LMP-1) that was induced by glucocorticoid treatment early during *in vitro* osteoblast differentiation (Boden et al., 1999). Marrow cells transfected with LMP-1 and delivered *in vivo* using a DBM carrier successfully induced spine fusion in athymic rats, whereas the carrier alone was ineffective (Fig. 12.5) (Boden et al., 1998).

Osteoinductive strategies for bone regeneration involving local delivery of proteins or genes have immense potential, but several important obstacles remain to clinical implementation. Issues such as the optimal dose and scaffold carrier matrix remain to be settled. The initial enthusiasm for osteoinductive proteins has been tempered somewhat by observations that the required concentration for bone induction in nonhuman primates is an order of magnitude higher than that required in rodents (Boden, 1999). Another major barrier has been identifying a resorbable carrier matrix that optimally delivers osteoinductive factors while providing mechanical support at load-bearing sites. As the design of bone regeneration constructs becomes more sophisticated, the interactions of the various construct elements must be considered. For example, the release kinetics of certain scaffold carriers may allow lower total doses of osteoinductive proteins to be used to achieve functional bone regeneration. Including osteoprogenitor cells within the scaffold carrier may also reduce the required therapeutic protein dose (Hollinger and Winn, 1999). Site-specific variations in response and patient variability due to age, disease, and other factors will almost certainly impact construct design parameters. While technical challenges remain, osteoinductive factors are certainly poised be an integral component of emerging strategies to engineer bone regeneration for a wide variety of clinical applications.

Test Beds for Bone Tissue Engineering Technologies

Numerous *in vivo* and *in vitro* test bed systems have been utilized to evaluate bone tissue engineering technologies. Mineralized nodule formation in two-dimensional cell culture has been used extensively to identify the osteogenic capacity of various cell types and osteoinductive properties of growth factors (Boden et al., 1997). Recent studies have reported culturing bone cells on three-dimensional substrates (Du et al., 1999; Ishaug-Riley et al., 1998). Glowacki et al. (1998) introduced a novel culture system providing media perfusion through three-dimensional collagen sponges seeded with murine marrow cells. Other recent work has confirmed the intuitive notion that three-dimensional cellular interactions differ significantly from those observed in two-dimensional culture systems (Kale et al., 2000). *In vitro* model systems have some significant advantages over *in vivo* model systems in terms of experimental efficiency and control. However, future efforts should prioritize development of novel *in vitro* model systems that account for the three-dimensional, load-bearing functional requirements of bone as well as validation of which aspects of the *in vivo* response may be predicted by *in vitro* systems.

In vivo studies of bone regeneration have typically utilized critical size defects in animal models ranging from rodents to nonhuman primates (Hollinger and Kleinschmidt, 1990). A requirement for any *in vivo* model used to test bone repair constructs is that there be little or no healing in the untreated control defects (Einhorn, 1999). The model should also be well controlled, reproducible, and provide quantitative outcome variables. In addition to defect models in the calvaria and long bones of various species, spine fusion and fracture healing models have also been used with success. Bone regeneration technologies designed for a specific application must ultimately be tested in a model system that closely mimics the clinical situation (Einhorn, 1999).

Bone chamber models have long been used to study the bone repair process under controlled *in vivo* conditions (Sandison, 1924). A recently developed bone chamber model that closely resembles spine fusion cages incorporates a large internal sampling volume for bone regeneration and the ability to apply a cyclic mechanical stimulus to the tissue forming within the chamber (Guldberg et al., 1997). This model has been utilized to investigate mechanisms of adaptation during bone repair and is now being applied to evaluate bone regeneration constructs under controlled *in vivo* mechanical loading conditions (Fig. 12.6). Although the influence of the local *in vivo* mechanical environment on bone repair and remodeling is well recognized, this potentially important factor has not been previously studied in the context of tissue engineered constructs. An understanding of how mechanical signals influ-

No Load Control   Cyclic Mechanical Loading

FIGURE 12.6. Hydraulic bone chamber model (*top*) and microtomography images showing the effects of 12 weeks of cyclic compressive mechanical loading on de novo bone formation (*bottom*). Loaded samples possessed increased volume fraction, trabecular thickness, connectivity, orientation with respect to the axis of loading, and apparent modulus compared to no load controls. (Reprinted with permission from Guldberg RE. *J. Bone Min. Res.* 12:1295–1302.)

ence construct integration may provide microstructural objectives for designing and manufacturing three-dimensional scaffold architectures. Furthermore, it may be possible to exploit the adaptive potential of bone by mechanically preconditioning constructs or cells prior to implantation.

Conclusions and Future Perspectives

Although technical challenges certainly remain, the future of bone tissue engineering technologies is very promising with tangible benefits to pa-

tients on the near horizon. Numerous questions remain regarding the relative cost, efficacy, and safety of the various approaches. However, several of the strategic approaches described above will likely prove to have significant advantages over the current clinical gold standard of autologous bone grafting. Advancements in biology have perhaps surged ahead of engineering over the last decade. As bone regeneration constructs become more sophisticated, critical design trade-offs will need to be made in the selection and integration of parameters. For example, the scaffold material and architecture that provide optimal mechanical properties may not be ideally suited for delivery of cells and proteins or facilitating the invasion of host cells and vascularity from the surrounding tissue bed. The design of bone regeneration constructs is a very challenging optimization problem that will require an even greater level of collaboration between biologists, material scientists, engineers, and clinicians than currently exists. Well-characterized test beds will be needed ranging from efficient initial screening models for proof of concept evaluations to well-controlled experimental systems for optimization of dosage levels and elucidation of interactions and potential synergisms between design parameters to validated *in vivo* models that closely mimic human clinical responses. More than one strategy will likely be needed to treat the wide variety of conditions and patients requiring restoration of functional bone. The approaches that eventually achieve clinical realization will not only need to consistently regenerate bone, but will also need to be cost effective, easy to use by surgeons, and flexible to meet the substantial demands of biological variability.

References

Archambault M.P., Peter S.J., Young R.G. 2000. Effects of cyclic tension on matrix synthesis and tissue formation in a three-dimensional culture system. Orthopedic Research Society Annual Meeting.

Aro H.T., Chao E.Y.S. 1993. Bone healing patterns affected by loading, fracture fragment stability, fracture type, and fracture site compression. *Clin. Orthop.* 293:8–17.

Ashton B.A., Eaglesom C.C. Bab I., Owen M.E. 1984. Distribution of fibroblastic colony-formiing cells in rabbit bone marrow and assay of their osteogenic potential an *in vivo* diffusion chamber method. *Calcif. Tissue Int.* 36:83–86.

Aubin J.E. 1999. Osteoprogenitor cell frequency in rat bone marrow stromal populations: role for heterotypic cell-cell interactions in osteoblast differentation. *J. Cell. Biochem.* 72:396–410.

Awad H.A., Butler D.L., Harris M.T., Ibrahim R.E., Wu Y., Young R.G., Kadiyala S., Boivin G.P. 2000. *In vitro* characterization of mesenchymal stem cell-seeded collagen scaffolds for tendon repair: effects of initial seeding density on contraction kinetics. *J. Biomed. Mater. Res.* 51:233–240.

Beresford J.N. 1989. Osteogenic stem cells and the stromal system of bone and marrow. *Clin. Orthop.* 240:270–280.

Boden S.D. 1999. Bioactive factors for bone tissue engineering. *Clin. Orthop.* 367S:S84–S94.

Boden S.D., Hair G., Titus L., Racine M., McCuaig K., Wozney J.M., Nanes M.S. 1997. Glucocorticoid-induced differentiation of fetal rat calvarial osteoblasts is mediated by bone morphogenetic protein-6. *Endocrinology* 138:2820–2828.

Boden S.D., Liu Y., Hair G.A., Helms J.A., Hu D., Racine M., Nanes M.S., Titus L. 1998a. LMP-1, a LIM-domain protein, mediates BMP-6 effects on bone formation. *Endocrinology* 139:5125-5134.

Boden S.D., Titus L., Hair G.A., Liu Y., Viggeswarapu M., Nanes M.S., Baranowski C. 1998b. Lumbar spine fusion by local gene therapy with a cDNA encoding a novel osteoinductive protein (LMP-1). *Spine* 23:2486–2492.

Boden S.D., Zdeblick T.A. Sandhu H.S., Heim S.E. 2000. The use of rhBMP-2 in interbody fusion cages. Definitive evidence of osteoinduction in humans: a preliminary report. *Spine* 25:376–381.

Boyan B.D., Lohmann C.H., Romero J., Schwartz Z. 1999. Bone and cartilage tissue engineering. *Tissue Eng.* 94:627–645.

Breitbart A.S., Grande D.A., Kessler R., Ryaby J.T., Fitzsimmons R.J., Grant R.T. 1998. Tissue engineered bone repair of calvarial defects using cultured periosteal cells. *Plast. Reconstr. Surg.* 101:567–574.

Breitbart A.S., Grande D.A., Mason J.M., Barcia M., James T., Grant R.T. 1999. Gene-enhanced tissue engineering: applications for bone healingusing cultured periosteal cells transduced retrovirally with the BMP-7 gene. *Ann. Plast. Surg.* 42:488–495.

Bruder S.P., Fox, B.S. 1999. Tissue engineering of bone. *Clin. Orthop.* 367S:S68–S83.

Bruder S.P., Jaiswal N., Haynesworth S.E. 1997. Growth kinetics, self-renewal, and the osteogenic potential of purified human mesenchymal stem cells during extensive subcultivation and following cryopreservation. *J Cell Biochem.* 64:278–294.

Bruder S.P., Kraus K.H., Goldberg V.M., Kadiyala S. 1998. The effect of implants loaded with autologous mesenchymal stem cells on the healing of canine segmental bone defects. *J. Bone Joint Surg. Am.* 80:985–996.

Carlevaro M.F., Cermelli S., Cancedda R., Cancedda F.D. 2000. Vascular endothelial growth factor (VEGF) in cartilage neovascularization and chondrocyte differentiation: auto-paracrine role during endochonral bone formation. *J. Cell. Sci.* 113:59–69.

Carter D.R., Blenman P.R., Beaupre G.S. 1988. Correlations between mechanical stress history and tissue differentiation in initial fracture healing. *J. Orthop. Res.* 6:736–748.

Case N.D., Duty A.O., Edison L.J., Ratcliffe A., Guldberg, R.E., 2000 Osseointegration of tissue-engineered cartilage constructs *in vivo.* Tissue Engineering Workshop, Davos, Switzerland. Also presented at the 46th Annual Meeting of the Orthopaedic Research Society, Orlando, Florida.

Connolly J.F., Guse R, Tiedman J., Dehne R. 1991. Autologous marrow injection as a substitute for operative grafting of tibial nonunions. *Clin. Orthop.* 266:259–270.

Cook S.D., Baffes G.C., Wolfe M.W., Sampath T.K., Rueger D.C. 1994. Recombinant human bone morphogenetic protein-7 induces healing in a canine long-bone segmental defect model. *Clin. Orthop.* 301:302–312.

Cornell C.N. 1999. Osteoconductive materials and their role as substitutes for autogenous bone grafts. *Orthop. Clin. North Am.* 30:591–598.

Daniels A.U., Turner J.L., Trieu J.L., Kemnitzer J.E. 2000. Effects of cyclic loads in saline on mechanical properties of bioabsorbable polymers. Society for Biomaterials, Sixth World Biomaterials Congress Transactions.

Davy D.T. 1999. Biomechanical issues in bone transplantation. *Orthop. Clin. North Am.* 30:553–563.

D'Ippolito G., Schiller P.C., Ricordi C., Roos B.A., Howard G.A. 1999. Age-related osteogenic potential of mesenchymal stromal stem cells from human vertebral bone marrow. *J. Bone Miner. Res.* 14:1115–1122.

Du C., Cui F.Z., Zhu D.X., de Groot K. 1999. Three-dimensional nan-Hap/collagen matrix loading with osteogenic cells in organ culture. *J. Biomed. Mater. Res.* 44:407–415.

Ducheyne P., El-Ghannam A., Shapiro I. 1994. Effect of bioactive glass templates on osteoblast proliferation and *in vitro* synthesis of bone-line tissue. *J. Cell. Biochem.* 56:162–167.

Ducy P., Zhang R., Geoffroy V., Ridall A.L., Karsenty G. 1997. Osf2/Cbfa1: a transcriptional activator of osteoblast differentiation. *Cell* 89: 747–754.

Einhorn T.A. 1999. Clinically applied models of bone regeneration in tissue engineering research. *Clin. Orthop.* (367 Suppl):S59–67.

Elgendy H.M., Norman M.E., Keaton A.R., Laurencin C.T. 1993. Osteoblast-like cell (MC3T3-E1) proliferation on bioerodible polymers: an approach towards the development of a bone-bioerodible polymer composite material. *Biomaterials* 14: 263–269.

Fang J., Zhu Y.Y., Smiley E., Bonadio J., Rouleau J.P., Goldstein S.A., McCauley L.K., Davidison B.L., Roessler B.J. 1996. Stimulation of new bone formation by direct transfer of osetogenic plasmid genes. *Proc. Natl. Acad. Sci. U.S.A.* 93:5753–5758.

Garcia A.J., Ducheyne P., Boettiger D. 1998. Effect of surface reaction stage on fibronectin-mediated adhesion of osteoblast-like cells to bioactive glass. *J. Biomed. Mater. Res.* 40(1):48–56.

Garcia A.J., Vega M.D., Boettiger D. 1999. Modulation of cell proliferation and differentiation through substrate-dependent changes in fibronectin confirmation. *Mol. Biol. Cell* 10:1–14.

Gebhart M., Lane J. 1991. A radiographical and biomechanical study of demineralized bone matrix implanted into a bone defect of rat femurs with and without bone marrow. *Acta Orthop. Belg.* 57: 130–143.

Glowacki J., Mizuno S., Greenberger J.S. 1998. Perfusion enhances functions of bone marrow stromal cells in three-dimensional culture. *Cell Transplant.* 3:319–326.

Goodship A.E., Kenwright J. 1985. The influence of induced micromovement upon the healing of experimental tibial fractures. *J. Bone Joint Surg. Am.* 67B(4):650–655.

Goodship A.E., Cunningham J.L., Kenwright J. 1998. Strain rate and timing of stimulation in mechanical modulation of fracture healing. *Clin. Orthop.* (355 Suppl):S105–115.

Grimsrud C.D., Romano P.R., D'Souza M., Puzas J.E., Reynolds P.R., Rosier R.N., O'Keefe R.J. 1999. BMP-6 is an autocrine stimulator of chondroycte differentation. *J. Bone Miner. Res.* 14: 475–486.

Guldberg R.E., Caldwell N.J., Guo X.E., Goulet R.W., Hollister S.J., Goldstein S.A. 1997 Mechanical stimulation of tissue repair in the hydraulic bone chamber. *J. Bone Miner. Res.* 12:1295–1302.

Haddock S.M., Debes J.C., Nauman E.A., Fong K.E., Arramon Y.P., Keaveny T.M. 1999. Structure-function relationships for coralline hydroxyapatite bone substitute. *J. Biomed. Mater. Res.* 47:71–78.

Haynesworth S.E., Baber M.A., Caplan A.I. 1992. Cell surface antigens on human marrow-derived mesenchymal cells are detected by monoclonal antibodies. *Bone* 13:69–80.

Hollinger J.O., Kleinschmidt J.C. 1990. The critical size defect as an experimental model to test bone repair materials. *J. Craniofac. Surg.* 1:60–68.

Hollinger J.O., Winn S.R. 1999. Tissue engineering of bone in the craniofacial complex. *Ann. N.Y. Acad. Sci.* 875:379–385.

Hollister S.J., Levy R.A., Chu T.-M., Halloran J.W., Feinberg S.E. 2000. An image-based approach for desgining and manufacturing craniofacial scaffolds. *Int. J. Oral Maxillofac. Surg.* 29:67–71.

Huibregtse B.A., Johnstone B., Goldberg V.M., Caplan A.I. 2000. Effect of age and sampling site on the chondro-osteogenic potential of rabbit marrow-derived mesenchymal progenitor cells. *J. Orthop. Res.* 18:18–24.

Hutmacher D.W. 2000. Scaffolds in tissue engineering bone and cartilage. *Biomaterials* 21:2529–2543.

Inada M., Yasui T., Nomura S., Miyake S., Deguchi K., Himeno M., Sato M., Yamagiwa H., Kimura T., Yasui N, Ochi T., Endo N., Kitamura Y., Kishimoto T., Komori T. 1999. Maturational disturbance of chondrocytes in Cbfa1-deficient mice. *Dev. Dyn.* 214:279–290.

Ishaug S.L., Crane G.M., Miller M.J., Yasko A.W., Yaszemski M.J., Mikos A.G. 1997. Bone formation by three-dimensional stromal osteoblast culture in biodegradable polymer scaffolds. *J. Biomed. Mater. Res.* 36:17–28.

Ishaug-Riley S.L., Crane-Kruger G.M., Yaszemski M.J., Mikos, A.G. 1998. Three-dimensional culture of rat calvarial osteoblasts in porous biodegradable polymers. *Biomaterials* 19:1405–1412.

Jaiswal N., et al., 1997. Osteogenic differentiation of purified, culture-expanded human mesenchymal stem cells *in vitro*. *J. Cell. Biochem.* 64:295–312.

Kadiyala S., Neelam J., Bruder S.P. 1997. Culture expanded, bone marrow-derived mesenchymal stem cells can regenerate a critical-sized segmental bone Defect. *Tissue Eng.* 3:173–185.

Kale S., Biermann S., Edwards C., Tarnowski C., Morris M., Long M.W. 2000. Three-dimensional cellular development is essential for ex vivo formation of human bone. *Nat. Biotechnol.* 18:954–958.

Kavalkovich K.W., Boynton R., Murphy J.M., Barry F.P. 2000. Effect of cell density on chondrogenic differentiation of mesenchymal stem cells. Orthopedic Research Society Annual Meeting, Orlando, FL.

Keaveny T.M., Wachtel E.F., Kopperdahl D.L. 1999. Mechanical behavior of human trabecular bone after overloading. *J. Orthop. Res.* 17:346–353.

Kellomaki M., Niiranen H., Puumanen K., Ashammakhi N., Waris T., Tormala P. 2000. Bioabsorbable scaffolds for guided bone regeneration and generation. *Biomaterials* 21:2495–2505.

Kenley R.A., Yim K., Abrams J., Ron E., Turek T., Marden L.J., Hollinger J.O. 1993. Biotechnology and bone graft substitutes. *Pharm. Res.* 10:1393–1401.

Khan S.N., Tomin E, Lane J.M. 2000. Clinical applications of bone graft substitutes. *Orthop. Clin. North Am.* 31:389–398.

Komori T., Yagi H., Nomura S., Yamaguchi A., Sasaki K., Deguchi K., Shimizu Y., Bronson R.T.,

Gao Y.H., Inada M., Sato M., Okamoto R., Kitamura Y., Yoshiki S., Kishimoto T. 1997. Targeted disruption of Cbfa1 results in a complete lack of bone formation owing to maturational arrest of osteoblasts. *Cell* 89:755–764.

Lane J.M., Yasko A.W., Cole T.E., Waller B.J., Brown M., Turek M., Gross J. 1999. Bone marrow and recombinant human bone morphogenetic protein-2 in osseous repair. *Clin. Orthop.* 361:216–227.

Laurencin C.T., Attawia M.A., Elgendy H.E., Herbert K.M. 1996. Tissue engineered bone-regeneration using degradable polymers: the formation of mineralized matrices. *Bone* 19(1 Suppl):93S–99S.

Lian J.B., McKee M.D., Todd A.M., Gerstenfeld L.C. 1993. Induction of bone-related proteins, osteocalcin and osteopontin, and their matrix ultrastructural localization with development of chondrocyte hypertrophy *in vitro*. *J. Cell Biochem.* 52:206–219.

Lieberman J.R., Le L.Q., Wu L., Finerman G.A., Berk A., Witte O.N., Stevenson S. 1998. Regional gene therapy with a BMP-2 producing murine stromal cell line induces heterotopic and orthotopic bone formation in rodents. *J. Orthop. Res.* 16:330–339.

Lieberman J.R., Daluiski A., Stevenson S., Wu L., McAllister P., Lee Y.P., Kabo J.M., Finerman G.A., Berk A.J., Witte O.N. 1999. The effect of regional gene therapy with bone morphogenetic protein-2-producing bone-marrow cells on the repair of segmental fermoral defects in rats. *J. Bone Joint Surg. Am.* 81:905–917.

Mackay A.M., Beck S.C., Murphy J.M., Barry F.P., Chichester C.O., Pittenger M.F. 1998. Chondrogenic differentiation of cultured human mesenchymal stem cells from marrow. *Tissue Eng.* 4:415–428.

Marra K.G., Szem J.W., Prashant N.K., DiMilla P.A., Weiss L.E. 1999. *In vitro* analysis of biodegradable polymer blend/hydroxyapatite composites for bone tissue engineering. *J. Biomed. Mater. Res.* 47:324–355.

Michel M.C., Guo X.D., Gibson L.J., McMahon T.A., Hayes W.C. 1993. Compressive fatigue behavior of bovine trabecular bone. *J. Biomech.* 26:453–463.

Mikos A.G., Sarakinos G., Leite S.M., Vacanti J.P., Langer R. 1993. Laminated three-dimensional biodegradable foams for use in tissue engineering. *Biomaterials.* 14(5):323–330.

Mow V.C., Hayes W.C. 1997. Biomechanics of cortical and trabecular bone: implications for assessment of fracture risk. In: *Basic Orthopaedic Biomechanics,* 2nd ed., pp. 69–111.

Nade S., Burwell R.G. 1977. Decalcified bone as a substrate for osteogenesis. An appraisal of the interrelation of bone and marrow in combined grafts. *J. Bone Joint Surg. (Br)* 59:189–196.

Nishida S., Endo N., Yamagiwa H., Tanizawa T., Takahashi H.E. 1999. Number of osteoprogenitor cells in human bone marrow markedly decreases after skeletal maturation. *J. Bone Miner Metab.* 17:171–177.

Peter S.J., Miller M.J., Yasko A.W., Yaszemeski, Mikos A.G. 1998. Polymer concepts in tissue engineering. *J. Biomed. Mater. Res.* 43:422–427.

Peter S.J., Livingston T., Gordon S., Yagami M., Klyushnenkova E., McIntosh K., Wagner J., Elkalay M., Young R.G., Kadiyala S. 2000. Bone formation via allogeneic mesenchymal stem cell implantation. In: World Congress of Biomaterials Conference. Vol. 1, Kamuela, Hawaii. p. 104.

Pittenger M.F., Mackay A.M., Beck S.C., Jaiswal R.K., Douglas R., Mosca J.D., Moorman M.A., Simonetti D.W., Craig S., Marshak D.R. 1999. Multilineage potential of adult human mesenchymal stem cells. *Science* 284:143–147.

Porter B.D., Oldham J.B., He S.L., Zorbitz M.E., Payne R.G., Currier B.L., Mikos A.G., Yaszemski M.J. 2000. Mechanical properties of a biodegradable bone regeneration scaffold. *J. Biomech. Eng.* 122:286–288.

Richards M., Goulet R., Schaffler M.B., Goldstein S.A. 1999 Marrow-derived progenitor cell injections enhance new bone formation during distraction. *J. Orthop. Res.* 17:900–908.

Riew K.D., Wright N.M., Cheng S., Avioli L.V., Lou J. 1998. Induction of bone formation using a recombinant adenoviral vector carrying the human BMP-2 gene in a rabbit spinal fusion model. *Calcif. Tissue Int.* 63:357–360.

Rogers J.J., Young H.E., Adkison L.R., Lucas P.A., Black A.C. 1995. Differentiation factors induce expression of muscle, fat, cartilage, and bone in a clone of mouse pluripotent mesenchymal stem cells. *Am. Surg.* 61:231–236.

Rubin C.T., Lanyon L.E. 1987. Osteoregulatory nature of mechanical stimuli: function as a determinant for adaptive remodeling. *J. Orthop. Res.* 5:300–310.

Salama R., Burwell R.D., Dickson I.R. 1973. Recombined grafts of bone and marrow. The benfficial effect upon osteogenesis of impregnating xenograft (heterograft) bone with autologous red marrow. *J. Bone Surg. (Br.)* 55:401–417.

Sandison J.C. 1924. A new method for the microscopic study of living growing tissues by the introduction of a transparent chamber in the rabbit's ear. *Anat. Rec.* 28:281–287.

Sasaki T., Watanabe C. 1995. Simulation of osteoinduction in bone wound healing by high-molecular hyaluronic acid. *Bone* 16:9–15.

Smultz W.P., Daniels A.U., Andriano K.P., France E.P., Heller J. 1991. Mechanical test methodology for environmental exposure testing of biodegradable polymers. *J. Appl. Biomat.* 2:13–22.

Solchaga L.A., Dennis J.E., Goldberg V.M., Caplan A.I. 1999. Hyaluronic acid-based polymers as cell carriers for tissue-engineered repair of bone and cartilage. *J. Orthop. Res.* 17:205–213.

Spector M. 1994. Anorganic bovine bone and ceramic analogs of bone mineral as implants to facilitate bone regeneration. *Clin. Plast. Surg.* 21:437–444.

Thomas C.H., McFarland C.D., Jenkins M.L., Rezania A., Steele J.G., Healy K.E. 1997. The role of vitronectin in the attachment and spatial distribution of bone-derived cells on materials with patterned surface chemistry. *J. Biomed. Mater. Res.* 37:81–93.

Thomson R.C., Yaszemski M.J., Powers J.M., Mikos A.G. 1998. Hydroxyapatite fiber reinforced poly(alpha-hydroxy ester) foams for bone regeneration. *Biomaterials* 19:1935–1943.

Thomson R.C., Mikos A.G., Beahm E., Lemon J.C., Satterfield W.C., Aufdemorte T.B., Miller M.J. 1999. Guided tissue fabrication from periosteum using preformed biodegradable polymer scaffolds. *Biomaterials* 20:2007–2018.

Urist M.R. 1965. Formation by autoinduction. *Science* 150:893–899.

Vacanti C.A., Kim W. Upton J., Mooney D., Vancanti J.P. 1995. The efficacy of periosteal cells compared to chondroycytes in the tissue engineered repair of bone defects. *Tissue Eng.* 1:301–308.

Wakitani S., Goto T., Pineda S.J., Young R.G., Mansour J.M., Caplan A.I., Goldberg V.M. 1994. Mesenchymal cell-based repair of large, full-thickness defects of articular cartilage. *J. Bone Joint Surg. Am.* 76:579–592.

Wang E.A., Rosen V., D'Alessandro J.S., Bauduy M., Cordes P., Harada T., Israel D.I., Hewick R.M., Kerns K.M., LaPan P., Luxemberg D.P., McQuaid D., Moutsatsos I.K., Nove J., Wozney J.M. 1990. Recombinant human bone morphogenetic protein induces bone formation. *Proc. Natl. Acad. Sci. U.S.A.* 87:2220–2224.

Whang K., Healy K.E., Elenz D.R., Nam E.K., Tsai D.C., Thomas C.H., Nuber G.W., Glorieux F.H., Travers R., Sprague S.M. 1999. Engineered bone regeneration with bioabsorbable scaffolds with novel microarchitecture. *Tissue Eng.* 5:35–51.

Wozney J.M., Rosen V. 1998. Bone morphogenetic protein and bone morphogenetic protein gene family in bone formation and repair. *Clin. Orthop.* 346:26–37.

Wozney J.M., Rosen V., Celeste A.J., Mitsock L.M., Whitters M.J., Kriz R.W., Hewick R.M., Wang E.A. 1988. Novel regulators of bone formation: molecular clones and activities. *Science* 242: 1528–1534.

Yaszemski M.J., Payne R.G., Hayes W.C., Langer R., Mikos A.G. 1996. Evolution of bone transplantation: molecular, cellular and tissue strategies to engineer human bone. *Biomaterials* 17:175–185.

Yoshikawa T., Ohgushi H., Tamai S. 1996. Immediate bone forming capability of prefabricated osteogenic hydroxyapatite. *J. Biomed. Mater. Res.* 32:481–492.

Zhang R., Ma P.X. 1999a. Poly(alpha-hydroxyl acids) hydroxyapatite porous composites for bone-tissue engineering. I. Preparation and morphology. *J. Biomed. Mater. Res.* 44:446–455.

Zhang R., Ma P.X. 1999b. Porous poly (L-lactic acid)/apatite composites created by biomimetic process. *J. Biomed. Mater. Res.* 45:285–293.

13

Tissue Engineering of Bone: The Potential Use of Gene Therapy for Difficult Bone Loss Problems

Brett Peterson and Jay R. Lieberman

Introduction

Bone has a unique ability to regenerate and heal without scar. In contrast, muscles, tendons, and ligaments heal by formation of scar tissue which is inferior to native tissue with respect to mechanical and biological properties. Both skeletogenesis and fracture healing involve a complicated cascade of bone-specific inductive factors and other signals that orchestrate cell migration, differentiation, and angiogenesis. Investigators have proposed that fracture healing recapitulates embryonic skeletal development (1). In both cases, a scaffold, whether a mesenchymal cartilaginous anlage or a fracture callus with periosteum and bone fragments, serves to guide vascular ingrowth and migration of osteogenic precursor cells. The biological matrix delivers growth factors and provides a surface for cell adhesion molecules to promote cell-matrix and cell-cell interactions. Osteogenic precursor cells proliferate, migrate, and eventually differentiate into osteoblasts. The osteoblasts then synthesize their own matrix and are incorporated into bone tissue. The goal of tissue engineering is to mimic these processes of either skeletogenesis or fracture healing and to produce a bone that has appropriate biological and biomechanical properties. This review summarizes the current progress toward the use of gene transfer techniques to facilitate the tissue engineering of bone.

Recently, solutions to bone loss problems have included the use of autograft bone, allograft bone, or Ilizarov bone transport techniques (2). Each of these techniques has a fairly high success rate in specific clinical situations as well as significant limitations. Autograft bone is currently the gold standard for bone graft material. Autologous bone contains osteogenic precursor cells, multiple growth factors, and its collagen and hydroxyappatite extracellular matrix are osteoconductive. However, autograft bone exists in limited supply. The harvesting of autograft bone has a significant associated morbidity (3, 4), and increases operating room time. Allograft bone avoids all of these limitations but has the inherent risk of disease transmission and the possibility of inducing an immune response. Allograft bone has been used successfully to treat small bone defects such as those that result after the elevation of the joint surface in an impacted tibial plateau fracture or in a defect resulting from the curettage of a benign bone tumor (5). However, allograft bone has limited osteoinductive potential and presently cannot be recommended for the treatment of large bone defects or in an environment that has limited regenerative potential (Figs. 13.1 and 13.2). Finally, the Ilizarov technique of bone transport is an innovative and effective treatment for major trauma to long bones (6, 7). Using the Ilizarov technique, the surgeon makes a corticotomy in the bone that is to be lengthened and applies an expandable ex-

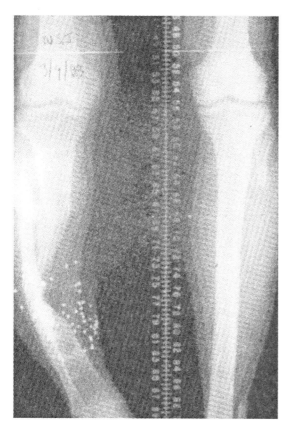

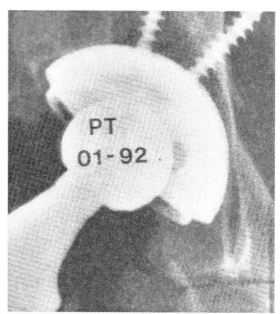

FIGURE 13.2. Example of a bone defect resulting from osteolysis secondary to wear debris from a total hip replacement.

FIGURE 13.1. Example of a bone defect in a tibia resulting from a shotgun blast. This large defect would require a more sophisticated method of treatment.

ternal fixator. When callus forms at the osteotomy site, the external fixator is expanded, the corticotomy callus is stretched, and ultimately the bone is lengthened (Fig. 13.3). This technique is highly specialized, requires extensive experience, and is limited by factors such as pin tract infections, patient compliance, and the fact that the callotasis takes a significant amount of time.

The need for tissue-engineered bone or a universal bone graft substitute is frequently encountered in orthopedic surgery. Difficult bone loss problems are associated with trauma, revision total joint arthroplasty, or oncologic resections. In these cases, the bone stock loss may be so extensive that conventional autologous bone grafting alone may not provide an adequate biological or mechanical solution. In addition, often times the patient's ability to induce an osteoinductive response is limited because of age (8), nicotine use (9), glucocorticoid use (10), or systemic conditions such as diabetes and peripheral vascular disease (11, 12). There is a wide spectrum of bone loss problems that require treatment. Some bone repair problems may be treated with autogenous bone graft or a bone graft substitute. However, more sophisticated tissue engineering strategies need to be developed to treat patients with extensive bone stock loss, particularly cases where there is compromise of the inherent mechanical structure of the bone and the vascularity of the surrounding soft tissue.

Bone Growth Factors

The components required for successful bone repair include osteoprogenitor cells, an adequate blood supply, a biological scaffold or matrix, and an osteogenic signal. The pioneering experiments by Urist were an attempt to isolate and characterize bone inducing molecules (13, 14).

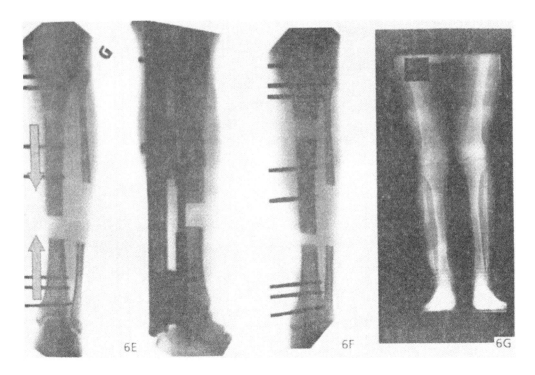

FIGURE 13.3. Distraction osteogenesis by the Ilizarov method. This method is being used to treat a bone defect resulting from trauma to the distal tibia. A proximal osteotomy is made in the tibia, and the resulting callus is stretched. (Adapted with permission from The British Council; Saleh M et al. Limb reconstruction after high energy trauma. British Medical Bulletin 1999. 55(4) p 870 figure 6.)

These experiments lead to the discovery of a family of bone morphogenetic proteins and to the cloning of bone morphogenetic proteins (BMPs) (15–17). The addition of certain bone growth factors such as BMP2 to demineralized bone matrix (DBM) enhances bone formation in a dose-dependent fashion (18). As more BMPs have been cloned, the recombinant proteins of other members such as BMP-7, also known as osteogenic protein-1 (OP-1), have also been shown to induce bone formation (19, 20). Both BMP-2 and OP-1 have been used to heal critical sized defects in a variety of animal models including rats (21), rabbits (22), canine (23), sheep (24), and non human primates (25) (Fig. 13.4). In humans, the success of recombinant proteins for bone healing has been more variable (26). In two clinical trials, significantly larger doses of BMP than would be expected were required to induce adequate bone repair (27, 28).

The reason for the discrepancy between the animal studies and the clinical trials is unclear.

In the clinical trials, a collagen sponge served as the carrier of the recombinant BMP-2 and OP-1. The collagen carrier was chosen because it was already approved for use in humans by the Food and Drug Administration (FDA), and surgeons were familiar with the material. However, collagen does not provide any biomechanical stability at the bone defect site. Furthermore, the preclinical studies were carried out in young, healthy animals with well-vascularized soft tissue beds. In patients, formation of bone may be limited by an inadequate blood supply, insufficient bone stock, or excessive fibrous or scar tissue. Furthermore, a single exposure to a recombinant protein may not produce an adequate osteogenic signal to induce bone repair (29).

Recently, the FDA approved the use of BMP-2 in a metallic cage for anterior spinal fusions, and the manufacturer of OP-1 has been granted a humanitarian device exemption to treat recalcitrant non-unions in long bones. The use of these recombinant proteins is an appealing alternative to

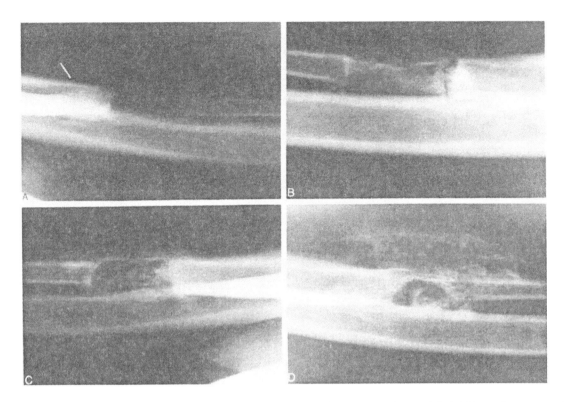

FIGURE 13.4. Recombinant human OP-1 used to treat ulnar defects in a canine model. (*A*) Zero micrograms of rhOP-1. (*B*) 625 micrograms of rhOP-1 (*C*) 1200 micrograms of rhOP-1. (*D*) 2500 micrograms of rhOP-1. All x-rays were taken at 16 weeks. (Adapted with permission from Lippincott, Williams and Wilkins; *Clinical Orhopedics and Related Research* 1994. p 305 figure 1.)

autogenous bone graft. However, the large doses of growth factor that are currently required to induce bone in humans suggests that growth factor delivery still needs to be optimized.

Carriers and the Delivery of Osteoinductive Factors

Osteoinductive factors are delivered on a carrier that functions to limit the diffusion of the factor from the desired site and to provide an osteoconductive matrix (30, 31). The requirements for a carrier are multiple and a complete discussion is beyond the scope of this review, but, in general, a good carrier must be biocompatible with the host and have similar biomechanical properties to the tissue being replaced (32). In addition, the carrier must be porous to allow for vascular ingrowth and for the transplant to be incorporated into the host. Theoretically, an ideal carrier would also be osteoconductive and provide a temporary extracellular matrix for cell ingrowth (33). Finally, the carrier must biodegrade at a rate that does not inhibit the repair process, and the breakdown products must not be toxic to the surrounding cells or soft tissue.

Most experience has been with collagen-based scaffolds such as demineralized bone matrix or collagen sponges. Krebsbach et al. have shown that carrier types such as polyvinyl sponges, hydroxyapatite, tricalcium phosphate, and poly (L-lactic acid) as well as the type of cell implanted on the carrier are critical factors with respect to whether bone formation occurs or not (34). In addition to the chemical structure of the carrier, the physical properties of the carrier such as the particle size affect the amount of bone that is formed (35). Finally, the pharmacokinetics of protein release, particularly the initial retention time, significantly influence the amount of bone that is produced for a given dose of growth factor (36). The ideal carrier for delivery of bone growth factors has not been determined, but once delivery systems are optimized, recombinant growth factors will become an even more effective option for treating bone defects.

Cell-Based Strategies

As an alternative to using large doses of recombinant proteins, some investigators have pursued cell-based strategies for the tissue engineering of bone. In cell-based strategies, bone precursor cells are seeded onto a carrier and placed into a bone defect. The most effective cell type appears to be bone marrow-derived mesenchymal stem cells (MSC) (37). Bone marrow is known to contain osteogenic precursor cells (38), and the success of autograft bone may stem from the fact that it supplies osteogenic precursor cells as well as osteoconductive and osteoinductive factors. The use of autologous bone graft is an example of a cell-based tissue engineering strategy that has been used for years. Traditionally, the use of autologous bone marrow has required open surgery for both the harvesting of the graft and for the insertion of the bone marrow into the bone defect site. Investigators have developed percutaneous techniques for the harvesting of bone marrow as well as its delivery to bone defects (39). Connolly et al. reported the healing of atrophic non-unions by aspirating bone marrow from the iliac crest and injecting it into tibial non-union sites (40).

However, unprocessed autologous marrow often does not offer an adequate quantity of osteoprogenitor cells, particularly in sick or elderly patients (41, 42). To increase the number of progenitor cells, bone marrow cells can be harvested, and techniques have been developed to produce a purified population of mesenchymal cells (43). These progenitor cells can be expanded over one billion-fold without losing pluripotency (44). The MSCs have demonstrated the ability to induce sufficient bone formation without any additional osteoinductive treatment. Bruder et al. have reported the healing of critically sized bone defects in rats (45) and dogs (46) by using a ceramic carrier seeded

with purified bone marrow-derived MSCs (Fig. 13.5). Culture-expanded MSCs from humans were implanted on ceramic blocks and also resulted in the healing of critically sized defects in nude rats (47). However, it remains to be determined whether the mesenchymal cells alone will be able to heal a bone defect in a clinical situation, particularly in elderly patients or those with compromised soft tissues.

It is possible that these cell-based therapies will be sufficient for the tissue engineering of bone in clinical situations. As stated previously, autologous bone grafts, whether delivered by open or percutaneous techniques, are currently important techniques for treating bone defects. The amount of bone graft can be augmented by mixing it with carriers such as demineralized bone matrix (48), but the limited supply of bone marrow is still a significant problem. The more recently developed technology of culture-expanded MSCs is quite promising. It is possible that these bone marrow-derived stem cells, particularly if combined with recombinant osteoinductive proteins, will enhance bone formation in patients. Such strategies may represent a new, safe, and effective means to treat bone defect problems clinically.

Gene Therapy

An even more sophisticated strategy for bone tissue engineering would be to combine growth factor delivery with a cell-based strategy via genetic modification. The goal of gene therapy is to convert the target cell's own biochemical machinery into *in vivo* bioreactors for recombinant protein production (49). It is postulated that an endogenously synthesized protein may have greater biological activity because of appropriate post-translational modifications (50). An osteoinductive signal delivered by this approach might be more physiologic and more effective. In addition, bone repair is particularly amenable to gene transfer strategies when compared with the treatment of diseases associated with genetic mutations such as cystic fibrosis. The transgene production is usually required only for a period of weeks to months until the bone defect is healed, and not for the life of the patient.

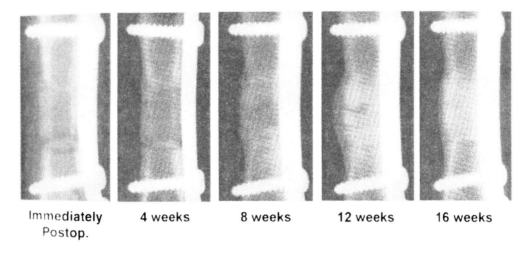

Immediately   4 weeks   8 weeks   12 weeks   16 weeks
Postop.

FIGURE 13.5. Canine segmental defect treated with a ceramic cylinder loaded with bone marrow-derived mesenchymal stem cells. (Adapted with permission from *J. Bone Joint Surg.* -Bruder S.P. et al. *J. Bone Joint Surg.* 1998. 80A p 989 figure 4.)

Depending on the vector chosen, gene delivery allows a brief or a more sustained production of the osteoinductive protein. Vectors used for gene therapy may be simple, such as a plasmid, which contains the cDNA of the gene of interest, or the vectors may be derived from a virus, which generally have higher transduction efficiencies. The most common viral delivery vehicles for gene therapy include adenoviruses, adeno-associated viruses, retroviruses, and lentiviruses.

Recombinant adenoviral vectors can be produced at high titers, efficiently infect a wide range of cell types, and remain episomal. In the treatment of cystic fibrosis, attempts at achieving a clinically useful gene therapy have been disappointing because of transient expression of the gene of interest (51). The reason for this transient expression is that the immune system recognizes native viral proteins that are produced along with the transgene. Thus, any cells that are infected by the recombinant adenovirus become targeted by the host immune system. However, in fracture non-unions and other bone defects, transient expression of the exogenous gene may be all that is required (52–54). Because the immune response to adenoviral vectors remains a potentially problematic issue, investigators have developed second-generation adenoviral vectors. These second-generation vectors produce fewer native adenoviral proteins, and

helper dependent adenoviral vectors have had even more of the native adenoviral proteins deleted (55). Experience with helper-dependent adenoviruses for bone formation is limited at this time, but these vectors can successfully deliver osteoinductive signals (56).

Retroviral vectors have also been used to deliver osteoinductive proteins (57, 58) and to heal bone defects (59). Retroviruses insert themselves into the genome of host cells and thus potentially have a longer duration of expression than vectors that remain episomal. Expression of the transgene that has been introduced by retrovirus does decrease over time (60), but this should not be a significant problem for applications requiring localized bone healing. The limitation of retroviral vectors lies in their transduction efficiency. Experiments involving hematopoietic stem cells have demonstrated that retroviral vectors that are all based on the Moloney murine leukemia virus (MMLV) are most efficient at infecting mouse cells (61). The transduction of larger animal and human cells remains problematic (62). Recently, higher transduction efficiencies have been achieved by treating stem cells with fibronectin and various cytokines such as thrombopoietin to increase the frequency of stem cells active in the cell cycle (63, 64). If higher transduction efficiencies with retroviral vectors become possible, it will make the vectors more attractive. However, the problem of random insertion into the host genome and the risk of tumorigenesis will need to be addressed.

In contrast to retroviral vectors, HIV-1-based lentiviruses are able to enter the nucleus of non-dividing cells (65). This ability allows lentiviral vectors to transduce quiescent cells, and efficiencies of approximately 60% in hematopoietic stem cells from human cord blood have been reported (66). Lentiviral vectors show a great deal of promise, but further studies are needed to investigate their efficacy and safety.

In addition to viral vectors, investigators have used nonviral methods such as liposomes, gene guns, electroporation, and naked DNA techniques to deliver recombinant genes into cells. These different gene deliver methods vary with respect to their efficiency of transduction as well as the duration of transgene production. A gene-activated matrix is a novel nonviral gene delivery technique that has been used successfully to enhance bone repair in animal models. The technique involves a biological carrier that has been impregnated with a plasmid carrying the cDNA of an osteoinductive molecule. Gene-activated matrices will be discussed in more detail later in this chapter.

Although the techniques and vectors used for gene therapy vary, the two general approaches are the direct or *in vivo* approach and the indirect or ex vivo approach. In the direct approach, a recombinant vector carrying the transgene of interest is injected at a specific anatomic site with the resulting transduction of the local surrounding cells. In the ex vivo approach, precursor cells are harvested, expanded in culture, transduced *in vitro,* and finally reimplanted in the host at the desired anatomic location. The direct approach is simpler and potentially less expensive. However, the direct approach is associated with a lower tranfection efficiency, and this technique does not allow for the selection or the expansion of a specific target cell population. Furthermore, the *in vivo* approach carries a greater risk of systemic migration of the virus. There are also concerns about the inflammatory response to the native adenoviral proteins that are produced in addition to the transgene product.

The ex vivo approach involves the harvesting and growing of target cells in culture. The cell culture step may be expensive and time consuming, but this step allows for the transfection of a specific target cell population. In addition, the efficiency of infection is enhanced with this approach (Fig. 13.6)

Gene Therapy for the Tissue Engineering of Bone In Vivo

Direct injection of adenovirus containing BMP-2 has induced bone formation in a number of animal models (67, 68). Using the *in vivo* approach, Baltzer et al. healed femoral defects in New Zealand white rabbits with a first-generation adenovirus containing the BMP-2 cDNA (69). They noted bone healing in the experimental group by 7 weeks and were able to detect extended expression of the transgene. Control femurs showed no bony union by 12 weeks, and all samples were assessed by x-ray, histologic, and biomechanical analysis.

The ex vivo gene therapy approach has also been used successfully in a femoral defect model. Lieberman et al. used a BMP-2 containing adenovirus to enhance bone marrow cell osteogenic potential (70). Healing rates and quality of healing of the femoral defects were compared between groups treated with either recombinant human BMP-2 versus groups treated with bone marrow cells infected with a BMP-2 containing adenovirus. Histomorphometric analysis revealed a more robust bone formation in femoral defects treated with BMP-2 producing bone marrow cells compared with recombinant BMP-2 alone (Figs. 13.7 and 13.8) Although biomechanical testing of the healed femurs did not show any differences between these groups, the histologically more robust bone formed by the BMP-2 producing cells might have been the result of a more continuous or physiologic release of the BMP-2 protein. It is also possible that autocrine loops might be important in producing a more vigorous osteogenic response (71). Bone marrow cells transduced with the adenovirus can respond to the BMP that they are overexpressing.

In spine surgery, a common complication is inadequate bone formation after attempted

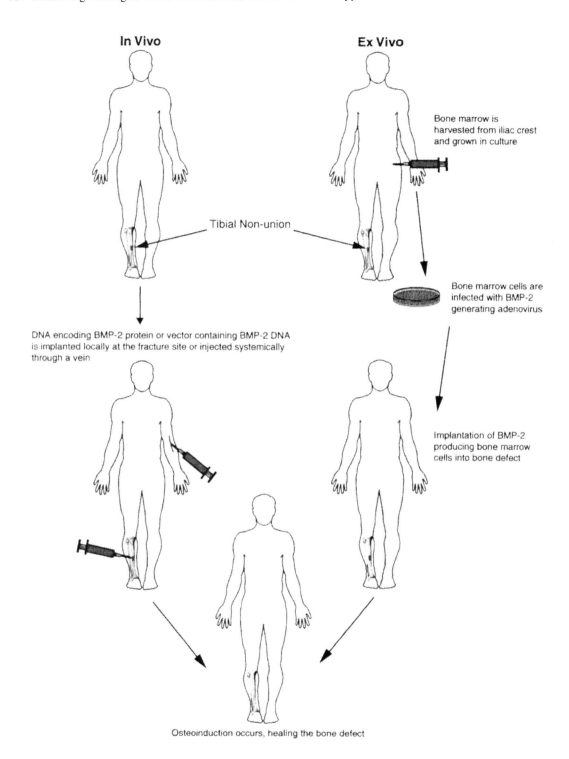

FIGURE 13.6. Ex vivo versus *in vivo* gene therapy strategies. (Adapted with permission from W.B. Saunders Co.; Scaduto et al. Orthop. *Clinics of North America* 1999. 30(4) p 627 figure 1.)

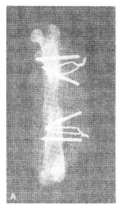 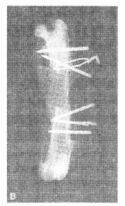

RBM-Ad-BMP-2   rh BMP-2 (20μg)

 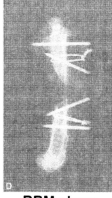 

RBM AD-LacZ   RBM alone   DBM alone

FIGURE 13.7. Critical sized femoral defects in Lewis rats. (*A*) Defect treated with rat bone marrow cells infected with BMP-2 containing adenovirus. (*B*) Defect treated with 20 micrograms of recombinant human BMP-2. (*C*) Defect treated with rat bone marrow cells infected with β-galactosidase containing adenovirus. (*D*) Defect treated with rat bone marrow cells alone. (*E*) Defect treated with demineralized bone matrix carrier alone. (Adapted with permission from *J. Bone Joint Surg.;* Liberman J.R. et al. *J. Bone Joint Surg.* 1999 81(7) p. 908. figure 3.)

arthrodesis. This results in pseudoarthrosis rates that have been reported to be as high as 40% (72). These clinical difficulties make the goal of using gene therapy to augment spine fusion even more desirable. In an animal model of spine fusion, Alden et al. first demonstrated that an *in vivo* approach could be used to induce bone formation in the paraspinous muscle of rats (73). The BMP-2 containing adenovirus, induced bone when it was percutaneously injected into nude rats. This treatment did not result in solid fusions. The same group were later able to demonstrate solid fusions at 16 weeks after in-

jecting a BMP-9 containing adenovirus into the paraspinous musculature of nude rats (74).

In an ex vivo approach with a BMP-2 containing adenovirus, Riew et al. reported bone formation in the lumbar spine region of New Zealand White rabbits. Autologous bone marrow cells infected with a BMP-2 containing adenovirus and implanted around the intertransverse process area resulted in new bone formation but no fusion (75). BMP-2 overexpressing bone marrow cells have also been used in our laboratory to successfully achieve intertransverse spine fusions in rats (76). Autologous bone marrow cells were

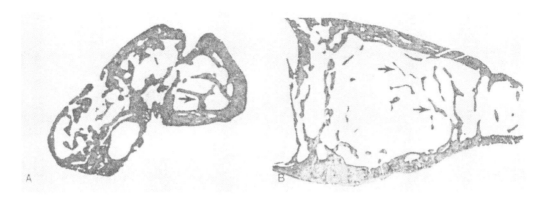

FIGURE 13.8. Histologic specimens of bone formed by BMP-2 producing bone marrow cells (*A*) and bone formed from rhBMP-2. Note the thicker trabeculae and more robust bone formation in femoral defects treated with BMP-2 producing bone marrow cells. (Adapted with permission from *J. Bone Joint Surg.;*—Liberman J.R. et al. *J. Bone Joint Surg* 1999 81(7) p. 911. figure 4.)

harvested from Lewis rats, cultured, and infected with a second-generation recombinant adenovirus. Demineralized bone matrix or collagen sponges were then impregnated with the infected cells and implanted into rats at the fusion site. The marrow cells infected with the BMP-2 containing adenovirus resulted in fusions in 15 of 15 spines that were implanted. None of the negative controls fused.

Boden et al. used an ex vivo approach with the cDNA for a novel osteoinductive transcription factor, LIM mineralization protein-1 (LMP-1). LMP-1 is an intracellular signaling molecule that is a putative transcription factor and has been shown to activate a cascade resulting in the secretion of bone morphogenetic proteins (77). The cDNA of LMP-1 was transfected into bone marrow cells and fused 9 of 9 lumbar spines in athymic rats whereas in the control group, with the anti-sense cDNA of LMP-1, none of the spines fused (78) (Fig. 13.9). More recently, the LMP-1 cDNA was subcloned into an adenoviral vector and used to infect bone marrow derived cells. Eight of eight rabbits treated with the bone marrow cells infected with the recombinant adenovirus fused while none of the negative controls developed a bone fusion mass (79). Although this strategy shows great clinical potential, further investigation of the biology of the LMP-1 protein is necessary. No knockout mouse or transgenic mouse has been reported,

and little is known of the possible side-effects of LMP-1 overexpression.

Bone marrow is not the only source of osteogenic precursor cells. Genetically altered fibroblasts, myoblasts, and fat derived mesenchymal cells have been able to produce bone *in vivo*. Krebsbach et al. harvested human gingival fibroblasts and rat skin fibroblasts, infected them with a BMP-7 containing adenovirus. With this technique, the group demonstrated *in vivo* bone formation in SCID mice (80). They also demonstrated that the bone formed was a chimera of the genetically altered fibroblasts and host osteoprogenitor cells induced by the recombinant BMP-7. Muscle can also be a source of osteoprogenitor cells. Lee et al. infected muscle derived cells with a BMP-2 containing adenovirus and healed calvarial defects in SCID mice (81). Using fluorescent in situ hybridization techniques, the group showed that a small percentage of the muscle derived cells that were implanted into the calvarial defects had differentiated into osteoblasts *in vivo*. Finally, Zuk et al. obtained human adipose tissue by liposuction. They demonstrated a fibroblast-like population of cells which behave like mesenchymal stem cells (MSC) (82). The fat-derived MSC, grown in the appropriate media, differentiated into bone, cartilage, fat, and muscle phenotypes *in vitro*. The full osteogenic potential of these cells has not yet been characterized *in vivo*.

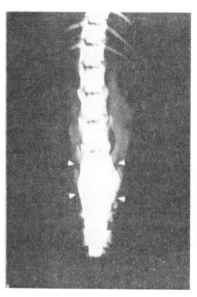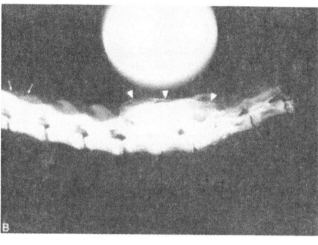

FIGURE 13.9. Nude rat spine fusion model. X-ray taken 4 weeks after implantation of LMP-1 transfected bone marrow cells. AP (*A*) and lateral (*B*) views. (Adapted with permission from Lippincott Williams and Wilkins; Boden S.D. et al. *Spine* 1998(23) p. 2486–2492 figure 1.)

However, it is known that the fat-derived MSC form robust bone *in vivo* when they are infected with BMP-2 containing adenovirus and implanted in a SCID mouse hindlimb (83).

Overall, the preclinical results of these gene therapy strategies are encouraging. But there have been problems involving the immune response to adenoviral vectors. Musgrave et al. used an adenovirus to deliver BMP-2 into the muscles of mice. Injection of the BMP-2 adenovirus into the triceps of immunocompetent mice produced significantly less bone than when injected into SCID mice (84). Analogously, another group induced ectopic bone by injecting a BMP-2 containing adenovirus into thigh muscles of nude rats, but bone formation was inhibited in immunocompetent Sprague-Dawley rats (85). Finally, in Wistar rats, Okubo et al. demonstrated bone induction after injecting a BMP-2 containing adenovirus but only when their rats were immunosuppressed with cyclophosphamide (86).

Another problem with the use of viral vectors for gene therapy relates to the safety of the host. While a full discussion is beyond the scope of this chapter, there is a concern of tumorigenesis from viral vectors that integrate into the host genome. This poses potential problems for the use of retroviruses or lentiviruses. In the case of adenoviruses that remain episomal, there is concern with the potential for systemic spread and toxic side effects. Even though there are over 300 gene therapy trials registered with the FDA with a few of these being clinical phase III trials (87), safety concerns are crictical as evinced by the "Gelsinger Case" in Philadelphia. At the University of Pennsylvania, a teenager with ornithine-transcarbamylase defiency died 4 days after being given a high dose of a recombinant adenovirus systemically (88). If gene therapy is to be used to treat a bone defect, an extremely high margin of safety must be assured.

To avoid the immune response as well as possible systemic toxicity one alternative strategy is the use of a gene-activated matrix (GAM) as a gene delivery vehicle. Each GAM consists of a collagen sponge or other biodegradeable carrier that has been impregnated with plasmid encoding the cDNA for an osteoinductive protein. The GAM is placed into a specific anatomic site where the goal is to turn the local fibroblasts into bioreactors that secrete the specific protein product (89). Fang et al. reported the healing of

5-mm femoral defects in Sprague-Dawley rats using a collagen sponge impregnated with a plasmid containing the BMP-4 cDNA or a fragment of parathyroid hormone cDNA (coding for amino acids 1–34) (90). Bony union occurred when GAMs containing either of these plasmids were implanted. Healing occurred more rapidly (4 weeks versus 9 weeks) if both cDNAs were included in the GAM suggesting a synergistic effect between BMP-4 and PTH1-34. In further studies, a GAM containing PTH1-34 was tested in a critically sized defect in a canine model (91). New bone formation was noted, and gene expression was detected at 6 weeks after implantation. However, the critical sized defects did not heal. The limited effectiveness of the GAM in this preclinical trial could have stemmed from the choice of PTH1-34 as the osteoinductive molecule; its anabolic effects on bone occurs when it is administered in pulsed doses (92). Also, most cells take up and express plasmid DNA inefficiently. It is likely that fibroblast- or myoblast-like precursors are the cell type responsible for the efficacy of gene activated matrices (93). If these cells can be better identified, they might be more specifically targeted and improve the efficiency of this nonviral strategy.

Conclusion

Bone loss and inadequate bone repair remain significant clinical problems. These problems exist as a spectrum ranging from small defects that can be treated by traditional methods such as autograft bone, recombinant osteoinductive protein, or a bone graft substitute. The recombinant proteins will be even more effective if carrier systems can be developed to optimize delivery of the growth factors. Cell-based strategies involving mesenchymal stem cells will also likely be used to treat bone defects in the near future. The cell based strategies may be synergistic when combined with recombinant proteins to heal bone defects.

There is currently no consistently satisfactory solution for the treatment of large bone defects. Perhaps a tissue engineering strategy that employs gene transfer technology will provide the answer to such difficult clinical problems. Gene

therapy offers the most sophisticated but also the most challenging biologically based treatment for bone defects. For gene therapy to become a viable clinical option, we will need to learn more about the biology of bone repair, the contribution of transduced progenitor cells to bone healing, the duration of recombinant protein production *in vivo*, and the short-term and long-term risks of using recombinant viral vectors in humans. In the future, orthopedic surgeons will have a number of options to treat bone repair problems including autologous bone graft, bone graft substitutes, recombinant proteins, cell-based strategies, and gene transfer strategies. The treatment selected will be based on the age of the patient, the patient's overall medical health, the severity of the bone loss, and the quality of the surrounding soft tissues. The ultimate goal is to develop a comprehensive tissue engineering strategy to enhance bone repair.

References

1. Ferguson C., Alpern E., Miclau T., Helms J.A. 1999. Does adult fracture repair recapitulate embryonic skeletal formation? *Mech. Dev.* 87:57–66.
2. Paley D., Maar D.C. 2000. Ilizarov bone transport treatment for tibial defects. *J. Orthop. Trauma.* 14:76–85.
3. Colterjohn N.R., Bednar D.A. 1997. Procurement of bone graft from the iliac crest. An operative approach with decreased morbidity. *J. Bone Joint Surg. Am.* 79:756–759.
4. Seiler J.G. 3rd, Johnson J. 2000. Iliac crest autogenous bone grafting: donor site complications. *J. South Orthop. Assoc.* 9:91–97.
5. Wilkins R.M. 2000. Unicameral bone cysts. *J. Am. Acad. Orthop. Surg.* 8:217–224.
6. Einhorn T.A., Glowacki J., Brighton C.T., Lane J.M., Friedlander G. eds. 1994. In: *Bone Formation and Repair.* American Academy of Orthopedic Surgeons. Distraction osteogenesis. pp. 235–270.
7. Saleh M., Yang L., Sims M. 1999. Limb reconstruction after high energy trauma. *Br. Med. Bull.* 55:870–884.
8. Buckwalter J.A. 1997. Maintaining and restoring mobility in middle and old age: the importance of the soft tissues. *Instr. Course Lect.* 46:459–469.
9. Porter S.E., Hanley E.N. Jr. 2001. The musculoskeletal effects of smoking. *J. Am. Acad. Orthop. Surg.* 9:9–17.

10. Sawin P.D., Dickman C.A., Crawford N.R., Melton M.S., Bichard W.D., Sonntag V.K. 2001. The effects of dexamethasone on bone fusion in an experimental model of posterolateral lumbar spinal arthrodesis. *J Neurosurg.* 94(1 Suppl):76–81.

11. Kagel E.M., Majeska R.J., Einhorn T.A. 1995. Effects of diabetes and steroids on fracture healing. *Curr. Opin. Orthop.* 6(5):7–13.

12. Perlman M.H., Thordarson D.B. 1999. Ankle fusion in a high risk population: an assessment of nonunion risk factors. *Foot Ankle Int.* 20:491–497.

13. Urist M.R., Nogami H., Mikulski A. 1976. A bone morphogenetic polypeptide. *Calcif. Tissue Res.* 21(Suppl):81–87.

14. Urist M.R. 1965. Bone: formation by autoinduction. *Science* 150:893–899.

15. Celeste A.J., Iannazzi J.A., Taylor R.C., Hewick R.M., Rosen V., Wang E.A., Wozney J.M. 1990. Identification of transforming growth factor β family members present in bone-inductive protein purified from bovine bone. *Proc. Natl. Acad. Sci. U.S.A.* 87:9843–9847.

16. Wozney J.M., Rosen V., Celeste A.J., Mitsock L.M., Whitters M.J., Kriz R.W., Hewick R.M., Wang E.A. 1988. Novel regulators of bone formation: molecular clones and activities. *Science.* 242:1528–1534.

17. Ozkaynak E., Rueger D.C., Drier E.A., Corbett C., Ridge R.J., Sampath T.K., Oppermann H. 1990. OP-1 cDNA encodes an osteogenic protein in the TGF-beta family. *EMBO J.* 9:2085–2093.

18. Schwartz Z., Somers A., Mellonig J.T., Carnes D.L. Jr., Wozney J.M., Dean D.D., Cochran D.L., Boyan B.D. 1998. Addition of human recombinant bone morphogenetic protein-2 to inactive commercial human demineralized freeze-dried bone allograft makes an effective composite bone inductive implant material. *J. Periodontol.* 69:1337–1345.

19. Cook S.D., et al. 1994. Recombinant human bone morphogenetic protein-7 induces healing in a canine segmental defect model. *Clin. Orthop.* 301:302–312.

20. Wang E.A. 1990. Recombinant human morphogenetic protein induces bone formation. *Proc. Natl. Acad. Sci. U.S.A.* 87:2220–2224.

21. Yasko A.W., Lane J.M., Fellinger E.J., Rosen V., Wozney J.M., Wang E.A. 1992. The healing of segmental bone defects, induced by recombinant human bone morphogenetic protein (rhBMP-2). A radiographic, histological, and biomechanical study in rats. *J. Bone Joint Surg. Am.* 74:659–670.

22. Cook S.D., Baffes G.C., Wolfe M.W., Sampath T.K., Rueger D.C., Whitecloud T.S. 3rd. 1994. The effect of recombinant human osteogenic protein-1

on healing of large segmental bone defects. *J. Bone Joint Surg. Am.* 76:827–838.

23. Cook S.D., Baffes G.C., Wolfe M.W., Sampath T.K., Rueger D.C. 1994. Recombinant human bone morphogenetic protein-7 induces healing in a canine long-bone segmental defect model. *Clin. Orthop.* (301):302–312.

24. Gerhart T.N., Kirker-Head C.A., Kriz M.J., Holtrop M.E., Hennig G.E., Hipp J., Schelling S.H., Wang E. 1993. Healing segmental femoral defects in sheep using recombinant human bone morphogenetic protein. *Clin. Orthop.* (293):317–326.

25. Cook S.D., Wolfe M.W., Salkeld S.L., Rueger D.C. 1995. Effect of recombinant human osteogenic protein-1 on healing of segmental defects in non-human primates. *J. Bone Joint Surg. Am.* 77:734–750.

26. Sakou T. 1998. Bone morphogenetic proteins: from basic studies to clinical approaches. *Bone.* 22:591–603.

27. Friedlaender G.E., Perry C.R., Cole J.D., Cook S.D., Cierny G., Muschler G.F., Zych G.A., Calhoun J.H., LaForte A.J., Yin S. 2001. Osteogenic protein-1 (bone morphogenetic protein-7) in the treatment of tibial nonunions. *J. Bone Joint Surg. Am.* 83-A Suppl 1(Pt 2):S151–158.

28. Boden S.D., Zdeblick T.A., Sandhu H.S., Heim S.E. 2000. The use of rhBMP-2 in interbody fusion cages. Definitive evidence of osteoinduction in humans: a preliminary report. *Spine.* 25:376–381.

29. Cook S.D., et al. 1995. In vivo evaluation of recombinant osteogenic protein-1 on healing of segmental defects in non-human primates. *J. Bone Joint Surg. Am.* 77:734.

30. Oakes D.A., Lieberman J.R. 2000. Osteoinductive applications of regional gene therapy: ex vivo gene transfer. *Clin. Orthop.* 379S:101–112.

31. Takakoa K., et al. 1988. Ectopic bone induction on and in porous hydroxyapatite combined with collagen and bone morphogenetic protein. *Clin. Orthop.* 234:250–254.

32. Mankani M.H., Kuznetsov S.A., Fowler B., Kingman A., Robey P.G. 2001. In vivo bone formation by human bone marrow stromal cells: effect of carrier particle size and shape. *Biotechnol. Bioeng.* 72:96–107.

33. Kadiyala S., Jaiswal N., Bruder S.P. 1997. Culture expanded bone marrow derived mesenchymal stem cells can regenerate a critical sized segmental bone defect. *Tissue Eng.* 3:173–185.

34. Krebsbach P.H., Kuznetsov S.A., Satomura K., Emmons R.V., Rowe D.W., Robey P.G. 1997. Bone formation in vivo: comparison of osteogenesis by transplanted mouse and human marrow stromal fibroblasts. *Transplantation* 63:1059–1069.

35. Mankani M.H., Kuznetsov S.A., Fowler B., Kingman A., Robey P.G. 2001. In vivo bone formation by human bone marrow stromal cells: effect of carrier particle size and shape. *Biotechnol. Bioeng.* 72:96–107.

36. Uludag H., D'Augusta D., Golden J., Li J., Timony G., et al. 2000. Implantation of recombinant human bone morphogenetic proteins with biomaterial carriers: a correlation between protein pharmacokinetics and osteoinduction in the rat ectopic model. *J. Biomed. Mater. Res.* 50:227–238.

37. Bruder S.P., Fox B.S. 1999. Tissue engineering of bone. Cell based strategies. *Clin. Orthop.* (367 Suppl):S68–83.

38. Beresford J.N. 1989. Osteogenic stem cells and the stromal system of bone and marrow. *Clin. Orthop.* 240:270–280.

39. Connolly J.F. 1998. Clinical use of marrow osteoprogenitor cells to stimulate osteogenesis. *Clin. Orthop.* (355 Suppl):S257–266.

40. Connolly J.F., Guse R., Tiedeman J., Dehne R. 1991. Autologous marrow injection as a substitute for operative grafting of tibial nonunions. *Clin. Orthop.* (266):259–270.

41. Haynesworth S.E., Goshima J., Goldberg V.M., Caplan A.I. 1992. Characterization of cells with osteogenic potential from human bone marrow. *Bone* 13:81–88.

42. Quarto R., Thomas D., Liang T. 1995. Bone progentior cell deficits and the age-associated decline in bone repair capacity. *Calcif. Tissue Int.* 56:123–129.

43. Bruder S.P., Fox B.S. 1999. Tissue engineering of bone. Cell based strategies. *Clin. Orthop.* (367 Suppl):S68–83.

44. Bruder S.P., Jaiswal N., Haynesworth S.E. 1997. Growth kinetics, self-renewal and the osteogenic potential of purified human mesenchymal stem cells during extensive subcultivation and following cryopreservation. *J. Cell Biochem.* 64:278–294.

45. Bruder S.P., Kurth A.A., Shea M., et al. 1998. Bone regeneration by implantation of purified, culture expanded human mesenchymal cells. *J. Orthop. Res.* 17:155–162.

46. Bruder S.P., Kraus K.H., Godberg V.M., Kadiyala S. 1998. The effect of implants loaded with autologous mesenchymal stem cells on the healing if canine segmental bone defects. *J. Bone Joint Surg.* 80A:985–996.

47. Bruder S.P., Kurth A.A., Shea M., et al. 1998. Bone regeneration by implantation of purified, culture expanded human mesenchymal cells. *J. Orthop. Res.* 17:155–162.

48. Tiedeman J.J., Garvin K.L., Kile T.A., Connolly J.F. 1995. The role of a composite, demineralized bone matrix and bone marrow in the treatment of osseous defects. *Orthopedics* 18:1153–1158.

49. Franceschi R.T., Wang D., Krebsbach P.H., Rutherford R.B. 2000. Gene therapy for bone formation: *in vitro* and *in vivo* osteogenic activity of an adenovirus expressing BMP-7. *J. Cell. Biochem.* 78:476–486.

50. Niyibizi C., et al. 1998. Potential role for gene therapy in the enhancement of fracture healing. *Clin. Orthop.* (355S):148–153.

51. Robbins P.D., Tahara H., Ghivizzani S.C. 1998. Viral vectors for gene therapy. *Tibtech.* 16:35–40.

52. Musgrave D.S., Bosch P., Ghivizzani S., Robbins P.D., Evans C.H., HuardJ. 1999. Adenovirus mediated direct gene therapy with bone morphogenetic protein-2 produces bone. *Bone* 24:541–547.

53. Goldstein S.A., Bonadio J. 1998. Potential role for gene therapy in the enhancement of fracture healing. *Clin. Orthop.* 355(Suppl)S148–153.

54. Riew K.D., Wright N.M., Cheng S., Aviolo L.V., Lou J. 1998. Induction of bone formation using a recombinant adenoviral vector carrying the human BMP-2 gene in a rabbit spinal fusion model. *Calcif Tissue Int.* 63:357–360.

55. Morsy M.A., Caskey C.T. 1999. Expanded-capacity adenoviral vectors—the helper-dependent vectors. *Mol. Med. Today.* 5:18–24.

56. Abe N., Lee Y., Sato M., Zhang J., Mitani K., Lieberman J. 2002. Enhancement of bone repair with a helper dependent BMP2 adenoviral vector. *Trans. Ortho Res. Soc.* March.

57. Engstrand T., et al. 2000. Transient production of bone morphogenetic protein 2 by allogeneic transplanted transduced cells induces bone formation. *Hum Gene Ther.* 11:205–211.

58. Laurencin C.T., Attawia M.A., Lu L.Q., Borden M.D., Lu H.H., Gorum W.J., Lieberman J.R. 2001. Poly(lactide-co-glycolide)/hydroxyapatite delivery of BMP-2-producing cells: a regional gene therapy approach to bone regeneration. *Biomaterials* 22:1271–1277.

59. Breitbart A.S., Grande D.A., Mason J.M., Barcia M., James T., Grant R.T. 1999. Gene-enhanced tissue engineering: applications for bone healing using cultured periosteal cells transduced retrovirally with the BMP-7 gene. *Ann. Plast. Surg.* 42:488–495.

60. Lund A.H., Duch M., Pedersen F.S. 1996. Transcriptional silencing of retroviral vectors. *J. Biomed. Sci.* 3:365–378.

61. Bodine D.M., Karlsson S., Nienhuis A.W. 1989. Combination of interleukins 3 an 6 preserves stem cell function in culture and enhances retrovirus-mediated gene transfer into hematopoietic stem cells. *Proc. Natl. Acad. Sci. U.S.A.* 86:8897–901.

62. Kohn D.B., Nolta J.A., Crooks J.A. 1999. Clinical trials of gene therapy using hematopoietic stem cells. In: Thomas E.D., Blume K.G., Forman S.J., eds. *Hematopoietic Stem Cell Transplantation*, 2nd ed. Boston, MA: Blackwell Scientific Publications, pp 97–102.

63. Kiem H.P., Andrews R.G., Morris J., et al. 1998. Improved gene transfer into baboon marrow repopulating cells using recombinant human fibronectin fragment CH-296 in combination with IL-6, stem cell factor, FLT-3 ligand, and megakaryocyte growth and development factor. *Blood* 92:1878–1886.

64. Dao M.A., Taylor N., Nolta J.A. 1998. Reduction in levels of the cyclin dependent kinase inhibitor p27kip-1 coupled with transforming growth factor neutralization induces cell-cycle entry and increasesretroviral transduction of primitive human hematopoietic cells. *Proc. Natl. Acad. Sci. USA* 95:13006–13011.

65. Naldini L., Blomer U., Gallay P. et al. 1996. In vivo gene delivery and stable transduction of nondividing cells by a lentiviral vector. *Science* 272:263–267.

66. Evans J.T., Kelly P.F., O'Neil E., Garcia J.V. 1999. Human cord blood CD34+CD38− cell transduction via lentivirus based gene transfer vectors. *Hum. Gene Ther.* 10:1479–1489.

67. Baltzer A.W., Lattermann C., Whalen J.D., Wooley P., Weiss K., Grimm M., Ghivizzani S.C., Robbins P.D., Evans C.H. 2000. Genetic enhancement of fracture repair: healing of an experimental segmental defect by adenoviral transfer of the BMP-2 gene. *Gene Ther.* 7:734–739.

68. Alden T.D., Pittman D.D., Hankins G.R., Beres E.j., et al. 1999. In vivo endochondral bone formation using a bone morphogenetic protein 2 adenoviral vector. *Hum. Gene Ther.* 10:2245–2253.

69. Baltzer A.W., Lattermann C., Whalen J.D., Wooley P., Weiss K., Grimm M., Ghivizzani S.C., Robbins P.D., Evans C.H. 2000. Genetic enhancement of fracture repair: healing of an experimental segmental defect by adenoviral transfer of the BMP-2 gene. *Gene Ther.* 7:734–739.

70. Lieberman J.R., Daluiski A., Stevenson S., Wu L., McAllister P., Lee Y.P., Kabo J.M., Finerman G.A., Berk A.J., Witte ON. 1999. The effect of regional gene therapy with bone morphogenetic protein-2-producing bone-marrow cells on the repair of segmental femoral defects in rats. *J Bone Joint Surg. Am.* 81:905–917.

71. Gazit D., et al. 1999. Engineered pluripotent mesenchymal cells integrate and differentiate in regenerating bone: a novel cell-mediated gene therapy. *J. Gene Med.* 1:121–133.

72. Boden S.D. 2000. Biology of lumbar spine fusion and use of bone graft substitutes: present, future, and next generation. *Tissue Eng.* 6:383–399.

73. Alden T.D., Pittman D.D., Beres E.J., Hankins G.R., Kallmes D.F., Wisotsky B.M., Kerns K.M., Helm G.A. 1999. Percutaneous spinal fusion using bone morphogenetic protein-2 gene therapy. *J. Neurosurg.* 90(1 Suppl):109–114.

74. Helm G.A., Alden T.D., Beres E.J., Hudson S.B., Das S., Engh J.A., Pittman D.D., Kerns K.M., Kallmes D.F. 2000. Use of bone morphogenetic protein-9 gene therapy to induce spinal arthrodesis in the rodent. *J. Neurosurg.* 92(2 Suppl):191–196.

75. Riew K.D., Wright N.M., Cheng S., Avioli L.V., Lou J. 1998. Induction of bone formation using a recombinant adenoviral vector carrying the human BMP-2 gene in a rabbit spinal fusion model. *Calcif. Tissue Int.* 63:357–360

76. Wang J., Lee Y., Kanim L., Lieberman J.R. 2002. Intertransverse process fusion in Lewis rats using bone marrow cells infected with a BMP-2 containing adenovirus. *J. Bone Joint Surg.* (In Press).

77. Boden S.D., Liu Y., Hair G.A., Helms J.A., Hu D., Racine M., Nanes M.S., Titus L. 1998. LMP-1, a LIM-domain protein, mediates BMP-6 effects on bone formation. *Endocrinology* 139:5125–5134.

78. Boden S.D., Titus L., Hair G., Liu Y., Viggeswarapu M., Nanes M.S., Baranowski C. 1998. Lumbar spine fusion by local gene therapy with a cDNA encoding a novel osteoinductive protein (LMP-1). *Spine* 23:2486–2492

79. Viggeswarapu M., et al. 2001. Adenoviral delivery of LIM mineralization protein-1 induces new-bone formation *in vitro* and *in vivo*. *J. Bone Joint Surg. Am.* 83-A(3):364–376.

80. Franceschi R.T., Wang D., Krebsbach P.H., Rutherford R.B. 2000. Gene therapy for bone formation: *in vitro* and *in vivo* osteogenic activity of an adenovirus expressing BMP-7. *J. Cell. Biochem.* 78:476–486.

81. Lee J.Y., Musgrave D., Pelinkovic D., Fukushima K., Cummins J., Usas A., Robbins P., Fu F.H., Huard J. 2001. Effect of bone morphogenetic protein-2-expressing muscle-derived cells on healing of critical-sized bone defects in mice. *J. Bone Joint Surg. Am.* 83-A(7):1032–1039.

82. Zuk P.A., et al. 2001. Multilineage cells from human adipose tissue: implications for cell-based therapies. *Tissue Eng.* 7:211–228.

83. Dragoo J.L., Choi J.Y., Hedrick M., Benhaim P. Lieberman, J.R. 2001. Bone induction of genetically manipulated stem cells derived from human fat. *Trans ORS* 27:21.

84. Musgrave D.S., et al. 1999. Adenovirus-mediated direct gene therapy with bone morphogenetic protein-2 produces bone. *Bone* 24:541–547.

85. Alden T.D., Pittman D.D., Hankins G.R., Beres E.J., Engh J.A., Das S., Hudson S.B., Kerns K.M., Kallmes D.F., Helm G.A. 1999. In vivo endochondral bone formation using a bone morphogenetic protein 2 adenoviral vector. *Hum Gene Ther.* 10:2245–2253.

86. Okubo Y., Bessho K., Fujimura K., Iizuka T., Miyatake S.I. 2000. Osteoinduction by BMP-2 via adenoviral vector under transient immunosuppression. *Biochem. Biophys. Res. Comm.* 267:382–387.

87. Scully, D. 2000. Gene therapy. *Clin. Orthop.* 379S:91–107.

88. Rubanyi G.M. 2001. The future of human gene therapy. *Mol Aspects Med.* 22:113–142.

89. Bonadio J. 2000. Tissue engineering via local gene delivery. *J. Mol. Med.* 78:303–311.

90. Fang J. et al. 1996. Stimulation of new bone by direct transfer of osteogenic plasmid genes. *Proc. Natl. Acad. Sci. U.S.A.* 93:5753–5758.

91. Bonadio J., et al. 1999. Localized direct plasmid gene delivery in vivo: prolonged therapy results in reproducible tissue regeneration. *Nat. Med.* 5: 753–759.

92. Morley P., Whitfield J.F., Willick G.E. 1999. Design and applications of parathyroid hormone analogues. *Curr. Med. Chem.* 6:1095–1106.

93. Goldstein S.A., Bonadio J. 1998. Potential role for direct gene transfer in the enhancement of fracture healing. *Clin. Orthop.* 355S:S154–S162.

14

Engineered Skeletal Muscle: Functional Tissues, Organs, and Interfaces

Robert G. Dennis

Introduction

The mission of the functional tissue engineer is to apply the engineering disciplines and information from the basic sciences to produce functional tissues and organs. Tissues and organs have emergent properties that arise from the organization of many cells, often of different type, transcending the functions of individual cells. It is incorrect to assert that a disorganized mass of living cells is a functional tissue. In this regard, the term *functional tissue engineering* suggests an emphasis on the functionality of engineered tissues and organs.

In any complex system, function arises at least in part from high-level organization, or structure. Often, the critical physiological function of a tissue arises from the organization and mechanical properties of the extracellular matrix, and is thus only indirectly related to cellular functions per se. A structure composed of disorganized, undifferentiated cells within a scaffold will not exhibit the complex functionality of a tissue or organ (Okano and Matsuda, 1998a,b; Okano et al., 1997; van Wachem et al., 1996, 1999). As an example, a box full of car parts is not an automobile. Analogously, a gel filled with myoblasts is not muscle tissue.

The engineer should begin with a clear understanding of the desired function of the tissue. To be meaningful for an engineer, a function must be quantifiable, controllable, and useful.

Most tissues perform many functions, and in general it is not yet possible to engineer a tissue that replaces 100% of normal function. In addition, it may be possible to modify an engineered tissue to perform a function that may not be its role under normal circumstances. An example would be to engineer skeletal muscle tissue to produce hormones, such as insulin, to replace the function of a nonmuscle tissue (Vandenburgh et al., 1996, 1998). Thus, engineered tissues include all engineered constructs that perform a useful function arising from the organization of living cells, even if that function is not native to the original tissue from which the cells were harvested.

Skeletal muscle tissue comprises nearly half of the body mass of adult mammals. It generates heat, produces large amounts of intracellular and extracellular proteins, and generates controlled force, work, and power for nearly all movements of the body. The development of skeletal muscle tissue from isolated myogenic precursor cells is reviewed in this text by Faulkner and Dennis (Chapter 5), and the potential use of muscle tissue as a protein factory is discussed by Kosnik, Dennis, and Vandenburgh (Chapter 28). In the context of this chapter, the function of skeletal muscle will be defined as excitability and contractility. Furthermore, the discussion will be limited to self-organizing tissue, in which the cells themselves reorganize in culture to produce a three-dimensional structure without the preexistence of a scaffold in the contractile region of the tissue.

Background

A general background in the area of functional skeletal muscle tissue engineering is presented in Chapter 28 by Kosnik, et al. Briefly, spontaneous contractions of cultured skeletal muscle cells were first reported by Lewis (1915). Though contractile function of cultured skeletal muscle may have been implicitly assumed since that report, it was not until 1991 that the first quantitative measures of the contractility of cultured skeletal muscle were reported (Vandenburgh et al., 1991a). Since that time, several approaches have been developed to promote the organization of muscle satellite cells and fibroblasts in culture into three-dimensional structures that exhibit tissuelike function in terms of excitability and isometric contractility. For any of the current approaches to skeletal muscle tissue engineering, the resulting tissue contractility ranges from 1% to 10% of control values for tissue from adult animals (Kosnik and Dennis,

2000). The low contractility of engineered muscle may be explained by the fact that the muscle fibers are not of adult phenotype, expressing chiefly developmental isoforms of myosin (LaFramboise et al., 1990; Reiser et al., 1988), and the structure, both within the muscle fibers and in the extracellular space, is not as well organized as that of control adult muscle (Fig. 14.1).

Current Research

At present, *in vitro* engineered functional skeletal muscle can be classified as either scaffold based or self-organizing. The scaffold-based

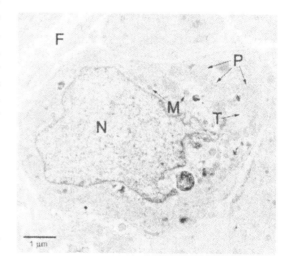

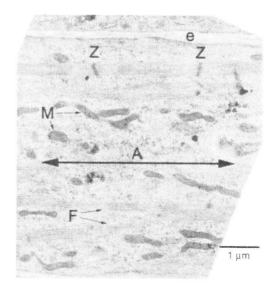

FIGURE 14.1. Electron micrographs of functional engineered skeletal muscle tissue after 31 days in culture. (*a*) Cross-section showing a single myotube with a centrally located nucleus **N**, peripheral assembly of myofibrils **P**, mitochondria **M**, and the formation of a tubule network **T**. Also visible are other less mature myotubes and fibroblasts **F** (plate x89). (*b*) Longitudinal section, showing the orientation of the long axis of the myotube indicated by the double-headed arrow **A**. Force is generated principally along the long axis. Also visible are mitochondria **M**, myofibrils in the process of organization **F**, and extracellular space between the myotubes **e**. The Z lines, which define individual sarcomeres within a myofibril, are also visible **Z**. The indicated sarcomere ranges in length from 2.46 to 2.62 μm. The muscle constructs are shortened by 5% before evaluation, and are fixed at this length, therefore the actual sarcomere length in this case is 2.58–2.75 μm (plate d69). Evidence for the arrested state of development and failure to achieve adult phenotype of the myotubes in the engineered muscle includes the centrally located nuclei and the sparse distribution of myofibrils, as well as the poorly defined Z lines in many sarcomeres.

functional muscle constructs, often referred to as BAMs (Powell et al., 1999; Vandenburgh et al., 1996), employ primary cells from mammals including humans, and can be maintained in culture for several weeks. The current state of research on scaffold-based functional muscle constructs is presented in Chapter 28 by Kosnik, Dennis, and Vandenburgh elsewhere in this text. Self-organizing muscle constructs, termed *myooids* (Dennis and Kosnik, 2000; Kosnik and Dennis, in press) employ mammalian primary cells and laminin—coated suture anchor materials, which serve as artificial tendons. Detailed methods for the cell culture and functional evaluation of myooids have been published (Dennis & Kosnik, 2000, in press; Dennis, et al., in press), and are summarized in Fig. 14.2. Co-culture of fibroblasts with the muscle cells is required to provide adequate extracellular material to mechanically support the cell monolayer during delamination and reorganization into a cylindrical structure (Dennis and Kosnik, 2000; Mayne et al., 1989; Vandenburgh et al., 1991a&b). After reorganization into a three-dimensional structure, the skeletal muscle construct may be handled by the silk suture 'tendons,' attached to a force transducer, and electrically stimulated to produce force.

Technical Challenges

Cell Source and Tissue Structure

In general, tissue engineers must strive to produce self-organizing tissues from primary cells. Established cell lines present many problems, both for implantation and the ultimate function of the engineered tissue. Generally, established cell lines will be of different genetic origin than the host into which they would be implanted, creating the serious problem of rejection by the host immune system. It is possible to learn much from established cell lines, but ultimately all engineered tissues will use primary cells from the eventual recipient organism. In addition, many of the currently available myogenic cell lines exhibit abnormal cellular function (Irintchev et al., 1998; Miller et al., 1991; Turner, 1986), and three-dimensional constructs engineered from

these cells have excitability and contractility inferior to that of tissues engineered from primary cells (Dennis et al., in press).

The use of primary cells is preferable for several reasons. It is generally assumed that both the number and proliferative potential of satellite cells is reduced with age (Schultz and McCormick, 1994). However, adequate numbers of myogenic precursor cells can be readily harvested from the tissues of neonatal, adult, or even extremely aged mammals to produce functional engineered skeletal muscles (Dennis et al., in press). Based upon the resulting tissue mass from muscles engineered from primary cells (Dennis and Kosnik, 2000; Dennis et al., in press), a 10-mg sample of tissue typically yields 1.5–2.0 mg of engineered muscle tissue. Although this at first appears to be an inefficient yield of cells since less tissue is produced than existed originally, it opens the possibility of harvesting muscle from areas of the body with plentiful tissue, such as the legs or torso, reengineering it to the appropriate architecture, and reimplanting the engineered tissue into a site requiring smaller muscle mass, such as the facial musculature. As for all other areas within the field of tissue engineering, the problem of an adequate source of primary cells will be relieved by the eventual development of technologies for harvesting pluripotent stem cells, with subsequent amplification and control of differentiation. Until such a time, skeletal muscle tissue engineers share an advantage with skin tissue engineers: the ready availability of large amounts of easily harvested tissues that require minimally invasive surgical techniques.

Scaffolds

The use of synthetic scaffolds for engineered skeletal muscle tissue has several key advantages and disadvantages. The advantages of scaffolds include the provision of a means by which the engineer may define the initial geometry of the muscle construct, provision of mechanical support for the cells, and the possibility to control cell attachment by the use of adhesion molecules affixed to the scaffold at controlled concentrations (Putnam and Mooney, 1996). In addition, growth factors and other bioactive compounds

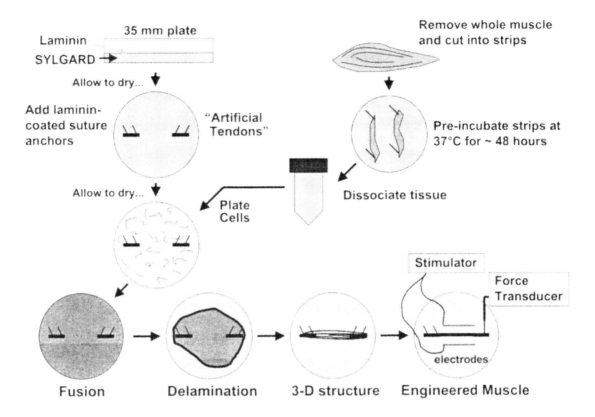

FIGURE 14.2. Schematic of the general process for preparing self-organizing functional engineered skeletal muscles *in vitro*. Cell adhesion to the SYLGARD substrate is controlled by the density of laminin on the surface. After removal from the animal, muscles are cut into strips and preincubated for ~48 hours to activate the satellite cells prior to tissue dissociation. Within 2 to 4 weeks, the monolayer of myotubes and fibroblasts delaminates from the substrate surface and self-organizes into a three-dimensional structure, with segments of surgical suture serving as artificial tendons.

can be delivered to the cells by means of the scaffold (Shea et al., 1999).

Artificial scaffolds have several important disadvantages. Unlike many other musculoskeletal tissues, skeletal muscle is composed of only a very small amount of extracellular matrix (ECM) material, generally less than 2%, whereas other musculoskeletal tissues such as tendon, ligament, and bone, are composed of 30–90% extracellular material (Nordin and Frankel, 1989). The presence of any significant amount of scaffold material in engineered muscle could seriously interfere with the normal organization of the tissue. For example, the presence of artificial extracellular material in an engineered skeletal muscle construct could easily result in internal shunting of the forces generated by the activated muscle cells, making it difficult or impossible to measure the contractility of engi-

neered muscle during development. One approach to the problem of scaffold interference is to employ biodegradable scaffold materials (Putnam and Mooney, 1996). In time, the scaffold would be resorbed, gradually being replaced by the cells and the ECM that they produce.

During differentiation of skeletal muscle, mononuclear myocytes fuse to form polynuclear myotubes, which go on to develop into adult muscle fibers that may be several centimeters in length and contain hundreds of nuclei (Engel and Franzini-Armstrong, 1994). For this to occur, the myocytes must be permitted considerable mobility, to come into physical contact with adjacent cells, align, and fuse. When embedded in currently available scaffold materials, the myocytes are generally prevented from fusing to form long, parallel myotubes. Future developments in scaffold materials and processing techniques may reduce this current limitation, or perhaps may even physically promote the formation of long, parallel myotubes.

One final limitation to the use of current artificial scaffold technologies is that, unlike other musculoskeletal tissues which normally are subjected to mechanical strain below 1%, skeletal muscle regularly experiences mechanical strains >10% during normal function. Over these physiologic strains, the passive force within control muscles arising from the naturally occurring extracellular material is generally quite low, often approximately zero. These special mechanical requirements for the ECM in functional muscle tissue severely restrict the number of candidate scaffold materials. Artificial scaffolds for functional tissue engineered skeletal muscle should occupy less than 2% of the tissue volume, permit a high level of cell mobility, tolerate repeated fully reversed mechanical strain >10% without failing, and have an extremely low elastic modulus.

Tissue Interfaces

The principle goal of functional tissue engineering for skeletal muscle is to produce highly organized muscle organs that exhibit the normal or desired function of skeletal muscle in the body, while providing interfaces for the engineered tissue to interact with other tissues and organ systems. Often, it is the tissue-to-tissue interfaces that are overlooked when considering engineered tissues. Organism-level function, as well as normal tissue function, growth, and maintenance depend critically upon tissue interfaces. The three major tissue interfaces for each muscle organ in the body are 1) the tendon muscle interface, including the myotendinous junction; 2) the nerve interface, including the neuromuscular junction and afferent nerves from muscle spindles and the Golgi tendon organs, and general afferent nerves for sensation of pain, etc.; and 3) the circulatory interface, including the fluid component (blood) as well as the architecturally fixed components such as the capillary bed, arterioles, venules, and extended to include the lymphatic system, which serves the important role of returning plasma capillary filtrate proteins (albumin) to the general circulation to prevent accumulation in the muscles and severe edema.

Tendon Muscle Interface

The mechanical interface between muscle and bone is tendon. At the interface between muscle and tendon (the myotendinous junction or MTJ), muscle fibers form very specific membrane structures and express MTJ-specific proteins (Engel and Franzini-Armstrong, 1994; Swasdison and Mayne, 1991, 1992). These MTJ structures are necessary for the transmission of the mechanical force and work, generated by the contractile apparatus within each muscle fiber, to the skeletal system and then to the outside world. In the absence of these structures, it would not be possible for muscle to exhibit normal contractility without damage to the cell membrane. The transition from muscle to tendon to bone is characterized by a gradual increase in mechanical stiffness to minimize stress concentration at the interface between the hard and soft tissues (Nordin and Frankel, 1989). Even in the case where muscle appears to interface with bone tissue directly, there is a transition zone of mechanical stiffness. By this mechanism, injuries to the soft tissues during normal muscle contractions are prevented. For any application in which functional skeletal muscle is to be employed, attention must be given to the tendon-muscle interface.

Nerve-Tissue Interface

Nerve-muscle interaction is critical to the development, phenotype, and control of skeletal muscle. When denervated, the excitability of mammalian skeletal muscle drops dramatically (unpublished data), such that longer stimulus pulses of greater amplitude are required to elicit any given level of activation of the muscle. Muscle contraction is controlled at the level of the motor unit, which is a single motor nerve and all of the muscle fibers that it innervates. The innervation ratio (number of muscle fibers innervated by each motor nerve) can vary over three orders of magnitude, and it is this ratio that defines the ability of a muscle to deliver controlled power and graded force. Muscle fiber phenotype (fast or slow fiber type) is controlled by the motor neuron innervating the fiber. Also, the localization and concentration of junctional receptors is highly dependant on innervation. Long-term denervation also results in dramatic and undesirable changes in the cellular structure and contractility of skeletal muscle (Adams et al., 1995; Goldman et al., 1988; Walke et al., 1996). Electrical stimulation can be used *in vivo* to activate denervated skeletal muscle and exert control over the subunit expression in the acetylcholine receptors (Adams et al., 1995). Using a nearly identical electrical stimulation pattern, muscle mass, excitability, and contractility can be maintained in denervated rat muscles for several months (Dow et al., 1999, 2000), but the same stimulation pattern applied to myooids in culture does not result in significant expression of adult myosin isoforms or control levels of contractility (unpublished data). It is probable that engineered muscle will be unable to fully achieve adult phenotype and normal excitability in the absence of innervation.

Vascular Tissue Interface

In the absence of a capillary bed and forced perfusion, engineered muscle will be limited in cross-section to sizes of approximately 0.5 mm in diameter (Dennis and Kosnik, 2000). Diffusion distances greater than a few fiber diameters prevent efficient delivery of nutrients and removal of metabolic waste products. *in vivo*, blood also delivers soluble signals (e.g., hormones) to muscle tissue, which certainly influence muscle development and the balance between tissue regeneration and breakdown during functional adaptation. In muscles of large cross-section, blood flow also acts to control the core temperature of heavily worked muscles by forced convection over a large surface area (the entire capillary bed), allowing the heat to be dissipated elsewhere in the body. A typical approach to vascularize engineered tissue constructs is to implant the constructs *in vivo*, to promote capillary infiltration (Okano and Matsuda, 1998b).

Excitability

The classical physiological measurements of excitable tissues (nerve and muscle) include quantitative measures of the stimulus parameters required to depolarize the cell membrane. The classical terms rheobase and chronaxie describe

the excitability of individual cell membranes of excitable tissues. Rheobase refers to the stimulus strength, or amplitude, chronaxie to the stimulus duration. A tissue specimen will respond to an arbitrary combination of stimulus pulse parameters, as shown in Figure 14.3. Most of the quantitative work in this area relates to nerve tissue. For nerve tissue, the determination of the thresholds of excitability are in a sense simplified, because the resulting action potential is an all-or-none phenomenon. For skeletal muscle as a bulk tissue, the twitch force depends upon both the stimulus amplitude as well as the stimulus duration. This is due in part to the fact that many cell membranes are depolarized during activation of the muscle, and each fiber can be stimulated to contract submaximally, generating peak force over a wide dynamic range depending on the value of the pulse amplitude and duration. The response of an engineered muscle to single stimulus pulses with a range of amplitudes and durations is shown in Figure 14.4. Be

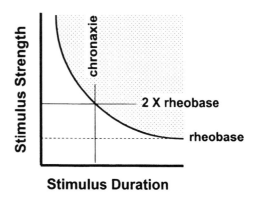

FIGURE 14.3. Schematic representation of the classical definition of the two quantitative parameters of tissue excitability: rheobase and chronaxie. An electrical stimulus pulse is characterized by the stimulus amplitude (strength), and the stimulus duration. The cell membrane of excitable tissues such as muscle and nerve will depolarize if the combination of stimulus strength and duration are adequate (shaded region of the graph). A stimulus pulse with a combination of amplitude and duration that falls outside of the shaded region will not depolarize the cell membrane. The quantification of the bulk excitability of engineered muscle tissue is based upon the classical definitions of rheobase and chronaxie.

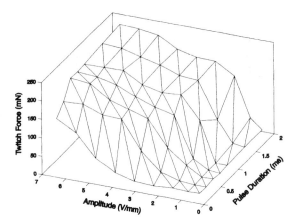

FIGURE 14.4. Mesh plot illustrating the interdependence of stimulus amplitude and stimulus intensity during electrically elicited twitch pulses for an engineered muscle (myooid). The twitch force (μN) for each combination of pulse amplitude (V/mm) and duration (ms) is plotted (plate a138).

cause of the range of possible submaximal responses of muscle to stimulation, modified definitions of rheobase and chronaxie are presented (Dennis and Kosnik, 2000). The variable R_{50} defines the stimulus amplitude required to elicit a half-maximal twitch force using a "long" stimulus pulse duration. A "long" pulse duration is defined as a pulse at least five times longer, and preferably 10 times longer, than the estimated chronaxie of the tissue. Thus, R_{50} is a measure of the rheobase of a skeletal muscle tissue specimen, either natural or engineered, and can be considered a bulk measure of the excitability of the tissue. To normalize the measure of R_{50} for different experimental setups, the stimulus amplitude (volts) is divided by the electrode separation (mm). The variable C_{50} (bulk muscle chronaxie) is simply measured as the pulse width required to elicit a half maximal twitch force when the stimulus amplitude is fixed at 2 times R_{50}. Increased values of either R_{50}, C_{50}, or both, indicate reduced excitability, because greater stimulus amplitude or pulse duration are required to elicit the desired contractile response. Of the two measures of excitability, R_{50} is dependent upon electrode configuration, whereas C_{50} is a more universal measure of excitability, being less subject to changes in experimental setup. Thus, relative differences in R_{50} can only be mea-

sured by using the same experimental electrode and tissue configuration, whereas comparisons of values of C_{50} for different specimens will be accurate provided only that the procedure for first measuring R_{50} is carried out consistently.

The excitability of engineered skeletal muscle tissue is of great importance in any application involving contractility. For muscle tissue *in vitro*, electrical stimulation is applied to elicit contractions. The total electrical energy required for each contraction depends upon the values of R_{50} and C_{50} for the muscle specimen in question. A simple calculation will demonstrate the importance of tissue excitability. Consider for the moment only the real component of the impedance (the DC resistance) of the electrode system used for stimulating the muscle specimens. During a single stimulation pulse, the voltage (V), current (I), and resistance (R) of the system are related by Ohm's law: V = IR. The electrical power (P) of each pulse is defined as P = IV. Solving for I from Ohm's Law, and substituting into the equation for power, we find that P = V²/R. From the physical definition of power, we know that power is the rate of work, P = W/t, where W is the work done in a period of time, t. Work is equivalent to energy (E), thus E = Pt = V²t/R. In the case of single electrical pulses applied to muscle specimens, using the values for pulse amplitude (R_{50}) and pulse duration (C_{50}), we find that the normative measure for the electrical energy in each electrical pulse is proportional to R_{50}^2 times C_{50}.

It is important to bear in mind that R_{50} and C_{50} are normative values for excitability, and pulses of greater amplitude and duration will be required to elicit maximal contractions from any given muscle specimen. However, for any given level of activation of a muscle specimen in a defined experimental configuration, the relative electrical energy per pulse will be proportional to the square of the rheobase (R_{50}), and directly proportional to the chronaxie (C_{50}). To assign tangible values, in preliminary experiments, we have determined that the value of R_{50} for engineered muscles is approximately 10 times greater than that of whole soleus muscles removed from a rat. The same is true for C_{50}. The practical consequence of the reduced excitability of engineered skeletal muscle tissues is that the electri-

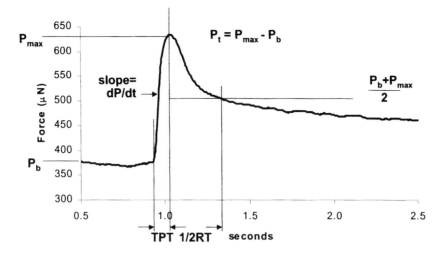

FIGURE 14.5. Raw data trace illustrating the contractility of a myooid during a single maximal twitch. Force traces such as this are quantified by a number of parameters that give insight into the function as well as the structure of muscle, both natural and engineered. In this figure, note that the baseline force is significantly above zero, being greater than half the value of the maximal twitch force, P_{max}. From this raw isometric contractility data, a number of important dynamic parameters can be measured, including the rate of force development (dP/dt), the time to peak tension (TPT), and the half relaxation time (½RT). Twitch force (P_t) is calculated by subtracting the baseline passive force (P_b) from the maximal twitch force (P_{max}).

cal energy required to elicit a contraction is approximately 1000-fold greater ($10^2 \times 10$) than that required to elicit a contraction in intact whole muscle. Tissue in general does not tolerate chronic exposure to electrical current, and a difference of 1000-fold in the total electrical energy requirement for tissue stimulation is extremely important. Thus, due to the low excitability of engineered muscle it is difficult to apply adequate electrical stimulation to measure contractility without damaging the tissue. This problem is exacerbated when electrical stimulation is employed for long periods during the culture of the engineered skeletal muscle construct, either to measure changes in contractility over time, or simply as an experimental intervention. In addition to tissue damage, the electrical currents tend to denature the culture medium. In any event, improved tissue excitability should be counted among the most important objectives of functional skeletal muscle tissue engineering.

Contractility

The fundamental quantitative assessment of muscle tissue function is contractility. Contrac-

tility includes all isometric force measurements, as well as measurements during lengthening and shortening contractions, submaximal contractions, and measurements of work and power. *In vitro* contractility measurements require electrical stimulation of the muscle, and measurement of force and muscle length (Dennis and Kosnik, in press). Muscle may be electrically stimulated *in vitro* in several ways. The two most commonly employed stimulation types are twitches, elicited by single electrical pulses, and tetani, elicited by a train of rapid stimulation pulses at fixed frequency. The force trace from an engineered muscle subjected to a single twitch pulse is shown in Figure 14.5. Both for twitches and tetani, the force trace reveals many important aspects of the contractility of the muscle, and it is possible to infer specific alterations in phenotype, structure, or even subcellular function on the basis of measured values from force traces (Close, 1964, 1972; Drachman and Johnson, 1973; Gordon et al., 1966). For example, the prolonged half relaxation time (½RT) of muscle engineered using the C2C12 cell line corroborates earlier findings that the calcium reuptake system in this cell line is impaired (Dennis et al., in press)

Several other types of contractility data from engineered skeletal muscle can be employed to directly assess the function of the tissue. Length-tension relationships (Fig. 14.6) can be determined, with the muscle in both the active and passive state (Gordon et al., 1966). Such quantitative relationships indicate important interactions between the contracting muscle and the extracellular matrix, and can be used to identify deviations from normal functionality that arise from structural characteristics of the engineered muscle. The length-tension relationship for myooids illustrated in Figure 14.6 clearly demonstrates that the passive baseline force in engineered muscles is much higher than that of control muscles at any position on the length-tension curve. The passive baseline force results from the tractile forces of fibroblasts in culture (Delvoye, 1991; Grinnell, 1994; Harris, 1991; Harris, et al., 1980, 1981; Higton and James, 1964; James and Taylor, 1969; Kolodney and Wysolmerski, 1992), which are organized around the periphery of each myooid as shown schematically in Figure 14.7, and are necessary for the self-organization of the muscle constructs (Dennis and Kosnik, 2000; Sanderson et al., 1986; Vandenburgh et al., 1991a).

Measures of contractility are essential in the engineering of functional skeletal muscle tissue. The contractility of muscle can not be determined by examination of histological sections, protein extraction, or any other nonmechanical method (Phillips et al., 1993). Muscle tissue may appear perfectly normal histologically, yet exhibit significantly reduced contractility. The only way to establish the contractility of engineered skeletal muscle is to measure it directly (Faulkner et al., 1997). Currently, the contractility of engineered skeletal muscle is at best 1–10% of that of control muscles from adult mammals (Dennis and Kosnik, 2000).

Muscle Architecture

Organ-level muscle function arises from the organization of individual muscle fibers, fibroblasts, and extracellular material into a higher-order structure. The muscle architecture plays an important role in the biomechanics of muscle organs, and how each muscle organ functions

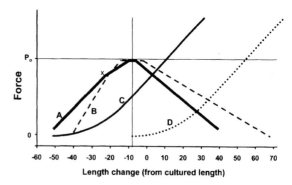

FIGURE 14.6. Length tension data for engineered muscles (myooids), superimposed upon data from control muscles (adapted from Gordon et al., 1966). The active (A) and passive (C) length tension relationship for myooids is shown with solid lines. The active (B) and passive (D) length-tension relationships for control muscles are scaled and shifted so that the center of the plateau of active force of the control muscle is aligned with that of the myooid. Length is expressed in terms of percentage length change from the cultured length for the myooid. Negative length changes indicate shortening of the specimen. Note that the passive curve for the engineered tissue (C) is very significantly left-shifted from what is normally observed in control muscles (D). Also note that the center of the plateau of the active curve for myooids occurs between −5% and −10% of the length at which the tissue self-organized. The point designated with an x on the ascending limb of both active curves designates the point at which the thin filaments (actin) begin to mechanically interfere within each sarcomere.

within a whole organism. The fiber orientation can be parallel, pennate, or more complex, including layers of fibers radiating from a central tendon at different orientations, such as in the diaphragm muscle. Muscle mass and fiber orientation, or the fiber:muscle length ratio (L_fL_o) dictate the physiologic cross section of the muscle. Muscles of the same apparent mass and shape can have very different physiologic cross-sectional areas, depending on the average fiber length and number of fibers in parallel. These factors account for major differences in muscle biomechanics, such as maximum shortening velocity and maximum tetanic force. Also of importance is the architecture of the connective tissue, which is important in the organization of the muscle fibers into fascicles and in the trans-

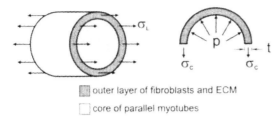

outer layer of fibroblasts and ECM
core of parallel myotubes

FIGURE 14.7. Schematic representation of the organization of fibroblasts and extracellular material around the core of myotubes in each myooid. The tractile forces generated by the outer layer of fibroblasts results in a longitudinal strain (σ_L) and a circumferential strain (σ_C). The σ_L is responsible for the high levels of baseline force in engineered muscle. The σ_C results in an internal pressure (**p**) within the myooid which can be estimated by calculating the *hoop stress*, assuming $\sigma_L = \sigma_C$ and a thin wall of fibroblasts of uniform thickness **t**.

mission of force from the muscle fibers to the skeletal system. Depending upon the biomechanics of the muscle organ and the arrangement of the muscle fibers within, the tendon can take the form of flattened sheets (aponeuroses) or thick sections.

Potential Applications

In the context of the broadest definition of the term function, engineered muscle tissue has tremendous potential for use in many capacities. The solution of each technical challenge in the field of functional muscle tissue engineering potentiates new applications with ever-increasing levels of functionality. It is certain that in some applications engineered skeletal muscle will displace the current need to use whole muscles explanted from living animals.

Basic research would benefit greatly from an *in vitro* model of skeletal muscle development and function. By utilizing functional engineered muscle, the number of animals used each year for basic scientific research could be greatly reduced. For example, using our current technology, a single laboratory rat can provide perhaps at most six or eight muscles for physiologic research. This limit is based both upon the number of anatomi-

cally suitable muscles in a single rodent, as well as the practical limitations imposed by the time required to surgically remove and evaluate each muscle. Additionally, each whole muscle remains viable for a maximum of 2 or 3 hours *in vitro*, limiting the time over which experiments can be carried out. In contrast, a single adult rat can easily provide enough myogenic precursor cells to produce 500–1000 myooids, each of which can be maintained in culture for several months (Dennis et al., in press). Experimental interventions can be applied to myooids in culture, and the contractility can be monitored at regular intervals using an automated bioreactor system. Thus, with the development of the technology to promote adult phenotype, functional engineered skeletal muscle constructs could replace whole muscle explants from living animals for longitudinal *in vitro* studies of muscle, thereby reducing the demand for laboratory animals by ~ 99% for many types of basic research.

New areas of basic research will be opened with the availability of an *in vitro* model of developing, functional skeletal muscle. These areas include the study of synaptogenesis in nerve–muscle co-cultures with the formation of functional motor units, the *in vitro* study of the development and biomechanics of muscle-tendon-bone joint systems, the study of the basic mechanisms of mechanotransduction of isolated fully functional muscle organs in culture (Vandenburgh et al., 1999), the influence of controlled mechanical, chemical, genetic, and electrical stimulatory signals on the architecture of muscle organs, the fiber phenotype and orientation, fiber hypertrophy and hyperplasia, contraction-induced injury and fiber regeneration, the plasticity of myogenic precursor cells from animals of all ages, and the study of the basic mechanisms of muscle aging and sarcopenia. The availability of a functional *in vitro* model of skeletal muscle will allow each of these areas of research to be carried out with the use of fewer animals and a commensurate reduction in the cost of the research. It is a general truth in engineering that we gain an understanding of physical systems by isolating and studying individual components of the system in question, then assembling these components into increasingly complex systems while quantifying the system behavior at each step. By providing the technology to isolate, assemble, and study muscles *in vitro* over long periods of time, our ability to directly study the fundamental mechanisms of muscle development, functional adaptation, injury, disease, and aging will be greatly enhanced.

There are several potential clinical applications for functional engineered skeletal muscle on the near horizon. Engineered muscle can serve as a protein factory. The engineered muscle tissue can contain genetically modified cells from a patient, with the express purpose of reimplantation of the tissue construct as a protein factory (Vandenburgh et al., 1996, 1998). Functional muscle constructs could also be used in diagnostic screening for congenital diseases such as malignant hyperthermia (Fig. 14.8), a life-threatening congenital disease of skeletal muscle that is one of the leading causes of death resulting from the administration of general anesthesia.

With the advent of the technology to produce functional engineered skeletal muscle with control levels of contractility, the constructs could be used to directly test the effectiveness of specific gene therapy strategies for human muscle. An example would be to use the available fetal tissue banks to engineer functional dystrophic muscle tissue, then to test gene/promoter constructs directly on this tissue, rather than on an animal model, such as the dystrophic (mdx) mouse. There are many advantages to this approach, including the ability to manufacture large numbers of engineered muscles *in vitro,* and for rapidly evaluating the functional outcome of many different gene/promoter constructs. The engineered muscle could then be tested for contractility and resistance to contraction-induced injury, to functionally quantify the attenuation of the dystrophic phenotype resulting from each therapeutic construct. The principle advantage of this strategy is that the genetic constructs would be tested directly on human tissues without putting any humans at risk. There is the added advantage that the genetic material could be easily introduced into the culture system, thereby avoiding many of the complexities and reducing the costs associated with the controlled introduction of genetic material into living animals. Adult phenotype skeletal muscle is postmitotic; in principle the genetic modifications would persist as long as the mus-

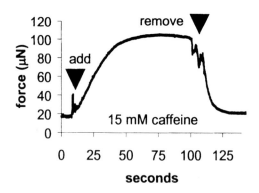

FIGURE 14.8. Illustration of one potential clinical application of myooids: the contractile response of myooids to the introduction and removal of 15 mM caffeine to the culture medium. This is similar to the contractility assay used for the diagnosis of malignant hyperthermia, during which a large tissue explant from the vastus lateralis muscle of a patient is assayed *in vitro* for contractility under exposure to caffeine and other agents.

cle fibers, which may be for the entire life of the animal (Decary et al., 1997; Vandenburgh et al., 1996).

The same strategy could be applied to drug screening and testing. Many drugs are tested for their potential collateral effects on muscle development and contractility, even if the expected clinical application of the drug is unrelated to diseases of muscle. Of course, all myoactive agents must also be thoroughly characterized in animal models and *in vitro* if possible, before clinical trials on humans can begin. There is currently a strong commercial interest in the development of *in vitro* models of functional skeletal muscle for evaluating the effectiveness of new compounds to combat muscle wasting disease. For any of these potential applications of functional engineered skeletal muscle to be realized, the technology must develop to the point that the engineered muscle exhibits control values of contractility. Without functional measures of contractility, three-dimensional skeletal muscle constructs offer few advantages over the more conventional two-dimensional tissue culture systems.

An additional potential commercial application for functional engineered skeletal muscle, undoubtedly on the distant technical horizon, is the large-scale production of edible animal proteins. Artificially farmed meat products could simultaneously reduce the cost while increasing the quality and purity of meat of all kinds. Presumably, the fat content, fiber type and architecture, and other factors of importance to food quality could all be controlled by the application of the correct interventions during tissue development in large scale commercial bioreactors.

Equally distant on the technical horizon, but ultimately one of the most important applications for functional engineered muscle tissue, is the creation of engineered muscle organs for use in the surgical correction of traumatic injury or congenital defects. The potential medical benefits of the availability of functional engineered muscle constructs from cells harvested directly from the patient are tremendous. For this vision to be realized, most, if not all, of the technical barriers listed above will need to be addressed and solved. The production of muscle organs for surgical implantation will remain perhaps the ultimate application for functional engineered skeletal muscle tissue.

Muscle is an extraordinarily versatile mechanical actuator. Considering the variety of muscles that have evolved in the animal kingdom, it is difficult to imagine a material more well suited to use as an actuator. Muscle is scalable, powering the motions of the smallest dust mite to the largest blue whale. Muscle force is increased by adding fibers in parallel, or by simply increasing the diameter of individual fibers, or both. The range of displacement of the muscle actuator is controlled by the number of sarcomeres in series, or effectively by increasing fiber length. Muscle works nearly silently, is biodegradable, can store its own fuel, can adapt to the demands placed upon it through the mechanisms of functional adaptation, and can heal when injured. In terms of power production, when including the energy source, muscle compares extremely favorably with all other actuator technologies in terms of power output (watts/kilogram) and chemomechanical transduction efficiency (work out/energy in) (Hollerbach et al., 1991). The chemomechanical transduction efficiency of muscle is even more impressive when one considers that it occurs at nearly equilibrium conditions, whereas the most efficient synthetic me-

chanical actuator systems operate at high temperatures and pressures, for example, diesel engines. Muscle has built-in mechanisms for self regulation at the cellular level, and the muscle-tendon unit contains internal sensors for force (Golgi tendon organs), displacement, and rate of displacement (muscle spindles). The architectural plasticity of muscle is superb. By arranging fibers in parallel and series, and by varying the fiber overlap and angle of pennation, muscle occurs in nature as a mechanical actuator in the form of long thin rods, short thick rods, fans, cones, thin flat sheets, plates, hollow tubes, and hollow spheres. Within a mechanical system, a muscle can produce positive work to serve as a motor, absorb energy to act as both a brake and a dashpot, as a spring to store energy, and as a strut or linkage to transmit force from one part of the skeleton to another (Dickinson et al., 2000). When acting in any of these capacities, the mechanical characteristics of the muscle can be tuned in tens of milliseconds to adjust important parameters such as stiffness and free length. Combinations of synergist (or agonist) and antagonist muscles about a joint can provide a virtually limitless range of options for control of the joint. With the built-in sensors and motor unit level of control exerted by the nervous system, muscle can operate in open or closed loop control, with force, position, or velocity feedback. In addition, muscle can generate excess heat to help maintain body temperature in cold environments. And, when properly prepared, muscle is an excellent source of dietary protein. Overall, these characteristics make muscle the most versatile actuator ever conceived, and there is growing interest in the use of living skeletal muscle tissue as an actuator for robotic and prosthetic applications.

Future Directions

Based upon the technical challenges and potential applications listed above, it is possible to roughly map out a reasonable scenario for the development of the field of functional skeletal muscle tissue engineering. The first and most important development will be to understand the signals that are required to promote expression of adult phenotype. Currently available muscle constructs are severely limited by the inability to generate control values of contractility and excitability. Our current understanding of the mechanical environment during the growth and development of muscle *in vivo* is incomplete (Harris, 1991). Several approaches to remediate this problem are possible. One is to embark on a more thorough study of the environmental milieu within developing organisms, with an eye toward elucidating those mechanisms directly related to musculoskeletal development. In addition to this approach, it is in principle now possible to actually reverse engineer the conditions that promote muscle growth and development, by applying combinations of interventions to muscle constructs *in vitro,* and monitoring the effects on excitability and contractility. Very many factors are involved, including gene expression, soluble chemical signals, tissue interactions, and mechanical signals. The problem is further complicated by the importance of not only the presence of each signal, but the sequencing of each signal at the appropriate time points during development. Searching for the optimal set of interventions in this large, *n*-dimensional solution space is itself a daunting task. It is logical to begin with known ranges of parameters, such as those measured during development of the tissue in question, but a large number of iterations will still be required to establish optimal protocols. Each iteration will require weeks or months for the functional outcome to be known, and it is therefore impractical to search for optimal solutions using inefficient search algorithms. Highly efficient searches of the solution space will be required, and advanced search techniques, such as "genetic search algorithms" (Hugh Herr, personal communication) can be advantageously employed in this and other tissue engineering systems, to find sets of optimal interventions to maximize tissue function.

Because of the innately mechanical function of skeletal muscle, it is likely that engineered skeletal muscles will require the development of a new bioreactor technology, to provide the necessary signals to the cells to induce full differentiation and adult phenotype. The first bioreactors of this type (Hatfaludy et al., 1989; Vandenburgh 1982;

Vandenburgh and Karlisch, 1989; Vandenburgh and Kaufman, 1979; Vandenburgh et al., 1989, 1990; 1991a,b) focused principally on the application of controlled mechanical strain to passive skeletal muscle organoids during culture, to simulate limb growth and movement during development. Because of the complex milieu of signals involved in muscle development, the next generation of bioreactors will apply a broad spectrum of mechanical, electrical, and chemical (soluble and insoluble) signals, to promote muscle development (Chromiak et al., 1998; Dennis and Kosnik, in press; Perrone et al., 1995; Shansky et al., 1997; Vandenburgh, 1992). The new class of bioreactors will incorporate electrical stimulation, force sensors, displacement transducers, and servo motors to emulate both the active and passive mechanical environment of the developing muscle. Quantitative measures of muscle excitability and contractility will be used as signals for closed-loop control of muscle activity in the bioreactors. A means to apply and control perfusion and the application of exogenous bioactive chemicals at prescribed times will also be required. Electrical and mechanical activity will be used to control the expression of activity-dependant genes, which in turn will be used to promote the formation of functional tissue interfaces *in vitro,* including formation of myotendinous junctions and synaptogenesis. Nerve co-culture will be employed, and electrode arrays will be used to monitor spontaneous depolarization of individual motor axons, as well as to assert motor unit-level control over the engineered constructs (Mensinger et al., 2000). Functional engineered muscle actuators will eventually be employed in both robotic and prosthetic applications. The synthetic device will act as a bioreactor for the muscle tissue, and the tissue will in turn provide a useful function for the synthetic device. For applications related to robotics and prosthetics, skeletal muscle tissue is in principle an excellent actuator (Hollerbach et al., 1991). As our ability to engineer more and better functional tissues improves, the hybrid prosthetic devices will be engineered with increasingly greater living tissue content, until eventually the entire prosthetic device is fully biological.

The ultimate goals of the functional skeletal muscle engineer must include the development of the technology to promote expression of normal adult muscle phenotype, contractility, and excitability. The engineered muscle constructs should be self-organizing tissue from primary mammalian cells, with intact and functional interfaces with the nervous system, the circulatory system, and the skeletal system. The technology must also allow control over the muscle architecture including fiber orientation, muscle cross-section, fiber length to muscle length ratio, organization of connective tissue fascicles, extracellular matrix type and architecture, and distribution of fibers within a motor unit throughout the muscle cross-section. This high-level structural organization is critical to the function of skeletal muscle.

References

Adams L., Carlson B.M., Henderson L., Goldman D. 1995. Adaptation of nicotinic acetylcholine receptor, myogenin, and MRF4 gene expression to long-term muscle denervation. *J Cell. Biol.* 131:1341–1349.

Chromiak J.A., Shansky J., Perrone C., Vandenburgh H.H. 1998. Bioreactor perfusion system for the long-term maintenance of tissue-engineered skeletal muscle organoids. *In Vitro Cell. Dev. Biol. Anim.* 34:694–703.

Close R. 1964. Dynamic properties of fast and slow skeletal muscles of the rat during development. *J. Physiol.* 173:74–95.

Close R.I. 1972. Dynamic properties of mammalian skeletal muscles. *Physiol. Rev.* 52:129–197.

Decary S., Mouly V., Hamida C.B., Sautet A., Barbet J.P., Butler—Browne G.S. 1997. Replicative potential and telomere length in human skeletal muscle: implications for satellite cell-mediated gene therapy. *Hum. Gene Ther.* 8:1429.

Delvoye P., Wiliquet P., Leveque J., Nusgens B.V., Lapiere C.M. 1991. Measurement of mechanical forces generated by skin fibroblasts embedded in a three-dimensional collagen gel. *J. Invest. Dermatol.* 97:898–902.

Dennis R.G., Kosnik P. Excitability and isometric contractile properties of mammalian skeletal muscle constructs engineered *in vitro*. *In Vitro Cell. Dev. Biol. Anim.* 36(5):327–335, 2000.

Dennis, R.G., Kosnik, II, P.E., Gilbert, M.E., and Faulkner, J.A. Excitability and contractility of skeletal muscle engineered from primary cultures

and cell lines. *Am J Physiol Cell Physiol* 280: C288–C295, 2001.

Dennis R.G., Kosnik P.E. 2000. Excitability and isometric contractile properties of mammalian skeletal muscle constructs engineered *in vitro*. *In Vitro Cell. Dev. Biol. Anim.* 36:327–335.

Dennis R.G., Kosnik P. 2002. Mesenchymal Cell Culture: Instrumentation and Methods for Evaluating Engineered Muscle. Anthony Atala, Robert P. Lanza, Eds. Academic Press San Diego In *Methods in Tissue Engineering.* Chapter 24 pages 307–315.

Dickinson M.H., Farley C.T., Full R.J., Koehl M.A.R., Kram R. Lehman S. 2000. How animals move: an integrative view. *Science* 288:100–106.

Dow D.E., Dennis R.G., Hassett C.A., Faulkner J.A. 1999. Electrical stimulation protocol to maintain mass and contractile force in denervated muscles. BMES-EMBS 1st Joint Conference, Session 6.1.2 Functional Neuromuscular Stimulation, Paper # 573.

Dow D.E., Dennis R.G., Hassett C.A., Faulkner J.A. 2000. Electrical stimulation to maintain functional properties of denervated EDL muscles of rats. 31st Annual Neural Prosthesis Workshop, National Institutes of Health, Lister Hill Center, Oct. 25–27.

Drachman D.B., Johnston D.M. 1973. Development of a mammalian fast muscle: dynamic and biochemical properties correlated. *J. Physiol. (Lond.)* 234:29–42.

Engel A.G., Franzini-Armstrong C. 1994. *Myology: Volume I, Basic and Clinical.* New York, McGraw-Hill, Inc.

Faulkner J.A., Brooks S.V., Dennis R.G. 1997. Measurement of recovery of function following whole muscle transfer, myoblast transfer, and gene therapy. In: *Methods in Tissue Engineering, Vol. 18: Tissue Engineering Methods and Protocols.* J.R. Morgan, M.L. Yarmush, eds. Totowa, NJ: Humana Press Inc., pp. 155–172.

Goldman D., Brenner H.R., Heinemann, S. 1988. Acetylcholine receptor α, β, γ, and δ-subunit mRNA levels are regulated by muscle activity. *Neuron* 1:329–333.

Gordon A.M., Huxley A.F., Julian F.J. 1966. The variation in isometric tension with sarcomere length in vertebrate muscle fibres. *J. Physiol. (Lond.)* 184:170–192.

Grinnell F. 1994. Fibroblasts, myofibroblasts, and wound contraction. *J. Cell. Biol.* 124:401.

Harris A.K. 1991. Physical forces and pattern formation in limb development. In: *Developmental Patterning of the Vertebrate Limb.* J.R. Hinchliffe, J.M. Hurle, D. Summerbell, eds. New York: Plenum Press, pp. 203–210.

Harris A.K., Wild P., Stopak D. 1980. Silicone rubber substrata: a new wrinkle in the study of cell locomotion. *Science* 208:177.

Harris A.K., Stopak D., Wild P. 1981. Fibroblast traction as a mechanism for collagen morphogenesis. *Nature* 290:249.

Hatfaludy S., Shansky J., Vandenburgh H.H. 1989. Metabolic alterations induced in cultured skeletal muscle by stretch-relaxation activity. *Am. J. Physiol.* 256(1 Pt 1):C175–181.

Higton D.I.R., James D.W. 1964. The force of contraction of full-thickness wounds of rabbit skin. *Br. J. Surg.* 51:462.

Hollerbach J.M., Hunter I.W., Ballantyne J. 1991. A comparative analysis of actuator technologies for robotics. In: *Robotics Review 2*. Cambridge, MA: MIT Press, pp. 299–342.

Irintchev A., Rosenblatt J.D., Cullen M.J., Zweyer M., Wernig A. 1998. Ectopic skeletal muscles derived from myoblasts implanted under the skin. *J. Cell Sci.* 111(Pt 22):3287–3297.

James D.W., Taylor J.F. 1969. The stress developed by sheets of chick fibroblasts *in vitro*. *Exp. Cell Res.* 54:107–110.

Kolodney M.S., Wysolmerski R.B. 1992. Isometric contraction by fibroblasts and endothelial cells in tissue culture: a quantitative study. *J. Cell Biol.* 117:73–82.

Kosnik P., Dennis R.G. 2002. Mesenchymal Cell Culture: Functional Mammalian Skeletal muscle constructs. In: *Methods in Tissue Engineering*. Anthony Atala and Robert P. Lanza, editors Academic Press, San Diego Chapter 23 pp. 299–306.

Kosnik P. Jr., Dennis R.G., Faulkner J.A. Functional Development of engineered skeletal muscle from adult and neonatal rats. *Tissue Engineering* 7(5): 573–584, 2001.

LaFramboise W.A., Daood M.J., Guthrie R.D., Butler-Browne, G.S., Whalen R.G., Ontell M. 1990. Myosin isoforms in neonatal rat extensor digitorum longus, diaphragm, and soleus muscles. *Am. J. Physiol.* 259:L116–L122.

Lewis M.R. 1915. Rhythmical contraction of the skeletal muscle tissue observed in tissue cultures. Am. J. Physiol. 38:153–161.

Lewis W.H., Lewis M.R. 1917. Behavior of cross striated muscle in tissue cultures. *Am. J. Anat.* 22:169.

Mayne R., Swasdison S., Sanderson R.D., Irwin M.H. 1989. Extracellular matrix, fibroblasts and the development of skeletal muscle. In: *Cellular and Molecular Biology of Muscle Development*. New York: Alan R. Liss, Inc., pp. 107–116.

Mensinger A.F., Anderson D.J., Buchko C.J., Johnson M.A., Martin D.C., Tresco P.A. Silver R.B., Highstein, S.M. 2000. Chronic recording of regenerating

VIIIth nerve axons with a sieve electrode. *J. Neurophysiol.* 83:611–615.

Miller, R.R., Rao J.S., Burton W.V., Festoff B.W. 1991. Proteoglycan synthesis by clonal skeletal muscle cells during *in vitro* myogenesis: differences detected in the types and patterns from primary cultures. *Int. J. Dev. Neurosci.* 9:259–267.

Nordin M., Frankel V.H. 1989. *Basic Biomechanics of the Musculoskeletal System* 2nd ed. Philadelphia: Lea & Febiger.

Okano T., Matsuda T. 1998a. Tissue engineered skeletal muscle: preparation of highly dense, highly oriented hybrid muscular tissues. *Cell Transplant.* 7:71–82.

Okano T., Matsuda T. 1998b. Muscular tissue engineering: capillary-incorporated hybrid muscular tissues *in vivo* tissue culture. *Cell Transplant,* 7:435–442.

Okano T., Satoh S., Oka T., Matsuda T. 1997. Tissue engineering of skeletal muscle: highly dense, highly oriented hybrid muscular tissues biomimicking native tissues. *ASAIO J.* 43:M749–M753.

Perrone C.E., Fenwick-Smith D., Vandenburgh H.H. 1995. Collagen and stretch modulate autocrine secretion of insulin-like growth factor-1 and insulin-like growth factor binding proteins from differentiated skeletal muscle cells. *J. Biol. Chem.* 270:2099–2106.

Phillips S.K., Wiseman R.W., Woledge R.C., Kushmerick M.J. 1993. Neither changes in phosphorus metabolite levels nor myosin isoforms can explain the weakness in aged mouse muscle. *J. Physiol. (Lond.)* 463:157–167.

Powell C., Shansky J., Del Tatto M., Forman D. E., Hennessey J., Sullivan K., Zielinski B.A., Vandenburgh H.H. 1999. Tissue-engineered human bioartificial muscles expressing a foreign recombinant protein for gene therapy. *Hum. Gene Ther.* 10:565–577.

Putnam A.J., Mooney D.J. 1996. Tissue engineering using synthetic extracellular matrices. *Nat. Med.* 2:824–826.

Reiser P.J., Kasper C.E., Greaser M.L., Moss R.L. 1988. Functional significance of myosin transitions in single fibers of developing soleus muscle. Am. J. Physiol. 254:C605–C613.

Sanderson R.D., Fitch J.M., Linsenmayer T.R., Mayne R. 1986. Fibroblasts promote the formation of a continuous basal lamina during myogenesis *in vitro*. *J. Cell Biol.* 102:740–747.

Schultz E., McCormick K.M. 1994. Skeletal muscle satellite cells. *Rev. Physiol. Biochem. Pharmacol.* 123:213–257.

Shansky J., Chromiak J., Del Tatto M., Vandenburgh H. 1997. A simplified method for tissue engineering skeletal muscle organoids *in vitro. In Vitro Cell. Dev. Biol. Anim.* 33:659–661.

Shea L.D., Smiley E., Bonadio J., Mooney D.J. 1999. DNA delivery from polymer matrices for tissue engineering. *Nat. Biotechnol.* 17:551–555.

Swasdison S., Mayne R. 1991. *In vitro* attachment of skeletal muscle fibers to a collagen gel duplicates the structure of the myotendinous junction. *Exp. Cell Res.* 193:227–231.

Swasdison S., Mayne R. 1992. Formation of highly organized skeletal muscle fibers *in vitro:* comparison with muscle development *in vivo. J. Cell. Sci.* 102:643–652.

Turner D.C. 1986. Cell–cell and cell–matrix interactions in the morphogenesis of skeletal muscle. *Dev. Biol.* 3:205–224.

van Wachem P.B., van Luyn M.J., da Costa M.L. 1996. Myoblast seeding in a collagen matrix evaluated *in vitro. J. Biomed. Mater. Res.* 30: 353–360.

van Wachem P.B., Brouwer L.A., van Luyn M.J. 1999. Absence of muscle regeneration after implantation of a collagen matrix seeded with myoblasts. *Biomaterials* 20:419–426.

Vandenburgh H.H. 1982. Dynamic mechanical orientation of skeletal myofibers *in vitro. Dev. Biol.* 93:438–443.

Vandenburgh H.H. 1988. A computerized mechanical cell stimulator for tissue culture: effects on skeletal muscle organogenesis. *In Vitro Cell. Dev. Biol.* 27:609–619.

Vandenburgh H.H. 1992. Mechanical forces and their second messengers in stimulating cell growth *in vitro. Am. J. Physiol.* 262(3 Pt 2):R350–355.

Vandenburgh H.H., Karlisch P. 1989. Longitudinal growth of skeletal myotubes *in vitro* in a new horizontal mechanical cell stimulator. *In Vitro Cell. Dev. Biol.* 25:607–616.

Vandenburgh H., Kaufman S. 1979. *In vitro* model for stretch-induced hypertrophy of skeletal muscle. *Science* 203:265–268.

Vandenburgh H.H., Hatfaludy S., Karlisch P., Shansky J. 1989. Skeletal muscle growth is stimulated by intermittent stretch-relaxation in tissue culture. *Am. J. Physiol.* 256(3 Pt 1):C674–682.

Vandenburgh H.H., Hatfaludy S., Sohar I., Shansky J. 1990. Stretch-induced prostaglandins and protein turnover in cultured skeletal muscle. *Am. J. Physiol.* 259(2 Pt 1):C232–240.

Vandenburgh H.H., Swasdison S., Karlisch P. 1991a. Computer-aided mechanogenesis of skeletal muscle organs from single cells *in vitro. FASEB J.* 5:2860–2867.

Vandenburgh H.H., Hatfaludy S., Karlisch P., Shansky, J. 1991b. Mechanically induced alterations in cultured skeletal muscle growth. *J. Biomech* 24 (Suppl 1):91–99.

Vandenburgh H., Del Tatto M., Shansky J., LeMaire J., Chang A., Payumo F., Lee P., Goodyear A., Raven L. 1996. Tissue-engineered skeletal muscle organoids for reversible gene therapy. *Hum. Gene Ther.* 7:2195–2200.

Vandenburgh H., Del Tatto M., Shansky J., Goldstein L., Russell K., Genes N., Chromiak J., Yamada S. 1998. Attenuation of skeletal muscle wasting with recombinant human growth hormone secreted from a tissue-engineered bioartificial muscle. *Hum. Gene Ther.* 9:2555–2564.

Vandenburgh H., Chromiak J., Shansky J., Del Tatto M., LeMaire J. 1999. Space travel directly induces skeletal muscle atrophy. *FASEB J.* 13:1031–1038.

Walke W., Xiao G., Goldman D. 1996. Identification and characterization of a 47 base pair activity-dependant enhancer of the rat nicotinic acetylcholine receptor δ-subunit promoter. *J. Neurosci.* 16:3641–3651.

15

Bioengineering the Growth of Articular Cartilage

Stephen M. Klisch, Michael A. DiMicco, Anne Hoger, and Robert L. Sah

Introduction

Articular cartilage is a thin layer of connective tissue located within joints and on the ends of Articular cartilage functions as a low-friction, wear-resistant, load-bearing material that facilitates joint motion (Maroudas, 1979; Mow and Ratcliffe, 1997). The articular cartilage of diarthrodial joints experiences a high level of biomechanical stress over many decades (Hodge et al., 1986) and, in many cases, can tolerate years of repetitive loading. However, cartilage damage and degeneration occur often with traumatic joint injury and advancing age at particular sites, such as the knee and hip.

The poor intrinsic healing capacity of articular cartilage has been described for centuries, and continues to be a problem today (Buckwalter and Mankin, 1998). This inadequate healing response is likely related to the relatively low cellularity and low metabolic activity of the chondrocytes within the tissue. Additionally, because of the avascularity of cartilage, cells that participate in systemic wound-healing responses are not able to access the injury site and promote tissue repair. When cartilage damage penetrates the subchondral bone, however, a cell-based repair response is initiated. Nevertheless, such a response results in the generation of a fibrocartilaginous repair tissue that does not match the composition, structure, or function of healthy articular cartilage.

The attainment of a number of specific design goals is likely to be critical to the development of a consistently successful strategy for the repair of cartilage defects (Hjertquist and Lemperg, 1969; Sah et al., 2001). One goal is to ultimately form tissue that has the normal site-specific bio- and subchondral bone. A second goal is that the formed tissue should integrate firmly to both the adjacent host cartilage and the underlying bone. To attain these goals, biological processes of repair and regeneration need to occur under the influence of postoperative biomechanical demands, which can be affected markedly by rehabilitation regimens. Thus, the time-dependent processes of cell-mediated matrix remodeling and matrix-dependent alteration of cartilage biomechanical properties need to be determined and controlled.

Current clinical strategies for treating defects of eroded or degenerated cartilage often involve the surgical clearing of the affected tissue area followed by the implantation of grafts that replace or form cartilage or, alternatively, both cartilage and bone. At one extreme is the approach of implanting preformed cartilage and bone; for example, an osteochondral fragment can be transplanted from an allogenic (Fig. 15.1A) or autogenic (Fig. 15.1B) source. At the other extreme is the approach of filling the bulk of the defect with cells that can facilitate the growth of appropriate cartilage and/or bone tissue. An example of this is autologous chondrocyte implantation (Fig. 15.1C), wherein a high-density cell suspension is introduced into the defect site under a tissue flap, and cartilaginous repair tissue is formed (Brittberg et al., 1994;

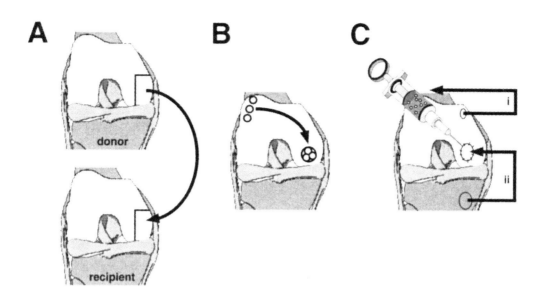

FIGURE 15.1. Current strategies for articular repair. (*A*) Allografting. After removing the injured tissue, an osteochondral fragment from a healthy individual is implanted. (*B*) Autografting osteochondral fragments from a healthy portion of a joint to a cleared defect site. (*C*) Autologous chondrocyte implantation, in which (*i*) cells are biopsied from healthy tissue, expanded in culture, and resuspended at high density for reimplantation into a surgically-cleared defect covered with (*ii*) a sutured autologous periosteal flap.

Grande et al., 1987). Experimentally, growth factors have been applied in order to stimulate the metabolic activity of cells in a repair situation (Fujimoto et al., 1999; Nixon et al., 1999). In between these extremes are the experimental strategies of forming immature cell- and matrix-laden tissue constructs *in vitro* and implanting the constructs into defects *in vivo*, after which maturation occurs.

The United States National Committee on Biomechanics goals for promoting functional tissue engineering include the identification of structural and mechanical requirements for engineered tissue constructs (Butler et al., 2000). To achieve this goal, *in vivo* mechanical loads need to be assessed in order to characterize the environment that a tissue-engineered construct will experience upon implantation. Patient-specific factors, such as age, health, and expectation for future use, may be key determinants of this mechanical loading environment. In addition, the mechanical properties of the tissue, whether a construct fabricated *in vitro* or a tissue

formed *in vivo*, must be determined within the framework of a mechanical model. Perhaps most importantly, the biological response and adaptation of the construct to the *in vivo* environment must be determined. An implant primarily composed of cells and relatively little matrix may evolve into functional tissue if it is mechanically protected during the early stages of growth, but may fail if overloaded. On the other hand, an implant that is composed of fully functional cell-laden cartilage may require early motion to maintain tissue homeostasis and attain a successful outcome.

Growth, resorption, and remodeling are fundamental processes that influence the size, shape, and properties of biological organs and tissues. Growth is "a normal process of increase in size of an organism as a result of accretion of tissue similar to that originally present" (Dorland, 1981). Here, volumetric growth of a constituent is interpreted as the deposition of constituent mass that has the same mechanical properties as the existing material. Volumetric

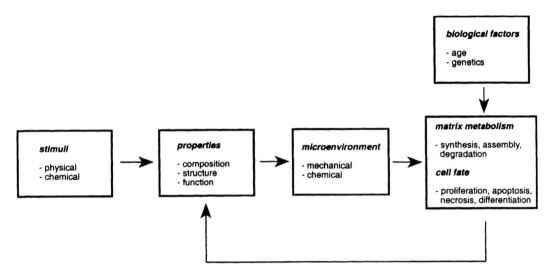

FIGURE 15.2. There exists a complex relationship between stimuli, tissue properties, the cellular microenvironment, and cell metabolism and fate.

growth, at either the constituent or at the tissue level, may change the residual stress field and, consequently, the overall structural and mechanical properties of the tissue (Fung, 1993; Skalak et al., 1982; Taber, 1998; Taber and Eggers, 1996). Resorption, "the loss of substance through physiologic or pathologic means" (Dorland, 1981), is, roughly, the opposite of growth. On the other hand, to remodel is "to alter the structure of; remake" (*Merriam-Webster Collegiate Dictionary*, 2001). Here, constituent remodeling is interpreted as a change in the structure of the constituent that alters its mechanical properties.

Mechanical and biochemical factors regulate the growth and remodeling of cartilage tissue matrix. When a stimulus is applied to cartilage *in vivo* or *in vitro*, the properties of the tissue, such as composition, structure, and biophysical function, determine how this stimulus translates into a mechanical and chemical change in the cellular and extracellular microenvironments (Fig. 15.2). This microenvironment, along with biological factors such as age and genetics, govern both cell metabolism and cell fate and, consequently, can lead to changes in the composition and the mechanical properties of the tissue. Models that describe this complex relationship between mechanical and biochemical stimuli,

cell metabolism, and tissue growth may lead to a better understanding of the processes of cartilage growth, degeneration, and repair.

Composition, Structure, and Function of Articular Cartilage

Adult cartilage is composed of a relatively small fraction of cells, called chondrocytes, within a fluid-filled extracellular matrix. Two of the molecular components of the solid matrix (Fig. 15.3), proteoglycan and collagen, appear to be predominantly responsible for the functional mechanical properties of the tissue (Grodzinsky, 1983; Maroudas, 1979; Mow and Ratcliffe, 1997). Aggrecan is the major proteoglycan of articular cartilage, and is composed of a protein core to which chondroitin sulfate and keratan sulfate glycosaminoglycan (GAG) chains are attached. The N-terminal of the core protein is able to associate with link protein, and both the core protein and the N-terminal of aggrecan have a strong binding affinity for hyaluronan. The coupling of many aggrecan molecules to hyaluronan results in macromolecular proteoglycan aggregates (total M.W. ~300 MDa), and these aggregates account for approximately

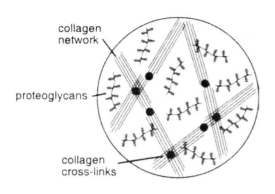

FIGURE 15.3. Schematic of the major components of the solid matrix of articular cartilage.

25–35% of the dry weight of adult articular cartilage (Buckwalter and Mankin, 1997). Approximately 60–80% of the proteoglycans in cartilage are in the form of these stable aggregates, while ~20–40% are soluble and mobile in the tissue matrix (Pottenger et al., 1985; Sajdera and Hascall, 1969). The GAG constituent provides the tissue with a fixed negative charge that increases the tissue's propensity to swell and to resist compressive loading (Basser et al., 1998; Lai et al., 1991). The collagen is mostly present (~95%) as fibrils effectively immobilized in the tissue matrix by cross-links (Furuto and Miller, 1987) and forming a collagen network. The cross-linked collagen network resists the swelling tendency of the proteoglycan, and provides the tissue with tensile and shear stiffness and strength (Mow and Ratcliffe, 1997; Venn and Maroudas, 1977; Woo et al., 1976).

Numerous studies have shown that the tissue's mechanical properties depend on its composition and structure. For example, the aggregate modulus of adult cartilage has been positively correlated with GAG content (Mow and Ratcliffe, 1997) and, to a lesser extent, with collagen content (Sah et al., 1996). Also, the permeability of adult articular cartilage has been inversely related to proteoglycan content, measured as fixed charge density (Maroudas et al., 1968). Throughout the maturation of bovine articular cartilage, from the fetal stage, through newborn calf, and to skeletal maturity, there is decrease in cellularity, an increase in the collagen content and crosslink density, and little or no change in the GAG

content (Pal et al., 1981; Strider et al., 1975; Thonar and Sweet, 1981). These biochemical changes from fetal to adult specimens are accompanied by an increase in the compressive modulus and a decrease in the permeability (Williamson et al., 2001; Wong et al., 2000), and each of these mechanical properties is significantly correlated with both GAG and collagen content. In addition, the tensile strength generally increases from the fetal to more advanced stages of development, and is positively correlated with collagen crosslink density (Williamson et al., 2001).

The transport properties of cartilage also depend on tissue composition. The transport of large solutes is hindered mainly by the aggregated proteoglycans. Since the spacing between the GAG chains of proteoglycans create the smallest "pores" in the cartilage extracellular matrix [(~2 nm, (Maroudas, 1979)] and are important for determining the water content of cartilage, variations in proteoglycan content can affect the transport of large solutes. For example, enzymatic removal of 93% of the proteoglycan increased the diffusivities of inulin (5000 Da) by 60% and that of dextran 70 (70,000 Da) by 100% (Torzilli et al., 1997) in cartilage explants.

Cartilage Metabolism and Growth

The cartilage extracellular matrix is synthesized, maintained, and degraded by chondrocytes. Tissue growth occurs when the normal balance between anabolic and catabolic processes is shifted toward matrix deposition. Such a shift may occur either by an increase in matrix synthesis by the chondrocytes or a decrease in the rate of loss of matrix constituents. A key feature of cartilage growth is that cell and matrix metabolism can be regulated by mechanical stimuli. Mechanical stimuli can cause growth abnormalities in pathological conditions such as hip dysplasia (Pauwels, 1976) and traumatic injury near the growth plate (Ogden, 1988). In vivo models of experimentally induced arthritis and joint immobilization indicate that abnormal loading conditions cause tissue remodeling and subsequent changes in matrix composition and mechanical properties (Guilak et al., 1997). *In vitro* experiments have quantified the metabolic response to mechanical stimuli such as hydrostatic pressure, dynamic compressive stress, and fluid shear (Guilak et al., 1997).

A second key feature of cartilage growth is that biochemical factors influence chondrocyte metabolism. Cytokines and other bioactive molecules may be produced within cartilage by chondrocytes, or they may reach chondrocytes via transport through the articular surface from the synovial fluid. Anabolic agents, such as insulin-like growth factor-1 (IGF-1), transforming growth factor-β (TGF-β), and basic fibroblast growth factor (bFGF) stimulate the biosynthesis of proteoglycans and/or collagens (Trippel, 1995). Catabolic agents including interleukin-1 (IL-1) and tumor necrosis factor-α (TNF-α) are known to promote matrix degradation (Fosang et al., 1991, Hardingham et al., 1992), possibly by increasing the production of endogenous proteases (Hutchinson et al., 1992; Kandel et al., 1990). Other molecules have more complicated actions on cartilage; nitric oxide, for example, is thought to simultaneously suppress matrix biosynthesis while protecting the matrix from breakdown (Evans et al., 1996).

Models of Cartilage Metabolism and Growth

The primary goal of models of cartilage metabolism and growth is to quantitatively describe the dynamics of the growth process in cartilage. These models can be used in a number of ways to develop a better understanding of the key features and mechanisms of cartilage growth. They can be used to predict parameters that change during growth and that are often difficult to measure, such as tissue composition and biomechanical properties. Also, they can describe and predict the outcome of therapies aimed at the successful repair of growth and degenerative abnormalities. This can be accomplished by applying the model with simple geometries such as those corresponding to an *in vitro* specimen, or with complex geometries such as those corresponding to a joint's articular surface where

computational models of the *in vivo* growth, degeneration, and repair processes can be developed. Finally, these models can be used to design experiments in order to identify the manner in which cartilage constructs may be stimulated *in vitro* for the fabrication of better implants.

Here, two types of models for tissue metabolism and growth are presented. The first group of models describes the metabolism of cartilage components. These models include single-compartmental models as well as spatially varying continuum models, the latter of which can explicitly describe biosynthesis, diffusion, and reaction processes. The second group of models describes the relationship between mechanical stimuli and tissue growth. These models include a cartilage growth mixture model with growth laws that define the amount and the orientation of deposition for various molecules in the tissue.

Compartmental Models of Tissue Metabolism

The steady-state composition of articular cartilage is established by a balance between the production of matrix molecules and the loss of these components from the tissue. Single-compartment kinetic models have been developed to describe the spatially averaged concentration (or content) of matrix components within cartilage tissue explants resulting from this metabolic balance, usually during long-term tissue culture. These models have proven useful in the determination of the effects of growth factor stimulation on the biosynthesis and loss (degradation) of proteoglycans and collagens (Hascall et al., 1990; Morales and Roberts, 1988; Sah et al., 1994). In these studies, the time courses of generation and loss of metabolically radiolabeled molecules were analyzed, and parameters were fit to an exponential model to derive rate constants. Similar compartmental models have also been used to investigate the kinetics of metabolic processes, such as intracellular proteoglycan synthesis and processing in a cell monolayer model (Lohmander and Kimura, 1986), and enzymatic collagen cross-linking in the cartilage extracellular matrix (Ahsan et al., 2000).

Compartmental models are governed by the concept of mass balance. The rate of change of material concentration (c) in a compartment is given by the difference between the rate of gain of molecules, r_g, and the rate of loss, r_l:

$$\frac{dc}{dt} = r_g - r_l \qquad (1)$$

In tissue explant studies, these rates are often calculated by measuring the concentration of a component at various times, both within the tissue and released into the medium. The resulting rates can then be fit to a model to determine rate constants. As an example, in one such model, the rate of change of proteoglycan concentration (c_p) is given by the difference in the synthesis rate k_s (constant in time) and the rate of loss, described as a first-order function of proteoglycan concentration (Hascall et al., 1990):

$$\frac{dc_p}{dt} = k_s - k_l c_p, \qquad (2)$$

where k_l is a rate constant.

In models of this type, only spatially averaged concentrations can be calculated, so that spatial concentration profiles are not possible. Additionally, the lumped parameters of a compartmental model are determined empirically, and depend on the geometry of the system under study. The translation of parameters derived for one geometry to another configuration is not clear.

Continuum Metabolic Model of Cartilage Growth

In vivo, matrix is distributed nonuniformly from the articular surface to the deeper layers (Maroudas, 1979). It is likely that these concentration profiles arise from synthesis, transport, binding, and degradation processes that occur in a spatially varying manner. Newly formed matrix molecules are secreted from cells into the extracellular matrix, where they may undergo binding interactions with the existing matrix, becoming effectively immobilized (Fig. 15.4). Alternately, they may remain soluble and be transported out of the matrix via diffusive or convective (flow-driven transport) mechanisms. Furthermore, matrix molecules can be degraded by a variety of endogenous proteolytic enzymes (Dean et al., 1989; Hollander et al., 1994), creat-

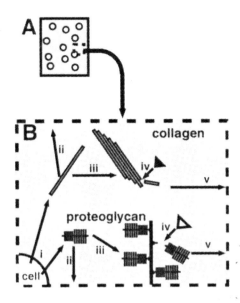

FIGURE 15.4. Cells in cartilage explants (*A*) produce matrix components (*B*). These newly formed molecules are secreted from the cells (*i*) and migrate into the matrix, where they may continue to diffuse away from the cell (*ii*) or become immobilized within the existing matrix (*iii*). The action of proteolytic enzymes (*iv*) can cause degradation of the matrix, leading to the release and transport (*v*) of degraded matrix molecules.

ing molecular fragments that may also diffuse through, and possibly out of, the tissue (Bolis et al., 1989; Sandy and Verscharen, 2001). These processes can result in spatially varying profiles of concentrations of newly synthesized, bound, and degraded matrix components, which are not described by a compartmental model.

The development of models involving spatial distributions of matrix molecules within cartilage builds on the large body of existing work in which the diffusion and reaction of exogenous, high-molecular-weight tracer molecules into and through devitalized cartilage has been studied, both experimentally (Bhakta et al., 2000; Schneiderman et al., 1995; Maroudas, 1976b) and mathematically (Torzilli et al., 1987). Recently, continuum models of cartilage metabolism have been developed to describe the relationships of nutrient transport (Galban and Locke, 1997, 1999) and matrix deposition (Obradovic et al., 2000) to the growth of tissue-engineered cartilage constructs. The following is a simplified form of this model, describing diffusion and reaction phenomena in free-swelling cartilage. A simple example of an application of this model in the context of an impermeable barrier (e.g., representing subchondral bone or a solid tissue culture surface) is given, and shows the development of a steady-state composition profile. The underlying hypothesis of the model is that processes of molecular biosynthesis, transport, binding, and degradation contribute to the dynamic regulation of cartilage composition.

Model Equations

In the derivation of this model, cartilage is treated as a continuum, in which its physical properties are defined by continuous functions of space. As with the compartmental model, a continuum model depends on mass balance descriptions of pools of newly formed molecules, bound molecules, and degraded molecules. In a continuum model, these parameters can be specified as functions of time and space, using knowledge of the system under study. The solution of these equations for each pool, subject to appropriate initial and boundary conditions, represents a description of the temporal and spatial variation of matrix components within a tissue explant. The following system of equations is an example of a one-dimensional continuum formulation allowing distribution of molecules within newly synthesized, bound, and degraded pools:

$$\text{newly synthesized:} \quad \frac{\partial c_s}{\partial t} = D_s \frac{\partial^2 c_s}{\partial x^2} + r_s - r_b, \quad (3)$$

$$\text{bound:} \quad \frac{\partial c_b}{\partial t} = r_b - r_d, \quad (4)$$

$$\text{degraded:} \quad \frac{\partial c_d}{\partial t} = D_d \frac{\partial^2 c_d}{\partial x^2} + r_d. \quad (5)$$

In these equations, c_s, c_b, and c_d refer to the concentrations of newly synthesized, bound, and degraded molecules. The mass balance allows diffusive transport of soluble components, and processes of synthesis (r_s), binding (r_b), and degradation (r_d). Each of these rates can be more exactly specified, and may depend on space, time, and concentrations.

Example: Steady-State Concentration of a Matrix Component in a Tissue Explant

The following example will be solved first using a compartmental model, then with a continuum model, so that the results may be compared. Consider the steady-state concentration of a newly synthesized matrix component within the cartilage portion of an osteochondral explant floating in an infinite, well-stirred bath of culture medium containing no soluble matrix molecules (Fig. 15.5A). In the compartmental model, setting the rate-of-change of the concentration in (2) to zero, the steady-state concentration $\langle c_{\circ,comp} \rangle$ is calculated as

$$\langle c_{\circ,comp} \rangle = k_s / k_l. \quad (6)$$

A continuum model can also describe the steady-state matrix concentration in the tissue, with boundary conditions defined by the culture configuration. Since the medium in contact with the articular surface ($z = 0$) contains no matrix molecules, the boundary condition at the articular surface is that the matrix concentration vanishes:

$$c(0) = 0. \quad (7)$$

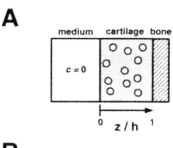

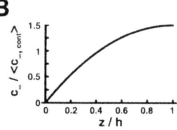

FIGURE 15.5. Continuum model of cartilage metabolism. (*A*) Cartilage explant culture configuration. (*B*) Resulting steady-state profile of matrix component within a cultured cartilage explant, as determined by a continuum model.

Assuming the bone-cartilage interface ($z = h$) to be impermeable to mobile matrix molecules, there is a no-flux boundary condition at this point, which is described as

$$\frac{\partial c}{\partial z}(h) = 0. \tag{8}$$

To simplify this example of a continuum model, assume that matrix component is synthesized in a soluble form, and is transported through the tissue solely by a diffusive process. If the biosynthetic rate is defined as the constant k_s, and the diffusivity D_s is also constant with respect to z and t, the mass-balance equation describing the spatial distribution of the newly synthesized matrix is derived from (3) as

$$\frac{\partial c}{\partial t} = D_s \frac{\partial^2 c}{\partial z^2} + k_s. \tag{9}$$

Then, the steady-state concentration profile $c_\infty(z)$ is given by setting the left side of equation (9) equal to zero and solving for c, subject to the boundary conditions (7,8), giving

$$c_\infty(z) = \frac{k_s z}{D_s}(h - \frac{z}{2}). \tag{10}$$

The average steady-state concentration within the tissue, $\langle c_{\infty,cont} \rangle$, is the integral of $c_\infty(z)$ with respect to z, divided by the thickness h:

$$\langle c_{\infty,cont} \rangle = \frac{k_s h^2}{3 D_s}. \tag{11}$$

The resulting concentration profile depicts the highest concentration at the deep layers of the tissue, with a decreasing concentration at the articular surface, as dictated by the boundary conditions (Fig. 15.5B). Comparison of equations (6) and (11) indicates that the average steady-state concentrations predicted by the compartmental and continuum models are similar in mathematical form. In the continuum model, however, the lumped parameter k_l is given by $3D_s/h^2$, which explicitly describes the loss rate in terms of transport and geometric parameters.

Discussion

The continuum model described above is able to predict spatial and temporal variations in matrix composition. Nonuniform spatial distributions of matrix components are a compositional feature of articular cartilage, since proteoglycan content and, to a lesser extent, collagen content, vary with distance from the articular surface toward the subchondral bone. These spatial variations in composition are thought to be important in determining depth-varying mechanical properties (e.g., Schinagl et al., 1997).

Studies on cartilage growth may also benefit from a continuum formulation in which temporal and spatial variations are included. During cartilage maturation from the fetus to the adult, tissue metabolism slows while matrix degradation increases, leading to the cessation of growth and maintenance of a steady-state composition. Furthermore, the metabolic responses of immature and mature cartilage to external stimuli are distinct (Brand et al., 1991; Li et al., 2001; Sandy and Plaas, 1986). Developmental changes can be modeled as long-term, time-varying relations between applied stimuli and metabolic responses.

A continuum model is valuable in its ability to allow coupling between internal or external stimuli, whether mechanical or biological, and the resulting metabolic shifts within the tissue. Those stimuli that have been shown to increase cell metabolism of certain matrix components (growth factors, dynamic mechanical compression, and fluid shear) can be included within the synthesis term in the expression for the newly synthesized pool [r_s in (3)]. Similarly, factors that promote degradation can be reflected as an increase in the r_d term in (5). Also, interactions between cytokine stimulation and the application of mechanical loads, such as that described recently for IGF-1 and static mechanical loading of cartilage explants (Bonassar et al., 2000) can be incorporated into this type of model. In the next section, biomechanical models that explicitly describe how mechanical stimuli regulate tissue growth are presented.

Biomechanical Models of Tissue Growth

The development of models that are capable of quantifying the complex relationship between mechanical factors and tissue growth and remodeling has been an important goal of bioengineering. A number of investigators have studied the remodeling of bone tissue. Cowin and

Hegedus (1976) developed a bone-remodeling theory for the solid, porous matrix. Numerous studies have proposed stress-strain relations and remodeling equations using this model, and analytical and numerical solutions have been developed (Cowin, 1993). In a different approach, Carter and colleagues (Beaupre et al., 1990; Carter and Wong, 1988) have developed a bone-remodeling theory in which the change in apparent density depends on the difference between a daily remodeling stimulus and a tissue homeostatic state (Beaupre et al., 1990). In both of these bone-remodeling theories, growth is represented by a change in a scalar parameter (volume fraction or apparent density).

In contrast to those approaches, other models have been proposed in which growth is represented by a tensor. Skalak and colleagues (Skalak et al., 1982, 1996, 1997) made several important contributions in the area of growth mechanics; in particular, in the report of Skalak et al. (1996) it was observed that nonuniform growth can lead to the development of residual stresses in the tissue, which affects its overall mechanical response. A general theory of growth for incompressible soft tissues was introduced by Rodriguez et al. (1994), which has been further developed by Chen and Hoger (2000) and Klisch et al. (2001c).

In (Chen and Hoger, 2000; Klisch et al., 2001c; Rodriguez et al., 1994), the deformation gradient due to growth was decomposed into two parts: a growth tensor that describes the amount and orientation of mass deposition, and an elastic accommodation tensor that ensures the continuity of the growing body. The introduction of the growth tensor requires a growth law, which defines the time rate of change of the growth tensor as a function of mechanical stimuli. Taber and colleagues have used models based on (Rodriguez et al., 1994) to study the stress-modulated growth of both the heart (Lin and Taber, 1995) and the aorta (Taber, 1998; Taber and Eggers, 1996) using several growth laws. Van Dyke and Hoger (2000s) further extended the theory and presented an implementation in which the growth law is defined on the current configuration of the growing material; this implementation has been used to model the growth of aortic tissue.

Cartilage Growth Mixture Model

The use of a biomechanical model of cartilage growth as a tool for quantifying the evolution of the composition, structure, and function of tissue-engineered cartilage may aid the successful repair of cartilage defects. Since the proteoglycan and collagen constituents of cartilage serve distinct mechanical roles, a model of cartilage growth that can predict the evolution of the tissue's mechanical properties during growth should allow these constituents to grow independently of each other. Recently, Klisch and Hoger (2001) derived a general growth mixture theory of an arbitrary number of growing elastic materials and a fluid, from which a cartilage growth mixture model was proposed (Klisch et al., 2000). The development in (Klisch and Hoger, 2001; Klisch et al., 2000) generalized previous work on the growth of elastic materials (Chen and Hoger, 2000; Klisch et al., 2001c; Rodriguez et al., 1994). In the cartilage growth mixture model, the proteoglycan and collagen constituents are allowed to develop distinct elastic strains during a growth process, which leads to an evolution in the composition and the mechanical properties of the solid matrix. Thus, the cartilage growth mixture model is capable of predicting remodeling effects associated with changes in tissue composition. In this section, the essential ingredients of the growth mixture theory of (Klisch and Hoger, 2001), along with a generic growth problem presented in (Klisch et al., 2000), are discussed.

Model Equations

The cartilage is modeled as a continuum mixture of two growing elastic materials, proteoglycan (p) and collagen (c), and a fluid (f). Here, in order to simplify the presentation, only equilibrium configurations are considered. The deformation gradient $\underset{\sim}{\mathbf{F}}^\alpha$ (superscript α = p or c) describes the overall deformation for each growing solid relative to a fixed reference configuration, and is decomposed as (Fig. 15.6)

$$\underset{\sim}{\mathbf{F}}^\alpha = \underset{\sim}{\mathbf{M}}^\alpha_e \, \underset{\sim}{\mathbf{M}}^\alpha_g . \qquad (12)$$

The tensor $\underset{\sim}{\mathbf{M}}^\alpha_e \underset{\sim}{\mathbf{M}}^\alpha_g$ describes the deformation due to growth relative to the fixed reference configuration, where the amount and orientation of

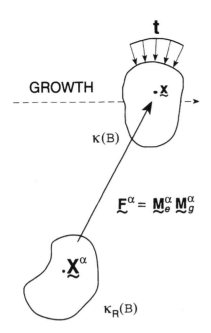

FIGURE 15.6. Decomposition of the deformation gradient tensors for the growing materials.

$$\rho^{\alpha} \det \underset{\sim}{\mathbf{F}}^{\alpha} = \rho_0^{\alpha} \exp \left(\int_{\tau=t_0}^{t} c^{\alpha} d\tau \right) \quad (14)$$

Here, ρ_0^{α} is the density in the reference configuration and c^{α} is the constituent mass growth function (rate of mass deposition per unit current mass). Due to the introduction of c^{α}, additional equations are required. It is assumed that the change in density is only due to the elastic part of the deformation, so

$$\rho^{\alpha} \det \underset{\sim}{\mathbf{M}}_{e}^{\alpha} = \rho_0^{\alpha}. \quad (15)$$

Combined with (14), this leads to the growth continuity equations

$$\det \underset{\sim}{\mathbf{M}}_{g}^{\alpha} = \exp \left(\int_{\tau=t_0}^{t} c^{\alpha} d\tau \right). \quad (16)$$

The equilibrium equations are

$$\operatorname{div} \underset{\sim}{\mathbf{T}}^{\alpha} = 0, \quad (17)$$

where $\underset{\sim}{\mathbf{T}}^{\alpha}$ is the Cauchy stress tensor. For non-equilibrium conditions, there exist additional terms, such as diffusive forces for each constituent and balance equations for angular momentum and energy (Klisch and Hoger, 2001). Also, constitutive equations are required for the stresses that appear in (17). In Klisch and Hoger (2001), general restrictions for the stress constitutive equations were derived from thermodynamical considerations.

To obtain a complete theory, equations that describe the time-rate of change of the growth tensors, $\dot{\underset{\sim}{\mathbf{M}}}_{g}^{\alpha}$, need to be specified. The growth law provides a description of the rate at which material is deposited (or resorbed) at a point, the orientation in which material deposition occurs, and the way in which mechanical factors influence mass deposition. In Van Dyke and Hoger (2001), it is argued that the physically intuitive growth law must be defined relative to the present configuration of a growing body. Since $\underset{\sim}{\mathbf{M}}_{g}$ is defined relative to a fixed reference configuration, a kinematic relationship between $\dot{\underset{\sim}{\mathbf{M}}}_{g}^{\alpha}$ and the growth law was established in Van Dyke and Hoger, (2001). For cartilage, the growth law for both the proteoglycan and collagen constituents may depend on a combination of mechanical stimuli; such as stress, strain, strain energy, strain rate, and interstitial fluid flow. This requires a constitutive equation of the form

mass deposition are described by $\underset{\sim}{\mathbf{M}}_{g}^{\alpha}$. The tensor $\underset{\sim}{\mathbf{M}}_{e}^{\alpha}$ is the elastic accommodation tensor that ensures continuity of the growing body, and may include a contribution arising from a superposed elastic deformation. In a tissue-engineering application, the fixed reference configuration may represent some early state of the growing tissue in which a protoeglycan-collagen matrix has formed. The internal constraint of intrinsic incompressibility (Mills, 1966) that is commonly employed in the study of cartilage mechanics (Mow et al., 1980) is used. Also, it is assumed that all of the proteoglycan and collagen molecules are bound to the extracellular solid matrix, so that their deformation gradients are equal. These constraints are expressed as

$$(\rho^{p}/\rho^{pT}) + (\rho^{c}/\rho^{cT}) + (\rho^{f}/\rho^{fT}) = 1, \quad \underset{\sim}{\mathbf{F}}^{p} = \underset{\sim}{\mathbf{F}}^{c}, \quad (13)$$

where ρ^{α} is the apparent density (per tissue volume) and $\rho^{\alpha T}$ is the true density (per constituent volume).

For the proteoglycan and collagen constituents, the balance of mass at time t can be written as

$$\dot{\underset{\sim}{\mathbf{M}}}{}^{\alpha}_{g} = \hat{\underset{\sim}{G}}\ (\underset{\sim}{\mathbf{M}}{}^{\alpha}), \qquad (18)$$

where $\hat{G}$ is some experimentally determined function of mechanical stimuli M^{α}.

Example: Homogeneous, Isotropic Growth of Unloaded Tissue Explants

In this section an example is presented (Klisch et al., 2000) that illustrates several of the main features of the cartilage growth mixture model. The particular example considers the uniform and isotropic growth of a sphere of cartilage for which the initial and final configurations are in equilibrium in the absence of external loads. Although external loading may serve to drive the growth process in biological tissues, in this example we restrict ourselves to studying the evolution of the solid matrix in an unloaded configuration and take the growth tensors to be independent of mechanical quantities. Therefore, we select specific forms for the growth tensors and use a no-load boundary condition. Although this example is restrictive and does not accurately reflect the actual growth process of cartilage *in vivo* or *in vitro,* it is simple enough to obtain exact solutions and to illustrate the capabilities of the cartilage growth mixture model.

Thus, consider homogeneous, isotropic growth tensors for the proteoglycan and collagen constituents which appear as spherical tensors

$$\underset{\sim}{\mathbf{M}}{}^{a}_{g} = g^{a}\,\underset{\sim}{\mathbf{1}}, \qquad (19)$$

where g^{p} and g^{c} are the *growth stretches*. The growth stretches are specified independently of each other, and the growth ratio $g^{p/c}$ is defined as the ratio of the proteoglycan to collagen growth stretches. The tensors $\underset{\sim}{\mathbf{M}}{}^{a}_{e}$ represent the elastic deformations required to maintain the constraint $(13)_2$; homogeneous, isotropic solutions exist of the form

$$\underset{\sim}{\mathbf{M}}{}^{a}_{e} = \lambda^{a}\underset{\sim}{\mathbf{1}}\ , \qquad (20)$$

where λ^{p} and λ^{c} are called the elastic stretches. For this special problem, the no-load boundary condition requires that the solid matrix stress $\underset{\sim}{\mathbf{T}}$ vanishes everywhere:

$$\underset{\sim}{\mathbf{T}} = \underset{\sim}{\mathbf{T}}{}^{p} + \underset{\sim}{\mathbf{T}}{}^{c} = 0. \qquad (21)$$

The proteoglycan stress constitutive equation $\underset{\sim}{\mathbf{T}}{}^{p}$ was derived using the model of (Basser et al., 1998) and compositional data for an adult bovine articular cartilage specimen (Williamson et al., 2001). An isotropic strain energy function for the solid matrix (Almeida and Spilker, 1997) was used to derive the stress-strain equation for the solid matrix, $\underset{\sim}{\mathbf{T}}$. Since the solid matrix stress is equal to the sum of the proteoglycan and collagen stresses, the collagen stress constitutive equation $\underset{\sim}{\mathbf{T}}{}^{c}$ was formed by subtracting $\underset{\sim}{\mathbf{T}}{}^{p}$ from $\underset{\sim}{\mathbf{T}}$ and evaluating the resulting equation at $\underset{\sim}{\mathbf{M}}{}_{e}$.

To solve a particular problem, $g^{p/c}$ is specified. The boundary condition (21) then yields an algebraic equation

$$f(\lambda^{p}, g^{p/c}) = 0, \qquad (22)$$

which is solved numerically to determine λ^{p}. For each equilibrium state following a prescribed growth, the aggregate modulus H_{A0} of the solid matrix is calculated by superimposing identical confined compression deformations on the proteoglycan and collagen constituents.

For example, consider the special case of growth where only additional proteoglycans are deposited (Fig. 15.7). In this special example, the stress supported by the proteoglycan becomes more compressive due to the increase in fixed charge density. In order to achieve a new equilibrium state for which there is a balance of stresses, the stress supported by the collagen network must become more tensile.

For various values of $g^{p/c}$, the calculated stress components, aggregate modulus, and contents are shown in Figure 15.8. When $g^{p/c} > 1$, more proteoglycan is deposited than collagen, and the proteoglycan stress becomes more compressive while the collagen stress becomes more tensile. Also, the aggregate modulus H_{A0} increases substantially as $g^{p/c} > 1$.

This example illustrates several features of the cartilage growth model. First, the model allows the constituents in the mixture to grow independently of each other. Second, the model describes the evolution of the constituent stresses due to growth. Third, the model can quantify the evolution of the mechanical properties of the tissue during growth. Last, the model can predict the evolution of the tissue composition due to growth.

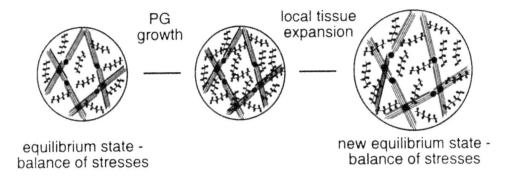

FIGURE 15.7. Schematic of a special case of cartilage growth, where only additional proteoglycans are deposited.

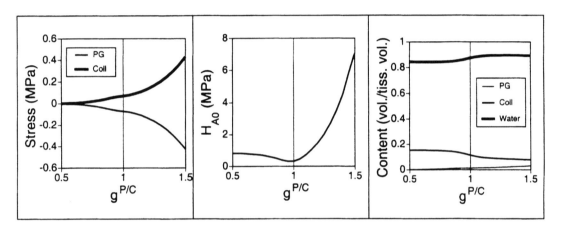

FIGURE 15.8. Results for the constituent stresses, solid matrix aggregate modulus, and tissue composition for the equilibrium configurations following isotropic growth of the proteoglycan and collagen constituents.

Discussion

The model developed here describes how the solid matrix of cartilage grows and remodels as a result of volumetric mass deposition of the proteoglycan and collagen constituents. A key assumption of this cartilage growth mixture model is that the material added during growth is assumed to have the same mechanical properties as the original material (Klisch and Hoger, 2001; Klisch et al., 2000). When the proteoglycan and collagen constituents grow without remodeling, the composition of the solid matrix may evolve over time and, consequently, the mechanical properties of solid matrix may change. Alternatively, when the proteoglycan and the collagen constituent experience remodeling without mass deposition or resorp-

tion, the structural and mechanical properties of the solid matrix may also change.

In Klisch et al. (2001a,b), the cartilage growth mixture model was extended to allow the collagen material constants to change during remodeling. In particular, an isotropic, linear collagen stress-strain equation was proposed with material constants that depended on the collagen crosslink density. In their work (Klisch et al., 2001a,b), this cartilage growth/remodeling mixture model was used to describe the changes observed in the composition and the mechanical properties of developing bovine cartilage. The results indicated that the collagen remodels during development; as it appears that the cross-link density increases to produce a stiffer collagen network. Also, that study outlined how the cartilage growth mixture model can be refined so that alterations in molecular structure, such as crosslink density, can be included in a quantitative analysis of the complex relationship between mechanical and biochemical stimuli, cell metabolism, and tissue growth.

The growth mixture theory developed by Klisch and Hoger (2001) is general enough so that it can be used in a variety of future applications. For example, the cartilage growth/remodeling mixture model developed in (Klisch et al., 2001a,b) can be used to study the development of osteoarthritis. Experimental measurements with healthy and osteoarthritic cartilage have shown that osteoarthritis is accompanied by an increase in tissue hydration and a decrease in proteoglycan content (Maroudas, 1976a), leading to the hypothesis that collagen damage increases in osteoarthritis (Basser et al., 1998). The cartilage growth/remodeling mixture model may be used to study the evolution of the tissue during osteoarthritis by specifying a proteoglycan growth tensor that leads to a decrease in proteoglycan content, and by including a collagen damage parameter as the remodeling parameter in the collagen stress.

The growth mixture theory is capable of modeling several phenomena observed in cartilage growth and remodeling that are not described by the model presented here. Recall that a small fraction of the proteoglcyans and the collagens are mobile and not bound to the extracellular matrix. Using the theory presented in the work of Klisch and colleagues (Klisch and Hoger, 2001; Klisch et al., 2000), the solid matrix can be modeled as being composed of four constituents: two that represent the proteoglycan and collagen molecules that are bound to the extracellular matrix, and two that represent the mobile proteoglycan and collagen molecules. Also, the role played by growth factors in the regulation of tissue growth can be modeled by allowing the growth factors to be represented by scalar parameters that parameterize the growth law. Then, through a series of experiments, a growth law with coefficients that depend on the growth factor may be formulated.

A formidable challenge in cartilage biomechanics is the determination of accurate stress-strain constitutive equations. For the cartilage growth mixture model, these equations need to be determined for the solid matrix, the proteoglycan constituent, and the collagen constituent. This is a crucial aim in developing the cartilage growth mixture model; because accurate constitutive equations must be used in order to determine the other parameters in the theory: for example, the growth tensors, the growth law, the remodeling parameters, etc.

The determination of the growth law may require experiments that quantify the *in vitro* growth of tissue-engineered constructs. For example, experiments may be performed that apply known levels of tissue stress, strain, and fluid diffusion. Also, the effect of various types of growth hormones on the cell metabolism during tissue engineering experiments can be quantified. These types of experiments will aid in identifying the form of the growth law; that is, the quantitative relationship between the evolution of the growth tensor and the mechanical/biochemical stimuli.

Conclusion

Many tissue engineering experiments have been, and continue to be, conducted with the aim of bioengineering the growth of articular cartilage. In order to achieve a better understanding of the complex relationship between mechanical and biochemical stimuli, cell metabolism, and tissue growth, these experimental approaches may be supplemented with quantitative models. At the present time, these types of quantitative models are sparse in the field of cartilage tissue engineering, although similar models have been used in the field of bone mechanics.

In this chapter, two types of models are reviewed that are capable of predicting the evolution of cartilage tissue composition, structure, and function as regulated by biochemical and mechanical stimuli. The continuum metabolic model of cartilage growth describes the dependence of cartilage matrix composition on matrix formation, binding, degradation, and molecular transport within and from the cartilage tissue. The cartilage growth mixture model, derived from the more general continuum growth mixture theory, describes the relationship between mechanical stimuli and cartilage composition, structure, and function during growth. In the future, it may prove useful to combine these models. For example, the growth mixture theory allows the definition of both newly synthesized and degraded mobile constituents. Then, the equations derived in the continuum metabolic model of cartilage growth would be used in a generalized cartilage growth mixture model.

These models may be useful in quantifying the growth and remodeling of tissue-engineered cartilage constructs, both before and after implantation. They can predict the manner in which constructs should be stimulated *in vitro* for the fabrication of better repair implants. Also, they can be used to design computer models of the *in vivo* growth, degeneration, and repair processes, which can be used as predictive tools for the design of therapies aimed at the successful repair of growth and degenerative abnormalities.

References

Ahsan T., Harwood F.H., Amiel D., Sah R.L. 2000. Kinetics of collagen crosslinking in adult bovine articular cartilage. *Trans. Orthop. Res. Soc.* 25:111.

Almeida E.S., Spilker R.L. 1997. Mixed and penalty finite element models for the nonlinear behavior of biphasic soft tissues in finite deformation. Part I— alternative formulations. *Comp. Meth. Biomech. Biomed. Eng.* 1:25–46.

Basser P.J., Schneiderman R., Bank R., Wachtel E., Maroudas A. 1998. Mechanical properties of the collagen network in human articular cartilage as measured by osmotic stress technique. *Arch. Biochem. Biophys.* 351:207–219.

Beaupre G.S., Orr T.E., Carter D.R. 1990. An approach for time-dependent bone modeling and remodeling—theoretical development. *J. Orthop. Res.* 8:651–661.

Bhakta N.R., Garcia A.M., Frank E.H., Grodzinsky A.J., Morales T.I. 2000. The insulin-like growth factors (IGFs) I and II bind to articular cartilage via the IGF-binding proteins. *J. Biol. Chem.* 275: 5860–5866.

Bolis S., Handley C.J., Comper W.D. 1989. Passive loss of proteoglycan from articular cartilage explants. *Biochim. Biophys. Acta* 993:157–167.

Bonassar L.J., Grodzinsky A.J., Srinivasan A., Davila S.G., Trippel S.B. 2000. Mechanical and physicochemical regulation of the action of insulin-like growth factor-I on articular cartilage. *Arch. Biochem. Biophys.* 379:57–63.

Brand H.S., de Koning M.H., van Kampen G.P., van der Korst J.K. 1991. Age related changes in the turnover of proteoglycans from explants of bovine articular cartilage. *J. Rheumatol.* 18:599–605.

Brittberg M., Lindahl A., Nilsson A., Ohlsson C., Isaksson O., Peterson L. 1994. Treatment of deep cartilage defects in the knee with autologous chondrocyte transplantation. *N. Engl. J. Med.* 331:889–895.

Buckwalter J.A., Mankin H.J. 1997. Articular cartilage. Part I: tissue design and chondrocyte-matrix interactions. *J. Bone Joint Surg.* 79-A:600–611.

Buckwalter J.A., Mankin H.J. 1998. Articular cartilage: degeneration and osteoarthritis, repair, regeneration, and transplantation. *Instr. Course Lect.* 47:487–504.

Butler D.L., Goldstein S.A., Guilak F. 2000. Functional tissue engineering: the role of biomechanics. *J. Biomech. Eng.* 122:570–575.

Carter D.R., Wong M. 1988. Mechanical stresses and endochondral ossification in the chondroepiphysis. *J. Orthop. Res.* 6:148–154.

Chen Y.C., Hoger A. 2000. Constitutive functions for elastic materials in finite growth and deformation. *J. Elast.* 59:175–193.

Cowin S.C. 1993. Bone stress adaptation models. *J. Biomech. Eng.* 115:528–533.

Cowin, S.C., Hegedus D.M. 1976. Bone remodeling I: A theory of adaptive elasticity. *J. Elast.* 6:313–325.

Dean D.D., Martel-Pelletier J., Pelletier J.P., Howell D.S., Woessner J.F.J. 1989. Evidence for metalloproteinase and metalloproteinase inhibitor imbalance in human osteoarthritic cartilage. *J. Clin. Invest.* 84:678–685.

Dorland W.A. 1981. *Dorland's Illustrated Medical Dictionary.* W.B. Saunders Company, Philadelphia.

Evans C.H., Watkins S.C., Stefanovic-Racic M. 1996. Nitric oxide and cartilage metabolism. *Methods Enzymol.* 269:75–88.

Fosang A.J., Tyler J.A., Hardingham T.E. 1991. Effect of interleukin-1 and insulin like growth factor-1 on the release of proteoglycan components and hyaluronan from pig articular cartilage in explant cultures. *Matrix* 11:17–24.

Fujimoto E., Ochi M., Kato Y., Mochizuki Y., Sumen Y., Ikuta Y. 1999. Beneficial effect of basic fibroblast growth factor on the repair of full-thickness defects in rabbit articular cartilage. *Arch. Orthop. Trauma Surg.* 119:139–145.

Fung Y.C. 1993. *Biomechanics: Mechanical Properties of Living Tissues.* Springer-Verlag, New York.

Furuto D.K., Miller E.J. 1987. Isolation and characterization of collagens and procollagens. *Methods Enzymol.* 144:41–139.

Galban C.J., Locke B.R. 1997. Analysis of cell growth in a polymer scaffold using a moving boundary approach. *Biotechnol. Bioeng.* 56:422–432.

Galban C.J., Locke B.R. 1999. Analysis of cell growth kinetics and substrate diffusion in a polymer scaffold. *Biotechnol. Bioeng.* 65:121–132.

Grande D.A., Singh I.J., Pugh J. 1987. Healing of experimentally produced lesions in articular cartilage following chondrocyte transplantation. *Anat. Rec.* 218:142–148.

Grodzinsky A.J. 1983. Electromechanical and physicochemical properties of connective tissue. *CRC Crit. Rev. Bioeng.* 9:133–199.

Guilak F., Sah R.L., Setton L.A. 1997. In Physical regulation of cartilage metabolism *Basic Orthopaedic Biomechanics.* V.C. Mow, W.C. Hayes, eds. Raven Press, New York, pp. 179–207.

Hardingham T.E., Bayliss M.T., Rayan V., Noble D.P. 1992. Effects of growth factors and cytokines on proteoglycan turnover in articular cartilage. *Br. J. Rheum.* 31S1:1–6.

Hascall V.C., Luyten F.P., Plaas A.H.K., Sandy J.D. 1990. In *Methods in Cartilage Research.* A. Maroudas, K. Kuettner, eds. Academic Press, San Diego, p. 108–112.

Hjertquist S.O., Lemperg R. 1969. Transplantation of autologous costal cartilage to an osteochondral defect on the femoral head. Histological and autoradiographical studies in adult rabbits after administration of 35S-sulphate and 3H-thymidine. *Virchows Arch. Pathol. Anat. Physiol. Klin. Med.* 346:345–360.

Hodge W.A., Fijan R.S., Carlson K.L., Burgess R.G., Harris W.H., Mann R.W. 1986. Contact pressures in the human hip joint measured in vivo. *Proc. Natl. Acad. Sci. U.S.A.* 83:2879–2883.

Hollander A.P., Heathfield T.F., Webber C., Iwata Y., Bourne R., Rorabeck C., Poole A.R. 1994. Increased damage to type II collagen in osteoarthritic articular cartilage detected by a new immunoassay. *J. Clin. Invest.* 93:1722–1732.

Hutchinson N.I., Lark M.W., MacNaul K.L., Harper C., Hoerrner L.A., McDonnell J., Donatelli S., Moore V., Bayne E.K. 1992. In vivo expression of stromelysin in synovium and cartilage of rabbits injected intraarticularly with interleukin-1β. *Arthritis Rheum* 35:1227–1233.

Kandel R.A., Pritzker K.P.H., Mills G.B., Cruz T.F. 1990. Fetal bovine serum inhibits chondrocyte collagenase production: interleukin 1 reverses this effect. *Biochim. Biophys. Acta.* 1053:130–134.

Klisch S.M., Hoger, A. 2001. A theory of volumetric growth for compressible elastic biological materials. *Math. Mech. Solids.* 6:551

Klisch S.M., Sah R.L., Hoger A. 2000. A growth mixture theory for cartilage. *ASME Mech. Biol.* BED-46:229–242.

Klisch S.M., Chen S.S., Hoger A., Sah R.L. 2001a. Modeling the compositional and mechanical changes in developing bovine articular cartilage using a cartilage growth theory. *J. Biomech. Eng.* (in revision)

Klisch S. M., Chen, S.S., Masuda, K., Thonar E.J.-M.A., Hoger A., Sah R.L. 2001b. Application of a growth and remodeling mixture theory to developing articular cartilage. *Trans. Orthop. Res. Soc.* 26:316.

Klisch S.M., Van Dyke T., Hoger A. 2001c. A theory of volumetric growth for compressible elastic materials. *Math. Mech. Solids* 6:551–575.

Lai W.M., Hou J.S., Mow V.C. 1991. A triphasic theory for the swelling and deformation behaviors of articular cartilage. *J. Biomech. Eng.* 113:245–258.

Li K.W., Williamson A.K., Wang A.S., Sah R.L. 2001. Growth responses of cartilage to static and dynamic compression. *Clin. Orthop.* 391 (Supp): S34–48.

Lin I.E., Taber L. 1995. A model for stress-induced growth in the developing heart. *J. Biomech. Eng.* 117:343–349.

Lohmander L.S., Kimura J. 1986. Biosynthesis of cartilage proteoglycan. In: *Articular Cartilage Biochemistry.* K. Kuettner, R. Schleyerbach, V.C. Hascall, eds. Raven Press, New York.

Maroudas A. 1976a. Balance between swelling pressure and collagen tension in normal and degenerate cartilage. *Nature* 260:808–809.

Maroudas A. 1976b. Transport of solutes through cartilage: permeability to large molecules. *J. Anat.* 122:335–347.

Maroudas A. 1979. Physiochemical properties of articular cartilage. In: *Adult Articular Cartilage.* M.A.R. Freeman, ed Pitman Medical, Tunbridge Wells, U.K., pp. 215–290.

Maroudas A., Bullough P., Swanson S.A.V., Freeman M.A.R. 1968. The permeability of articular cartilage. *J. Bone Joint Surg.* 50-B:166–177.

Merriam-Webster Collegiate Dictionary. 2001. Merriam-Webster, Springfield, MA.

Mills N. 1966. Incompressible mixtures of Newtonian fluids. *Int. J. Eng. Sci.* 4:97–112.

Morales T.I., Roberts A.B. 1988. Transforming growth factor-β regulates the metabolism of proteoglycans in bovine cartilage organ cultures. *J. Biol. Chem.* 263:12828–12831.

Mow V.C., Ratcliffe A. 1997. In: *Basic Orthopaedic Biomechanics*. V.C. Mow, W.C. Hayes, eds. Raven Press, New York, pp. 113–178.

Mow V.C., Kuei S.C., Lai W.M., Armstrong C.G. 1980. Biphasic creep and stress relaxation of articular cartilage in compression: theory and experiment. *J. Biomech. Eng.* 102:73–84.

Nixon A.J., Fortier L.A., Williams J., Mohammed H. 1999. Enhanced repair of extensive articular defects by insulin-like growth factor-I-laden fibrin composites. *J. Orthop. Res.* 17:475–487.

Obradovic B., Meldon J.H., Freed L.E., Vunjak-Novakovic G. 2000. Glycosaminoglycan deposition in engineered cartilage: experiments and mathematical model. *AIChE J* 46:1860–1871.

Ogden J.A. 1988. In: *Behavior of the Growth Plate*. H.K. Uhthoff, J.J. Wiley, ed. Raven Press, New York, pp. 85–96.

Pal S., Tang L.-H., Choi H., Habermann E., Rosenberg L., Roughley P., Poole A.R. 1981. Structural changes during development in bovine fetal epiphyseal cartilage. *Collagen Rel. Res.* 1:151–176.

Pauwels, F. 1976. *Biomechanics of the Normal and Diseased Hip; Theoretical Foundations, Technique and Results of Treatment; An Atlas.* Springer-Verlag, New York.

Pottenger L.A., Webb J.E., Lyon N.B. 1985. Kinetics of extraction of proteoglycans from human cartilage. *Arthritis Rheum.* 28:323–330.

Rodriguez E.K., Hoger A., McCulloch A.D. 1994. Stress-dependent finite growth in soft elastic tissues. *J. Biomech.* 27:455–467.

Sah R.L., Chen A.C., Grodzinsky A.J., Trippel S.B. 1994. Differential effects of IGF-I and bFGF on matrix metabolism in calf and adult bovine cartilage explants. *Arch. Biochem. Biophys.* 308:137–147.

Sah R.L., Trippel S.B., Grodzinsky A.J. 1996. Differential effects of serum, IGF-I, and FGF-2 on the maintenance of cartilage physical properties during long-term culture. *J. Orthop. Res.* 14:44–52.

Sah R.L., Chen A.C., Chen S.S., Li K.W., DiMicco M.A., Kurtis M.S., Lottman L.M., Sandy J.D. 2001. In: *Arthritis and Allied Conditions. A Textbook of Rheumatology.* W.J. Koopman, ed. Lippincott, Williams & Wilkins, Philadelphia; pp. 2264–2278.

Sajdera S.W., Hascall V.C. 1969. Proteinpolysaccharide complex from bovine nasal cartilage. A comparison of low and high shear extraction procedures. *J. Biol. Chem.* 244:77–87.

Sandy J.D., Plaas A.H.K. 1986. Age-related changes in the kinetics of release of proteoglycans from normal rabbit cartilage explants. *J. Orthop. Res.* 4:263–272.

Sandy J., Verscharen C. 2001. Analysis of aggrecan in human knee cartilage and synovial fluid indicates that aggrecanase (ADAMTS) activity is responsible for the catabolic turnover of whole aggrecan whereas MMP-like activity is required primarily for C-terminal processing of the molecule. *Biochem. J.* 358:615–626

Schinagl R.M., Gurskis D., Chen A.C., Sah R.L. 1997. Depth-dependent confined compression modulus of full-thickness bovine articular cartilage. *J. Orthop. Res.* 15:499–506.

Schneiderman R., Snir E., Popper O., Hiss J., Stein H., Maroudas A. 1995. Insulin-like growth factor-I and its complexes in normal human articular cartilage: studies of partition and diffusion. *Arch. Biochem. Biophys.* 324:159–172.

Skalak R., Gasgupta G., Moss M., Otten E., Dullemeijer P., Vilmann H. 1982. Analytical description of growth. *J. Theor. Biol.* 94:555–577.

Skalak R., Zargaryan S., Jain R.K., Netti P.A., Hoger A. 1996. Compatability and the genesis of residual stress by volumetric growth. *J. Math. Biol.* 34:889–914.

Skalak R., Farrow D.A., Hoger A. 1997. Kinematics of surface growth. *J. Math. Biol.* 35:869–907.

Strider W., Pal S., Rosenberg L. 1975. Comparison of proteoglycans from bovine articular cartilage. *Biochim. Biophys. Acta.* 379:271–281.

Taber L. 1998. A model for aortic growth based on fluid shear and fiber stresses. *J. Biomech. Eng.* 120:348–354.

Taber L.A., Eggers D.W. 1996. Theoretical study of stress-modulated growth in the aorta. *J. Theor. Biol.* 180:343–357.

Thonar E.J.-M., Sweet M.B.E. 1981. Maturation-related changes in proteoglycans of fetal articular cartilage. *Arch. Biochem. Biophys.* 208:535–547.

Torzilli P.A., Adams T.C., Mis R.J. 1987. Transient solute diffusion in articular cartilage. *J. Biomech.* 20:203–214.

Torzilli P.A., Arduino J.M., Gregory J.D., Bansal M. 1997. Effect of proteoglycan removal on solute mobility in articular cartilage. *J. Biomech.* 30:895–902.

Trippel S.B. 1995. Growth factor actions on articular cartilage. *J Rheumatol* 43S:129–132.

Van Dyke T.J., Hoger A. 2001. Should the growth law be defined on the current or initial configuration? *J. Theor. Biol.* (submitted).

Venn M. F., Maroudas A. 1977. Chemical composition and swelling of normal and osteoarthritic femoral head cartilage I: chemical composition. *Ann. Rheum. Dis.* 36:121–129.

Williamson A.W., Chen A.C., Sah R.L. 2001. Compressive properties and structure-function relationships of developing bovine articular cartilage. *J. Orthop. Res.* 19:1113–1112.

Wong M., Ponticiello M., Kovanen V., Jurvelin J.S. 2000. Volumetric changes of articular cartilage during stress relaxation in unconfined compression. *J. Biomech.* 33:1049–1054.

Woo S.L.-Y., Akeson W.H., Jemmott G.F. 1976. Measurements of nonhomogeneous directional mechanical properties of articular cartilage in tension. *J. Biomech.* 9:785–791.

Part IV

Assessment of Junction in Engineered Tissues

16

Functional Tissue Engineering: Assessment of Function in Tendon and Ligament Repair

David L. Butler, Matthew Dressler, and Hani Awad

Introduction

It is estimated that over two million people visit a physician for musculoskeletal injuries each year (Praemer et al., 1999). Over 43% of these visits are for soft tissue injuries that occur to ligaments (connective tissues that link bone to bone at a joint), tendons (connective tissues that link muscle to bone), or joint capsule. Some of the most commonly injured ligaments and tendons are the anterior cruciate and medial collateral ligaments and the Achilles, patellar, and rotator cuff tendons.

Certain ligaments in the knee, like the medial collateral ligament, heal rather well, whether or not they are repaired (Frank et al., 1983; Weiss et al., 1991). These tissues respond particularly well if the injuries are partial rather than complete ligament tears. Other soft connective tissues like the Achilles tendon respond well to surgical repair by forming a collagen callus to tolerate the large *in vivo* forces (Grood et al., 1985).

However, some soft tissue structures must be reconstructed or replaced rather than directly repaired. For example, the anterior cruciate ligament heals incompletely when surgically repaired (Arnoczky et al., 1979; Cabaud et al., 1979, 1980). Thus, orthopedic surgeons typically reconstruct the ligament using autogenous or allogeneic grafts (Butler et al., 1989; Jackson et al., 1987a; Noyes et al., 1983a, 1984; Paulos et al., 1983). The repair tissue that fills the defects created in the tendon due to graft reconstruction is also weakened by the surgery (Awad et al., in press; Beynnon et al., 1995; Burks et al., 1990; Proctor et al., 1997).

Tissue engineering is an attractive solution for treating many of these problem injuries. However, tissue-engineered constructs must also be capable of resisting the large and impulsive forces transmitted by muscle and by ground reaction forces. Functional tissue engineering (FTE) accounts for these *in vivo* forces and seeks to determine those parameters that tissue engineers can use in designing implants that more effectively repair tendon and ligament injuries.

Summary of Current Fields of Tissue Engineering and Functional Tissue Engineering Epidemiology of Musculoskeletal Soft Tissue Injury and Repair

Injuries to musculoskeletal soft tissues like tendon and ligament continue to escalate. Over 12 million (44%) of the estimated 28 million patients who experience a musculoskeletal injury in the United States each year sustain soft tissue sprains, strains, and dislocations (Praemer et al., 1992, 1999). Tissues about the knee, shoulder, and ankle are among the most commonly injured. The Achilles tendon, for example, is one of the most frequently ruptured tendons (Ballas et al., 1998; Maffulli, 1998), with estimates approaching 100,000 injuries each year (Praemer, 1996). In addition, occupational injuries to the shoulder, primarily to the rotator cuff, rank second to low back and neck pain in clinical frequency (Sommerich et al., 1993).

Attempts to repair these frequently injured tissues can pose long-term consequences for the patient. Epidemiologists estimate that surgeons replace the anterior cruciate ligament in the knee between 42,000 (Praemer et al., 1992) and 102,000 (AAOS, 2000) times each year. Patients receiving patellar tendon grafts can experience limited range of motion as well as anterior knee pain and muscle weakness in the region where the graft was procured. Though surgeons repair other ligaments less frequently, the medial collateral ligament in the knee is still repaired about 4000 times each year (AAOS, 2000). The American Academy of Orthopaedic Surgeons also estimates that the rotator cuff tendons are repaired over 50,000 times a year (AAOS, 2000). It is not surprising, therefore, that all soft tissue injuries account for 47% of all work-loss days and over 36% of all school-loss days every year (Praemer et al., 1992).

Natural Healing and Surgically Assisted Repair and Reconstruction of Ligaments

Over the years, investigators have created a rather large database from animal studies about the biomechanical and histological changes that occur to ligaments during natural healing, surgical repair, and soft tissue reconstruction. Understanding how these specific tissues respond to injury and repair is essential to establishing benchmarks for judging the effectiveness of tissue engineering methodologies.

Medial Collateral Ligament Healing and Repair

The geometry, biomechanics, and biology of the normal medial collateral ligament (MCL) are

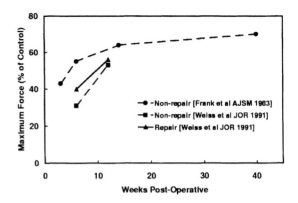

FIGURE 16.1. Maximum force generated for natural healing and surgical repair of rabbit medial collateral ligament injuries as a percentage of unoperated control. Note that the repaired tendon (*solid line*) fares no better than the non-repaired tendon (*dashed line*). Adapted from Frank et al. (1983) and Weiss et al. (1991).

complicated but well studied. The collagen fibers of the MCL originate as a zonal insertion on the medial femoral epicondyle and insert through a periosteal insertion on the proximal aspect of the tibia (Bartel et al., 1977; Marieb, 1998). The MCL serves as a primary restraint to medial joint opening (Grood et al., 1981) and to external tibial rotation (Haimes et al., 1994; Inoue et al., 1987). This multifunctional role in the knee is accomplished by differential recruitment of collagen fibers of different lengths that work synergistically with changes in joint position.

The MCL heals rather well without surgical repair (Frank et al., 1983; Weiss et al., 1991). Frank et al. (1983) found that 14 weeks after surgical transection, the failure load of the unrepaired MCL achieved about 65% of contralateral, normal ligament values. However, repair strength then showed little additional improvement thereafter up to 40 weeks (Fig. 16.1). These below-normal failure forces were accompanied by increased medial knee laxity and temporal changes in failure modes, which occurred in the midsubstance at 3 weeks post surgery, changed to failure at the tibial insertion at 6 weeks, and then changed back to failure in the midsubstance at 14 weeks (Frank et al., 1983). The cross-sectional areas of the repairs, while consistently larger than their corresponding unoperated controls, decreased over time up to 14 weeks and then showed no further improvement.

In an attempt to improve the strength of these naturally healing tissues, Weiss et al. (1991) used a mattress suture technique to directly repair a surgically induced injury in the rabbit MCL and compared the biomechanical results with an identical injury in the contralateral ligament that received no augmentation or assistance. This direct repair produced no significant improvement in structural or material properties over natural (unassisted) healing. Maximum force values for both models were 30–40% of normal values at 6 weeks post surgery and 50% of normal at 12 weeks (Fig. 16.1). The maximum stress and modulus remained relatively constant between 6 and 12 weeks, achieving 25–34% of normal values. This study, therefore, confirmed that directly repairing the ligament provides no added benefit over natural healing in the rabbit MCL model.

Anterior Cruciate Ligament

The anterior cruciate ligament (ACL) is an intra-articular but extrasynovial structure that is the primary restraint to anterior-posterior translation of the tibia at 30° and 90° of knee flexion (Butler et al., 1980). Despite bleeding or hemarthrosis in the knee after injury (Arnoczky, 1985; Arnoczky et al., 1979; Noyes et al., 1983b), the ACL is generally assumed to lack a vigorous blood supply, relying instead upon synovial nutrition. Attempts to surgically reattach the failed ends have been problematic due to the extent of injury throughout the ligament, the difficulty in identifying and reconnecting the correct bundles to restore the complicated helical structure, the lack of collateral protection from adjacent tissues, the limited blood supply, and the harsh synovial environment (Grood et al., 1985).

Direct repair of the ruptured ACL has been controversial. One clinical study reported that ACL repair improved rating scores and decreased anterior tibial translation at 20° flexion, the so-called Lachman test (Marshall et al., 1982). However, another clinical study showed significantly decreased knee function at late follow-up (Odensten et al., 1984). Early animal studies by O'Donoghue and co-workers demonstrated long-term declines in canine ACL repair strength (O'Donoghue et al., 1966, 1971). Later work by Cabaud et al. revealed very weak ACL repairs in the canine model (2–10% of normal maximum force) but better outcomes in the primate (47–63% of normal), which the group attributed to difficulties in controlling immobilization in the dog model after surgery (Cabaud et al., 1979). Cabaud concluded that augmented repair with the patellar tendon was required to protect these ACL repairs (Cabaud et al., 1980).

Instead, surgeons use autogenous and allogeneic tissue grafts to reconstruct the damaged ACL. Until the 1980s, however, orthopedists had no basis for choosing one graft type over another, sometimes selecting semitendinosus, gracilis, or patellar tendons, and other times electing to use patellar retinaculum, fascia lata, or iliotibial band (Noyes et al., 1984). Then in 1984, Noyes, Butler and co-workers directly contrasted the biomechanics of these tissues by reporting the strength (i.e., maximum force), stiffness, and material properties of these human cadaveric tissues relative to the normal anterior cruciate ligament. They showed that if strength and stiffness were used as the primary criteria for graft selection, then only the properties of the patellar tendon (PT) with attached patellar and tibial bone blocks exceeded corresponding values for the normal anterior cruciate ligament. In fact, a 14-mm-wide graft developed 159–168% of normal ACL strength (Noyes et al., 1984). Unfortunately, these biomechanical advantages were lost when the PT-bone graft was subsequently implanted in the ACL-deficient knee in animals (Butler et al., 1983, 1989; Clancy et al., 1981; Ishibashi et al., 1997; Jackson et al., 1987a,b, 1993; Kusayama, 1994; Ng et al., 1995). Graft strength declined to only 10–15% of normal ACL values within weeks after surgery, presumably due to incomplete fixation into bone (Butler, 1987; Butler et al., 1983, 1989). However, even one year after surgery, graft strength was still less than 50% of normal ACL maximum force (Butler et al., 1989; Clancy et al., 1981). The "envelope" of graft strength as a percentage of ACL strength from many of these studies is adapted from Bush-Joseph et al (1996) in Figure 16.2.

While ACL reconstruction using the patellar tendon remains a gold standard for graft selection in the young active patient, grafts still do

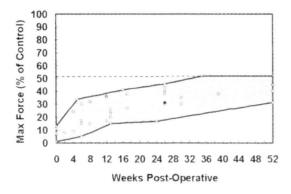

FIGURE 16.2. Envelope of maximum force response for patellar tendon reconstructions as a percentage of the maximum force for the unoperated anterior cruciate ligament. Note the low strength early after surgery (due to weak fixation strength) and the modest 40–50% strength of the reconstructions one year after surgery. The points represent data taken from a series of studies utilizing autogenous patellar tendon reconstructions. Adapted from Bush-Joseph et al. (1996).

fail with a frequency that is problematic. Consequently, revision surgery is becoming ever more frequent. Couple these clinical findings with the marginal biomechanical strength from animal studies, and it is clear that other options will need to be pursued. In developing these options, including tissue-engineered ACL replacements, investigators will need to select which parameters (strength, stiffness, etc.) should be used to judge whether the graft is successful in the patient. Establishing these measures for autografts, allografts, and tissue-engineered repairs, especially relative to expected *in vivo* forces after surgery, remains a top priority.

Surgical Repair of Injured Tendon

Achilles Tendon

Conventional repair of the Achilles tendon has not been uniformly successful. Spanning between the gastrocnemius muscle and the calcaneus at the ankle, the Achilles tendon transmits up to eight times body weight during running (Soma and Mandelbaum, 1995). It is generally agreed

that injuries that are treated surgically result in a lower incidence of re-rupture than more conservative nonoperative treatments. In animal studies, however, injuries treated by nonoperative means have actually fared quite well. For example, Reddy et al. (1999) found that rabbits regained 48% of normal maximum force and 30% of normal strain energy only 15 days after surgical tenotomy. Murrell et al. (1994) found no benefit to surgical repair of rat Achilles ruptures up to 15 days after injury, and Roberts et al. (1983) found no biomechanical benefit to using a polyglactin mesh instead of a conventional suture. Whether nonoperative approaches yield suitable strength and stiffness in the longer term is not well known.

Patellar Tendon

The patellar tendon transmits forces from the quadriceps muscle to the proximal tibia to effect knee extension. These forces can approach 15 times body weight before failure (Zernicke et al., 1977). The tissue is particularly sensitive to changes in the *in vivo* forces as demonstrated in the studies by Yasuda and Hayashi (Ohno et al., 1993, 1995; Yamamoto et al., 1993, 1995) who found patellar tendon properties to decline rapidly when the tissue was stress shielded and to recover slowly when re-stressed *in vivo*. This tendon has also received particular attention due to its popular use in ACL reconstruction (Noyes and Barber-Westin, 1997; Noyes et al., 1983a, 1984). After removing its central third to create the graft, the entire patellar tendon (repair tissue and adjacent struts) recovers only 17% of normal linear modulus, 60–70% of normal maximum force, and 73% of normal stiffness by 6 months post surgery (Beynnon et al., 1995; Burks et al., 1990). Assuming the defect has undergone some level of repair, the medial and lateral struts must have remodeled as well. Recently, investigators have isolated the repair tissue (scar) from the normal medial and lateral struts, allowing direct comparison of repair tissue quality (Awad et al., In press; Proctor et al., 1997). Proctor et al. (1997) showed that 21 months after graft removal, the cross-sectional area was 42% greater than normal. However, the maximum force and maximum stress of the repair site alone were only 49% and 35% of normal patellar tendon values, respectively. The fact that the repair tissue was still not normal almost two years after surgery suggests that tissue-engineering methods might prove useful in enhancing tissue quality.

Rotator Cuff

The four muscle-tendon units of the rotator cuff (subscapularis, supraspinatus, infraspinatus, teres minor muscle) help provide stability to the glenohumeral joint of the shoulder (Marieb, 1998). Although most injuries to the rotator cuff tendons are partial thickness tears that respond well to nonsurgical treatment, some injuries develop into chronic problems. The mechanisms/factors that determine which injuries will heal and which will persist are not well known. Carpenter, Soslowsky, and co-workers (Carpenter et al., 1998) studied repair after injury to the rat supraspinatus tendon. In situ freezing was used to lyse the local cells, compromising the tendon's intrinsic ability for repair. The healing tissue was evaluated biomechanically at 3, 6, and 12 weeks after surgery. The investigators found that tendon with debilitated healing fared no worse than the controls, both achieving only 42–46% of normal tendon maximum force and 38–43% of normal stiffness. The repair quality, as assessed by measuring material properties like maximum stress and modulus, achieved only 9–15% of normal values. The cross-sectional areas of these repairs were also six to seven times larger than those of the normal tendon (Carpenter et al., 1998). Clearly, the biomechanics of these repair tissues were not restored to normal values. While this attempt at hindering the intrinsic healing capacity of the tendon did not yield a significant decrease in repair outcome, a tissue-engineered therapy might inhance the healing capacity and result in a stronger and better organized repair.

Conventional Tissue Engineering

While additional animal studies are needed to determine the optimal repair potential of various ligaments and tendons, it is clear that many months after surgery, most of these soft tissues still have not achieved biomechanical values even close to normal. Given that patients sustain far more devastating injuries than investigators can create in the laboratory, new strategies are now needed to both speed the repair process and produce a tissue with more normal mechanical properties.

The inability of natural healing and surgical repair to truly "regenerate" normal soft tissue has been the impetus for tissue engineering (TE). Combining principles of engineering and biology, TE endeavors to fabricate new tissues in the laboratory that are designed to rapidly restore tissue form (i.e., three-dimensional architecture and composition of the normal tissue) and function (normal structural and material properties) to permit the patient to return to his/her preinjury status (Butler and Awad, 1999; Butler et al., 2000).

The TE process typically involves introducing living cells and "carrier" into a wound site or mixing these cells and carrier with a natural or man-made scaffold material. Collagen gels often serve as carriers for these cells (Awad et al., 1999, 2000; Young et al., 1998), but over the years, scaffolds have included carbon fibers (Makisalo et al., 1988; Weiss et al., 1985), collagen fibers (Chvapil et al., 1993; Dunn et al., 1993, 1994; Jackson et al., 1996) and polylactic acid (PLA) (Sato et al., 2000), polyglycolic acid (PGA) (Cabaud et al., 1982), and Dacron (Wredmark and Engstrom, 1993) sutures. When these constituents are combined, the implants should encourage cellular recruitment and tissue ingrowth. The stiffer scaffold material should protect the cells and newly forming repair tissue from high forces during the early phases of repair. To avoid stress shielding, however, the scaffolds should also ideally degrade at the same rate as newly formed tissue begins to bear significant load. The degrading scaffold should pose no threat of rejection, however (Awad et al., 1999; Butler and Awad, 1999).

Cellular-based implants for tendon and ligament repair have used differentiated fibroblasts and undifferentiated mesenchymal stem cells (MSCs). Following earlier studies that used collagenous matrix bioprostheses to repair ruptured ligaments (Dunn et al., 1988, 1992, 1993, 1994), Dunn et al. (1995) seeded collagen scaffolds with differentiated fibroblasts to repair the

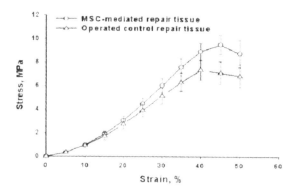

FIGURE 16.3. Average stress-strain curve for MSC-mediated repair tissue and operated control tissue from a window defect model in the rabbit patellar tendon at 4 weeks post surgery. Error bars indicate standard error of the mean. Printed with permission from Awad et al. (1999).

ACL. While they found that the scaffolds improved attachment of the fibroblasts and increased collagen synthesis *in vitro*, the source of the cells (PT vs. ACL fibroblasts) and the type of matrix influenced synthetic activity. In an attempt to reduce this variability in repair outcome due to cell source and potentially improve cell vitality, Awad et al. (1997) filled patellar tendon window defects with MSCs that had been suspended in a type I collagen gel purified from bovine skin. They reasoned that these pluripotential cells, with the capacity to form cartilage, bone, or other tissues (Caplan, 1991, 1994; Caplan and Boyan, 1994; Caplan et al., 1993), would take their cues from the local environment. Awad et al. (1997) showed that the material properties of the repair sites were 18–33% greater than corresponding properties for natural healing. The average stress-strain curves for the treated and operated control tissue repairs are shown in Figure 16.3.

Despite these improvements in repair outcome, many questions remain. How strong and stiff must the repairs be to adequately resist the *in vivo* force environment? What are these *in vivo* force regimes? What biomechanical parameters are most important in judging these improvements? What might be done in culture before surgery to improve the mechanical integrity of the implants to better meet these functional demands?

Functional Tissue Engineering

To address some of these questions, the U.S. National Committee on Biomechanics in 1998 adopted the concept of functional tissue engineering (Butler et al., 2000). Functional tissue engineering extends the goals of tissue engineering by considering the actual functional demands placed on the implants that are intended to repair the damaged tissue and by incorporating these requirements into meaningful parameters that tissue engineers should design into their constructs before surgery.

What follows are a series of studies conducted by our group that apply the principles of functional tissue engineering to the design of tendon and ligament implants. These studies are simply examples of studies that could be performed to develop improved constructs. As such, these studies should be expanded in order to develop more functionally improved tissue engineered constructs.

Determination of *In Vivo* Force Patterns and Thresholds for Tendon and Ligament

Tissue engineers face a major challenge in measuring force patterns and force thresholds acting on their constructs. Little information is available for most tissue systems. Fortunately, however, several groups, including our own (Korvick et al., 1992, 1996; Malaviya et al., 1998), have determined force patterns and thresholds in normal tendons and ligaments. Early work by Korvick et al. (1992, 1996) and Holden et al. (1994) showed that *in vivo* forces, as a percentage of failure force, differed for tendon and ligament. After inserting a modified pressure transducer in the goat's ACL, these investigators showed that the ACL generated dual or bimodal loading peaks during the stance phase of gait. The first peak was higher if the animal walked up a ramp and the second peak was higher if the animal walked down a ramp. The peak forces increased as the speed of the activity increased (Holden et al., 1994; Korvick et al., 1992). However, these peak forces never exceeded 10% of the ACL failure force, regardless of the activity (Fig. 16.4). By contrast, the PT in the goat generated more

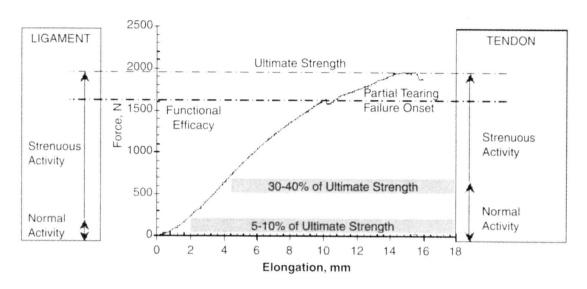

FIGURE 16.4. Functional requirements of ligaments and tendons. *In vivo* loads in general achieve no more than 10% of their ultimate force for ligaments and about 30–40% for tendons. Adapted from Noyes et al. (1984).

unimodal peaks than the ACL (Holden et al., 1994; Korvick et al., 1996) and these PT forces were initiated earlier in time than those in the ACL (Korvick et al., 1992). The peak forces in the PT reached 32–40% of failure force as the speed of the activity increased (Korvick et al., 1996) (Fig. 16.4). Subsequent studies in the rabbit model showed that the flexor digitorum profundus tendon achieved 29% of failure force during inclined hopping activities (Malaviya et al., 1998).

It is likely that force thresholds will change for different tendons and for different ligaments. However, knowing these thresholds and the general patterns of force developed for several activity levels can be useful to the tissue engineer. While the engineer will likely be unable to protect the implant from traumatic forces that are generated with injury, at least the implants will be designed to meet the goal of resisting the forces associated with normal activities of daily living. If the tissue engineer also builds a safety factor into the design, the likelihood of premature damage during the early remodeling phase is reduced.

Aligning the Cells in the Cell-Based Tissue Engineered Implants: Biomechanical Assessment Following Surgery

Recently, Young et al. (1998) postulated that tendon repair properties would be improved if the cells could be aligned before implantation. By manipulating the construct before surgery using an alignment scaffold, they reasoned that cellular alignment would lead to improved extracellular matrix organization and thus improved tensile properties after surgery. They removed bone marrow aspirates from the ilium of the rabbit and isolated and propagated mesenchymal stem cells (MSCs). MSCs are noncommitted progenitor cells of musculoskeletal tissues with the ability to repair or regenerate soft and hard tissues (Caplan et al., 1993; Wakitani et al., 1994). The cells were suspended in a type I collagen gel and placed in glass "boats" through which a suture was suspended. The cells contracted the gel onto the suture and the composite was then used to span a 1-cm defect in the Achilles tendon of the same rabbit. A suture with neither cells nor gel spanned the contralateral defect. The load-

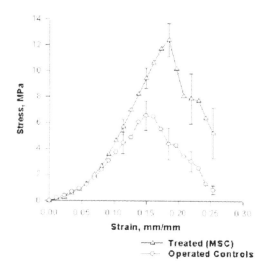

FIGURE 16.5. Average stress-strain curve for MSC-treated repair tissue and operated acellular controls in rabbit Achilles tendon at 12 weeks after surgery. Error bars indicate standard error of the mean. Printed with permission from Young et al. (1998).

related structural and material properties (all but elongation and strain) were significantly greater for the MSC-treated repairs than for the treated, control repairs at 4, 8, and 12 weeks after surgery. In fact, the values for the MSC-treated repairs were generally twice those of the control repairs (Fig. 16.5). Particularly encouraging was the fact that by 12 weeks post-op, the MSC-based repairs had achieved 63% of normal tendon stiffness and 66% of normal maximum force. These increases were primarily due to the larger (1.7 to 3.5 times the cross-sectional area) but poorly organized repair tissue compared with normal tendon. When these structural properties were normalized by area and length, the resulting material properties showed a continued improvement over time. However, even at 12 weeks after surgery, the maximum stress was still only 36% of the ultimate stress for the normal tendon. Using the *in vivo* force thresholds measured for tendon from the prior section

(40% of ultimate), one could thus conclude that the MSC treatment had adequately repaired the structural properties of the tendon (66%) but not the material properties (36%). Whether these properties would continue to improve over time, whether other variations on the treatment (such as cell seeding density) could further improve these results, and whether this treatment would be as good in other tendon systems were still not known.

Optimizing Cell Shape and Alignment in Culture: Effects of Varying Cell Seeding Density

One reason why the cell-gel-suture implant dramatically improved Achilles tendon repair after surgery (Young et al., 1998) might be that the presence of the suture aligned the cells *in vitro* and changed their phenotype. Much like the organization of normal tendon, these aligned and more spindle-shaped cells would be expected to synthesize more aligned extracellular matrix, which in turn would improve biomechanical properties. To test this concept, Awad et al. (2000) used an identical procedure as before (Young et al., 1998) to isolate and expand MSCs from rabbit bone marrow and then seed these cells in collagen gels. The group hypothesized that as the cells contracted the implants around the pre-tensioned suture, the rate and magnitude of the contraction as well as any changes in cell shape and alignment would be controlled by cell-seeding density. Thus, Awad et al. (2000) measured lateral contraction of the implants every 8 hours and used stereology to record changes in cell nuclear aspect ratio (AR) and orientation angle (OA) at three cell concentrations (1, 4, and 8 million cells/mL or 1 M, 4 M, and 8 M) and at three time points during incubation (24, 48, and 72 hours). Cell-seeding density significantly influenced both contraction kinetics and nuclear appearance. The 1 M constructs contracted more slowly than the 4 M or 8 M constructs, and their diameters at 72 hours were approximately 60–70% larger than those seeded at the higher densities. The cells in the higher density constructs also were more elongated and better aligned than the MSCs from the 1 M constructs (Awad et al., 2000).

Alignment strategies offer a "passive" approach to organizing the implants before surgery. Before implementing more complex dynamic approaches for implant loading in culture, however, the effects of cell seeding density were examined *in vivo* to determine their potential benefit after surgery.

Effects of Cell-Seeding Density on Repair Biomechanics after Surgery

The same suture alignment strategy was applied to filling in a full-length defect in the patellar tendon (Awad et al., In Press). The authors hypothesized that increasing cell-seeding density, which results in more contracted MSC implants with better-aligned and spindle-shaped cells *in vitro* (Awad et al., 2000), would result in improved repair biomechanics over time. Full-length defects were created along the central third of the patellar tendon in left and right knees of rabbits. Small defects were also created in the bone ends to simulate preparation of ligament replacement grafts. MSC-gel constructs were created as before at concentrations of 1 M, 4 M, and 8 M. The 1 M and 4 M constructs were surgically implanted into the wound sites of left and right knees of one group of animals and evaluated at 6, 12, and 26 weeks post surgery. The 8 M constructs were compared to unfilled defects in a second group of animals at the same time points. The MSC-based repairs were found to produce significantly improved material properties at all time points compared to the natural healing repairs. The material properties of the MSC-based repairs also increased at a significantly faster rate than the natural repairs. However, the investigators found no dose-dependent improvement in repair outcome for higher concentrations of MSCs and all parameters were no greater than about 25% of unoperated control tendon results. This surprising outcome indicates that simply adjusting cell concentration alone is not sufficient to dramatically increase the repair quality compared to normal. Other strategies need to be investigated.

Technical Challenges in Functional Tissue Engineering

Numerous technical challenges face researchers seeking to tissue engineer tendon and ligament. Three of the most significant challenges, in the opinion of the authors, are briefly described below.

Developing Minimally Invasive Methods for Measuring In Vivo Forces and Deformations

New and improved technologies must be developed that permit investigators to measure *in vivo* force and deformation data in the specific tissues that they intend to repair. These technologies should enable *in vivo* signals to be measured less invasively than with conventional strain gauges and transducers. Micro-Electro-Mechanical Systems (MEMS) technology offers one attractive approach. Smaller devices should be designed to be capable of being placed within the tissue rather on its surface when there is a possibility of interference from adjacent tissues. Transducer materials should be selected that minimize encapsulation and rejection by the body. More durable transducers should also be designed to permit signals to be recorded for longer periods of time after surgery, possibly using telemetry rather than hard wiring to prevent lead wire and connector damage. By measuring signals up to weeks rather than days after surgery, researchers can expect to obtain more realistic forces and deformations without the adverse effects of surgery.

In vivo forces must also be measured in intact tendons and ligaments in patients and in tissue engineered repairs in patients and in animals. Little *in vivo* tendon and ligament data is available from patients, and more will be needed if tissue engineered repairs are to be designed that can tolerate various activities of daily living in the real setting. Devising "smart" repairs that contain transducers that can provide actual *in vivo* forces and deformations after surgery also offer more realistic data that indicate whether the tissue engineered repairs are functioning as expected.

Devising More Robust FTE Repairs to Meet these *In Vivo* Force Requirements

A significant challenge that tissue engineers face is how to create an implant that will resist these large *in vivo* forces early after surgery while still stimulating the cells that must participate in the repair process. Implants that are too weak and compliant at surgery will stretch under the large forces applied to them *in vivo*. Reinforcing these implants with suture or other stiff scaffolds protects them during this early healing phase, but also stress-shields the cells that sense their force and deformation environment to synthesize appropriate extracellular matrix. As previously discussed, new technologies are required that yield new scaffold materials that can degrade at the same rate that new matrix is synthesized without producing adverse tissue reactions.

Mechanically Stimulating Functional Tissue Engineering Implants Before Surgery

New mechanical testing systems are required that can deliver more relevant forces or deformations to the implants during incubation. Given the weak and compliant nature of the implants, providing signals that mimic conditions measured *in vivo* offers an attractive way to "precondition" the cells before surgery. These systems must be capable of delivering precisely controlled signals, preferably under computer control, for up to weeks at a time. These signals will likely need to be tailored to the specific tissue application and may need to be increased over time so as to gradually condition rather than mechanically shock the cells in the implant. This "ramping up" process offers the opportunity to simulate increasing levels of activity.

Future Perspectives and Directions in Functional Tissue Engineering

Functional tissue engineering for tendon and ligament repair has an exciting future. *In vivo* data is already available for several tendons and ligaments that can guide the next generation of devices. These *in vivo* forces are being used to determine safety factors so as to better design these implants. Within these thresholds of expected activity, mechanical parameters are being identified that the tissue engineer can use to determine if "good is good enough." Early generations of mechanical stimulation systems are already available that will permit implants to be preconditioned before surgery.

Despite these advances, much research is still needed to translate the fundamental laboratory and animal studies "from the bench to the bedside." During the tissue engineering phase (prior to surgical implantation), appropriate cell sources, either differentiated or undifferentiated, must be selected for seeding tendon and ligament implants. Differentiated cells will already be of the right type but will require compromise of otherwise healthy tissue and methods for their removal from dense collagenous structures. Undifferentiated cells will be pluripotential for developing more complex tissues like ligament-bone composites but will require efficient extraction procedures and control of proliferation and differentiation for appropriate tissue to be synthesized. Decisions will be required as to whether autogenous or allogeneic cell types can be used in clinical situations. Presuming that tissue-engineered implants alone will not possess sufficient strength and stiffness before surgery, biomaterials experts must be identified to work closely with tissue engineers to formulate scaffolds with the right mechanical integrity, degradation characteristics, and biocompatibility to facilitate aligned extracellular matrix of the correct composition. Various mechanical and chemical stimulation methods must also be developed during the tissue engineering phase. The effects of mechanical signals on early implant development will probably need to be determined separately from the effects of growth factor stimulation after which the synergistic nature of both treatments can be investigated.

Several significant research advances will be required after surgical implantation as well. Minimally invasive methods such as MRI will be needed to assess the maturation of the repairs. In the early stages, these methods can be coupled with histological evaluations in animal studies to try and develop quantitative imaging measures that will be useful in assessing results in patients who ultimately receive these tissue engineered implants. Functional "yardsticks" will also be needed to assess whether the repairs are working as intended. Quantitative or semi-quantitative response measures will need to be established during simple functional tests on the patient to determine whether the tissue engineering repair strategy has achieved the intended result. Finally, databases must be developed whereby *in vitro* treatments can be directly correlated with *in vivo* outcomes in animals and in patients. These *in vitro* to *in vivo* correlations are needed in order to make more rapid and positive changes in implant characteristics before surgery that can increase the likelihood of a successful result after surgery. Developing these *in vitro* and *in vivo* strategies for functional tissue engineered tendon and ligament repair should dramatically improve the chance of treating these problematic soft tissue injuries.

Acknowledgments

The authors wish to acknowledge support of the National Institutes of Health (AR46574) and a Merit Grant from the Veteran's Administration. The authors also thank Drs. Gregory Boivin, Arnold Caplan, Randall Young, David Fink, and Stephen Gordon for their contributions to research cited in this article.

References

AAOS. 2000. Arthroplasty and total joint replacement procedures: United States 1990 to 1997. Available at: http://www.aaos.org/wordhtml/press/arthropl.html

Arnoczky S.P. 1985. Blood supply to the anterior cruciate ligament and supporting structures. *Orthop. Clin. North Am.* 16:15–28.

Arnoczky S.P., Rubin R.M., Marshall J.L. 1979. Microvasculature of the cruciate ligaments and its response to injury. An experimental study in dogs. *J. Bone Joint Sug.* 61A:1221–1229.

Awad H.A., Butler D.L., Malaviya P., Boivin G.P., Huibregtse B., Caplan A.I. 1997. Autologous mesenchymal stem cell-impants can improve the repair of the injured rabbit patellar tendon. *43rd Annual Meeting of the Orthopaedic Research Society* 22:535.

Awad H.A., Butler D.L., Boivin G.P., Smith F., Malaviya P., Huibregtse B., Caplan A.I. 1999. Autologous mesenchymal stem cell-mediated repair of tendon. *Tissue Eng.* 5:267–277.

Awad H.A., Butler D.L., Harris M.T., Ibrahim R.E., Wu Y., Young R.G., Kadiyala S., Boivin G.P. 2000. *In vitro* characterization of mesenchymal stem cell-seeded collagen scaffolds for tendon repair: effects of initial seeding density on contraction kinetics. *J. Biomed. Mater. Res.* 51:233–240.

Awad H.A., Butler D.L., Boivin G.P., Dressler M.R., Smith F.N.L., Young R. In press. Repair of patellar tendon injuries using a cell-collagen composite. *J. Orthop. Res.*

Ballas M.T., Tytko J., Mannarino F. 1998. Commonly missed orthopedic problems. *Am. Fam. Phys.* 57:267–274.

Bartel D.L., Marshall J.L., Schieck R.A., Wang J.B. 1977. Surgical repositioning of the medial collateral ligament. An anatomical and mechanical analysis. *J. Bone Joint Surg.* 59A:107–116.

Beynnon B.D., Proffer D., Drez D.J., Jr., Stankewich C.J., Johnson R.J. 1995. Biomechanical assessment of the healing response of the rabbit patellar tendon after removal of its central third. *Am. J. Sports Med.* 23:452–457.

Burks R.T., Haut R.C., Lancaster R.L. 1990. Biomechanical and histological observations of the dog patellar tendon after removal of its central third. *Am. J. Sports Med.* 18:146–153.

Bush-Joseph C.A., Cummings J.F., Buseck M., Bylski-Austrow D.I., Butler D.L., Noyes F.R., Grood E.S. 1996. Effect of tibial attachment location on the healing of the anterior cruciate ligament freeze model. *J. Orthop. Res.* 14:534–541.

Butler D.L. 1987. Evaluation of fixation methods in cruciate ligament replacement. *Instr. Course Lect.* 36:173–178.

Butler D.L., Awad H.A. 1999. Perspectives on cell-collagen composites for tendon repair. *Clin. Orthop.* 367(Suppl):S324–332.

Butler D.L., Noyes F.R., Grood E.S. 1980. Ligamentous restraints to anterior-posterior drawer in the human knee. A biomechanical study. *J. Bone Joint Surg.* 62A:259–270.

Butler D.L., Hulse D.A., Kay M.D., Grood E.S. et al. 1983. Biomecanics of cranial cruciate ligament reconstruction in the dog. *Vet. Surg.* 12:113–118.

Butler D.L., Grood E.S., Noyes F.R., Zernicke R.F., Brackett K. 1984. Effects of structure and strain measurement technique on the material properties of young human tendons and fascia. *J. Biomech.* 17:579–596.

Butler D.L., Grood E.S., Noyes F.R., Olmstead M.L., Hohn R.B., Arnoczky S.P., Siegel M.G. 1989. Mechanical properties of primate vascularized vs non-vascularized patellar tendon grafts; changes over time. *J. Orthop. Res.* 7:68–79.

Butler D.L., Goldstein S.A., Guilak F. 2000. Functional tissue engineering: the role of biomechanics. *J. Biomech. Eng.* 122:570–575.

Cabaud H.E., Rodkey W.G., Feagin J.A. 1979. Experimental studies of acute anterior cruciate ligament injury and repair. *Am. J. Sports Med.* 7:18–22.

Cabaud H.E., Feagin J.A., Rodkey W.G. 1980. Acute anterior cruciate ligament injury and augmented repair. Experimental studies. *Am. J. Sports Med.* 8:395–401.

Cabaud H.E., Feagin J.A., Rodkey W.G. 1982. Acute anterior cruciate ligament injury and repair reinforced with a biodegradable intraarticular ligament. Experimental studies. *Am. J. Sports Med.* 10:259–265.

Caplan A.I. 1991. Mesenchymal stem cells. *J. Orthop. Res.* 9:641–650.

Caplan A.I. 1994. The mesengenic process. *Clin. Plastic Surg.* 21:429–435.

Caplan A., Boyan B. 1994. Endochondral bone formation: the lineage cascade. In: *Bone,* Vol. 8. B. Hall. ed. CRC Press.

Caplan A.I., Fink D.J., Goto T., Linton A.E., Young R.G., Wakitani S., Goldberg V.M., Haynesworth S.E. 1993. Mesenchymal stem cells and tissue repair. In: *The Anterior Cruciate Ligament—Current and Future Concepts.* D.W. Jackson, S.P. Arnoczky, S.L.-Y. Woo, C.B. Frank, T.M. Simon eds. Raven Press, New York, pp. 405–417.

Carpenter J., Thomopoulos S., Flanagan C., DeBano C., Soslowsky L. 1998. Rotator cuff defect healing: a biomechanical and histologic analysis in an animal model. *J. Shoulder Elbow Surg.* 7:599–605.

Chvapil M., Speer D.P., Holubec H., Chvapil T.A., King D.H. 1993. Collagen fibers as a temporary scaffold for replacement of ACL in goats. *J. Biomed. Mater. Res.* 27:313–325.

Clancy W.G., Jr., Narechania R.G., Rosenberg T.D., Gmeiner J.G., Wisnefske D.D., Lange T.A. 1981. Anterior and posterior cruciate ligament reconstruction in rhesus monkeys. *J. Bone Joint Surg.* 63A:1270–1284.

Dunn M.G., Doillon C.J., Berg R.A., Olson R.M., Silver F.H. 1988. Wound healing using a collagen matrix: Effect of dc electrical stimulation. *J. Biomed. Mater. Res.* 22:191–206.

Dunn M.G., Tria A.J., Kato Y.P., Bechler J.R., Ochner R.S., Zawadsky J.P., Silver F.H. 1992. Anterior cruciate ligament reconstruction using a composite collagenous prosthesis. A biomechanical and histologic study in rabbits. *Am. J. Sports Med.* 20:507–515.

Dunn M.G., Avasarala P.N., Zawadsky J.P. 1993. Optimization of extruded collagen fibers for ACL reconstruction. *J. Biomed. Mater. Res.* 27:1545–1552.

Dunn M.G., Maxian S.H., Zawadsky J.P. 1994. Intraosseous incorporation of composite collagen prostheses designed for ligament reconstruction. *J. Orthop. Res.* 12:128–137.

Dunn M.G., Liesch J.B., Tiku M.L., Zawadsky J.P. 1995. Development of fibroblast-seeded ligament analogs for ACL reconstruction. *J. Biomed. Mater. Res.* 29:1363–1371.

Frank C., Woo S.L., Amiel D., Harwood F., Gomez M., Akeson W. 1983. Medial collateral ligament healing. A multidisciplinary assessment in rabbits. *Am. J. Sports Med.* 11:379–389.

Grood E.S., Noyes F.R., Butler D.L., Suntay W.J. 1981. Ligamentous and capsular restraints preventing straight medial and lateral laxity in intact human cadaver knees. *J. Bone Joint Surg.* 63A:1257–1269.

Grood E.S., Butler D.L., Noyes F.R. 1985. Models of ligament repairs and grafts. In: *AAOS Symposium on Sports Medicine. The Knee.* C.V. Mosby Co., St. Louis, pp. 169–181.

Haimes J.L., Wroble R.R., Grood E.S., Noyes F.R. 1994. Role of the medial structures in the intact and anterior cruciate ligament-deficient knee. Limits of motion in the human knee. *Am. J. Sports Med.* 22:402–409.

Holden J.P., Grood E.S., Korvick D.L., Cummings J.F., Butler D.L., Bylski-Austrow, D.I. 1994. In vivo forces in the anterior cruciate ligament: direct measurements during walking and trotting in a quadruped. *J. Biomech.* 27:517–526.

Inoue M., McGurk-Burleson E., Hollis J.M., Woo S.L. 1987. Treatment of the medial collateral ligament injury. I: the importance of anterior cruciate ligament on the varus-valgus knee laxity. *Am. J. Sports Med.* 15:15–21.

Ishibashi Y., Rudy T.W., Livesay G.A., Stone J.D., Fu F.H., Woo S.L. 1997. The effect of anterior cruciate ligament graft fixation site at the tibia on knee stability: evaluation using a robotic testing system. *Arthroscopy* 13:177–182.

Jackson D.W., Grood E.S., Arnoczky S.P., Butler D.L., Simon T.M. 1987a. Cruciate reconstruction using freeze dried anterior cruciate ligament allograft and a ligament augmentation device (LAD). An experimental study in a goat model. *Am. J. Sports Med.* 15:528–538.

Jackson D.W., Grood E.S., Arnoczky S.P., Butler D.L., Simon T.M. 1987b. Freeze dried anterior cruciate ligament allografts. Preliminary studies in a goat model [published erratum appears in *Am. J.*

Sports Med. 1987 sep-oct;15(5):482]. *Am. J. Sports Med.* 15:295–303.

Jackson D.W., Grood E.S., Goldstein J.D., Rosen M.A., Kurzweil P.R., Cummings J.F., Simon T.M. 1993. A comparison of patellar tendon autograft and allograft used for anterior cruciate ligament reconstruction in the goat model. *Am. J. Sports Med.* 21:176–185.

Jackson D.W., Simon T.M., Lowery W., Gendler E. 1996. Biologic remodeling after anterior cruciate ligament reconstruction using a collagen matrix derived from demineralized bone. An experimental study in the goat model. *Am. J. Sports Med.* 24: 405–414.

Korvick D.L., Holden J.P., Grood E.S., Cummings J.F., Rupert M.P. 1992. Relationships between patellar tendon, anterior curciate ligament, and vertical ground reaction forces during gait: Preliminary studies in a quadruped. *Adv. Bioeng.* 22:99–102.

Korvick D.L., Cummings J.F., Grood E.S., Holden J.P., Feder S.M., Butler D.L. 1996. The use of an implantable force transducer to measure patellar tendon forces in goats [see comments]. *J. Biomech.* 29:557–561.

Kusayama T. 1994. Reconstruction of the rabbit anterior cruciate ligament using the autogenous patellar tendon. *Tokai J. Exp. Clin. Med.* 19:23–28.

Maffulli N. 1998. The clinical diagnosis of subcutaneous tear of the achilles tendon. A prospective study in 174 patients. *Am. J. Sports Med.* 26:266–270.

Makisalo S., Skutnabb K., Holmstrom T., Gronblad M., Paavolainen P. 1988. Reconstruction of anterior cruciate ligament with carbon fiber. An experimental study on pigs. *Am. J. Sports Med.* 16: 589–593.

Malaviya P., Butler D.L., Korvick D.L., Proch F.S. 1998. In vivo tendon forces correlate with activity level and remain bounded: Evidence in a rabbit flexor tendon model. *J. Biomech.* 31:1043–1049.

Marieb E. 1998. *Human Anatomy and Physiology,* 4th ed. Benjamin/Cummings Science Publishing, New York.

Marshall J.L., Warren R.F., Wickiewicz T.L. 1982. Primary surgical treatment of anterior cruciate ligament lesions. *Am. J. Sports Med.* 10:103–107.

Murrell G.A., Lilly E.G., Goldner R.D., Seaber A.V., Best T.M. 1994. Effects of immobilization on achilles tendon healing in a rat model. *J. Orthop. Res.* 12:582–591.

Ng G.Y., Oakes B.W., Deacon O.W., McLean I.D., Lampard D. 1995. Biomechanics of patellar tendon autograft for reconstruction of the anterior cruciate ligament. *J. Orthop. Res.* 13:602–608.

Noyes F.R., Barber-Westin S.D. 1997. A comparison of results in acute and chronic anterior cruciate ligament ruptures of arthroscopically assisted autogenous patellar tendon reconstruction. *Am. J. Sports Med.* 25:460–471.

Noyes F.R., Butler D.L., Paulos L.E., Grood E.S. 1983a. Intra-articular cruciate reconstruction. I: perspectives on graft strength, vascularization, and immediate motion after replacement. *Clin. Orthop.* 172:71–77.

Noyes F.R., Mooar P.A., Matthews D.S., Butler D.L. 1983b. The symptomatic anterior cruciate-deficient knee. Part I: the long-term functional disability in athletically active individuals. *J. Bone Joint Surg.* 65A:154–162.

Noyes F.R., Butler D.L., Grood E.S., Zernicke R.F., Hefzy M.S. 1984. Biomechanical analysis of human ligament grafts used in knee-ligament repairs and reconstructions. *J. Bone Joint Surg.* 66A:344–352.

Odensten M., Lysholm J., Gillquist J. 1984. Suture of fresh ruptures of the anterior cruciate ligament. A 5-year follow-up. *Acta Orthop. Scand.* 55:270–272.

O'Donoghue D.H., Rockwood C.A., Frank G.R., Jack S.C., Kenyon R. 1966. Repair of the anterior cruciate ligament in dogs. *J. Bone Joint Surg.* 48A:503–519.

O'Donoghue D.H., Frank G.R., Jeter G.L., Johnson W., Zeiders, J.W., Kenyon R. 1971. Repair and reconstruction of the anterior cruciate ligament in dogs. Factors influencing long-term results. *J. Bone Joint Surg.* 53A:710–718.

Ohno K., Yasuda K., Yamamoto N., Kaneda K., Hayashi K. 1993. Effects of complete stress-shielding on the mechanical properties and histology of in situ frozen patellar tendon. *J. Orthop. Res.* 11:592–602.

Ohno K., Pomaybo A.S., Schmidt C.C., Levine R.E., Ohland K.J., Woo S.L. 1995. Healing of the medial collateral ligament after a combined medial collateral and anterior cruciate ligament injury and reconstruction of the anterior cruciate ligament: comparison of repair and nonrepair of medial collateral ligament tears in rabbits. *J. Orthop. Res.* 13:442–449.

Paulos L.E., Butler D.L., Noyes F.R., Grood E.S. 1983. Intra-articular cruciate reconstruction. II: Replacement with vascularized patellar tendon. *Clin. Orthop.* 172:78–84.

Praemer A. 1996. Personal communications.

Praemer A., Furner S., Rice D.P. 1992. *Musculoskeletal Condition in the United States.* American Academy of Orthopaedic Surgeons, Park Ridge, IL.

Praemer A., Furner S., Rice D.P. 1999. *Musculoskeletal Condition in the United States,* 2nd ed. American Academy of Orthopaedic Surgeons, Rosemont, IL.

Proctor C.S., Jackson D.W., Simon T.M. 1997. Characterization of the repair tissue after removal of the central one-third of the patellar ligament. An experimental study in the goat model. *J. Bone Joint Surg.* 79A:997–1006.

Reddy G.K., Stehno-Bittel L., Enwemeka C.S. 1999. Matrix remodeling in healing rabbit achilles tendon. *Wound Repair Regeneration* 7:518–527.

Roberts J.M., Goldstrohm G.L., Brown T.D., Mears D.C. 1983. Comparison of unrepaired, primarily repaired, and polyglactin mesh-reinforced achilles tendon lacerations in rabbits. *Clin. Orthop.* 181:244–249.

Sato M., Maeda M., Kurosawa H., Inoue Y., Yamauchi Y., Iwase H. 2000. Reconstruction of rabbit achilles tendon with three bioabsorbable materials: histological and biomechanical studies. *J. Orthop. Sci.* 5:256–267.

Soma C.A., Mandelbaum B.R. 1995. Repair of acute achilles tendon ruptures. *Orthop. Clin. North Am.* 26:239–247.

Sommerich C.M., McGlothlin J.D., Marras W.S. 1993. Occupational risk factors associated with soft tissue disorders of the shoulder: a review of recent investigations in the literature. *Ergonomics* 36:697–717.

Wakitani S., Goto T., Pineda S.J., Young R.G., Mansour J.M., Caplan A.I., Goldberg V.M. 1994. Mesenchymal cell-based repair of large, full-thickness defects of articular cartilage. *J. Bone Joint Surg.* 76A:579–592.

Weiss A.B., Blazina M.E., Goldstein A.R., Alexander H. 1985. Ligament replacement with an absorbable copolymer carbon fiber scaffold—early clinical experience. *Clin. Orthop.* 196:77–85.

Weiss J.A., Woo S.L.Y., Ohland K.J., Horibe S., Newton, P.O. 1991. Evaluation of a new injury model to study medial collateral ligament healing—primary repair versus nonoperative treatment. *J. Orthop. Res.* 9:516–528.

Wredmark T., Engstrom B. 1993. Five-year results of anterior cruciate ligament reconstruction with the Stryker dacron high-strength ligament. *Knee Surg. Sports Traumatol. Arthrosc.* 1:71–75.

Yamamoto N., Ohno K., Hayashi K., Kuriyama H., Yasuda K., Kaneda K. 1993. Effects of stress shielding on the mechanical properties of rabbit patellar tendon. *J. Biomech. Eng.* 115:23–28.

Yamato M., Adachi E., Yamamoto K., Hayashi T. 1995. Condensation of collagen fibrils to the direct vicinity of fibroblasts as a cause of gel contraction. *J. Biochem.* 117:940–946.

Young R.G., Butler D.L., Weber W., Caplan A.I., Gordon S.L., Fink D.J. 1998. Use of mesenchymal stem cells in a collagen matrix for achilles tendon repair. *J. Orthop. Res.* 16:406–413.

Zernicke R.F., Garhammer J., Jobe F.W. 1977. Human patellar-tendon rupture: a kinetic analysis. *J. Bone Joint Surg.* 59A:179–183.

17

The Role of Mechanical Forces in Tissue Engineering of Articular Cartilage

Jethy C.Y. Hu and Kyriacos A. Athanasiou

Introduction

Articular cartilage is a specialized form of hyaline cartilage that covers the articulating ends of bones and serves as a wear-resistant, friction-reducing surface that evenly distributes forces onto the bone. Since chondrocytes do not sustain a healing response (Hunziker, 1999), injuries to the articular surface generally remain permanent. Injuries that penetrate down to the subchondral bone result in the formation of the mechanically inferior fibrocartilage, and this repair tissue eventually breaks down with usage (Buckwalter, 1998). Tissue engineering may be a solution to assist the healing process, and mechanical forces may be utilized to stimulate the chondrocytes to produce and organize matrix components to recreate the function of articular cartilage.

Structure and Function of Articular Cartilage

Articular cartilage is composed of a fluid phase and a solid phase. The interaction between these two phases determines the mechanical properties, and thus the function of cartilage. The fluid phase is water with physiological concentrations of ionic and non–ionic solutes, and makes up 75–80% of the wet weight (ww) of cartilage (Maroudas, 1979). By wet weight, the solid phase is composed of approximately 10% chondrocytes, 10–30% collagen, 3–10% proteoglycans, approximately 10% lipids, and minor amounts of glycoproteins (Muir, 1979).

Beginning with the articulating surface, cartilage can be separated into the superficial, middle, deep, and calcified zones (Fig. 17.1). These four distinct zones vary in their structure and function. The superficial zone makes up 10–20% of the full thickness of articular cartilage and contains densely packed collagen II fibrils and flattened, elongated cells oriented in the direction of shear stress (Kim et al., 1994). Serving as a transition from a tangential orientation of the superficial zone to the radial orientation in the deep zone, the middle zone contains round cells and collagen fibers that are randomly arranged and is 40–60% of the full thickness. In the deep zone, columns of ellipsoid cells and collagen fibers are oriented radially, and the collagen fibrils extend into the calcified zone to reinforce the bond between cartilage and bone (Maroudas, 1979). The deep zone is separated from the calcified zone by a distinct tidemark. The calcified zone contains cells trapped within the calcified matrix. Through the depth of cartilage, water content also falls linearly, from approximately 84% ww to 40–60% ww. The collagen content falls from 86% dry weight (dw) in the superficial zone to 67% dw in the deep zone. The proteoglycan content increases from around 15% dw in

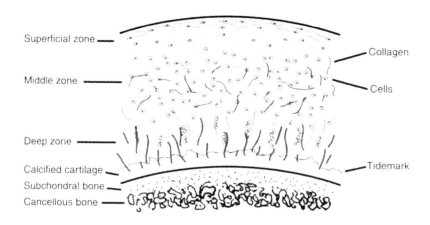

FIGURE 17.1. Articular cartilage is divided into four zones. The superficial zone contains flattened cells and collagen fibers aligned in the direction of shear. The middle zone contains randomly distributed round cells and collagen fibers that serve as a transition between the superficial and deep zones. The deep zone contains cells that are arranged in columns and collagen fibers that extend into the calcified zone of cartilage to reinforce the bond between cartilage and bone. Separated from the deep zone by a distinct tidemark, the calcified zone contains cells locked in the calcified matrix.

the superficial zone to a peak of 25% dw in the middle zone, then falls to 20% in the deep zone (Maroudas, 1979).

The chondrocyte is the basic metabolic unit of cartilage, and is responsible for matrix remodeling (Mow et al., 1992). Since articular cartilage is avascular, chondrocytes obtain nutrients by diffusion from the synovial fluid, facilitated during joint movement (Honner and Thompson, 1971). Type II collagen contributes to the tensile strength of cartilage and is the most abundant type of collagen in articular cartilage, accounting for 90–95% of the collagen in the matrix (Kempson et al., 1976). Its fibril thickness is affected by the other collagen types and varies from finest on the surface, to gradually becoming coarser in the deep zone. The functions of type X collagen are not entirely clear. It is classified as a network-forming collagen and is found mineralized in the calcified zone of cartilage. Proteoglycans are a special class of glycoproteins with long, un-branched, and highly negatively charged gly-cosaminoglycan (GAG) chains. The major type of proteoglycan found in cartilage is aggrecan, which provides, through water retention, the compressive strength of cartilage (Kempson, 1979). Aggrecan is formed by a core protein of high molecular weight (~250,000), with attached

GAG side chains, mostly chondroitin sulfate and keratin sulfate. Each aggrecan contains ~100 chondroitin sulfate chains and up to 60 keratin sulfate chains. Chondroitin sulfate is also larger, at around ~20 kDa each, while keratin sulfate chains are around 5–15 kDa each. The ratio of the two GAGs varies with the depth of cartilage. Expressed mostly in cartilaginous tissues, aggrecan aggregates with hyaluronan, an unbranched polysaccharide with molecular weight up to several million, and a link protein stabilizes this aggregation. The negative charges on aggrecan attracts cations into the matrix to result in a high osmotic pressure, called the Donnan osmotic pressure, causing the tissue to swell (Broom, 1990; Lai et al., 1993; Maroudas and Grushko, 1990). When cartilage is loaded, the load acts to displace the fluid from the cartilage, and force is thus dissipated, as modeled by the biphasic theory (Mow et al., 1980).

The biphasic theory models cartilage as consisting of an incompressible, porous, permeable solid and an incompressible viscous fluid that fills and interacts with the pores (Mow et al., 1980). When cartilage is loaded, the fluid flow drag of the interstitial fluid with the solid matrix balances out the applied forces (Mow et al., 1992). Modeling with the biphasic theory yields

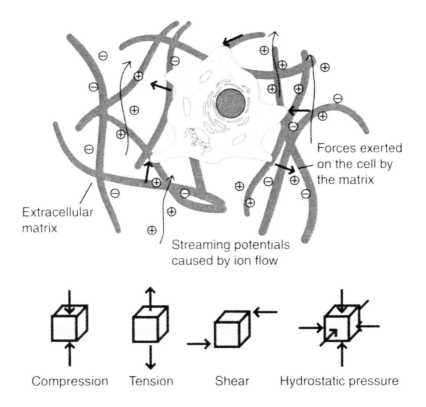

FIGURE 17.2. The types of forces a chondrocyte may experience during mechanical loading of the joint include direct compressive deformation, shear, and hydrostatic pressure. Chondrocytes may also experience streaming potentials caused by ion flow.

three independent variables from a creep indentation test, the compressive aggregate modulus, the Poisson's ratio, and the permeability. Human articular cartilage has thus been shown to have an aggregate modulus that ranges from 0.53 to 1.34 MPa, a Poisson's ratio which ranges from 0.00 to 0.14, and a permeability that ranges from 0.90×10^{-15} to 4.56×10^{-15} m^4/Ns (Athanasiou et al., 1991, 1994, 1995, 1998). The mechanical properties of cartilage are a direct result of its function, and thus vary from anatomical location to location. Even within a joint, high and low weight-bearing areas exist, and the compressive moduli, permeability, Poisson's ratio, and thickness may be different, as demonstrated by our experiments on the ankle (Athanasiou et al., 1995); hip (Athanasiou et al., 1994), knee (Athanasiou et al., 1991), and first metarsophalangeal joint (Athanasiou et al., 1998).

Forces Used in Engineering of Articular Cartilage

The goal of fabricating a mechanically functional tissue engineered articular cartilage is approached by manipulating four main variables: the cell type, the scaffold, chemical factors (peptides, growth factors), and mechanical forces. During joint loading, articular cartilage may experience compression and a plowing motion that exudes fluid at the leading edge of the motion and imbibes fluid at the trailing end (Mansour and Mow, 1976). Chondrocytes in the cartilage may thus experience direct compressive deformation, shear, and hydrostatic pressure (Fig. 17.2). Thus, chondrocytes have been cultured under these mechanical forces, and their responses have been assessed to understand their mechanotransduction pathway.

Compression

Articular cartilage undergoes cyclic compressive loading during joint movement. This cyclic compressive loading results in fluid exudation and reabsorption that facilitates diffusion of nutrients, growth factors, and wastes to and from the chondrocytes. In addition, the forces exerted onto the chondrocytes during this loading configuration also serve to modulate chondrocyte growth and activity. Numerous studies have been performed by compressing cartilage explants, while observing the expression and synthesis rates of the extracellular matrix (ECM) molecules. It has been shown that static compression of explants suppresses proteoglycan synthesis for up to 50% strain, and that cyclic compression has varied effects, depending on the loading regimen (Gray et al., 1988; Palmoski and Brandt, 1984; Sah et al., 1989; Torzilli et al., 1997). In tissue engineering efforts, Lee and Bader (1997) found that GAG synthesis at 15% compression was inhibited at 0.3 Hz, increased at 1 Hz, and remained the same as free-swelling controls at 3 Hz for chondrocytes seeded in agarose. Furthermore, Mauck and associates (2000) observed increased GAG synthesis and increased equilibrium aggregate modulus of cell-agarose constructs compressed at 10% strain at 1 Hz over free-swelling constructs. Though collagen synthesis was observed by Mauck and associates (2000) to increase, collagen synthesis was observed to decrease with increased frequency by Lee and Bader (1997). Still, cyclic compression at physiologic ranges, approximately 1 Hz and 10–15% strain, appears optimal for up-regulating GAG synthesis for tissue engineering. Cellular responses to compression include changes in proliferation, collagen synthesis, and GAG synthesis. These parameters were found to respond to dynamic strain regimens in distinct manners, suggesting that the signaling mechanisms involved are uncoupled (Lee and Bader, 1997). Modeling the chondrocyte response during compression has been separated into two categories, static and dynamic compression.

Static Compression

Static compression of cartilage explants has been shown to result mainly in the exudation of water (Armstrong et al., 1984). This fluid loss provides many possible mechanisms for transduction, such as an increase in the charge density of the matrix and a related change in the pH, an increase in the osmotic pressure, the creation of electric currents caused by ion movement, the facilitated diffusion of solutes, and the application of shear stress to the cells.

Fluid flow out of cartilage during compression results in increased matrix fixed charge density, which causes the counterion concentration to rise (Gray et al., 1988). Experiments have thus been preformed to observe the effects of ion concentrations on chondrocytes with and without compressive stress. By subjecting cartilage from the reserve zone of the epiphyseal plate to static compressive stresses of 0–3 MPa for 12 hr, SO_4^{2-}, K^+, and H^+ concentrations were altered to find that radiolabeled proline and sulfate incorporation were independent of SO_4^{2-} and K^+, but dependent on pH (Gray et al., 1988). For articular chondrocytes, Na^+ addition to medium resulted in decreased GAG synthesis (Schneiderman et al., 1986) that mirrors the effects of static compression. However, since Na^+ is the main component regulating the pericellular osmotic pressure, and since a decrease in GAG synthesis was also observed when sucrose was added to the medium to alter osmotic pressure (Schneiderman et al., 1986), the mechanism by which Na^+ acts to decrease GAG synthesis is still unclear. For tissue engineering purposes, the pericellular osmotic pressure should thus be kept as close to physiological conditions as possible to prevent a decrease in GAG synthesis (Bayliss et al., 1986; Schneiderman et al., 1986; Urban et al., 1993). The creation of electric currents caused by ion movement is called streaming potential. It has been modeled and examined in cartilage (Kim et al., 1995; Lai et al., 2000), but its direct effects are still unclear. Later, the role of diffusion and shear stress will be discussed together since current testing methodologies on chondrocytes have not been designed to observe the effects of one independently from the other.

Dynamic Compression

It is important to note that the mechanical environment of chondrocytes in explants is time dependent and inhomogeneous (Bachrach et al., 1995), thus the chondrocytes are subjected to different mechanical stimuli dependent on time and location.

Sah and associates (1989) observed increased proteoglycan production at 0.0001–0.001 Hz, which may be a result of increased diffusion, since significant fluid exudation occurs at this frequency. At higher frequencies, the hydrostatic pressure in the explant is increased, and strains of 1–5% stimulated radioisotope incorporation (Sah et al., 1989). The effects of hydrostatic pressure on chondrocytes have thus been investigated, and will be discussed later. In experiments that applied oscillatory compression of up to 10% to unconfined cartilage disk explants, Kim and associates (1994) observed more radiolabel incorporation in the outer ring of the explant than the inner core. Armstrong and associates (1984) have performed calculations to show that, in the case of unconfined compression, the outer ring experiences fluid flow, while the inner core mainly experiences hydrostatic pressure. This suggests that, in compression loading, the chondrocyte's response due to fluid flow or changes in cell shape may be greater than the response to hydrostatic pressure. Since fluid flow may also offer increased diffusion, the effect of diffusion was investigated by using disks with different surface area to volume ratios, for which no differences were observed (Kim et al., 1994). Thus, chondrocyte response in this case may be due to shear or changes in cell shape, and shear forces have been studied as a pathway for chondrocyte mechanotransduction.

Shear

When a pin is driven into an articular cartilage explant, stress fissures will form in the direction of shear stress because the collagen fibers in the lamina splendens are aligned lengthwise in this direction (Buckwalter et al., 1991). The cells in the superficial zone, directly beneath the lamina splendens, are also polar and flattened, with their axis along the direction of shear (Buckwalter et al., 1991). For this reason, and also because shear may act on chondrocytes during matrix compression, the effects of shear on chondrocytes have been investigated.

Smith and associates (1995) applied fluid-induced shear to normal human and bovine chondrocytes in culture using a cone viscometer. A shear of 1.6 Pa was applied for periods of 24, 48, and 72 hours. Both types of chondrocytes cultured under shear were observed to align in a major axis and an orthotropic minor axis. Glycosaminoglycan biosynthesis was found to increase two-fold for both human and bovine chondrocytes under shear stress, and increased chain lengths were also observed for both types of cells. Increased proteoglycan size was observed only in human chondrocytes, but the lack of change in the bovine population was attributed to the absence of serum in these cultures. The release of prostaglandin E2 was also increased 10 times in bovine and 20 times in human chondrocytes under shear (Smith et al., 1995). A reason for the increased GAG chain size was attributed to the release of nitric oxide by chondrocytes under shear, and that this nitric oxide then influences GAG metabolism (Das et al., 1997). In this case, shear forces may have detrimental effects of chondrocytes since experimental osteoarthritis has shown increased chondroitin sulfate chain lengths (Carney et al., 1985), and prostaglandin E2 is associated with proinflammatory mediators.

It is important to note, however, that efforts at tissue engineering of articular cartilage in bioreactors require a replication in the function of the tissue, and the persistence of chondrocytic release of prostaglandin E2 and nitric oxide due to shear should be assessed. If prostaglandin E2 is only released in culture, but not present at the time of implantation, shear may still be applied to promote synthesis of ECM *in vitro*, as long as effects associated with osteoarthritis do not remain in the tissue engineered product.

Hydrostatic Pressure

Chondrocytes in articular cartilage may experience hydrostatic pressure during compression of the matrix during loading, as modeled by Armstrong and associates (1984). In an experimental set-up, direct compression of the fluid medium is preferable over compression of the gas above it due to altered solubility of gases during compression, though some methodologies employ an inert gas (e.g., He) for this purpose.

It has been shown that chondrocytes can survive at hydrostatic pressures as high as 30 MPa (Parkkinen et al., 1995). At the low range, Lippiello and associates (1985) observed suppression of matrix synthesis when 0.52–2.06 MPa was applied continuously to cartilage explants for 24 hours. At 2.59 MPa, though, continuous hydrostatic pressure resulted in 10–15% increase in $^{35}SO_4$ incorporation (Lippiello et al., 1985). Continuous pressures above 20 MPa lowered the incorporation of radiolabels (Hall et al., 1991; Lammi et al., 1994), but this decrease was reversible (Hall et al., 1991). A "rebound" response when the cartilage was unloaded was observed with 60% increase in metabolic rate (Lippiello et al., 1985), suggesting that cyclic pressures may be more beneficial in up-regulating ECM production. Cyclic hydrostatic pressure at physiological ranges (5–15 MPa) applied for 20 seconds or for 5 minutes has been shown to increase chondrocyte matrix production, as indicated by the increased incorporation of radiolabeled sulfates and proline (Hall et al., 1991). Smith and associates (2000) have shown that an intermittent loading regimen of 10 MPa at 1 Hz at times less than one day showed increases for type II collagen mRNA and aggrecan mRNA levels. It was found that collagen mRNA signals increased five-fold at 4–8 hours loading periods and then decreased, whereas the aggrecan mRNA signal increased to three-fold throughout this loading period. When the loading profile was changed to 10 MPa at 1 Hz for 4 hours per day for 4 days, the type II collagen mRNA signal was increased by nine-fold, and the aggrecan mRNA signal was increased by 20-fold when compared to unloaded cultures (Smith et al., 2000).

It has been shown that the stimulation of histamine H1 receptors cause prostaglandin E2 production, while the stimulation of the H2 receptor causes cAMP synthesis (Taylor and Woolley, 1987). That the two pathways are separate has also been observed by Uchida and associates (1988), who observed an increase in cAMP synthesis but not in prostaglandin synthesis in cultured rib chondrocytes exposed to tension (Uchida et al., 1988). Since blocking either channel does not affect the membrane hyperpolarization of chondrocytes in response to cyclic hydrostatic pressure, the effects of hydrostatic pressure has been suggested to not include the production of prostaglandins (Wright et al., 1992), unlike shear.

Hydrostatic pressure may be transduced to the cell through the cytoskeleton. For instance, the Golgi apparatus has been shown to respond to hydrostatic pressure through microtubules (Parkkinen et al., 1993). Continuous application of hydrostatic pressure at 30 MPa to chondrocytes resulted in compaction of the Golgi apparatus and a decrease in proteoglycan synthesis (Parkkinen et al., 1993). When the microtubules are disrupted with nocodazole, the Golgi apparatus did not deform, and loading did not stimulate the incorporation of radiolabels (Parkkinen et al., 1993). Both nocodazole and taxol inhibited proteoglycan synthesis under hydrostatic pressure, though the structure and function of the synthesized proteoglycans were not altered (Jortikka et al., 2000). This would suggest that the effects of hydrostatic pressure are not through microtubule-dependent vesicle traffic (Jortikka et al., 2000). Chondrocyte stress fibers nearly completely disappeared under continuous pressure at 30 MPa, while intermittent pressure resulted in thinning of the stress fibers (Parkkinen et al., 1995). The actin cytoskeleton has been suggested to play a role in chondrocyte response to mechanical stimuli by modulating membrane potential since chondrocytes, which depolarize their membranes under hydrostatic pressure, did not depolarize their membranes when exposed to cytochalasin B, used to disrupt actin (Wright et al., 1992). This is in contrast to chondrocytes maintaining their response to hydrostatic pressure in the presence of colchicine, a substance known to distrupt microtubules, thus suggesting that microtubules are not involved in the depolarization response (Wright et al., 1992).

Continuous hydrostatic pressure results in depolarization of the chondrocyte cell membrane (Wright et al., 1992). Since depolarization occurred in the presence of verapamil, which blocks Ca^+ channels, it is thought that depolarization occurs due to Na^+ channels. This observation is further supported with the observation that addition of tetrodotoxin, shown to block Na^+ channels in fibroblasts, also blocked depolarization in chondrocytes subjected to hydrostatic pressure (Wright et al., 1992). Quinidine blocks Ca^+-dependent K^+ channels, and its addition prevented the hyperpolarization of the chondrocyte cell membrane as a result to pressure (Wright et al., 1992), suggesting that these channels are also involved. Other ion transport mechanisms affected by hydrostatic pressure include the Na/K pump (supressed at 2.5–50 MPa) and the Na/K/2Cl cotransporter (inhibited at 7.5 MPa), to result in a pressure dependent inhibition of K flux (Hall, 1999). A change in ion concentration may result in the reorganization of stress fibers (Otter et al., 1987). Chondrocyte communication may occur via ion exchange, since micropipette distortion of the plasma membrane induced a wave of increased calcium ion concentration that was spread to surrounding cells by phospholipase C permeating through gap junctions (D'Andrea et al., 2000). Ion channels can also cause a change in intracellular pH. It has been shown that the resting intracellular pH in chondrocytes is 7.10 ± 0.04 (Wilkins and Hall, 1992), and that this pH is not altered by hydrostatic pressure at 30.3 MPa (Browning et al., 1999), though the rate of recovery from ammonium-induced intracellular acidification was increased. It has been shown that the transporter responsible for the regulation of pH in chondrocytes is the $Na^+ \times H^+$ exchanger (Hall et al., 1996), and its activation under hydrostatic pressure involves direct phosphorylation of the transporter protein itself (Browning et al., 1999).

Zonal Variations Pertinent to Tissue Engineering

Though chondrocytes have been categorized as all belonging to the same phenotype, transient metabolic differences between the chondrocytes of different sizes (Trippel et al., 1980) and zonal affiliations (Aydelotte et al., 1988; Aydelotte and Kuettner, 1988; Archer et al., 1990; Siczkowski and Watt, 1990; Zanetti et al., 1985) have been observed *in vitro.* For instance, superficial zone chondrocytes were found to attach to tissue culture plastic slower than those from the deeper zones (Siczkowski and Watt, 1990). Keratin sulfate synthesis has been observed to gradually in-

crease through cartilage depth (Archer et al., 1990; Aydelotte and Kuettner, 1988; Aydelotte et al., 1988; Siczkowski and Watt, 1990; Zanetti et al., 1985). Just as many cells types lose their phenotypes *in vitro,* these zonal differences in cell morphology and metabolic product decreased in time as chondrocyte subpopulations were cultured in monolayers (Siczkowski and Watt, 1990). However, chondrocytes cultured in agarose retained morphological and proteoglycan synthesis differences for at least a month (Archer et al., 1990; Aydelotte and Kuettner, 1988; Aydelotte et al., 1988).

In addition to metabolic differences, other variations pertinent to mechanotransduction have also been found with depth, such as the expression of integrin receptors. Integrin receptors attach the chondrocyte to its ECM, and thus may pass mechanical forces applied to the ECM on to the chondrocyte via the cytoskeleton or by causing the release of secondary messengers (Fig. 17.3). The cytoskeleton was observed to respond to mechanical stimuli across the cell membrane via the $\beta 1$ integrins, as shown by Wang and associates (1993) using endothelial cells. In this case, force-dependent stiffening of cells subjected to mechanical forces to adhesion (integrin $\beta 1$) and nonadhesion cell surface receptors was observed only when microtubules, intermediate filaments, and microfilaments were intact. Mechanical signals can also progress through the integrin receptor to result in the release of secondary messengers, and further progression through the cytoskeleton may result in the activation of gene transcription (Ben-Ze'ev, 1991; Ingber, 1991). In the case of chondrocytes, Millward-Sadler and associates (2000) observed increased levels of aggrecan mRNA and decreased levels of matrix metalloproteinase-3 mRNA in mechanically stimulated chondrocytes, a response that was dependent on integrins, stretch-activated ion channels, and interleukin-4. Integrins that link the chondrocyte to the ECM include $\beta 1$ and $\alpha 5$ subunits that mediate attachment to collagen, osteopontin, and fibronectin, among the many ECM components of cartilage (Attur et al., 2000; Loeser, 1993; Shimizu et al., 1997). Other integrin receptors observed by immunohistochemical staining on chondrocytes include $\alpha 1$, αV, $\beta 4$, and $\beta 5$ inte-

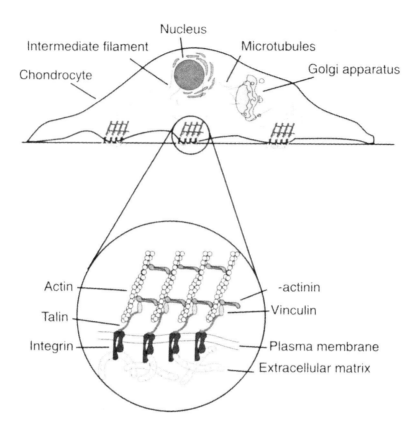

FIGURE 17.3. Integrin receptors link the cytoskeleton to the extracellular matrix (ECM), and mechanical forces may be transferred from the ECM to various organelles that may be targets of mechanotransduction.

grins (Ostergaard et al., 1998). Salter and associates (1992) have shown that integrin expression is heterogeneous throughout the depth of articular cartilage. For instance, the αV subunit is preferentially expressed by chondrocytes in the superficial zone than by cells in the deeper zones (Ostergaard et al., 1998).

The cytoskeletal components are believed to be part of the pathway of mechanotransduction and have also been found to vary through depth. In the case of intermediate filaments (IF), IFs are more prominent in weight-bearing than non-weight-bearing areas of articular cartilage; this is believed to be related to the high swelling pressure of the middle zone (Eggli et al., 1988). Ralphs and associates (1993) have found a number of IFs such as cytokeratins and vimentin that vary with the zonal position and the developmental stage of rat articular cartilage. Cytokeratins are associated in the superficial zone and

may contribute to the flattened morphology of these chondrocytes (Ralphs et al., 1993). Cells that experience large deformations (i.e., closer to the surface) contain less vimentin than those experiencing smaller deformations (i.e., the cells in the deep zone) (Durrant et al., 1999). Vimentin has thus been hypothesized to resist compression (Ghadially, 1983; Ralphs et al., 1992). Vimentin may also play a role in mechanotransduction since it is observed to associate with focal contacts and with the nucleus, and its assembly depends on the mechanical culturing conditions (Durrant et al., 1999). In static culture, where cartilage explants were free to swell, the vimentin cytoskeleton was disassembled after 1 hour. The vimentin cytoskeleton then reappeared after several hours. Under compression of greater than 0.5 MPa, where the loading counteracted the free swelling present in static culture, the vimentin cytoskeleton was preserved. The actin cytoskeleton was observed to be maintained in both static culture and under compressive loading (Durrant et al., 1999). The co-localization of actin and vimentin at focal points suggests that the actin cytoskeleton is indirectly linked to the pericellular matrix. Thus, these data suggest that the actin is responsible for sensing the load, since it was present in all cases, while the vimentin structure is a response to the load (Durrant et al., 1999). A link between actin and the nucleus has been shown, since nuclei of chondrocytes under compression only deformed with intact actin filaments (Guilak, 1995). Microtubules are found as cilium in many chondrocytes, and are thought to sense mechanical, physiochemical, or osmotic stimuli in the pericellular environment (Poole et al., 1987). For instance, when nocodazole and taxol were used to disrupt or stabilize microtubules in chondrocytes monolayers, disruption by nocodazole decreased [^{35}S]sulfate incorporation by 39–48%, while stabilization by taxol also decreased [^{35}S]sulfate incorporation, but by 17% (Jortikka et al., 2000). The variation of microtubules through depth has not been studied extensively; it is only known that microtubules are found in cells of the middle zone (Palfrey and Davies, 1966).

When using differentiated chondrocytes for tissue engineering, it is thus important to note differences of the chondrocytes from each zone, since they display metabolic differences, different expression of integrins, different cytoskeletal components, and reside in different mechanical environments. The response of different zones of chondrocytes to mechanical forces has been demonstrated to be different. The effects of hydrostatic pressure vary according to species, joint, and the topographical location in the joint (Hall et al., 1991). Hall and associates (1991) suspect that these variations are not due to the previous loading history of the tissue. The difference arising from topographical location may be attributed to the varied proportion of surface to deep zone cells from one topographical location to the next, and that cells from these zones may respond to hydrostatic pressures differently (Hall et al., 1991). Compressive loading of subpopulations of chondrocytes in agarose at 15% strain showed a general inhibition of GAG synthesis in superficial zone chondrocytes, while the

same strain applied at 1 Hz resulted in a 50% stimulation of GAG synthesis by deep zone chondroctes (Lee et al., 1998).

The ECM of the chondrocytes may also affect their response to hydrostatic pressure. Decreased sulfate incorporation was observed in cultured chondrocytes loaded at 0.05 Hz for 1.5 hours, but the same loading in explants resulted in an increased sulfate incorporation in explants (Parkkinen et al., 1993). Cell–cell contacts have also been found to resist the effects of hydrostatic pressure (Parkkinen et al., 1995). The structure/function relationship of cartilage means that material heterogeneity is directly translated into mechanical differences. Variations in collagen fibril diameter, density, and orientation, and variations in the type and amount of glycosaminoglycans with respect to depth also mean that the mechanical properties of cartilage vary over depth. Early compressive testing has mainly modeled cartilage as a homogenous tissue, but it has been shown recently that the compressive modulus increased from 0.079 ± 0.039 MPa in the superficial layer to 2.10 ± 2.69 MPa in the deepest layer (Schinagl et al., 1997). Zonal variations in the properties of articular cartilage have also been found using a microindentation technique (Shin et al., 1998). It has also been shown that the tensile properties of cartilage vary with depth, and the tensile modulus of the ECM correlated strongly with the collagen/proteoglycan ratio (Akizuki et al., 1986; Woo et al., 1976). Since chondrocytes respond to mechanical signals, these differences may be responsible for the synthetic differences of chondrocytes from zone to zone. For example, cartilage explants compressed at 1 MPa showed the greatest reduction in GAG synthesis in the superficial zone (Guilak et al., 1994). Since the upper layers of cartilage are compressed more than the lower layers, it was found that inhibition of biosynthetic activity was generally highest for areas undergoing the highest compressive axial strains (Wong et al., 1997).

Mechanics of Single Cells

Chondrocytes in explants are likely to experience numerous forms of mechanical forces simultaneously, but the mechanisms of mechanotransduction may be better understood from simple models involving fewer types of forces. The mechanisms of the chondrocyte response to deformation *in vivo* may be elucidated by *in vitro* experiments with single chondrocytes, and, in this case, the cytomechanical properties of the chondrocyte, that is, the stiffnesses of its cytoskeleton and plasma membrane, are important parameters that can be quantified and may be correlated to the chondrocyte response.

The earliest experiments in observing chondrocyte deformation were done in cartilage explants and visualized by microscopy (Broom and Myers, 1980), and, recently, the deformation of chondrocytes in a matrix has been mathematically modeled (Wu et al., 1999). Since tissue engineering of cartilage may require a matrix to carry the cells, chondrocyte deformation in agarose has also been observed (Lee and Bader, 1995). In both cases, chondrocytes were found to deform during compression of the matrix in which they resided. In the case where the cartilage explant was compressed, surface zone chondrocytes were found to deform anisotropically, expanding more in the direction perpendicular to the local split-line pattern. Also, these chondrocytes underwent more deformation than deep zone chondrocytes (Guilak et al., 1995), while compression of chondrocytes in agarose showed the reverse (Lee and Bader, 1995). In both cases, the mechanical properties of the matrices prevented the direct determination of chondrocyte cytomechanical properties. However, with these two experiments, the importance of matrix mechanical properties to influence mechanotransduction in tissue engineering has been illustrated. Chondrocyte deformation in explants may be influenced by the changing mechanical properties of the cartilage zones (Kempson et al., 1976; Narmoneva et al., 1999; Schinagl et al., 1997; Shin et al., 1998). Tissue engineering scaffolds that mimic the mechanical properties of cartilage zones may provide the correct mechanical signals required for chondrocyte matrix synthesis and remodeling to form a tissue that replicates the functions of these three zones.

Recent mathematical modeling of chondrocyte deformation requires not only knowledge of matrix mechanical properties, but also the mechanical properties of the chondrocyte to yield the stress-strain and fluid flow environment of the chondrocyte. Guilak and associates (1999) have directly measured chondrocyte cytomechanical properties using micropipette aspiration, and chondrocytes and their pericellular matrices have been found to have equilibrium elastic moduli of 0.6 kPa and 1.5 kPa, respectively, which are three orders of magnitudes smaller than the elastic modulus of the ECM. To observe the role of the cytoskeleton in this deformation process, Guilak (1995) disrupted actin cytoskeleton using cytochalasin D and found that the cell deformed in volume as before, but deformation of the nuclei was only observed in nondisrupted cells. Determination of the zonal variations of the stiffness of chondrocytes may further aid in modeling, and this effort may be accomplished using micropipette aspiration or a cytoindentation technique, developed by Shin and Athanasiou (1999). Briefly, a cantilever beam with an indenting probe is applied to cells seeded on glass slides. The force applied via this indentation may be calculated by measuring the deflection of this cantilever and its spring constant. Time-dependent deformation of a cell may thus be obtained and modeled by the biphasic theory to yield cytomechanical properties. Of particular interest is that a similar configuration may also be employed to directly determine a cell's adhesiveness to its substrate (Athanasiou et al., 1999; Yamamoto et al., 1998), an important parameter for chondrocyte mechanotransduction as evidenced by their anchorage dependence. Cytoskeletal reorganizations due to the formation of focal adhesions may change the cell stiffness, and, thus, using these methods, cell adhesion and cell stiffness may be correlated to observe the effect of different substrates on cellular mechanotransduction.

Bioreactors

Currently, direct compression of the matrix provides a complex array of mechanical signals, as previously discussed. The effects of shear on chondrocyte metabolism have been investigated using a cone viscometer (Das et al., 1997; Smith et al., 1995). In this case, the effects of fluid induced shear cannot be separated from the effects of increased mass transport, and secondary fluid flows such as eddies around the cells were not quantified. Direct cell distortion applied by seeding chondrocytes on a membrane and stretching it (Uchida et al., 1988) moves the chondrocytes relative to the medium and may result in shear and secondary flows. So far, only hydrostatic pressure has been applied independent of other forces. On the other hand, many bioreactors have been developed that apply a combination of forces simultaneously, and these are briefly discussed below.

Carver and Heath (1999) have performed several tests using a semi-perfusion bioreactor with intermittent pressure. Pressures of 3.44 and 6.87 MPa were applied intermittently at 5 seconds on and 15 seconds off for 20 minutes every 4 hours on chondrocytes seeded on PGA. The compressive modulus of the resulting constructs was higher for those cultured in the bioreactor as opposed to those cultured in spinner flasks. It is interesting to note that although constructs displayed native-level concentrations of GAG, their compressive moduli did not reach native levels, though the modulus did increase with GAG content (Carver and Heath, 1999). This suggests possible, though incomplete, organization of the secreted matrix due to mechanical forces. In addition to synthesis, the ECM must be structured to withstand loads to function as an implant. How mechanical forces may signal chondrocytes to organize its matrix is still a mystery. Matrix components may be organized actively by chondrocytes during production, or they may organize themselves after secretion by the chondrocytes. Thus, the current challenge in producing mechanically functional tissue using mechanotransduction is not only in the up-regulation of ECM synthesis, but also its retention in the matrix and its organization to replicate the structure/function relationship of cartilage.

Microgravity has also been investigated in the role of cartilage engineering, in conjunction with rotating wall bioreactors. Freed and associates (1997) when comparing constructs cultured on the Mir Space Station and those cultured on earth, found the constructs cultured on earth to be mechanically superior. Whether this difference may be completely attributed to the difference in gravity is uncertain since the spinning rates of the rotating wall bioreactor in space and on earth were different, and thus the effects of diffusion may also serve as a modulator of chondrocyte response.

Rotating wall bioreactors have been used to model "microgravity" conditions. During rotation, the centrifugal, fluid shear, and gravitational force vectors acting on a construct in this bioreactor sums to zero and the construct is suspended. Tissue-engineered constructs produced using this bioreactor have been shown to contain 68% as much GAG and 33% as much type II collagen (ww) after 40 days (Freed et al., 1998). Vunjak-Novakovic and associates (1999) compared tissue-engineered cartilage cultured in static flask, mixed flask, and rotating wall bioreactor to conclude that constructs cultured in rotating wall bioreactors contained more GAG than those from either static or mixed flasks and more collagen than those from static flasks. The GAG content of constructs from rotating wall bioreactors was lower, though not significantly, than native tissue. However, the equilibrium modulus of this construct is less than one third of that of native tissue. This presents a similar problem to that of the perfusion bioreactor with intermittent pressure, that is, synthesis of matrix components with incomplete organization. The increased ECM production in this case is attributed to laminar flow and facilitated diffusion. Whether the effect of laminar flow is to provide a certain level of shear stress or to provide as low of a shear stress as possible still needs to be investigated.

Challenges and Future Directions

The use of mechanical forces is only one component in our philosophy to tissue engineer cartilage. This component itself involves several steps. First, the mechanical function of the native tissue has to be defined with respect to the zonal variations. Secondly, salient mechanical forces with quantifiable parameters should be applied to single cells or a colony of cells in order to determine their response to this one mechanical stimulus. A model should then be developed to find optimal load regimens, and these load regimens may then be applied to tissue engineer cartilage. The biomechanical and biochemical properties of this tissue-engineered cartilage should then be compared with the native tissue, and the model should be modified accordingly.

Challenges in using mechanical forces in cartilage tissue engineering include finding the type of mechanical forces, amplitude, and duty cycle that are necessary to induce the chosen cell type to produce a functional tissue, and developing bioreactors that are capable of applying these forces. To do this, further work needs to be done in understanding the mechanisms of mechanotransduction, and models need to be developed to gain a predictive understanding of how chondrocytes behave under bioreactor conditions. Tissue engineering of cartilage should correlate mechanical properties of tissue-engineered constructs with mechanical forces applied, if these correlations exist. Efforts to determine the forces needed to tissue engineer articular cartilage have been approached from two directions, one by determining the mechanotransduction pathways at the cellular level, and another by directly assessing the resulting construct due to applied force(s). A convergence of the two models to predict how mechanical forces may contribute to matrix production and organization at the cellular level is a long way off and may be too complex to simulate for engineering purposes. Currently, ECM synthesis is correlated with mechanical forces, and then the overall biomechanical properties of the construct are correlated with the degree of ECM synthesis (Vunjak-Novakovic et al., 1999). However, one must note that tissue-engineered cartilage may be inhomogeneous, as was this case, and thus different zones of the cartilage will have different mechanical properties. What may be achieved now is to develop a simplified model for the forces in a bioreactor and the forces that cells experience given their matrix enclosure. From this, cells in regions in the matrix that experience similar forces may be delineated, and a model may be developed to correlate the structural organization and mechanical properties of these regions with the applied mechanical force(s).

Though articular chondrocytes have been classified as belonging to one phenotype, significant differences in their response to mechanical signals may warrant their separation in tissue engineering, as opposed to current methods were chondrocytes from all zones are seeded homogeneously. Since compression has been applied to chondrocyte subpopulations and shown that different metabolic responses result, we may now move on to observe these components separately. In our effort to reach this goal, the immediate approach may first be to further identify the cytomechanical properties of chondrocytes with respect to zonal and substrate variations, and under different mechanical stresses. Shear, hydrostatic pressure, and facilitated diffusion may be applied separately to chondrocyte subpopulations and their metabolic responses observed. If the structure/function relationship of cartilage is any indication of chondrocyte matrix remodeling, we may expect superficial cells to produce more collagen under shear, deep cells to produce more proteoglycans under hydrostatic compression, and the synthesis rates of both populations to increase with facilitated diffusion. Also, scaffolds with different mechanical properties may be incorporated into the same tissue engineering construct to aid in delineating and also in replicating the functions of the zones of cartilage. Concurrently, standards must be developed to evaluate the mechanical function of tissue-engineered constructs.

References

Akizuki S., Mow V.C., Muller F., Pita J.C., Howell D.S., Manicourt D.H. 1986. Tensile properties of human knee joint cartilage: I. Influence of ionic conditions, weight bearing, and fibrillation on the tensile modulus. *J. Orthop. Res.* 4:379–392.

Archer C.W., McDowell J., Bayliss M.T., Stephens M.D., Bentley G. 1990. Phenotypic modulation in sub-populations of human articular chondrocytes *in vitro*. *J. Cell Sci.* 97:361–371.

Armstrong C.G., Lai W.M., Mow V.C. 1984. An analysis of the unconfined compression of articular cartilage. *J. Biomech. Eng.* 106:165–173.

Athanasiou K.A., Agarwal A., Dzida F.J. 1994. Comparative study of the intrinsic mechanical properties of the human acetabular and femoral head cartilage. *J. Orthop. Res.* 12:340–349.

Athanasiou K.A., Liu G.T., Lavery L.A., Lanctot D.R., Schenck R.C., Jr. 1998. Biomechanical topography of human articular cartilage in the first metatarsophalangeal joint. *Clin. Orthop.* 348:269–281.

Athanasiou K.A., Niederauer G.G., Schenck R.C., Jr. 1995. Biomechanical topography of human ankle cartilage. *Ann. Biomed. Eng.* 23:697–704.

Athanasiou K.A., Rosenwasser M.P., Buckwalter J.A., Malinin T.I., Mow V.C. 1991. Interspecies comparisons of in situ intrinsic mechanical properties of distal femoral cartilage. *J. Orthop. Res.* 9:330–340.

Athanasiou K.A., Thoma B.S., Lanctot D.R., Shin D., Agrawal C.M., LeBaron R.G. 1999. Development of the cytodetachment technique to quantify mechanical adhesiveness of the single cell. *Biomaterials* 20:2405–2415.

Attur M.G., Dave M.N., Clancy R.M., Patel I.R., Abramson S.B., Amin A.R. 2000. Functional genomic analysis in arthritis-affected cartilage: yin-yang regulation of inflammatory mediators by alpha 5 beta 1 and alpha V beta 3 integrins. *J. Immunol.* 164:2684–2691.

Aydelotte M.B., Greenhill R.R., Kuettner K.E. 1988. Differences between sub-populations of cultured bovine articular chondrocytes. II. Proteoglycan metabolism. *Connect. Tissue Res.* 18:223–234.

Aydelotte M.B., Kuettner K.E. 1988. Differences between sub-populations of cultured bovine articular chondrocytes. I. Morphology and cartilage matrix production. *Connect. Tissue Res.* 18:205–222.

Bachrach N.M., Valhmu W.B., Stazzone E., Ratcliffe A., Lai W.M., Mow V.C. 1995. Changes in proteoglycan synthesis of chondrocytes in articular cartilage are associated with the time-dependent changes in their mechanical environment. *J. Biomech.* 28:1561–1569.

Bayliss M.T., Urban J.P., Johnstone B., Holm S. 1986. in vitro method for measuring synthesis rates in the intervertebral disc. *J. Orthop. Res.* 4:10–17.

Ben-Ze'ev A. 1991. Animal cell shape changes and gene expression. *Bioessays* 13:207–212.

Broom N.D. 1990. New experimental approaches to the understanding of structure-function relationships in articular cartilage. In: *Methods in Cartilage Research.* A. Maroudas and K. Kuettner, eds. Academic Press Limited, San Diego; pp. 70–73.

Broom N.D., Myers D.B. 1980. A study of the structural response of wet hyaline cartilage to various loading situations. *Connect. Tissue. Res.* 7:227–237.

Browning J.A., Walker R.E., Hall A.C., Wilkins R.J. 1999. Modulation of $Na^+ \times H^+$ exchange by hydrostatic pressure in isolated bovine articular chondrocytes. *Acta Physiol. Scand.* 166:39–45.

Buckwalter J.A. 1998. Articular cartilage: injuries and potential for healing. *J. Orthop. Sports Phys. Ther.* 28:192–202.

Buckwalter J.A., Hunziker E.B., Rosenberg L.C., Coutts R., Adams M., Eyre D. 1991. Articular cartilage: composition and structure. In: *Injury and Repair of the Musculoskeletal Soft Tissues,* 2nd ed. S. L. Woo and J. A. Buckwalter, eds. American Academy of Orthopaedic Surgeons, Park Ridge, pp. 405–425.

Carney S.L., Billingham M.E., Muir H., Sandy J.D. 1985. Structure of newly synthesised (^{35}S)-proteoglycans and (^{35}S)- proteoglycan turnover products of cartilage explant cultures from dogs with experimental osteoarthritis. *J. Orthop. Res.* 3:140–147.

Carver S.E., Heath C.A. 1999. Influence of intermittent pressure, fluid flow, and mixing on the regenerative properties of articular chondrocytes. *Biotechnol. Bioeng.* 65:274–281.

D'Andrea P., Calabrese A., Capozzi I., Grandolfo M., Tonon R., Vittur F. 2000. Intercellular Ca^{2+} waves in mechanically stimulated articular chondrocytes. *Biorheology* 37:75–83.

Das P., Schurman D.J., Smith R.L. 1997. Nitric oxide and G proteins mediate the response of bovine articular chondrocytes to fluid-induced shear. *J. Orthop. Res.* 15:87–93.

Durrant L.A., Archer C.W., Benjamin M., Ralphs J.R. 1999. Organization of the chondrocyte cytoskeleton and its response to changing mechanical conditions in organ culture. *J. Anat.* 194:343–353.

Eggli P.S., Hunziker E.B., Schenk R.K. 1988. Quantification of structural features characterizing weight- and less- weight-bearing regions in articular cartilage: a stereological analysis of medial femoral condyles in young adult rabbits. *Anat. Rec.* 222:217–227.

Freed L.E., Hollander A.P., Martin I., Barry J.R., Langer R., Vunjak-Novakovic G. 1998. Chondrogenesis in a cell-polymer-bioreactor system. *Exp. Cell Res.* 240:58–65.

Freed L.E., Langer R., Martin I., Pellis N.R., Vunjak-Novakovic G. 1997. Tissue engineering of cartilage in space. *Proc. Natl. Acad. Sci. U.S.A.* 94: 13885–13890.

Ghadially F. N. 1983. *Fine Structure of Synovial Joints: A Text and Atlas of the Ultrastructure of Normal and Pathological Articular Tissues.* Butterworths, London.

Gray M.L., Pizzanelli A.M., Grodzinsky A.J., Lee R.C. 1988. Mechanical and physiochemical determinants of the chondrocyte biosynthetic response. *J. Orthop. Res.* 6:777–792.

Guilak F. 1995. Compression-induced changes in the shape and volume of the chondrocyte nucleus. *J. Biomech.* 28:1529–1541.

Guilak F., Jones W.R., Ting-Beall H.P., Lee G.M. 1999. The deformation behavior and mechanical properties of chondrocytes in articular cartilage. *Osteoarthritis Cartilage* 7:59–70.

Guilak F., Meyer B., Ratcliffe A., Mow V. 1994. The effects of matrix compression on proteoglycan metabolism in articular cartilage explants. *Osteoarthritis Cartilage* 2:91–101.

Guilak F., Ratcliffe A., Mow V.C. 1995. Chondrocyte deformation and local tissue strain in articular cartilage: a confocal microscopy study. *J. Orthop. Res.* 13:410–421.

Hall A.C. 1999. Differential effects of hydrostatic pressure on cation transport pathways of isolated articular chondrocytes. *J. Cell. Physiol.* 178:197–204.

Hall A.C., Horwitz E.R., Wilkins R.J. 1996. The cellular physiology of articular cartilage. *Exp. Physiol.* 81:535–545.

Hall A.C., Urban J.P., Gehl K.A. 1991. The effects of hydrostatic pressure on matrix synthesis in articular cartilage. *J. Orthop. Res.* 9:1–10.

Honner R., Thompson R.C. 1971. The nutritional pathways of articular cartilage. An autoradiographic study in rabbits using 35S injected intravenously. *J. Bone Joint Surg. [Am.]* 53:742–748.

Hunziker E.B. 1999. Articular cartilage repair: are the intrinsic biological constraints undermining this process insuperable? *Osteoarthritis Cartilage* 7:15–28.

Ingber D. 1991. Integrins as mechanochemical transducers. *Curr. Opin. Cell. Biol.* 3:841–848.

Jortikka M.O. Parkkinen J.J., Inkinen R.I., Karner J., Jarvelainen H.T., Nelimarkka L.O., Tammi M.I., Lammi M.J. 2000. The role of microtubules in the regulation of proteoglycan synthesis in chondrocytes under hydrostatic pressure. *Arch. Biochem. Biophys.* 374:172–180.

Kempson G.E. 1979. Mechanical properties of articular cartilage. In: *Adult Articular Cartilage,* 2nd ed., M.A.R. Freeman, ed. Pitman Medical, Kent, England; pp. 333–414.

Kempson G.E., Tuke M.A., Dingle J.T., Barrett A.J., Horsfield P.H. 1976. The effects of proteolytic enzymes on the mechanical properties of adult human articular cartilage. *Biochim. Biophys. Acta* 428:741–760.

Kim Y.J., Bonassar L.J., Grodzinsky A.J. 1995. The role of cartilage streaming potential, fluid flow and pressure in the stimulation of chondrocyte biosynthesis during dynamic compression. *J. Biomech.* 28:1055–1066.

Kim Y.J., Sah R.L., Grodzinsky A.J., Plaas A.H., Sandy, J.D. 1994. Mechanical regulation of cartilage biosynthetic behavior: physical stimuli. *Arch. Biochem. Biophys.* 311:1–12.

Lai W.M., Mow V.C., Sun D.D., Ateshian G.A. 2000. On the electric potentials inside a charged soft hydrated biological tissue: streaming potential versus diffusion potential. *J. Biomech. Eng.* 122:336–346.

Lai W.M., Mow V.C., Zhu W. 1993. Constitutive modeling of articular cartilage and biomacromolecular solutions. *J. Biomech. Eng.* 115:474–480.

Lammi M.J., Inkinen R., Parkkinen J.J. Hakkinen T., Jortikka M., Nelimarkka L.O., Jarvelainen H.T., Tammi M.I. 1994. Expression of reduced amounts of structurally altered aggrecan in articular cartilage chondrocytes exposed to high hydrostatic pressure. *Biochem. J.* 304:723–730.

Lee D.A., Bader D.L. 1995. The development and characterization of an *in vitro* system to study strain-induced cell deformation in isolated chondrocytes. *In Vitro Cell. Dev. Biol. Anim.* 31:828–835.

Lee D.A., Bader D.L. 1997. Compressive strains at physiological frequencies influence the metabolism of chondrocytes seeded in agarose. *J. Orthop. Res.* 15:181–188.

Lee D.A., Noguchi T., Knight M.M., O'Donnell L., Bentley G., Bader D.L. 1998. Response of chondrocyte subpopulations cultured within unloaded and loaded agarose. *J. Orthop. Res.* 16:726–733.

Lippiello L., Kaye C., Neumata T., Mankin H.J. 1985. *in vitro* metabolic response of articular cartilage segments to low levels of hydrostatic pressure. *Connect. Tissue Res.* 13:99–107.

Loeser R.F. 1993. Integrin-mediated attachment of articular chondrocytes to extracellular matrix proteins. *Arthritis Rheum.* 36:1103–1110.

Mansour J.M., Mow V.C. 1976. The permeability of articular cartilage under compressive strain and at high pressures. *J. Bone. Joint Surg. [Am.]* 58:509–516.

Maroudas A. 1979. Physicochemical properties of articular cartilage. In: *Adult Articular Cartilage,* 2nd M.A.R. Freeman, ed. Pitman Medical, Kent, England pp. 215–290.

Maroudas A., Grushko G. 1990. Measurement of swelling pressure of cartilage. In: *Methods in Cartilage Research,* A. Maroudas and K. Kuettner, eds. New York, Academic Press; pp. 298–301.

Mauck R.L., Soltz M.A., Wang C.C., Wong D.D., Chao P.H., Valhmu W.B., Hung C.T. 2000. Functional tissue engineering of articular cartilage through dynamic loading of chondrocyte-seeded agarose gels. *J. Biomech. Eng.* 122:252–260.

Millward-Sadler S.J., Wright M.O., Davies L.W., Nuki G., Salter D.M. 2000. Mechanotransduction via integrins and interlukin-4 results in altered aggrecan and matrix metalloproteinase 3 gene expression in normal, but not osteoarthritic, human articular chondrocytes. *Arthritis Rheum.* 43:2091–2099.

Mow V.C., Kuei S.C., Lai W.M., Armstrong C.G. 1980. Biphasic creep and stress relaxation of articular cartilage in compression? Theory and experiments. *J. Biomech. Eng.* 102:73–84.

Mow V.C., Ratcliffe A., Poole A.R. 1992. Cartilage and diarthrodial joints as paradigms for hierarchical materials and structures. *Biomaterials* 13:67–97.

Muir I.H.M. 1979. Biochemistry. In: *Adult Articular Cartilage,* 2nd ed., M. Freeman, ed. Pitman Medical, Kent, England, pp. 145–214.

Narmoneva D.A., Wang J.Y., Setton L.A. 1999. Nonuniform swelling-induced residual strains in articular cartilage. *J. Biomech.* 32:401–408.

Ostergaard K., Salter D.M., Petersen J., Bendtzen K., Hvolris J., Andersen C.B. 1998. Expression of alpha and beta subunits of the integrin superfamily in articular cartilage from macroscopically normal and osteoarthritic human femoral heads. *Ann. Rheum. Dis.* 57:303–308.

Otter T., Bourns B., Franklin S., Reider C., Salmon E.D. 1987. Hydrostatic pressure effects on cytoskeletal organization and ciliary motility: a calcium hypothesis. In: *Current Perspectives in High Pressure Biology.* H.W. Jannasch, R.E. Marquis, A.M. Zimmerman, eds. Academic Press, London, pp. 75–93.

Palfrey A.J., Davies D.V. 1966. The fine structure of chondrocytes. *J. Anat.* 100:213–226.

Palmoski M.J., Brandt K.D. 1984. Effects of static and cyclic compressive loading on articular cartilage plugs *in vitro. Arthritis Rheum.* 27:675–681.

Parkkinen J.J., Ikonen J., Lammi M.J., Laakkonen J., Tammi M., Helminen H.J. 1993. Effects of cyclic hydrostatic pressure on proteoglycan synthesis in cultured chondrocytes and articular cartilage explants. *Arch. Biochem. Biophys.* 300:458–465.

Parkkinen J.J., Lammi M.J., Inkinen R., Jortikka M., Tammi M., Virtanen I., Helminen H.J. 1995. Influence of short-term hydrostatic pressure on organization of stress fibers in cultured chondrocytes. *J. Orthop. Res.* 13:495–502.

Parkkinen J.J., Lammi M.J., Pelttari A., Helminen H.J., Tammi M., Virtanen I. 1993. Altered Golgi apparatus in hydrostatically loaded articular cartilage chondrocytes. *Ann. Rheum. Dis.* 52:192–198.

Poole C.A., Flint M.H., Beaumont B.W. 1987. Chondrons in cartilage: ultrastructural analysis of the pericellular microenvironment in adult human articular cartilages. *J. Orthop. Res.* 5:509–522.

Ralphs J.R., Benjamin M., Lewis A., Archer C.W. 1993. Cytokeratin expression in articular chondrocytes. *Trans. Orthop. Res. Soc.* 18:616.

Ralphs J.R., Tyers R.N., Benjamin M. 1992. Development of functionally distinct fibrocartilages at two sites in the quadriceps tendon of the rat: the suprapatella and the attachment to the patella. *Anat. Embryol.* 185:181–187.

Sah R.L., Kim Y.J., Doong J.Y., Grodzinsky A.J., Plaas A.H., Sandy J.D. 1989. Biosynthetic response of cartilage explants to dynamic compression. *J. Orthop. Res.* 7:619–636.

Salter D.M., Hughes D.E., Simpson R., Gardner D.L. 1992. Integrin expression by human articular chondrocytes. *Br. J. Rheumatol.* 31:231–234.

Schinagl R.M., Gurskis D., Chen A.C., Sah R.L. 1997. Depth-dependent confined compression modulus of full-thickness bovine articular cartilage. *J. Orthop. Res.* 15:499–506.

Schneiderman R., Kevet D., Maroudas A. 1986. Effects of mechanical and osmotic pressure on the rate of glycosaminoglycan synthesis in the human adult femoral head cartilage: an *in vitro* study. *J. Orthop. Res.* 4:393–408.

Shimizu M., Minakuchi K., Kaji S., and Koga J. 1997. Chondrocyte migration to fibronectin, type I collagen, and type II collagen. *Cell Struct. Funct.* 22:309–315.

Shin D., Athanasiou K. 1999. Cytoindentation for obtaining cell biomechanical properties. *J. Orthop. Res.* 17:880–890.

Shin D., Lin J.H., Agrawal C.M., Athanasiou K.A. 1998. Zonal variations in microindentation properties of articular cartilage. *Transactions of the 44th Annual Meeting of the Orthopaedic Research Society.* G. B. J. Andersson, ed. 23:903.

Siczkowski M., Watt F.M. 1990. Subpopulations of chondrocytes from different zones of pig articular cartilage. Isolation, growth and proteoglycan synthesis in culture. *J. Cell. Sci.* 97:349–360.

Smith R.L., Donlon B.S., Gupta M.K., Mohtai M., Das P., Carter D.R., Cooke J., Gibbons G., Hutchinson N., Schurman D.J. 1995. Effects of fluid-induced shear on articular chondrocyte morphology and metabolism *in vitro. J. Orthop. Res.* 13:824–831.

Smith R.L., Lin J., Trindade M.C., Shida J., Kajiyama G., Vu T., Hoffman A.R., van der Meulen M.C., Goodman S.B., Schurman D.J., Carter D.R. 2000. Time-dependent effects of intermittent hydrostatic pressure on articular chondrocyte type II collagen and aggrecan mRNA expression. *J. Rehabil. Res. Dev.* 37:153–161.

Taylor D.J., Woolley D.E. 1987. Evidence for both histamine H1 and H2 receptors on human articular chondrocytes. *Ann. Rheum. Dis.* 46:431–435.

Torzilli P.A., Grigiene R., Huang C., Friedman S.M., Doty S.B., Boskey A.L., and Lust G. 1997. Characterization of cartilage metabolic response to static and dynamic stress using a mechanical explant test system. *J. Biomech.* 30:1–9.

Trippel S.B., Ehrlich M.G., Lippiello L., Mankin H.J. 1980. Characterization of chondrocytes from bovine articular cartilage: I. Metabolic and morphological experimental studies. *J. Bone Joint Surg. [Am.]* 62:816–820.

Uchida A., Yamashita K., Hashimoto K., Shimomura Y. 1988. The effect of mechanical stress on cultured growth cartilage cells. *Connect. Tissue. Res.* 17:305–311.

Urban J.P., Hall A.C., Gehl K.A. 1993. Regulation of matrix synthesis rates by the ionic and osmotic environment of articular chondrocytes. *J. Cell. Physiol.* 154:262–270.

Vunjak-Novakovic G., Martin I., Obradovic B., Treppo S., Grodzinsky A.J., Langer R., Freed L.E. 1999. Bioreactor cultivation conditions modulate the composition and mechanical properties of tissue-engineered cartilage. *J. Orthop. Res.* 17: 130–138.

Wang N., Butler J.P., Ingber D.E. 1993. Mechanotransduction across the cell surface and through the cytoskeleton. *Science* 260:1124–1127.

Wilkins R.J., Hall A.C. 1992. Measurement of intracellular pH in isolated bovine articular chondrocytes. *Exp. Physiol.* 77:521–524.

Wong M., Wuethrich P., Buschmann M.D., Eggli P., Hunziker E. 1997. Chondrocyte biosynthesis correlates with local tissue strain in statically compressed adult articular cartilage. *J. Orthop. Res.* 15:189–196.

Woo S.L., Akeson W.H., Jemmott G.F. 1976. Measurements of nonhomogeneous, directional mechanical properties of articular cartilage in tension. *J. Biomech.* 9:785–791.

Wright M.O., Stockwell R.A., Nuki G. 1992. Response of plasma membrane to applied hydrostatic pressure in chondrocytes and fibroblasts. *Connect. Tissue Res.* 28:49–70.

Wu J.Z., Herzog W., Epstein M. 1999. Modeling of location- and time-dependent deformation of chondrocytes during cartilage loading. *J. Biomech.* 32:563–572.

Yamamoto A., Mishima S., Maruyama N., Sumita M. 1998. A new technique for direct measurement of the shear force necessary to detach a cell from a material. *Biomaterials* 19:871–879.

Zanetti M., Ratcliffe A., Watt F.M. 1985. Two subpopulations of differentiated chondrocytes identified with a monoclonal antibody to keratin sulfate. *J. Cell Biol.* 101:53–59.

18

Biomechanics of Native and Engineered Heart Valve Tissues

Michael S. Sacks

Introduction

On the most basic functional level, the aortic heart valve is essentially a check-valve that serves to prevent retrograde blood flow from the aorta back into the left ventricle (Fig. 1a). This seemingly simple function belies the structural complexity, elegant solid-fluid mechanical interaction, and durability necessary for normal aortic valve function (Fung, 1984). For example, the aortic valve is capable of withstanding 30–40 million cycles per year, resulting in a total of ~3 billion cycles in single lifetime (Thubrikar, 1990). No valve made from nonliving materials has been able to demonstrate comparable functional performance and durability.

However, this staggering level of performance can be cut short by aortic valve disease, the most common form being stenosis resulting from calcification. Currently, the treatment of aortic valve disease is usually complete valve replacement. First performed successfully in 1960, surgical replacement of diseased human heart valves by valve prostheses is now commonplace and enhances survival and quality of life for many patients. The vast majority of prosthetic valve designs are either mechanical prosthesis and bioprosthetic heart valves (BHV). Mechanical prostheses are fabricated from synthetic materials, mainly pyrolytic carbon leaflets mounted in a titanium frame. BHV are fabricated from ei-

ther porcine aortic valve or bovine pericardium, chemically treated with glutaraldehyde to reduce immunogenecity and improve durability, and usually mounted onto a flexible metal frame (stent) that is covered with Dacron to facilitate surgical implementation.

Of these two major prosthesis designs, BHV are now used in approximately 40% of the estimated 75,000 U.S. and 275,000 worldwide valve replacements done annually (Schoen, 2001). The BHV aortic valve has the advantage of low rates of thromboembolic complications without chronic anticoagulation therapy (and its associated morbidity and mortality risks) required for mechanical prostheses. However, they suffer high rates of late structural dysfunction owing to tissue degradation (Schoen and Levy, 1999; Turina et al., 1993). The principal processes that account for BHV tissue degradation *in vivo* are widely considered to be 1) cuspal mineralization, causing cuspal stiffening with or without tearing, and 2) non calcific cuspal damage, including mechanical fatigue and possibly proteolytic degradation of the collagenous extracellular matrix, causing cuspal tears and perforations (Schoen and Levy, 1999; Turina et al., 1993). Approximately 90% of all BHV fabricated from porcine aortic valves fail with tearing, and some fail with little or no calcification (Schoen and Cohn, 1986; Schoen and Hobson, 1985; Schoen and Levy, 1999).

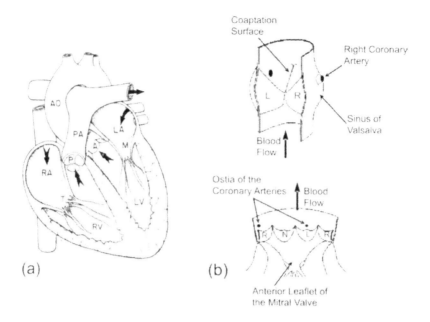

FIGURE 18.1. (*a*) Schematic of the heart showing its valves and chambers. RA, right atrium; RV, right ventricle; LA, left atrium; LV, left ventricle; T, tricuspid valve; P, pulmonary valve; M, mitral valve; A, aortic valve; PA, pulmonary artery; AO, aorta. Arrows show the path of blood flow. Illustrations of the aortic valve (*b*) in the closed configuration, and (*c*) cut open and laid flat. These schematics show the coaptation area, sinuses and ostia of the coronary arteries and highlight their relative positions. N, noncoronary (posterior) cusp; R, right coronary cusp; L, left coronary cusp. (Adapted from Thubrikar, 1990.)

While much effort in currently underway in both ours and other laboratories to improve current BHV (Billiar and Sacks, 2000; Gloeckner et al., 1999; Sacks and Billiar, 1997; Sacks and Smith, 1998; Sacks et al., 2001; Smith et al., 1999; Vyavahare et al., 1997, 199a, b), in the long-term new technologies will aortic valve to be developed. This is especially the case in pediatric applications, where growth of the replacement valve is essential to eliminate the need for reoperations. Further, repairs of congenital deformities require very small valve sizes that are simply not commercially available.

Tissue engineering (TE) offers the potential to create cardiac replacement structures containing living cells, which has the potential for growth and remodeling, overcoming the limitations of current pediatric heart valve devices (Hoerstrup et al., 1998, 1999, 2000; Mayer et al., 1997; Sodian et al., 2000a–c). Using autologous cells and biodegradable polymers, TE heart valves (TEHV) have been fabricated and have functioned in the pulmonary circulation of growing

lambs for up to five months, with the beginnings of a specialized layered structure (Hoerstrup et al., 2000). Despite these promising results, significant questions remain. For example, the role of initial scaffold structure and mechanical properties to guide the development of optimal extra cellular matrix (ECM) structure and strength are largely unexplored. While detailed biomechanical investigations of the *in vitro* incubation process could shed much light on optimizing TEHV designs, little work has been conducted to date.

Perhaps the current biomechanical challenges with TEHV are best exemplified by the principals of functional tissue engineering, as recently stated by Butler et al. (2000). While a long-term goal is duplication of the native valve, interim TE prostheses do not necessarily have to achieve this goal to be successful. There is thus a need to establish minimal functional parameters necessary to produce a functional valve replacement. Further, there is a need for models of long-term remodeling and implant survival to minimize costly ani-

(a)   (b)   (c)

FIGURE 18.2. (a) Histological cross-section of the aortic valve demonstrating the trilayered structure (F, fibrosa; S, spongiosa; and V, ventricularis). (b) Further illustration of the trilayered structure of the aortic cusp created by assembling several hundred contiguous histologic sections and fluorescent imaging, using a technique developed by Resolution Science Corporation, CA. (c) Same image as that shown in (b) but with only the interstitial cells remaining, demonstrating the high degree of cellularity of the cuspal tissue.

mal trials, particularly since conventional durability testing techniques are inapplicable.

The purpose of this chapter is to present a review of the structure-strength relationships for native, chemically treated, and engineered heart valve tissues. While not exhaustive reviews, highlights of the major works in the field have been included, focusing on studies performed in the author's laboratory. This includes the author's collaboration with Dr. John Mayer of Boston Children's Hospital on the development of a TEHV for pediatric surgical reconstruction. A final section of this review discusses how current biomechanical techniques can be applied to the development of TEHV and suggests future research directions.

The Structure of the Native Aortic Valve

Tissue Structure

The aortic valve is consists of three cusps located at the base of the aortic root (Fig. 18.1, b and c). The valve seals by coaptation of the three cusps, closing by deceleration of the blood flow in late systole (Bellhouse and Bellhouse, 1969; Bellhouse and Reid, 1969), and thus does not require backflow that would lead a reduced cardiac output. To achieve this elegant function,

low flexural rigidity necessary to allow normal valve opening and high tensile strength to resist transvalvular pressures in excess of 80 mm Hg are required.

To achieve these demanding design goals using available biological materials (e.g., collagens, elastin, proteoglycans, etc.), nature has evolved a trilayered cuspal structure (Fig. 18.2). These layers are the ventricularis, spongiosa, and fibrosa (Thubrikar, 1990). As its name implies, the ventricularis layer faces the left ventricular chamber and is composed of a dense network of collagen and elastin fibers. The spongiosa layer contains a high concentration of proteoglycans. The fibrosa layer is composed predominantly of a dense network of collagen fibers, and is thought to be the major stress-bearing layer. Interstitial cells (myofibroblasts) permeate the entire tissue structure, although are generally more numerous in the spongiosa (Fig. 18.2c). Vesely et al. (1992) have reported the occurrence of residual stresses in the aortic valve cusp, thought to be the result of passive contractile forces generated by the elastin fibers within the ventricularis layer acting on the fibrosa layer. These results suggest a complex structural mechanical interaction between the layers of the aortic valve cusp.

It is well known that collagen fibers can withstand high tensile forces, but have low torsional and flexural stiffness. Thus, directions in which the

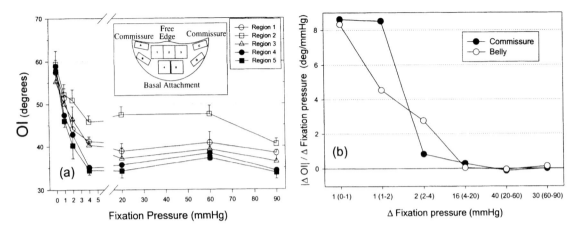

FIGURE 18.3. (a) OI at transvalvular pressures 0–90 mm Hg for belly regions 1–5, where OI values were statistically different from 0 mm Hg mean at all nonzero pressure levels and regions (n = 6, $p < 0.05$); inset, aortic cusp regions. The mean absolute value change in OI per mm Hg of applied transvalvular pressure ($|\Delta OI|/\Delta$ Fixation pressure) was also computed at each transvalvular pressure interval (b). The largest changes in OI (over 8°/mm Hg) occurred between 0 and 1 mm Hg, with all cuspal regions showed negligible rates of change past 4 mm Hg. (Adapted from Sacks et al., 1998.)

fibers are oriented can be identified with the directions in which the tissue is able to withstand the greatest tensile stresses. Gross fiber orientation thus leads to an understanding and predictability of the mechanical properties of the tissue. This is especially the case in the study of the structure of the aortic heart valve, which is uniquely suited for efficient transmission of mechanical stresses with the minimal use of material.

We have previously performed a quantitative study of the gross fiber architecture of the aortic valve and how it changes under increasing transvalvular pressure (Sacks et al., 1998). To quantify the gross fiber architecture of the valve cusp, we used small angle light scattering (SALS) (Sacks et al., 1997). In SALS, laser light is passed through a tissue specimen and the spatial intensity distribution of the resulting scattered light represents the sum of all structural information within the light beam envelope. The resulting angular distribution of scattered light intensity, $I(\Phi)$, about the laser axis represents the distribution of fiber angles within the light beam envelope at the current tissue location. The width of the $I(\Phi)$ distribution is indicative of the degree of fiber orientation; highly oriented fiber networks result in a very narrow peak, while more randomly distributed fibers yield a broader

peak. To quantify the degree of fiber orientation, we utilized a physically intuitive orientation index (OI) defined as the angle that contains one half of the total area under the $I(\Phi)$ distribution, representing 50% of the total number of fibers (Sacks et al., 1997). Thus, highly oriented fiber networks will have low OI values, while more randomly oriented networks will have larger values.

To simulate on the changes to aortic valve cuspal structure with increasing transvalvular pressure, fresh porcine aortic valves were fixed at transvalvular pressures ranging from 0 to 90 mm Hg. In addition, cusps were also dissected into their respective fibrosa and ventricularis layers and then remeasured by SALS. This allowed us to explore the interrelationships between the intact and separated layer structural responses at different levels of transvalvular pressure.

Overall, increasing transvalvular pressure induced the greatest changes in fiber alignment (i.e., reduction in OI value) between 0 and 1 mm Hg, and past 4 mm Hg there was no detectable improvement in fiber alignment (Fig. 18.3a). Interestingly, there were significant regional variations; the commissural regions (regions A–D, Fig. 18.3a, inset) exhibited a high rate of fiber alignment increase up to 4 mm Hg then abruptly halted (Fig. 18.3b). In contrast, the belly regions (regions 1–5, Fig. 18.3a, inset) exhibited a more gradual reduction in the rate of fiber alignment (Fig. 18.3b). When the fibrosa and ventricularis layers of the cusps were rescanned separately, the degree of orientation for both layers became more similar once the transvalvular pressure exceeded 4 mm Hg, and was almost indistinguishable by 60 mm Hg.

Our major finding was that the fibrous microstructure of the natural aortic valve's rate and amount of reorientation due to transvalvular pressure is both regionally and layer variant. These results further suggest that as transvalvular pressure increases, the ventricularis substantially stretches out along the circumferential direction, inducing a high degree of fiber alignment not present in the unloaded state. It is possible that, in addition to retracting the aortic cusp during systole, the ventricularis may mechanically contribute to the diastolic cuspal stiffness at high transvalvular pressures. This may help to prevent overdistention of the cusp at high transvalvular pressures.

Further evidence of an adaptive structure is the unique structure of the commissure region, which approximately corresponds to the coaptation region. The coaptation region is under no transvalvular pressure, but is loaded instead in a uniaxial-like manner due to tethering forces generated at the attachment of the commissures to the aortic root. Unlike the biaxially loaded belly region, the uniaxial loading of the commissures would tend to make their structure more highly aligned, that is, more like a tendon. Like tendons, a highly aligned fiber network would have a very short transition region from low to high stiffness, as evidenced by rapid fiber uncrimping with stress. The highly aligned nature of the commissure region at unloaded state and the more rapid realignment with transvalvular pressure in the commissure regions are consistent with the pretransition strain level behavior of tendonlike materials.

Based on these results, we see that the aortic valve cusp has evolved specialized structures adapted to the local stress state. It is interesting to speculate as to the mechanisms behind these specialized structures. Clearly, the local state of mechanical stress will be a major factor in determining local structure, for example, Wolff's law

for bone or residual stresses in arteries. This is supported by work by Peskin et al. (1994), who, using the equations of mechanical equilibrium, not only predicted circumferentially orientated fibers but also the dense fibrous cords of the fibrosa layer. Peskin's work suggests that structural variability found in the aortic valve may be due, in part, to slight differences in the final equilibrium state of the cuspal structure as it initially develops and adapts to the adult hemodynamic stress state.

Heart Valve Cells

Unlike the comparatively in-depth knowledge of vascular endothelial and interstitial cells, very little is known about cardiac valve cell phenotype and function. This lack of knowledge not only limits our understanding of the native valve, but also choice of the appropriate cell types for use in heart valve tissue engineering. Giachelli et al. (2000) have isolated and characterized bovine aortic valve endothelial cells (BVECs) and interstitial cells (BVICs). Adhesive characteristics of these cells were determined to evaluate candidate scaffold coating ligands. Recently, Batten et al. (2001) investigated T-cell response to valve endothelial and interstitial cells in vitro. It was found that valve endothelial and interstitial cells express similar levels of human leukocyte antigens and adhesion and co-stimulatory molecules, which are either induced or up-regulated after interferon gamma treatment. Although valve endothelial and interstitial cells express a similar range of cell-surface molecules, it is only the endothelial cells that are immunogenic. In addition, it was shown that these two cell types interact in a donor-specific manner to orchestrate the immune response and therefore may have clinical relevance in the allogeneic response of the heart valve recipients.

These studies provide some light on the much needed information about cardiac valve cells. Clearly, the relation between cuspal fiber architecture, in vivo strain history, cuspal geometry, cell loading/straining and their subsequent responses in the native aortic valve is complex. Regardless of the specifics of the approach (e.g., collagen, collagen/polymer combination), all TEHV will have to ultimately duplicate natural valve structure and biological responses in order to properly duplicate their *in vivo* function.

Biomechanical Behavior of the Native Aortic Valve Cusp

Planar Biaxial Mechanical Properties

The mechanics of soft tissues are complex: they exhibit a highly nonlinear stress-strain relationship, undergo large deformations, complex viscoelasticity, and complex axial coupling behaviors that defy simple experiments and material models. Much of this behavior is a direct result of changes in their internal structure with strain, which involves both straightening of highly crimped collagen fibers and rotation of these fibers toward the stretch axis.

Most previous work on the mechanical properties of the native and chemically treated aortic valve has relied on uniaxial mechanical testing (Lee et al., 1984a, b; Vesely and Noseworthy, 1992). These studies demonstrate that chemical fixation of intact valves, especially under pressure, alters the mechanical properties of the cusps. Marked decreases in the extensibility are generally attributed to "locking" the collagen fibers in the uncrimped state (Broom and Christie, 1982; Christie, 1992). Tests on thin tissue strips, however, cannot mimic the heterogeneous multiaxial deformation fields, combined loading sequences and native fiber kinematics found in the physiological environment. Mayne et al. (1989) and Christie et al. (1995) have performed equi-biaxial testing (i.e., equal levels of tension applied to each test axis) that overcomes many of the above limitations of uniaxial loading. However, derivation of a constitutive relationship solely from equi-biaxial test data is limited due to multiple co-linearities that confound the ability to obtain reliable, unique model parameter values (Brossollet and Vito, 1995).

Billiar and Sacks (2000) generated the first complete biaxial mechanical data necessary for constitutive modeling aortic valve cusp. Due to the small size and heterogeneous structure of the aortic valve cusp, new testing methods were developed and validated. Cuspal specimens were subjected to biaxial tests utilizing seven loading

protocols to provide a range of loading states that encompass the physiological loading state. The cusps demonstrated a complex, highly anisotropic mechanical behavior, including pronounced mechanical coupling between the circumferential and radial directions. Mechanical coupling between the axes produced negative strains along the circumferential direction and/or nonmonotonic stress-strain behavior in many samples subjected to equi-biaxial tension. This behavior was also noted by Mayne et al. (1989) but could not be explained. Clearly, a constitutive model is needed to truly understand the aortic cusp behavior and its implications on the mechanics of the intact valve.

The quantified fiber architecture (Sacks et al., 1998) and biaxial mechanical data (Billiar and Sacks, 2000) suggest that a structural approach is the most suitable method for the formulation of a constitutive model for the aortic valve cusp. Our related work on native and chemically treated bovine pericardium suggests that a structural approach is both feasible and attractive for bioprosthetic heart valve biomaterials (Sacks, 2000).

Details of the model have been previously presented (Billiar and Sacks, 2000), and follows the structural modeling approach by Lanir et al. (1979, 1983). In this approach, the tissues net or total strain energy is assumed to be the sum of the individual fiber strain energies, linked through appropriate tensor transformation from the fiber coordinate to the global tissue coordinates. For the aortic valve, we assume that the planar biaxial mechanical properties of the cusp can be represented as a planar array of collagen fibers. Anatomically, these fibers most closely represent the dense, highly aligned collagen fibers in the fibrosa layer. Next, the angular fiber distribution and the density of the fibers are assumed constant throughout the tissue. Based on our SALS results for the aortic valve cusp (Sacks et al., 1998), we utilize the fact that the angular distribution of the collagen fibers, $R(\theta)$, can be represented by a Gaussian distribution,

$$R(\theta) = \frac{1}{\sigma\sqrt{2\pi}} \exp\left[\frac{-(\theta - M)^2}{2\sigma^2}\right] \quad (1)$$

where θ is the direction with respect to the x_1 or circumferential axis (Fig. 18.4b), σ is the stan-

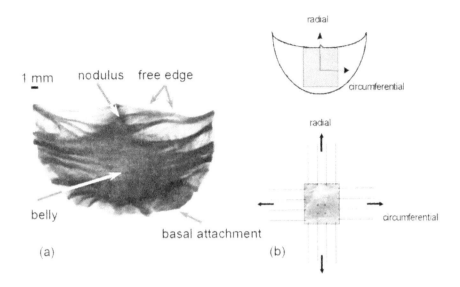

FIGURE 18.4. (a) A partially polarized image of an aortic cusp highlighting main anatomical features. (b) Schematic of the location of the biaxial test specimen taken from the aortic valve cusp. (Taken from Billiar and Sacks, 2000.)

dard deviation and M is the mean of the distribution. M was determined experimentally for each specimen by using the preferred fiber directions as determined by SALS (Sacks et al., 1998). The "effective" fiber stress-strain properties were represented using:

$$S_f = A\left[\exp\left(BE_f\right)-1\right] \qquad (2)$$

where S_f is the second Piola-Kirchhoff fiber stress, E_f is the fiber Green's strain. This formulation for the fiber stress-strain law avoids detailed descriptions of complex crimp distributions.

For valvular tissue, it is more convenient to work with membrane stresses due to considerations such as variable total and layer thickness, and heterogeneous layer structure (Billiar and Sacks, 2000). Further, since the biaxial mechanical tests are run using membrane stress control using the specimen's unloaded dimensions, a Lagrangian membrane stress measure is used in the constitutive formulation. We also assume that interspecimen variations in fiber volume fraction V_f and thickness h are negligible, so that the product hV_f can be conveniently absorbed into the material constant A. The resulting expressions for the Lagrangian membrane stresses T_{ij} are:

$$T_{11} = \int_{-\pi/2}^{\pi/2} S_f^*(E_f)R(\theta)\left(\lambda_1 \cos^2\theta + \kappa_1 \sin\theta\cos\theta\right)d\theta$$

$$\qquad (3)$$

$$T_{22} = \int_{-\pi/2}^{\pi/2} S_f^*(E_f)R(\theta)\left(\lambda_2 \sin^2\theta + \kappa_2 \sin\theta\cos\theta\right)d\theta$$

where $A^* = hV_f A$ and $S_f^* = A^* [\exp(BE_f) - 1]$. The pareters A^*, B, and σ were estimated by fitting Eqn. 3 to the complete biaxial data set (Billiar and Sacks, 2000).

The fit of the model to the data was good despite the complexity of the mechanical response over the broad range of biaxial loading states (Fig. 18.5). The model fit the data from all seven protocols well even though the data from the outer protocols (1 and 7, see inset in Fig. 18.5) were not used in the parameter estimation. Using only three material parameters, the quantitative "goodness of fit" was comparable to phenomenological models of other tissues (Choi and Vito, 1990; Humphrey et al., 1990; May-Newman and Yin, 1998).

Another important aspect of the structural approach is that the two distinguishing aspects of the aortic valve cusp biaxial behavior, namely the extreme mechanical anisotropy and the strong

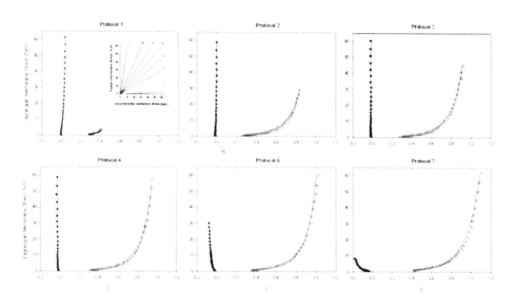

FIGURE 18.5. Stress-strain curves for six of the seven loading protocols for a 4 mm Hg fixed specimen (*open circles*) and the simulated stress-strain curves (*lines*). The structural constitutive model demonstrated an excellent fit to the data, including the presence of large negative circumferential strains due to strong axial coupling. Although protocol five was not shown for clarity of presentation, it also demonstrated an equally good agreement between theory and experiment. Inset: biaxial testing protocols with showing the corresponding protocol numbers shown for reference. (Taken from Billiar and Sacks, 2000.)

mechanical coupling between the axes, can be explained by the angular distribution of fibers. To more clearly demonstrate this effect, we generated simulations under equi-biaxial loading for a given set of A* and B values by letting σ vary (Fig. 6). These simulations indicate that the value of σ is the primary determinant of the biaxial stress-strain response, as shown for a) nearly random ($\sigma = 90°$), b) moderately anisotropic ($\sigma = 35°$), c) highly anisotropic, including contraction along one axis ($\sigma = 20°$), and d) extremely anisotropic ($\sigma = 10°$). Although we assumed a simplified tissue structure in the formulation of the model, the structural approach highlighted the importance of the angular orientation of the fibers in determining the complex anisotropic mechanical behavior of the tissue.

Flexural Mechanical Properties of Native and Porcine BHV

As previously stated, the aortic heart valve is essentially a check-valve that serves to prevent retrograde blood flow from the aorta back into the left ventricle (Fig. 18.1a). This function involves highly complex solid-fluid mechanical interactions, which to date have not been be accurately simulated with current computational approaches. In addition to understanding the mechanics of the native valve, there is a general consensus that stresses developed in the leaflets during valve operation play a significant role in both calcific- and noncalcific-related damage (Broom, 1978: Schoen and Hobson, 1985; Thubrikar et al., 1983). In particular, flexural stresses during valve opening and closing have been thought to play a considerable role in limiting the long-term BHV durability (Ishihara et al., 1981; Smith et al., 1999; Thubrikar et al., 1983; Vesely et al., 1988).

To quantify the complexity of the shape changes the native and BHV cusp can undergo during the cardiac cycle, we recently developed a novel optical-based heart valve imaging system (Iyengar et al., 2001). Here, three-dimensional surface reconstructions were generated, demonstrating that BHV cusps undergo a sequence of complex deformations, which was

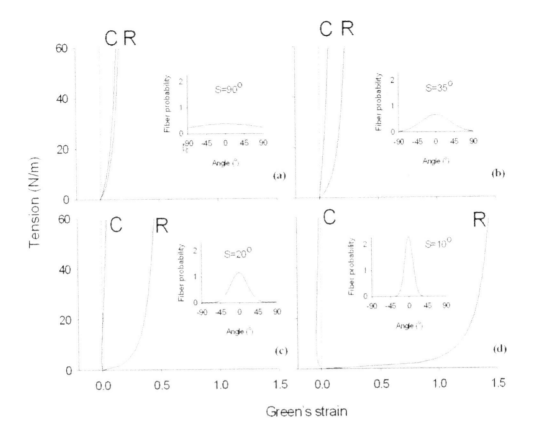

FIGURE 18.6. Simulation of the effect of misalignment of the biaxial test specimen to testing axes by setting μ = 10°, demonstrating a large change in peak extensibilities, especially in the radial direction. This result underscores the need for complete deformation state analysis and compensation of small misalignments when studying highly aligned fibrous tissues such as the aortic valve cusp. (Taken from Billiar and Sacks, 2000.)

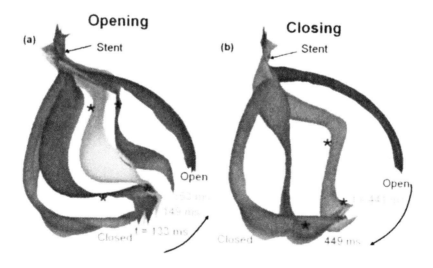

FIGURE 18.7. A sequence of three-dimensional fitted surfaces from a BHV at various times in the cardiac cycle, highlighting the differences in the (a) opening and (b) closing behavior. The most interesting aspect is the substantial differences in leaflet shape during the two phases. Note too the points of high flexure identified by a *.

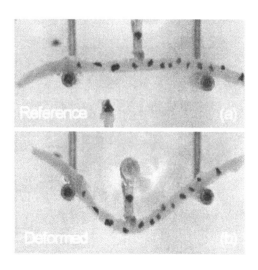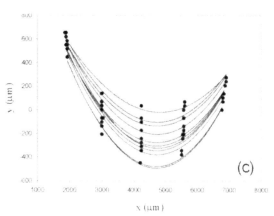

FIGURE 18.8. Photos of heart valve biomaterial strip before (*a*) and after (*b*) being subjected to three-point bending. As shown in (*c*), the small graphite particles are tracked and fit to a quadric spline to compute local curvatures.

different during the opening and closing phases. Moreover, points of severe flexion were directly visualized, which change in location both with time and between the opening and closing phases. This study underscores the need to study the flexure properties of both native and BHV cusps.

To quantify the complex flexure response of heart valve biomaterials, we have developed a specially designed flexure testing apparatus (Gloeckner et al., 1999). Details of the experimental methods and analyses have been described in previously (Gloeckner et al., 1999; Sacks et al., 2001). Briefly, 2 mm × 10 mm samples aligned to the circumferential direction were prepared from aortic valve cusps. Along one edge small (~0.3 mm), black graphite markers were attached with cyanoacrylate (Fig. 18.8). The markers were placed so that the shape of the sample could be measured while the specimen was subjected to three-point bending.

Video images of each flexure experiment were recorded on SVHS video tape, from which the applied load was determined. The spatial positions of the graphite markers were obtained by tracking their x-y positions, to which a quadratic curve was fit to the marker positions at each frame (Fig. 18.8b).

$$y = ax^2 + bx + c \qquad (4)$$

From this quadratic fit, the instantaneous curvature κ at each frame was computed using (Struik, 1961):

$$\kappa = y'' / (1 + (y')^2)^{3/2} \qquad (5)$$

where $y'' = d^2y/dx^2 = 2a$, and $y' = dy/dx = 2ax + b$.

To compute the effective stiffness (E_{eff}) we utilized the Bernoulli-Euler moment-curvature relation for beams undergoing large displacements (Timoshenko, 1955):

$$M = E_{eff} I \Delta\kappa \qquad (6)$$

where $\Delta\kappa$ is the change in curvature, I is the 2nd moment of inertia computed from the specimen geometry, and M is the applied bending moment determined from the applied loads and testing geometry. Physically, E_{eff} represents the net structural stiffness of the tissue. The value for E_{eff} for each specimen is reported as the value at the center of specimen where the maximum M occurred. In practice, data is presented as M/I vs. $\Delta\kappa$ so that the slope of the curve is equal to E_{eff}. Due to the specimen structural heterogeneity, the value of E_{eff} represents the effective bending stiffness only for the current bending direc-

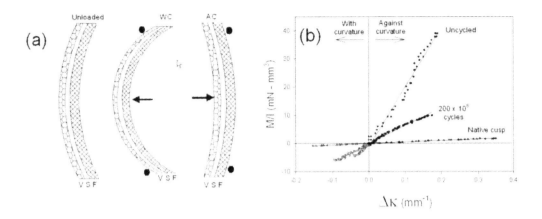

FIGURE 18.9. (*a*) A representative M/I vs. Δκ curve for fresh aortic valve cusp demonstrating a linear response over the full range of flexure, with E_{eff} = 5.3 kPa in both AC and WC bending directions. In contrast, an uncycled porcine BHV cusp demonstrated a profound, bending direction dependent increase in stiffness, with E_{eff} = 201.4 kPa in AC direction (a 20-fold increase from the native state), and E_{eff} = 71.6 kPa in the WC direction. After 200×10^6 cycles, the AC direction stiffness dropped substantially to E_{eff} = 127.3 kPa, while the WC direction dropped only slightly to 62.8 kPa.

tion. To subject the cuspal layers to either tension or compressive stresses, each specimen was subjected to three-point bending in both the against (AC) and with (WC) the natural curvature of the cusp (Fig. 18.9*a*).

Results demonstrated that, interestingly, fresh tissue had a nearly perfect linear response (r^2 = 0.99) over the entire bending curve, with E_{eff} = 5.3 kPa in both AC and WC bending directions (Fig. 18.9*b*). In contrast, porcine BHV demonstrated a profound, bending direction dependent increase in stiffness, with E_{eff} = 201.4 kPa in AC direction (a 20-fold increase from the native state), and E_{eff} = 71.6 kPa in the WC direction (Fig. 18.9*b*). When we examined porcine BHV subjected to 200×10^6 accelerated test cycles, the AC direction stiffness dropped substantially to E_{eff} = 127.3 kPa, while the WC direction dropped only slightly to 62.8 kPa. These results suggest layer-specific damage, and further demonstrate the sensitivity of flexure testing in assessing heart valve biomaterial mechanical properties.

We refer to "flexural properties" as those mechanical properties measured while a tissue is exposed to a precisely defined state of pure bending. We have adopted this approach because of its sensitivity and its ability to simulate the flex-

ural deformations of the functioning valve. This approach is in contrast to that of Talman and Boughner (1996), who subjected porcine BHV tissue to transverse shearing by applying forces tangent to the upper and lower (i.e., the fibrosa and ventricularis sides, respectively) surfaces of a circular cuspal specimen. This methodology does not, using our definition, directly address the flexural mechanical properties of the porcine BHV cusp. Rather, it measures mechanical properties of cuspal tissue when sheared between two parallel plates. While providing information on the heart valve tissue shear behavior, the correlation between these properties to both our three-point bending method and the functioning cusp is unclear.

Flexural Mechanical Properties of TE Heart Valve Cusp Biomaterials

Flexure testing can also be used as a highly sensitive method to assess TEHV biomaterials. In a recent study, Perry et al. (2001) fabricated trileaflet valve constructs (TEHV) from P4HB-coated PGA sheet, and seeded the TEHV using peripheral blood-derived mesenchymal cells. The cell-polymer construct was allowed to rotate in suspension for two weeks in a rotating biore-

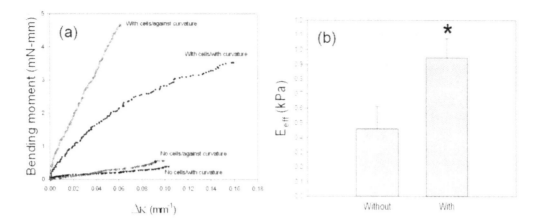

FIGURE 18.10. Flexure testing results for TEHV biomaterials. In (*a*), a dramatic increase in flexural rigidity in the incubated tissue can be observed. Moreover, while the no-cell controls demonstrated no directional dependence, the TEHV-incubated tissue demonstrated a directional-dependent flexural stiffness response, suggesting transmural variations in ECM structure. Effective tissue modulus, E_{eff}, computed from the EI data in (*a*), indicated that effective biomaterial stiffness increased two-fold, which could not be explained by the increase in thickness, but in fact *indicates that the incubated tissue was intrinsically stiffer than the no-cell (i.e., scaffold only) controls.*

actor before being transported to a pulse-duplicating bioreactor for an additional two weeks. Using the flexure mechanical testing methods described previously, mechanical properties were assessed at zero and four weeks. Results clearly indicated a marked reduction in flexural stiffness with incubation time, as well as nonlinearties in the moment-curvature relation. This latter finding suggests that nonlinear soft tissue properties may become more prominent as the polymer scaffold degrades and the deposited soft tissue mass increases.

We recently conducted a more complete study wherein tri-leaflet valve constructs were fabricated from P4HB-coated PGA seeded with ovine endothelial and carotid artery medial cells, cultured *in vitro* 18 days (4 days static/14 days dynamic) in a bioreactor (Guleserian et al., submitted). In addition, nonseeded controls were identically prepared and incubated as a means to determine the contribution of the de novo deposited ECM. Incubated tissue exhibited a doubling in tissue thickness after only 18 days in the bioreactor, while flexure testing indicated a dramatic increase in flexural rigidity in the incubated tissue (Fig. 18.10*a*). Moreover, while the no-cell controls demonstrated no directional dependence, the TEHV-incubated tissue demonstrated a directional-dependent flexural stiffness

response (Fig. 18.10*a*). As for the porcine BHV (Fig. 18.9), this result suggests transmural variations in ECM structure. Further analysis of the pilot studies indicated that the effective tissue modulus E_{eff} increased two-fold (Fig. 18.10*b*). This result is supported by the fact that in beam theory, the flexural rigidity (EI) is proportional to h^3, where h is the thickness (Frisch-Fay, 1962). Therefore, a doubling in thickness would only result in an 8-fold increase in flexural rigidity, compared with over 10-fold increase observed here. Thus, the incubated tissue with cells was intrinsically stiffer than no-cell (i.e., scaffold only) controls.

Other Engineered Heart Valve Structures

The chordae are bundles of collagenous fibers that emerge from papillary muscles, support the mitral valve leaflets, and are thus essential to the proper function of the mitral valve. When chordae stretch or tear, the mitral valve becomes regurgitant; mitral valve repair is the most common surgical technique used to correct mitral regurgitation. Vesely et al. (Shi and Vesely, Submitted) have attempted to create artificial chordae from cells and reconstituted type I collagen, using directed collagen gel shrinkage techniques (Auger et al., 1998; Girton et al., 1999; Tranquillo et al., 1992). Constructs were prepared and cultured for up to 8 weeks, demonstrating a highly nonlinear stress/strain response, pronounced hysteresis, and progressive elongation during preconditioning tests. Stiffness and failure strength were reported to be 5 MPa and 1 MPa, respectively. Although the mechanical properties of these constructs were qualitatively similar to these of native chordae, they were about an order of magnitude less stiff and less strong than native chordae. However, they were substantially stronger than the related approaches, likely due to the very high contraction ratios and collagen fiber densities achieved.

Closure

In the long term, chemically treated BHV will be supplanted by tissue-engineered valve prostheses. Regardless of the specifics of the biomaterial design (e.g., biologically derived collagen, de novo synthesized collagen, collagen/polymer combination), all tissue-engineered bioprostheses will have to duplicate natural tissue mechanics to some extent in order to properly simulate their *in vivo* function. Biomechanics can contribute significantly to this process by utilizing fundamental continuum mechanics principals and rigorous experimental techniques to bring together many of the disparate findings in biology into a unifying theory for growth and remodeling of engineered soft biological tissues.

For example, our constitutive model was able to describe the complete measured planar biaxial stress-strain of the aortic valve cusp using a structural approach. Only three parameters were needed to fully simulate the highly anisotropic, and nonlinear in-plane biaxial mechanical behavior. Although based upon a simplified cuspal structure, the model underscored the role of the angular orientation of the fibers that completely accounted for extreme mechanical anisotropy and pronounced axial coupling. Such knowledge of the mechanics of the aortic cusp derived from this model can help lay the groundwork for the design of tissue engineered scaffolds for replacement heart valves.

Ideally, a TEHV replacement will require the geometry, mechanical properties, and cell biology closely matched to the native valve and outflow track tissues. The polymeric scaffold is meant to provide many of these features during the initial period of tissue formation. However, because the scaffold is absorbed over time, these functions must be taken over progressively by the developing tissue. Ideally, the scaffold/tissue construct will be seeded and grown *in vitro* for a period of time to achieve sufficient functionality, prior to implantation and continued growth and remodeling *in vivo*. While much progress has been made in cell biology and the chemical composition of scaffolds, comparatively little work has been performed on the tissue mechanical aspects of TEHV biomaterials. This is in part due to a lack of methods to assess and model the degrading polymer/ECM biocomposite tissue equivalent, as well as knowledge of the biomechanical behavior of the natural valves. Computational models also needed to understand basic micromechanical phenomena of the biodegradation and ECM deposition process. Further, successful computational models can greatly aid in optimizing the incubation process by minimizing the current costly and inefficient trial-and-error approach, and in understanding the *in vivo* remodeling process and ultimately the long-term fate of the TEHV. While TE offers the potential to overcome limitations of current heart valve prosthesis (especially the very limited options for the pediatric population), the bioengineering challenge is determining how to optimize biological, structural, and mechanical factors of ECM formation for *in vivo* success. A major step in this long-term goal is developing a thorough understanding of the underlying structural and mechanical factors responsible for the initial formation of functionally optimal TEHV tissue structures.

Acknowledgments

Partial funding of the author's work by NIH grant HL-63026-02 and funding of the TEHV work of Dr. John Mayer by NIH grant HL-97-005 is gratefully acknowledged. The author is also an Established Investigator of the American Heart Association.

References

Auger F.A., Rouabhia M., Goulet F., Berthod F., Moulin V., Germain L. 1998. Tissue-engineered human skin substitutes developed from collagen-populated hydrated gels: clinical and fundamental applications. *Med. Biol. Eng. Comput.* 36:801–812.

Batten P., McCormack A.M., Rose M.L., Yacoub M.H. 2001. Valve interstitial cells induce donor-specific T-cell anergy. *J. Thorac. Cardiovasc. Surg.* 122:129–135.

Bellhouse B., Bellhouse F. 1969. Fluid mechanics of model normal and stenosed aortic valves. *Circ. Res.* 25:693–704.

Bellhouse B.J., and Reid K.G. 1969. Fluid mechanics of the aortic valve. *Br. Heart J.* 31:391.

Billiar K.L., Sacks M.S. 2000b. Biaxial mechanical properties of the natural and glutaraldehyde treated aortic valve cusp—Part I: Experimental results. *J. Biomech. Eng.* 122:23–30.

Billiar K.L., Sacks M.S. 2000a. Biaxial mechanical properties of fresh and glutaraldehyde treated porcine aortic valve cusps: Part II—A structurally guided constitutive model. *J. Biomech. Eng.* 122:327–335.

Broom N. 1978. Fatigue-induced damage in glutaraldehyde preserved heart valve tissue. *J. Cardiovasc. Surg.* 76:202–211.

Broom N., Christie G.W. 1982. The structure/function relationship of fresh and gluteraldehyde-fixed aortic valve leaflets. In: *Cardiac Bioprosthesis.* L.H. Cohn, V. Gallucci, eds. Yorke Medical Books, New York. pp. 477–491.

Brossollet L.J., Vito R.P. 1995. An alternate formulation of blood vessel mechanics and the meaning of the *in vivo* property. *J. Biomech.* 28:679–87.

Butler D.L., Goldstein S., Guilak F. 2000. Functional tissue engineering: the role of biomechancs. *J. Biomech. Eng.* 122:570–575.

Choi H.S., Vito R.P. 1990. Two-dimensional stress-strain relationship for canine pericardium. *J. Biomech. Eng.* 112:153–159.

Christie G.W. 1992. Anatomy of aortic heart valve leaflets: the influence of glutaraldehyde fixation on function. *Eur. J. Cardiothorac. Surg.* 6:S25–S33.

Christie G.W., Barratt-Boyes B.G. 1995. Age-dependent changes in the radial stretch of human aortic valve leaflets determined by biaxial stretching. *Ann. Thorac. Surg.* 60:S156–159.

Frisch-Fay R. 1962. *Flexible Bars.* Butterworths, Washington, DC.

Fung Y.C. 1984. *Biodynamics: Circulation.* Springer-Verlag, New York.

Giachelli C., Wiester L., Cuy J. 2000. *Isolation and Characterization of Cardiac Valve Interstitial Fibroblast and Endothelial Cells for use in Tissue Engineering.* Third Biennial Meeting of the Tissue Engineering Society, Tampa, FL.

Girton, T.S., Oegema T.R., Tranquillo R.T. 1999. Exploiting glycation to stiffen and strengthen tissue equivalents for tissue engineering. *J. Biomed. Mater. Res.* 46:87–92.

Gloeckner D.C., Billiar K.L., Sacks M.S. 1999. Effects of mechanical fatigue on the bending properties of the porcine bioprosthetic heart valve. *ASAIO J.* 45:59–63.

Guleserian, K.J., Sacks M.S., Ulrich J., Martin D.P., Mayer J.E. (submitted). The effects of cell seeding on the flexural mechanical properties of tissue engineered heart valve cusps. *Tissue Eng.*

Hoerstrup S.P., Zund G., Lachat M., Schoeberlein A., Uhlschmid G., Vogt P., Turina M. 1998. Tissue engineering: a new approach in cardiovascular surgery—seeding of human fibroblasts on resorbable mesh. *Swiss Surg.* Suppl(2):23–25.

Hoerstrup S.P., Zund G., Ye Q., Schoeberlein A., Schmid A.C., Turina M.I. 1999. Tissue engineering of a bioprosthetic heart valve: stimulation of extracellular matrix assessed by hydroxyproline assay. *ASAIO J.* 45:397–402.

Hoerstrup S.P., Sodian R., Daebritz S., Wang J., Bacha E.A., Martin D.P., Moran A.M., Guleserian K.J., Sperling J.S., Kaushal S., Vacanti J.P., Schoen F.J., Mayer, Jr. J.E. 2000. Functional living trileaflet heart valves grown *in vitro* Circulation 102(19 Suppl 3):III44–49.

Humphrey J.D., Strumpf R.K., Yin F.C. 1990. Determination of a constitutive relation for passive myocardium: II. Parameter estimation. *J. Biomech. Eng.* 112:340–346.

Ishihara T., Ferrans V.J., Boyce S.W., Jones M., Roberts W.C. 1981. Structure and classification of cuspal tears and perforations in porcine bioprosthetic cardiac valves implanted in patients. *Am. J. Cardiol.* 48:665–678.

Iyengar A., Sugimoto H., Smith D.B., Sacks M. 2001. Dynamic in-vitro 3D reconstruction of heart valve leaflets using structured light projection. *Ann. Biomed. Eng.* 29:963–973.

Lanir Y. 1979. A structural theory for the homogeneous biaxial stress-strain relationships in flat collageneous tissues. *J. Biomech.* 12:423–436.

Lanir Y. 1983. Constitutive equations for fibrous connective tissues. *J. Biomech.* 16:1–12.

Lee J.M., Boughner D.R., Courtman D.W. 1984a. The glutaraldehyde-stabilized porcine aortic valve xenograft. II. Effect of fixation with or without pressure on the tensile viscoelastic properties of the leaflet material. *J. Biomed. Mater. Res.* 18:79–98.

Lee J.M., Courtman D.W., Boughner D.R. 1984b. The glutaraldehyde-stablized porcine aortic valve xenograft. I. Tensile viscoelastic properties of the fresh leaflet material. *J. Biomed. Mater. Res.* 18:61–77.

Mayer J.E., Jr., Shin'oka T., Shum-Tim D. 1997. Tissue engineering of cardiovascular structures. *Curr. Opin. Cardiol.* 12:528–532.

Mayne A.S., Christie G.W., Smaill B.H., Hunter P.J., Barratt-Boyes B.G. 1989. An assessment of the mechanical properties of leaflets from four second-generation porcine bioprostheses with biaxial testing techniques [see comments]. *J. Thorac. Cardiovasc. Surg.* 98:170–180.

May-Newman K., Yin F.C. 1998. A constitutive law for mitral valve tissue. *J. Biomech. Eng.* 120:38–47.

Perry, T.E., Kaushal S., Nasseri B., Sutherland F.W.H., Wang J., Guleserian K.J., Bischoff J., Vacanti J.P., Sacks M.S., Mayer J.E. 2001. Peripheral blood as a cell source for tissue engineering heart valves. *Surg. Forum* LII:99–101.

Peskin C., McQueen D. 1994. Mechanical equilibrium determines the fractal fiber architecture of aortic heart valve leaflets. *Am. J. Physiol.* 266:H319.

Sacks M.S. 2000. A structural constitutive model for chemically treated planar connective tissues under biaxial loading. *Comput. Mech.* 26:243–249.

Sacks M.S., Smith D.B. 1998. Effects of accelerated testing on porcine bioprosthetic heart valve fiber architecture. *Biomaterials* 19:1027–1036.

Sacks M., Billiar K. 1997. Biaxial mechanical behavior of bioprosthetic heart cusps subjected to accelerated testing. In: *Advances in Anticalcific and Antidegenerative Treatment of Heart Valve Bioprostheses.* S. Gabbay, R. Frater, eds. Silent Partners, Austin, TX.

Sacks M., Gloeckner D., Vyavahare N., Levy R. 2001. Loss of flexural rigidity in bioprosthetic heart valves with fatigue: new findings and the relation to collagen damage. *J. Heart Valve Dis.* in-press.

Sacks M.S., Smith D.B., Hiester E.D. 1997. A small angle light scattering device for planar connective tissue microstructural analysis. *Ann Biomed Eng* 25:678–689.

Sacks M.S., Smith D.B., Hiester E.D. 1998. The aortic valve microstructure: effects of transvalvular pressure. *J. Biomed. Mater. Res.* 41:131–141.

Schoen F.J. 2001. Pathology of heart valve substitution with mechanical and tissue prostheses. In: *Cardiovascular Pathology.* M.D. Silver, A.I. Gotlieb, F.J. Schoen, eds. Livingstone, New York.

Schoen F.J., Cohn L.H. 1986. Explant analysis of porcine bioprosthetic heart valves: mode of failure and stent creep. In: *Biological and Bioprosthetic Valves.* E. Bodnar, M.H. Yacoub, eds. Yorke, New York, pp. 356–365.

Schoen, F.D., Hobson C.E. 1985. Anatomic analysis of removed prosthetic heart valves: causes of failure of 33 mechanical valves and 58 bioprostheses, 1980–1983. *Hum. Pathol.* 16:545–549.

Schoen F., Levy R. 1999. Tissue heart valves: Current challenges and future research perspectives. *J. Biomed. Mater. Res.* 47:439–465.

Shi Y., I. Vesely (Submitted). Fabrication of mitral valve chordae using directed collagen gel shrinkage. *Tissue Eng.*

Smith D.B., Sacks M.S., Pattany P.M., Schroeder R. 1999. Fatigue-induced changes in bioprosthetic

heart valve three-dimensional geometry and the relation to tissue damage. *J. Heart Valve Dis.* 8:25–33.

Sodian R., Hoerstrup S.P., Sperling J.S., Daebritz S., Martin D.P., Moran A.M., Kim B.S., Schoen F.J., Vacanti J.P., Mayer, Jr., J.E. 2000a. Early *in vivo* experience with tissue-engineered trileaflet heart valves. *Circulation* 102(19 Suppl 3):III22–29.

Sodian R., Hoerstrup S.P., Sperling J.S., Martin D.P., Daebritz S., Mayer, Jr. J.E., Vacanti J.P. 2000b. Evaluation of biodegradable, three-dimensional matrices for tissue engineering of heart valves. *ASAIO J.* 46:107–110.

Sodian, R., Sperling J.S., Martin D.P., Egozy A., Stock U., Mayer, Jr. J.E., Vacanti J.P. 2000c. Fabrication of a trileaflet heart valve scaffold from a polyhydroxyalkanoate biopolyester for use in tissue engineering. *Tissue Eng.* 6:183–188.

Struik D.J. 1961. *Lectures on Classical Differential Geometry.* Dover, New York.

Talman E., Boughner D. 1996. Internal shear properties of fresh aortic valve cusps: implications for normal valve function. *J. Heart Valve Dis.* 5:152–159.

Thubrikar M. 1990. *The Aortic Valve.* CRC, Boca Raton, FL.

Thubrikar M., Deck J., Aouad J., Nolan S. 1983. Role of mechanical stress in calcification of aortic bioprosthetic valves. *J. Thorac. Cardiovasc. Surg.* 86:115–125.

Timoshenko S. 1955. *Strength of Materials.* D. Van Nostrand, New York.

Tranquillo R.T., Durrani M.A., Moon A.G. 1992. Tissue engineering science: consequences of cell traction force. *Cytotechnology* 10:225–250.

Turina J., Hess O.M., Turina M., Krayenbuehl H.P. 1993. Cardiac bioprostheses in the 1990s. *Circulation* 88:775–781.

Vesely I., Boughner D., Song T. 1988. Tissue buckling as a mechanism of Bioprosthetic Valve failure. *Ann. Thorac. Surg.* 46:302–308.

Vesely I., Noseworthy R. 1992. Micromechanics of the fibrosa and the ventricularis in aortic valve leaflets. *J. Biomech.* 25:101–113.

Vyavahare N., Chen W., Joshi R.R., Lee C.H., Hirsch D., Levy J., Schoen F.J., Levy R.J. 1997. Current progress in anticalcification for bioprosthetic and polymeric heart valves. *Cardiovasc. Pathol.* 6:219–229.

Vyavahare N., Ogle M., Schoen F.J., Levy R.J. 1999a. Elastin calcification and its prevention with aluminum chloride pretreatment. *Am. J. Pathol.* 155:973–982.

Vyavahare N., Ogle M., Schoen F.J., Zand R., Gloeckner D.C., Sacks M., Levy R.J. 1999. Mechanisms of bioprosthetic heart valve failure: fatigue causes collagen denaturation and glycosaminoglycan loss. *J. Biomed. Mater. Res.* 46:44–50.

19

Assessment of Function in Tissue-Engineered Vascular Grafts

David N. Ku and Hai-Chao Han

Introduction

The development of tissue-engineered vascular grafts is beset with many practical and organiza- systematic approach to the development of a tissue-engineered vascular graft is to utilize design controls for medical devices required by the Food and Drug Administration (FDA, 21 CFR Part 820) and the International Organization for Standardization (ISO 9001). This process currently entails defining design inputs, verifying/ validating the design, and documenting design outputs. To this end, it may prove useful to follow this design method before a vessel design is implemented. Such a process of design inputs, design outputs, and validation testing is the basic premise currently endorsed by the FDA and ISO.

Laboratory tests for assessing the function of tissue engineered vascular grafts need to be developed to be consistent with FDA and ISO regulations on design controls.

Design Criteria/Inputs

In its most basic sense, a tissue-engineered vascular graft must function as an artery over long periods of time. As a result, it must exhibit significant mechanical strength to withstand the high blood pressures that may spike up to 300 mm Hg in the arterial system, transmit blood flow at low resistance, and have low potential for thrombogenicity.

As we consider vascular grafts with tissue engineering possibilities, we would like these grafts to exhibit several physiologic characteristics such as the ability to adjust diameter in response arteries will typically become stronger when subjected to hypertension and will change their diameters in response to flow. Therefore, design criteria governing the final tissue engineered graft design will be 1) high mechanical strength, 2) a lumen composed of living cells performing specific physiologic functions, 3) short- and long-term functional adaptive responses to varying hemodynamic conditions, 4) low thrombogenicity, and 5) very low immunoreactivity or host versus graft response.

Verification/Validation Test Methods

Many of these design criteria can be verified through a series of tests that can be categorized into four groups: *in vitro* bench-top testing, ex vivo bench-top testing, ex vivo animal testing, and *in vivo* animal testing. For successful design validation, we need to select the correct tests or their combination to assess the function of the vascular grafts in a cost-effective manner. Vascular grafts are designed to work in a physiologic environment. Since the environment significantly affects vascular graft performance, it is important to test vascular grafts in a physiological environment/conditions. Many physiologic functions can hardly be tested at nonphysiologic conditions *in vitro*. *In vivo* animal testing has good physiological and clinical relevance but is costly and has poor control of the testing conditions.

One effective way to assess performance of vascular grafts is to place the vascular grafts in a simulated physiological environmental test chamber, that is, to use an ex vivo bench-top system. In this chapter, we will discuss the details of the organ culture system used for verification and validation of tissue engineering vascular grafts.

Several issues may be addressed with tests using the ex vivo bench-top system (i.e., an organ culture system). For instance, mechanical strength, short-term adaptation in cells with physiologic functions to a variety of hemodynamic parameters, waste and nutrient control, and use of different cell phenotypes and cell sources can be tested. Because the system is perfused with media instead of whole blood, thromobogenic, and immunologic effects are removed from the overall testing at this point.

Ex Vivo Bench-Top Test System

Historically, physiologists have been using bench-top perfusion systems in vascular physiology studies. The system was then developed to provide a physiological pressure condition for vascular study (Brant et al., 1987) and later evolved to organ culture systems to maintain the viability of blood vessels for a period of days (Bardy et al., 1995; Labadie et al., 1996; Ligush et al., 1992). Organ culture with physiological perfusion has more physiologic relevance than traditional static tissue culture. Normal vascular structure and cellular functions have been reported for arteries and veins cultured for a few days up to a few weeks (Bardy et al., 1995; Koo et al., 1991) with vasomotor responses reported for up to 7 days (Han and Ku, 1999, 2001).

System Design

Though variations exist in systems used by different groups, the setup mostly consists of a vessel chamber, fluid reservoir, driven pump, and control and measurement units. Figure 19.1 illustrates the setup used in our laboratory. Vessel segments are mounted between stainless steel cannulae in their *in vivo* flow direction and can be stretched longitudinally to reestablish the original in situ length or to other designated stretch ratios. A wide range of perfusion flow rate (0–500 ml/min) is achievable, producing a physiological wall shear stress (15 dyn/cm^2) or pathological wall shear stress (0–30 dyn/cm^2). Mean pressure is adjusted with the resistance clamp while pulse amplitude is regulated through adjustment of the effective length of the T-end, which was designed as an elastic chamber. Further, the T-end can be replaced with a pulse-dampening dome to produce steady-state flow.

System Verification

In order to verify our system, we have tested normal porcine arteries using the system. We used segments (5–7 cm in length) of common carotid arteries harvested from farm pigs (6–7 months old, body weight 250–290 lb.) at a local abattoir. After culturing for 7 days at in situ axial length and under physiologic flow conditions with a pulsatile pressure of 100 ± 20 mm Hg, a mean lumen wall shear stress of 15 dyn/cm^2 at a flow rate of 190 ml/min, the arteries exhibited normal histology with hematoxylin and eosin stain, active cell mitochondria with tetrazolium stain, a basal tone, and strong contractile responses to agonists norepinephrine, carbachol, and sodium nitroprusside (Han and Ku, 1999, 2001). We found that the cell death rate in the 7-day cultured arteries was comparable to fresh arteries as previously reported by Schiotz et al., (2000). Studies in our lab and others have shown that cultured arteries remodel in a mode similar to *in vivo* (Han and Ku, 1998, 2001; Ligush et al., 1992; Matsumoto et al., 1999).

Ex Vivo Bench-Top Tests

The following tests can be conducted using ex vivo bench-top testing. The test results on natural arteries can serve as the standard to which future tissue engineered vascular grafts can be compared.

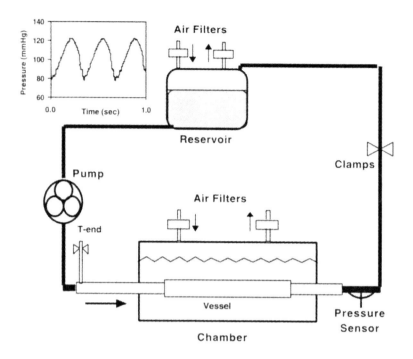

FIGURE 19.1. Schematic of the ex vivo bench-top testing system. The system consists of a vessel chamber, perfusion medium reservoir, and peristaltic pump. The artery is mounted on both ends to cannulae in the vessel chamber. A roller pump drives medium from the reservoir to perfuse the vessel. The resistance clamp controls pressure while the T-end tubing functions as an elastic chamber to adjust the pulse amplitude of pressure. Pressure is measured immediately downstream of the chamber with a manometer. The flow loop can be sterilized in an autoclave and the whole system is placed in an incubator for gas exchange and temperature control. Both perfusion and bath media were composed of Dulbecco's modified Eagles medium (DMEM) supplemented with sodium bicarbonate, L-glutamine, calf serum (10%), and antibiotics. The viscosity of the perfusion medium was increased to the level of human blood (4 cP) by adding dextran. From Han and Ku (2001) with permission.

Mechanical Properties

Mechanical properties and strength of vascular grafts can be evaluated using several tests including uniaxial or biaxial tensile test for stress-strain relation, pressure diameter relation test for compliance, burst test for burst pressure, flow impedance, and suture retention test.

Both tensile test and pressure diameter relation test are well suited to evaluate the mechanical properties and strength of natural blood vessels and vascular grafts. The tensile test uses intact cylindrical vessel segments or cut pieces of the vessel, either ring segments or rectangular strips. Some researchers prefer using intact cylindrical segments (Gupta and Kasyanov, 1997), while others prefer using rings (Seliktar et al.,

2000) or strips (Tanaka and Fung, 1974). Testing of vessel pieces (rings, strips) produces direct information on the material properties of the vascular wall while testing of the intact cylindrical segments incorporates information on the structural properties. Though results from both tests are convertible (Hayashi, 1993), each test has inherent advantages and limitations. The ring test requires a smaller sample size, but the intact cylinder tests are less traumatic to the vascular cells and have better clinical relevance. Therefore, we think the intact cylindrical segment test is better for evaluating the function and mechanical properties of the vascular grafts.

Using ex vivo bench-top testing, pressure-diameter curves are directly obtained by changing the perfusion pressure and measuring the

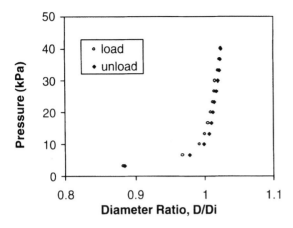

FIGURE 19.2. Pressure-diameter relation in porcine carotid arteries after cultured in the ex vivo bench-top system for 1 day under physiologic hemodynamic conditions (100 mm Hg, wall shear stress 15 dyn/cm^2).

struments (AAMI, 1994) has set up a standard for suture retention testing. Following this standard, a 6-0 polypropylene suture is sewn to one end of a vessel 2 mm from its edge and stretched at a displacement loading rate of 150 mm/min. The retention load for arteries is found to be in the range of 4–12 N with a mean of 7 N (Chin Quee et al., 2000). In comparison, the PVA grafts have a suture retention strength between 3 and 5 N with a mean of 4.5 N (Chin Quee et al., 2001). A much lower suture retention strength (less than 1 N) was reported for tissue engineered arteries (Niklason et al., 1999), while no data is available for other current tissue-engineered arteries.

Vascular Cell Viability and Proliferation

The ex vivo bench-top system also allows evaluation of the viability and function of living cells in vascular grafts. Cellular functions and their adaptations to various hemodynamical and biochemical environmental changes can be examined by putting vascular grafts under a physiological environment in the ex vivo bench-top system. Apoptotic/necrotic cells and new proliferating cells can be labeled, for example, with fluorescent dyes such as propidium or ethidium and bromodeoxylurindine, respectively, and examined with fluorescent microscopy. Cellular mitochondria activity can be demonstrated by methylthiazol terazolium (MTT) staining (Mosmann, 1983), a tetrazolium salt that cleaves to the mitochondria of viable cells and stains the tissue dark blue. Vascular wall architecture and components can be assessed histologically with light microscopy. These tests have been validated for natural arteries (Chin Quee et al., 2000; Han and Ku, 2001).

corresponding diameter changes (Fig. 19.2), thus producing data that closely correlates to data obtained from *in vivo* physiological conditions. Vascular stiffness and compliance can also be derived from the pressure-diameter relation curves. Vascular flow impedance is derived from flow rate and the pressure difference measured between the chamber's inlet and outlet (Chin Quee et al., 2000).

Mechanical Strength

Burst pressure is a very good parameter to describe the mechanical strength of a blood vessel. It is simple and direct. The burst pressure of an artery or a vascular graft can be measured by increasing lumenal pressure of the vessel that is placed in the ex vivo bench-top system until small leaks appear. For example, the burst pressure of six arteries tested in the system is over 3000 mm Hg (Chin Quee et al., 2000). Lower burst pressures have been reported for current tissue engineered arteries in the literature (L'Heureux et al., 1995; Niklason et al., 1999; Seliktar et al., 2000) though the test conditions were not well described. A Poly Vinyl Alcohol (PVA) hydrogel graft developed in our lab reached a burst pressure of 1000 mm Hg, which is the level of a vein graft (Chin Quee et al., 2001).

Suture retention is important for surgical handling. The American Association of Medical In-

Vasomotor Function

Contractile function can be assessed by conventional myography or vasomotion assay. The myography uses isolated vessel rings stimulated with various agonists and measures the contraction forces (Niklason et al., 1999; Somers et al., 2000). Advantages include small specimen size and high sensitivity while disadvantages include trauma to vessel specimens from handling. The diameter vasomotion assay challenges an intact vascular tube directly with agonists and measures the changes in vessel diameter (Han and Ku, 1999; Labadie et al., 1996; Matsumoto et al., 1999). Advantages include no handling-induced trauma and better physiological relevance.

For porcine carotid arteries cultured under physiological condition for 7 days, results from our studies showed that the diameter constricts nearly 20% in response to norepinephrine (NE). The endothelium dependent relaxation to Carbachol (carbamylcholine chloride, CCh) verifies endothelial function (Fig. 19.3) while demonstration of larger diameters after relaxation vasodilation by sodium nitroprusside (SNP) indicates the existence of basal tone in the vessel. Vasomotor function is a very important feature tissue-engineered vascular grafts try to achieve. However, current tissue engineered arteries have either very little or no vasomotor function largely due to the changes of smooth muscle cell from contractile phenotype to synthetic phenotype during *in vitro* cell culture (Owen, 1995). Rebuilding smooth muscle cells contractility has been a great challenge for tissue-engineered arteries (Stegemann et al., 2001). Pulsatile flow could be one of the important factors that modulate smooth muscle cell phenotype. Tissue-engineered arteries developed under pulsatile flow conditions using a similar system have been reported with better vasomotor function and microstructure (Niklason et al., 1999; Seliktar et al., 2000). The bench-top testing system would become a useful tool in developing and preconditioning tissue-engineered arteries.

Vascular Adaptation to Pressure and Flow

Adaptation in cells with physiologic function can be addressed using the ex vivo bench-top testing system, allowing assessment of vascular adaptation to a variety of hemodynamic parameters in a well-controlled environment.

Adaptation under Hypertensive Pressure: Opening Angle Changes

When cutting arterial rings with a radial cut, they pop open into C-shaped sectors (Fung, 1990). The opening angle of the C-shaped sector

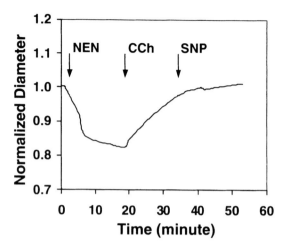

FIGURE 19.3. Diameter variations in response to agonists NE, CCh, and SNP in an artery after being maintained in culture for 7 days. The diameter is normalized with the initial diameter obtained immediately before any pharmacological challenge. The arrows marked by NE, CCH, SNP, and SNP4 show the time point when the agonists NE (10^{-5} M), CCh (10^{-4} M), SNP (10^{-5} M), and SNP (10^{-4} M) were applied, respectively.

is a concise parameter directly indicating the residual stress in the vessel and affects the mechanical behavior of the arterial wall (Han and Fung, 1996). Change in opening angles is a simple parameter that reveals wall remodeling proceeding across the wall thickness (Fung, 1990; Liu and Fung, 1989).

Significantly larger opening angles developed in arteries cultured under hypertensive pressure for 3 or 7 days in the ex vivo bench-top testing system (Fig. 19.4). These opening angle changes are consistent with results of previous hypertensive animal studies (Liu and Fung, 1989, Matsumoto and Hayashi 1996), providing evidence that the arteries are remodeling similarly to *in vivo* conditions. Overall, this phenomenon displays simple evidence of arterial wall adaptation in the bench-top testing system.

One interesting new finding is that smooth muscle contraction is intensified near the lumenal side of the wall for those arteries cultured under hypertensive pressure. This stronger contraction significantly decreases the opening angle of hypertensive arteries stimulated by NE while the opening angles of normotensive arter-

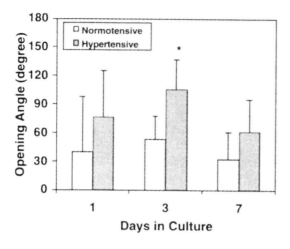

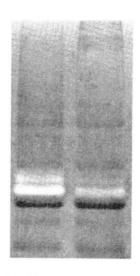

FIGURE 19.4. The opening angle in arteries cultured under normotensive pressure (100 ± 20 mm Hg) and hypertensive pressure (200 ± 30 mm Hg) for 1, 3, and 7 days. Values are mean $\pm$ SD, $n = 5$, 6, and 6 for 1-, 3-, and 7- day groups, respectively. $*p < 0.05$.

FIGURE 19.5. MMP activity of the endothelial cells in arteries cultured under static conditions (*left*) and physiologic pulsatile flow (*right*). The band shown is a 58-kDa. gelatinase on the SDS-page zymography.

ies remained relatively constant (Han and Ku, 1998). In fact, the total vascular wall contractile response to NE was increased as revealed by stronger diameter constriction in the intact vessel segment (Han and Ku, 2001). Equally interesting is that a better-organized microstructure was observed in tissue-engineered arteries cultured under pulsatile flow conditions using a system similar to the bench-top test system (Niklason et al., 1999; Seliktar et al., 2000). However, it is not clear whether intensified mechanical loading can promote or speed up the remodeling process for tissue-engineered arteries.

Artery MMP Activity

Changes in matrix metalloproteinase (MMP) activity provides further evidence of vascular wall adaptation to hemodynamic changes in the ex vivo environment. Matrix metalloproteinases, produced by either smooth muscle cells, macrophages, or fibroblasts, break down the extracellular matrix (degrade the collagens and elastin). Initiation of vascular adaptation comes with changes in MMP activity. Our previous work shows that MMP9 (gelatinase B) activity is elevated when arteries are subject to a mean hypertensive pressure of 200 mm Hg (Chesler et al., 1999). It is likely that increased pressure has a

large effect on collagenase activity thus affecting cell migration, cell hypertrophy, and proliferation. On the other hand, decrease in flow from physiological level to zero also causes MMP activity to increase (Han et al., 1997; Fig.19.5). In comparison, higher MMP activity has been reported in tissue engineered arteries cultured under pulsatile flow conditions (Seliktar et al., 1999).

Thus, in the ex vivo bench-top testing one can assess the structural and functional adaptation of tissue-engineered arteries in response to a variety of hemodynamic conditions. Further evidence of arterial wall adaptation in response to hypertensive pressure can be found in the literature (see, e.g., Bardy et al., 1996; Birukov et al., 1997; Holt et al., 1992; Soyombo et al., 1992).

Other Tests

Cellular and vascular responses to many pharmacological drugs as well as biomechanical and biochemical environment alterations can be studied using the ex vivo bench-top system. Other issues such as endothelial cell adhesion strength and migration rate, as well as mass transport and drug delivery in the vascular grafts, can also be addressed using ex vivo bench-top testing. Vascular grafts are subjected

TABLE 19.1. Issues that can be addressed using the *ex vivo* bench-top testing and the associate tests needed to evaluate the vascular grafts.

Issues	Tests
Mechanical strength	Burst pressure, flow resistance, compliance, and stress strain or pressure diameter curves
Surgical handling	Suture retention
Structural	Histology
Cell viability, proliferation	MTT stain, PI stain/TUNEL stain, BrdU/PCNA stain
Vasomotor function	Diameter response to norepinephrine and carbachol, etc.
Short-term adaptation/preconditioning	Changes of above measurements in response to mechanical or biochemical changes

to axial/longitudinal stress after implantation. Longitudinal tensile strength of vascular grafts also can be evaluated by axial stretch testing using the ex vivo bench-top system (Han et al., 2000). Many other cellular and vascular responses to pressure and flow changes (Davies 1995; Langille et al. 1995) may also be addressed using the ex vivo bench-top system. Further work needs to be done in these areas.

In summary, many design issues for vascular grafts can be addressed using the ex vivo bench-top testing. These issues include, but are not limited to, mechanical strength, cell viability, and contractile function, as well as cellular, structural, and functional adaptation of vascular grafts. Table 19.1 summarizes some design issues that can be tested using the ex vivo bench-top system. From the results of these bench-top tests, a tissue-engineered vascular graft can be evaluated in comparison with natural arteries. Different vascular grafts can be compared based on their scores in a score system that reflects the order of importance of these tests (Chin Quee et al., 2001).

Discussion

We have presented a series of ex vivo bench-top tests to verify the design of tissue-engineered arteries and to validate tissue-engineered arteries before *in vivo* testing.

The advantage of using organ culture compared with animal studies includes a well-defined controllable hemodynamic environment and decreased cost. The vessels are perfused under *in vivo* pressure, flow, and axial stretch to simulate physiology mechanical loads. Hemodynamic parameters can be controlled and changed accurately and consistently. Individual parameters can be varied independently without affecting other parameters, enabling changes that are hard to achieve *in vivo* (Herman et al., 1987). Specifically, the flow rate, mean pressure, and pulsatile amplitude can be varied independently; the wall shear stress and shear rate can also be altered separately by changing medium viscosity (Han and Ku, 2001). The limitations are the lack of circulating hormones and blood cells. It is not suitable for thromobogenic and immunologic studies and very long-term study.

In addition to its application in validation tests, a bench-top test system may also be used to provide suitable hemodynamic conditions in the development of tissue-engineered arteries. Other applications of the bench-top test system include tests for vascular devices such as stents, catheters, or endovascular imaging devices. The system may be adapted for the tissue engineering of other tissues and organs such as bone, cartilage, muscle, or heart. An application for trabecula culture was reported by Janssen and colleagues (Janssen et al., 1998).

One issue in validation tests is that the testing environmental conditions need to be standardized. For example, it is obvious that the temperature needs to be set at body temperature (37°C) and vascular grafts should be placed in either tissue culture mediums or physiologic solutions. However, issues such as the minimum number of tests required for new vascular grafts need to be determined. Professional standards should be established and adopted by the associated vascular graft research, development, manufacturing community as recommendations to the FDA, requiring the joint efforts of surgeons, engineers, manufacturers, and government agencies.

While the ex vivo bench-top system is suited to validate many of the criteria, further study is needed to resolve other outstanding issues, including standardization, thrombogencity, and immunoreactivity.

After assessment in the ex vivo bench-top system, two additional design criteria must be validated. Low thrombogenicity can be achieved through an ex vivo vascular shunt model in which blood is perfused through the vascular graft and returned to the animal. This system, developed by Hanson and colleagues (Harker et al., 1991) has the strong advantage of allowing a single animal to test multiple vascular grafts in a short period of time. Using radiolabeled platelets, one can quantify the amount of platelet adhesion as a function of time. As baseline numbers have been measured for Dacron, polytetrafluoroethylene (PTFE), and a variety of other biomaterial surfaces, one can quantitatively assess whether the new tissue-engineered vascular graft is adequate or not. Finally, low immunoreactivity will need to be assessed in an animal model. The assessment of immunoreactivity is beset with two major issues. One is species difference, and the second is individual-to-individual antigenicity. Both of these responses can elicit a large immunologic response. One can assess an initial screen of immunoreactivity by using a mixed lymphocyte culture. However, ultimately a full implantation will provide the most realistic data on immunoreactivity. If one starts with a tissue-engineered vascular graft using human cells, it is unlikely that a test in a dog or a sheep will provide much useful information. Thus, one may be forced into a human study without animal testing. However, if in the development of a vascular graft one is able to assess all of the other design inputs in a bench-top test system before one places this in a human, then the confidence for success can be increased.

In this chapter we have attempted to apply a design control approach to our design validation testing of tissue-engineered vascular grafts and demonstrate a bench-top system that may be used to validate these designs in the future.

Conclusion

Critical issues for vascular graft such as mechanical strength, cell viability, vasomotor function, and short-term adaptive response, etc. can be addressed by ex vivo bench-top testing. These tests are cost effective and would help preselect vascular graft designs suited for further *in vivo* animal testing. It would also be helpful to use design inputs for new vascular graft designs to improve existing vascular grafts. Overall, implementing these design inputs would lead to a tissue-engineered artery that is mechanically strong, structurally similar to natural arteries, physiologically active, and adaptable to hemodynamic changes.

Acknowledgment

This work was supported primarily by the ERC Program of the National Science Foundation under Award Number EEC-9731643. We thank Ms. Casey McIntosh for her comments and suggestions.

References

AAMI (Association for the Advancement of Medical Instrumentation). 1994. American National Standard for Cardiovascular Implants—Vascular Protheses. American National Standards Institute, AAMI, VP20, Arlington VA.

Bardy N, Karillon GJ, Merval R, Samuel J-L, Tedgui A. 1995. Differential effects of pressure and flow on DNA and protein synthesis and on fibronectin expression by arteries in a novel organ culture system. *Circ. Res.* 77:684–694.

Bardy N, Merval R, Benessiano J, Samuel J-L, Tedgui A. 1996. Pressure and angiotensin II synergistically induce aortic fibronectin expression in organ culture model of rabbit aorta. Evidence for a pressure-induced tissue renin-angiotensin system. *Circ. Res.* 79:70–78.

Birukov KG, Lehoux S, Birukova AA, Merval R, Tkachuk VA, Tedgui A. (1997). Increased pressure induces sustained protein kinase C-independent herbimycin A-sensitive activation of extracellular signal-related kinase 1/2 in the rabbit aorta in organ culture. *Circ. Res.* 81:895–903.

Brant AM, Teodori MF, Kormos RL, Borovetz HS. 1987. Effect of variations in pressure and flow on the geometry of isolated canine carotid arteries. *J. Biomech.* 20:831–838.

Chesler NC, Ku DN, Galis ZS. 1999. Transmural pressure induces matrix-degrading activity in porcine arteries ex vivo. *Am. J. Physiol.* 277(5 Pt 2):H2002–2009.

Chin Quee S, Han HC, Ku DN. 2000. Bench-top validation tests for tissue engineered arteries. *Adv. Bioeng.* BED 48:59–60.

Chin Quee S, Han HC, Ku DN. 2001. Verification tests for tissue engineered vascular grafts. *Ann. Biomed. Eng.* 29(S1):S148.

Davies PF. 1995. Flow-mediated endothelial mechanotransduction. *Physiol. Rev.* 75:519–560.

Fung YC. 1990. *Biomechanics: Stress, Strain and Growth.* Springer-Verlag, New York.

Gupta BS, Kasyanov VA. 1997. Biomechanics of human common carotid artery and design of novel hybrid textile. *J. Biomed. Mater. Res.* 34:341-349.

Han HC, Fung YC. 1996. Direct measurement of transverse residual strains in aorta. *Am. J. Physiol.* 270:H750–H759.

Han HC, Ku DN 1998. Smooth muscle function remodels nonuniformly in hypertensive arteries in organ culture. *Proceedings of the 3rd World Congress Biomechanics.* Sapporo, Japan. p. 423.

Han HC, Ku DN. 1999. Organ culture as a standard test for evaluation of tissue engineered arteries. *Proceedings of the First BMES-EMBS Conference,* Atlanta, GA. October 1999.

Han HC, Ku DN. 2001. Contractile responses in arteries subjected to hypertensive pressure in 7-day organ culture. *Ann. Biomed. Eng.* 29:467–475.

Han HC, Conklin BS, Hamilton G, Girard PR, Ku DN. 1997. Shear stress modulates the proteinase activity of endothelial cells in porcine carotid artery (Abstract). *Ann. Biomed. Eng.* 25(S1):S19.

Han HC, Vito RP, Michael K, Ku DN. 2000. Axial stretch increases cell proliferation in arteries in organ culture. *Adv. Bioeng.* BED 48:63–64.

Harker LA, Kelly AB, Hanson SR 1991. Experimental arterial thrombosis in nonhuman primates. *Circulation* 83(6 Suppl):IV41–55.

Hayashi K. 1993. Experimental approaches on measuring the mechanical properties and constitutive laws of arterial walls. *J. Biomech. Eng.* 115(4B):481–488.

Herman IM, Brant AM, Warty VS, Bonaccorso J, Klein, EC, Kormos, RL, Borovetz HS. 1987. Hemodynamics and the vascular endothelial cytoskeleton. *J. Cell. Biol.* 105:291–302.

Holt CM, Francis SE, Rogers S, Gadsdon PA, Taylor TCC, Soyombo A, Newby, AC, Angelini GD. 1992. Intimal proliferation in an organ culture of human internal mammary artery. *Cardiovasc. Res.* 26:1189–1194.

Janssen PML, Lehnart SE, Prestle J, Lynker JC, Salfeld P, Just H, Hasenfuss G. 1998. The trabecula culture system: a novel technique to study contractile parameters over a multiday time period. *Am. J. Physiol.* 274:H1481–H1488.

Koo EWY, Gotlieb AI. 1991. Neointimal formation in the porcine aortic organ culture: cellular dynamics over one month. *Lab. Invest.* 64:743–753.

Ku DN. 1997. Blood flow in arteries. *Ann. Rev. Fluid Mech.* 29:399–434.

Labadie RF, Antaki JF, Williams JL, Katyal S, Ligush J, Watkins SC, Pham SM, Borovetz HS. 1996. Pustile perfusion system for ex vivo investigation of biochemical pathways in intact vascular tissue. *Am. J. Physiol.* 270:H760–H768.

Langille BL. 1995. Blood flow-induced remodeling of the arterial wall. In: *Flow-Dependent Regulation of Vascular Function.* Bevan JA, Kaley G, Rubanyi GM, eds. Oxford Univ. Press, Oxford, p. 277–299.

L'Heureux N, Paquet S, Labbe R, Germain L, Auger FA. 1998. A completely biological tissue-engineered human blood vessel. *FASEB J.* 12:47–56.

Ligush J, Labadie RF, Berceli SA, Ochoa JB, Borovetz HS. 1992. Evaluation of endothelium-derived nitric oxide mediated vasodilatation utilizing ex vivo perfusion of an intact vascular tissue. *J. Surg. Res.* 52:416–421.

Liu SQ, Fung YC. 1989. Relationship between hypertension, hypertrophy, and opening angle of the zero-stress state of arteries following aortic constriction. *J. Biomech. Eng.* 111:325–335.

Matsumoto T, Hayashi K. 1996. Stress and strain distribution in hypertensive and normotensive rat aorta considering residual strain. *J. Biomech. Eng.* 118:62–73.

Matsumoto T, Okumura E, Miura Y, Sato M. 1999. Mechanical and dimensional adaptation of rabbit carotid artery cultured *in vitro.* *Med. Biol. Eng. Comput.* 37:252–256.

Mosmann T. 1983. Rapid colorimetric assay for cellular growth and survival: application to proliferation and cytotoxicity assay. *J. Immunol. Methods* 65:55–63.

Niklason LE, Gao J, Abbott WM, Hirschi KK, Houser S, Marini R, Langer R. 1999. Functional arteries grown *in vitro.* *Science* 284:489–493.

Owen GK. 1995. Regulation of differentiation of vascular smooth muscle cells. *Physiol Rev* 75:487–517.

Schiotz L, Buus CL, Hessellund A, Mulvany MJ. 2000. Effect of mitogens on growth and contractile responses of rat small arteries: *in vitro* studies. *Acta Physiol. Scand.* 169:103–113.

Seliktar D, Galis ZS, Nerem RM. 1999. Mechanical properties of tissue-engineered blood vessels: the role of MMPs in matrix degradation. *Workshop on Tissue Engineering, Gene Delivery, and Regenerative Healing,* Feb. 1999, Hilton Head, South Carolina.

Seliktar D, Black RA, Vito RP, Nerem RM. 2000. Dynamic mechanical conditioning of collagen-gel blood vessel. *Ann. Biomed. Eng.* 28:351–362.

Somers MJ, Mavromatis K, Galis ZS, Harrison DG. 2000. Vascular superoxide production and vasomotor function in hypertension induced by deoxycorticosterone acetate-salt. *Circulation* 101:1722–1728.

Soyombo A, Angelini GD, Bryan AJ, Newby AC. 1992. Surgical preparation induces injury and promotes smooth muscle cell proliferation in a culture of human saphenous vein. *Cardiovasc. Res.* 27: 1961–1967.

Stegemann JP, Nerem RM. 2001. Combined biochemical and mechanical stimulation of tissue engineered blood vessels. *Ann. Biomed. Eng.* 29(S1): S145

Tanaka TT, Fung YC. 1974. Elastic and inelastic properties of the canine aorta and their variation along the aortic tree. *J. Biomech.* 7:357–370.

20

Assessment of the Performance of Engineered Tissues in Humans Requires the Development of Highly Sensitive and Quantitative Noninvasive Outcome Measures

R.J. O'Keefe, S. Totterman, E.M. Schwarz, J.M. Looney, G.S. Seo, J. Tamez-Pena, A. Boyd, S. Bukata, J. Monu, and K. Parker

Introduction

Tissue engineering is one of the rapidly growing areas of biomedical research. The use of engineered cells and tissues to improve repair processes and treat degenerative conditions has gained enthusiasm in a number of organ systems, and promising animal studies have been performed to treat neurological disorders, such as Parkinson's disease and brain injury, as well as endocrine diseases, such as diabetes (Lanza and Cooper, 1998; Li et al., 1999). The musculoskeletal system is also an area of tremendous potential for tissue engineering. Musculoskeletal degenerative conditions are a major source of morbidity in the aging population. Currently, there are no effective therapies to slow or halt the degeneration of the articular surface, meniscus, inter vertebral disc, or muscle/tendinous units. Similarly, injury to these structures results in inadequate reparative processes and leads to long-term morbidity. Thus, the application of tissue engineering to orthopedic diseases has tremendous potential.

In keeping with the potential of tissue engineering to impact orthopedic diseases has been an increase in funding through National Institute of Arthritis and Musculoskeletal and Skin Diseases for projects investigating this application. As of 1998, more than 30 research programs investigating bone and cartilage tissue engineering were funded through NIAMS (Slavkin et al., 1999). While the majority of projects focus on bone and cartilage, additional projects investigating tendon/ligament, meniscus, and joint are also in progress. In addition, industry-initiated and sponsored tissue engineering programs are also increasing. Since the vast majority of current research activity is in the preclinical and developmental stages, the degree of research activity suggests that a large number of human clinical trials will be under way in the next 10–15 years.

Outcomes Measures in Preclinical Studies

Outcomes research has become extremely popular term in recent years (Wright and Weinstein, 1998). This is the mechanism through which the efficacy of an intervention can be assessed. An outcome can be judged in multiple ways, including patient and physician assessment, physical findings, laboratory assessment, and radiological assessment. For an outcome instrument to be valid, it must reproducibly measure a real and meaningful effect or difference (Wright and Weinstein, 1998). In many cases, a single instrument combines several modalities such as the Western Ontario and McMaster Universities Osteoarthritis Index, which uses three measures: stiffness, pain, and activities of daily living, which can be summed to give a single score (Boardman et al., 2000).

Both *in vivo* preclinical and clinical studies utilize specific outcomes. In the case of preclinical animal-based investigations, these endpoints frequently involve the histologic analysis of tissues harvested from sacrificed animals at specific time points (Holzer et al., 1999; Kanayama et al., 1999). This is particularly relevant for tissue engineering applications where the effects of growth factors, matrix components, or cells, either alone or in combination, present a large matrix of potential outcomes (Arnoczky, 1999; Grande et al., 1999). Thus, investigations designed to determine improved efficacy based upon relatively subtle differences in cell populations, growth factor concentrations, or matrix composition require a particularly sensitive outcome measure. This is especially true when considering the use of these three components in combination (Arnoczky, 1999; Grande et al., 1999).

Harvested tissues provide a sensitive measure of tissue repair and regeneration in a variety of orthopedic diseases. Histology assessment of bone union has been widely used and validated for fractures, non-unions, critical bone defects, and spine fusion (Holzer et al., 1999; Kanayama et al., 1999). This has been considered more sensitive than radiological measures, while biomechanical tests of strength further assess differences between histologically healed specimens (Athanasiou et al., 2000; Hoshino et al., 2000; Kanayama et al., 1999). Histologic endpoints have also permitted evaluation of the efficacy of different therapies for the treatment of cartilage defects, including growth factor treatment, and cell- and tissue-based therapies. These studies have permitted analysis of endpoints such as proteoglycan content, matrix organization, cellularity, healing to adjacent tissue, and biomechanical measurements. The ability to harvest

tissue provides for the use of sensitive outcome measures with the subsequent selection of the optimal materials.

Outcomes Measures in Human Clinical Trials

Human clinical trials also are critically dependent upon outcome assessment (Boardman et al., 2000; Wright and Weinstein, 1998). Similar to preclinical studies, histologic or tissue-based studies can be very meaningful. For example, a recent human clinical pilot study examining the use of gamma interferon in the treatment of pulmonary fibrosis showed a marked decrease in the expression of transforming growth factor (TGF)-beta in the lungs of treated patients, consistent with other measures of clinical improvement (Ziesche et al., 1999). However, for the most part, histologic or tissue-based endpoints are not traditionally used to measure outcome, and a subsequent large multicenter clinical trial examining gamma interferon for the treatment of pulmonary fibrosis is not using tissue samples. Obtaining histologic material from humans, even in experimental studies, is invasive and associated with potential risks and morbidity. This is particularly true for orthopedic conditions where removal or biopsy of tissue can result in structural weakness or tissue abnormalities that predispose to degenerative disorders.

For this reason, validated outcome measures for the assessment of musculoskeletal measures have focused primarily on noninvasive measurements. These include validated measures of functional outcome such as the WOMAC scoring system and SF-36, but have also included a variety of other scoring systems based upon both objective and subjective data (Boardman et al., 2000). These have included scoring systems designed to assess the outcome of joint replacements, fracture healing, spine fusions, and tendon and ligament reconstructions (Carragee, 1997; Christie et al., 1999; Plancher et al., 1998; Wright et al., 2000). While modalities such as subjective symptoms, joint motion, etc., are useful, they are indirect measures of tissue regeneration or repair.

In some cases, radiological imaging studies have supplemented outcome measures. Similar to histologic evaluation, radiological imaging has the advantage of directly assessing the tissue repair process (Holzer et al., 1999). Thus, x-rays can determine fracture union, bone regeneration, and evidence of fusion (Kanayama et al., 1999). Computerized tomography has increased sensitivity, as has been shown in animal studies with the use of micro-computerized tomography as well as in human studies (Holzer et al., 1999; Kanayama et al., 1999). Magnetic resonance imaging has the added advantage of being able to assess repair in soft tissues, and has been used successfully to examine the thickness of articular cartilage (van der Heijde, 2000). Importantly, these imaging modalities are noninvasive and can be used at multiple intervals to directly assess the performance of the engineered composites.

However, while these imaging modalities are sensitive, their use will need to be customized for specific orthopedic tissue engineering applications. Thus, quantitative radiological measures that are noninvasive, reproducible, sensitive, and meaningful will be essential adjuncts as tissue engineering research moves from preclinical to human trials. These types of measurements will range from the use and adaptation of existing technology, to the creation of new technologies. They will permit the direct quantitative assessment of the performance of tissue-engineered constructs and will supplement other indirect, but validated outcome measures.

Development of Imaging Modalities to Assess Periprosthetic Acetabular Bone Loss

We have recently initiated a clinical trial at the University of Rochester examining the efficacy of a biological reagent, Enbrel, to prevent progressive bone loss adjacent to a loose acetabular prosthesis (Schwarz et al., 2000b). Our ability to perform a meaningful clinical trial has been dependent upon the development of a quantitative methodology to assess the volume of periprosthetic bone loss in a sensitive and reproducible manner.

Similar to other human clinical trials, this trial has its basis in in vitro preclinical data based upon sound biochemical and molecular principles, as well as upon proven efficacy in accepted in vivo animal models. In vitro studies demonstrate the ability of the genetically soluble tumor necrosis factor (TNF)-alpha receptor to inhibit osteoclast formation and inflammation in response to titanium particles. On the basis of these findings, an in vivo animal model was used to further investigate the effects of Enbrel on bone resorption in a mouse calvarial model (Childs et al., 2001; Schwarz et al., 2000a). In this model, 30 mg of titanium particles are placed onto the mouse calvaria (Fig. 20.1A). In controls, this results in the rapid development of an inflammatory reaction with subsequent stimulation of osteoclastic bone resorption (Schwarz et al., 2000a).

The degree of bone resorption has been quantified through histological analysis. The bone resorption is related to the area of soft tissue located around the saggital suture. To standardize this measurement, harvested calvaria are fixed in methylmethacrylate and sections are cut longitudinally such that the midline saggital suture is in the middle of the slice (Schwarz et al., 2000a). The saggital suture area is measured in 3-micron sections just anterior to the occipital suture. Three separate sections are measured and a total of five animals are used in each group (Fig. 20.1B). Special stains can be used to examine additional cellular and tissue parameters, such as the number of osteoclasts and the amount of reactive bone formation (Schwarz et al., 2000a).

Thus, this methodology represents a sensitive and reproducible method to assess bone resorption in response to titanium particles. Experiments with Enbrel demonstrate a complete inhibition in bone resorption following administration of 0.2 mg/kg every other day beginning on the day of surgery (Fig. 20.1B) (Childs et al., 2001). However, as previously discussed, these preclinical animal studies are easy to interpret due to the ability to harvest tissue for quantitative analysis of histologic, biochemical, immunologic, histomorphometric, and molecular parameters. While these tissue-based studies

Animal Model

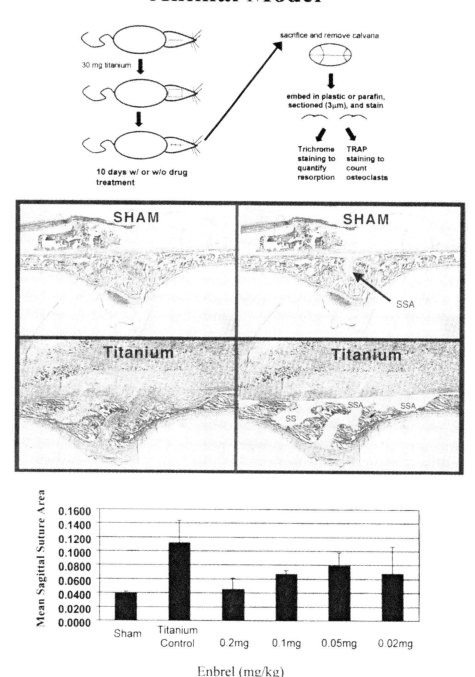

FIGURE 20.1. *An in vivo model of particle mediated bone resorption.* (*A*). Schematical representation of the mouse calvarial mode of bone resorption model. (*B*). Representative histological sections of the mouse calvarial bone in sham-operated and titanium-treated animals. The panels on the right demonstrate the area around the sagital sutures that is considered in the calculation of the sagittal suture area. (*C*). Intraperitoneal injection of Enbrel every other day inhibits bone resorption as demonstrated by a reduction in the mean sagittal suture area in animals harvested 10 days following surgery and drug treatment. [*B* is reproduced with the permission of the *J. Orthop. Res.* (Schwarz et al., 2000a); *C* is reproduced with permission of the *J. Bone Miner Res.* (Childs et al., 2001).]

provide direct measurement of osteolysis, they are not applicable to human investigations.

As we proceeded toward the development of a human clinical trial of prosthetic loosening, we realized that the development of noninvasive measures that are reproducible, sensitive, and quantitative for local bone loss was essential. Quantitative measurement of bone catabolism products (N-telopeptides) in urine and serum are valuable, but are limited in that they provide only an indirect measure of periprosthetic bone loss (Antoniou et al., 2000). Furthermore, they are systemic, and not localized measurements. Although X-ray analysis is a direct measure, the current methodology is relatively insensitive (Emerson et al., 1999). Plain radiographic analysis is a two-dimensional measurement of a process that occurs in three dimensions. Furthermore, it is not readily applicable to the measurement of periacetabular osteolysis due to the complex three-dimensional structure of the pelvic bone. Thus, while acetabular loosening is more clinically relevant and associated with a larger volumetric bone loss than that occuring in the proximal femur, pelvic osteolysis cannot be adequately assessed by current imaging techniques. Finally, the insensitivity inherent in a two-dimensional measurement means that additional time and increased numbers of patients are required to determine the effectiveness of a given intervention. More importantly, significant effects might be missed by a relatively insensitive outcome measure.

For this reason, we worked with an imaging company, VirtualScopic LLC, to develop a methodology to permit quantitative assessment of the efficacy of Enbrel to affect the progression of bone loss in patients with established bone loss in the periacetabular region. The imaging technique was adapted from standard computerized tomographic scanning for which streak artifact elimination and three-dimensional segmentation algorithms were developed. This permitted elimination of metal artifacts, segmentation of various tissues, and development of sensitive three-dimensional measurement parameters.

An example of the use of this methodology is presented in the case shown in Figure 20.2. The pelvic radiograph of this 60-year-old man demonstrates evidence of bone loss in the periacetabular region (Fig. 20.2*A*). More sensitive, is standard computer tomography, which demonstrates a much more profound bone loss than can be observed on plain radiographs (Fig. 20.2*B*). However, neither plane radiographs, nor the standard computer tomographic images are readily quantifiable. Further the computer tomographic images suffer from streak artifacts created by the prosthesis itself. The algorithmic measurement permits segmentation of the various tissues and quantification of the volume of bone loss (Fig. 20.2, *C* and *D*). Repeated measures have shown a sensitivity of 0.1 cm^3 and a reproducibility of ±4%.

This three-dimensional imaging, in association with other validated clinical and laboratory outcome measures, is the basis of the current clinical trial, and permits a comprehensive analysis of drug performance over time in comparison to placebo controls. The imaging component of this project provides an important basis for the development of future clinical trials examining intervention in this complex disease process and is an important advance in the assessment of periprosthetic osteolysis.

Summary

The treatment of human disease with tissue engineering constructs has tremendous potential and is an area of active research. However, the development of optimal combinations of matrices, cells, and growth factors/genes will require careful assessment of both preclinical as well as human clinical trials. Evaluation of human clinical trials will be markedly improved by the development of noninvasive methods designed to directly assess the performance of the engineered construct. Adaptation and development of imaging methods will enhance evaluation, and will play a central role in clinical trials, as evidenced by the role of imaging in the ongoing trial examining the effect of Enbrel on periprosthetic bone loss.

Acknowledgments

The authors would like to acknowledge the support of Public Health Services Awards AR44220 and AR46545.

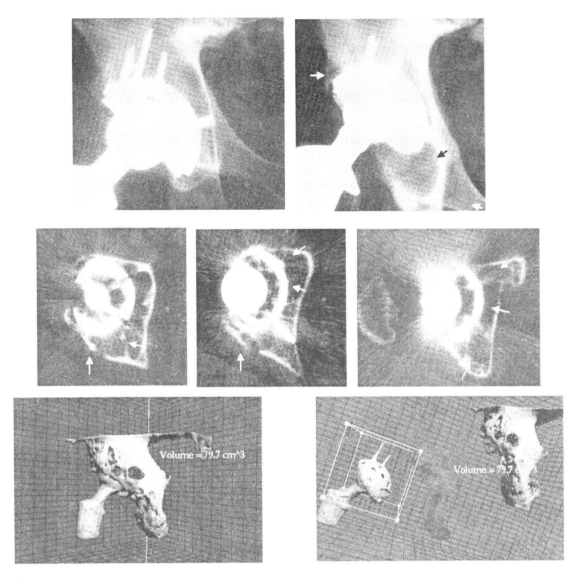

FIGURE 20.2. *Quantitative evaluation of periacetabular bone loss in humans.* (*A*). Plain radiographs at 4 years (*left panel*) and 12 years (*right panel*) following total hip arthroplasty in a 60-year-old man. (*B*). A computerized tomographic scan 12 years following arthroplasty. *Arrows* demonstrate areas of bone loss extending from the anterior to the posterior column. A cortical defect is present in the posterior column. (*C* and *D*). Application of a customized three-dimensional segmentational algorithm to the computerized tomogram results in segmentation of various tissues, eliminated metal artifacts, and permits development of sensitive three dimensional measurement parameters. The volumetric bone loss was calculated as 79.7 cm^3.

References

Antoniou J, Huk O, Zuko RD, Eyre D, Alini M. 2000. Collagen crosslinked N-telopeptides as markers for evaluating particulate osteolysis: a preliminary study. *J. Orthop. Res.* 18:64–67.

Arnoczky S. 1999. Building a meniscus. Biologic considerations. *Clin. Orthop.* 367:S244–253.

Athanasiou KA, Zhu C, Lanctot DR, Agrawal CM, Wang X. 2000. Fundamentals of biomechanics in tissue engineering of bone. *Tissue Eng.* 6:361–381.

Boardman DL, Dorey F, Thomas BJ, Lieberman JR. 2000. The accuracy of assessing total hip arthroplasty outcomes: a prospective correlation study of walking ability and 2 validated measurement devices. *J. Arthrop.* 15:200–204.

Carragee EJ. 1997. Single-level posterolateral arthrodesis, with or without posterior decompression, for the treatment of isthmic spondylolisthesis in adults. A prospective, randomized study. *J. Bone Joint Surg.* 79:1175–1180.

Childs LM, Goater JJ, O'Keefe RJ, Schwarz EM. 2001. Efficacy of Etanercept for wear debris-induced osteolysis. *J. Bone Miner. Res.* 16:338–47.

Christie MJ, DeBoer DK, Trick LW, Brothers JC, Jones RE, Vise GT, Gruen TA. 1999. Primary total hip arthroplasty with use of the modular S-ROM prosthesis. Four to seven-year clinical and radiographic results. *J. Bone Joint Surg.* 81:1707–1716.

Emerson RHJ, Sanders SB, Head WC, Higgins L. 1999. Effect of circumferential plasma-spray porous coating on the rate of femoral osteolysis after total hip arthroplasty. *J. Bone Joint Surg.* 81:1291–1298.

Grande DA, Breitbart AS, Mason J, Paulino C, Laser J, Schwartz RE. 1999. Cartilage tissue engineering: current limitations and solutions. *Clin. Orthop.* 367:S176–185.

Holzer G, Majeska RJ, Lundy MW, Hartke JR, Einhorn TA. 1999. Parathyroid hormone enhances fracture healing. A preliminary report. *Clin. Orthop.* 366:258–263.

Hoshino T, Muranishi H, Saito K, Notoya K, Makino H, Nagai H, Sohda T, Ogawa Y. 2000. Enhancement of fracture repair in rats with streptozotocin-induced diabetes by a single injection of biodegradable microcapsules containing a bone formation stimulant, TAK-778. *J. Biomed Mater Res.* 51:299–306.

Kanayama M, Cunningham BW, Sefter JC, Goldstein JA, Stewart G, Kaneda K, McAfee PC. 1999. Does spinal instrumentation influence the healing process of posterolateral spinal fusion? An *in vivo* animal model. *Spine* 24:1058–1065.

Lanza RP, Cooper DK. 1998. Xenotransplantation of cells and tissues: application to a range of diseases, from diabetes to Alzheimer's. *Mol. Med. Today* 4:39–45.

Li R, Williams S, White M, Rein D. 1999. Dose control with cell lines used for encapsulated cell therapy. *Tissue Eng.* 5:453–466.

Plancher KD, Steadman JR, Briggs KK, Hutton KS. 1998. Reconstruction of the anterior cruciate ligament in patients who are at least forty years old. A long-term follow-up and outcome study. *J. Bone Joint Surg.* 80A:184–197.

Schwarz EM, Benz EB, Lu AP, Goater JJ, Mollano AV, Rosier RN, Puzas JE, O'Keefe RJ. 2000a. A quantitative small animal surrogate to evaluate drug efficacy in preventing wear debris-induced osteolysis. *J. Orthop. Res.* 18:849–855.

Schwarz EM, Looney RJ, O'Keefe RJ. 2000b. Anti-TNFa therapy as a clinical intervention for periprosthetic osteolysis. *Arthritis Res.* 2:165–168.

Slavkin HC, Panagis JS, Kousvelari E. 1999. Future opportunities for bioengineering research at the National Institutes of Health. *Clin. Orthop.* 367:S17–30.

van der Heijde DM. 2000. Radiographic imaging: the "gold standard" for assessment of disease progression in rheumatoid arthritis. *Rheumatology* 39:9–16.

Wright JC, Weinstein MC. 1998. Gains in life expectancy from medical interventions—standardizing data on outcomes. *N. Engl. J. Med.* 339:380–386.

Wright JG, Young NL, Waddell JP. 2000. The reliability and validity of the self-reported patient-specific index for total hip arthroplasty. *J. Bone Joint Surg.* 82:829–837.

Ziesche R, Hofbauer E, Wittmann K, Petkov V, Block LH. 1999. A preliminary study of long-term treatment with interferon gamma-1b and low-dose prednisolone in patients with idiopathic pulmonary fibrosis. *N. Engl. J. Med.* 341:1264–1269.

Part V

Cell-Matrix Interactions in Functional Tissue Engineering

21

Functional Tissue Engineering and the Role of Biomechanical Signaling in Articular Cartilage Repair

Farshid Guilak and Lori A. Setton

Introduction

Articular cartilage is the thin layer of deformable, load-bearing material that lines the ends of long bones in diarthrodial joints. The primary functions of this tissue are to support and distribute forces generated during joint loading and to provide a lubricating surface to prevent wear or degradation of the joint (Mow et al., 1992). Under normal physiologic conditions, the unique composition and structure of cartilage allow it to perform these essential biomechanical functions with little wear or damage. The components and architecture of this tissue are maintained through a balance of the anabolic and catabolic activities of the chondrocytes, which comprise a small fraction of the tissue volume. However, articular cartilage possesses limited capacity for intrinsic repair, and even minor lesions or injuries may lead to progressive damage and joint degeneration. In recent years, investigators have developed novel approaches to promote cartilage repair or replacement using a variety of tissue engineering technologies through the implantation of a combination of cells, biomaterials, and biologically active molecules (Coutts et al., 1992; Grande et al., 1989; O'Driscoll et al., 1986; Sams and Nixon, 1995; Temenoff and Mikos, 2000; Wakitani et al., 1994).

Despite early successes, however, tissue engineering has faced significant challenges in the successful long-term repair or replacement of cartilage and other tissues that serve a primarily biomechanical role in the body. Within this text, we emphasize a new and evolving discipline termed *functional tissue engineering,* which seeks to meet these challenges by developing guidelines for addressing the fundamental challenges involved in the interaction of tissue-engineered constructs with mechanical forces prior to and following implantation. Recently, a series of formal principles of functional tissue engineering have been proposed in a format that is meant to apply generally to various tissues (Butler et al., 2000) and specifically to articular cartilage (Guilak et al., 2001). In the preface of this text, the general guidelines and recommendations for functional tissue engineering have been expanded to encompass several additional aspects of the interactions of mechanical stresses with engineered tissues. It is hoped that this approach will significantly improve the potential for successful tissue-engineered repair of load-bearing tissues. In this chapter, we focus specifically on the role of mechanical stresses in regulating cellular activity in engineered cartilage, within the context of functional tissue engineering. Many of the issues discussed in this chapter will be relevant to other tissues and organs of the body that are known to respond biologically to their mechanical environment and are targets for engineered tissue replacement (e.g., bone, muscle, blood vessels).

Mechanical Factors and Their Role in Cartilage Homeostasis

Chondrocyte activity is influenced by physical factors such as joint loading in combination with genetic and biochemical factors (e.g., growth factors, cytokines, and hormones). Chondrocytes are able to perceive and respond to signals generated by the normal load-bearing activities of daily living, such as walking and running (Helminen et al., 1987). Under abnormal conditions, however, mechanical factors may be responsible for initiating degradative processes that ultimately may lead to osteoarthritis and progressive joint degeneration. The physical mechanisms involved in the process of mechanical signal transduction potentially involve mechanical, chemical, and electrical signals that are generated secondary to joint loading. Physiological loading of the joint produces deformation of the cartilage layer, and associated changes in the mechanical environment of the cell within the extracellular matrix, such as spatially varying tensile, compressive and shear stresses and strains (Mow et al., 1992). In addition, the presence of a large fluid phase containing mobile ions, as well as a high density of negatively charged proteoglycans in the solid matrix (i.e., fixed charge density), gives rise to coupled electrical and chemical phenomena during joint loading (Grodzinsky, 1983; Lai et al., 1991). The ability of the chondrocytes to regulate their metabolic activity in response to the mechanical, electrical, or chemical signals of their physical environment provides a means by which articular cartilage can alter its structure and composition to the physical demands of the body. In this sense, the mechanical environment of the cells is believed to play an important role in the health and function of the diarthrodial joint. Therefore, it is presumed that the long-term success of cell-based repairs may require similar capabilities for cells to remodel, and possibly repair, the tissue in response to the functional demands of the joint. It would be important, therefore, to have detailed knowledge of the role of mechanical factors in regulating cell activity in native tissues as well as in engineered repair tissues.

Models of Cartilage Response to Loading

A number of different approaches have been used to investigate the role of physical stimuli in regulating cartilage and chondrocyte activity, ranging from *in vivo* studies to experiments at the cell and molecular level [see reviews in (Guilak et al., 1997; Helminen et al., 1987; Stockwell, 1987; van Campen and van de Stadt, 1987)]. Each level of study provides important advantages and disadvantages that often involve a tradeoff between the potential physiologic relevance of the experiment and the ability of the investigator to maximally control the test conditions. For example, *in vivo* animal studies based on emulating the physiological relevant loading conditions provide a means to study long-term (i.e., weeks to years) tissue changes associated with growth, remodeling or aging (Moskowitz, 1992; Pritzker, 1994), but are limited by difficulties in determining the precise loading history of the tissue as well as the effect of systemic factors or mediators (e.g., hormones, cytokines, enzymes). These confounding effects make it difficult to relate specific mechanical stimuli directly to the biological response within the joint. At the tissue level, *in vitro* studies of cartilage explants or isolated cells grown in three-dimensional matrices can provide model systems where both the applied loading and biochemical environment can be better controlled over time periods generally ranging from hours up to several months. Such systems can provide information on the relationships between matrix loading and cell metabolism. However, one potential limitation of tissue-level experiments is that the presence of the cartilage extracellular matrix generates a variety of biophysical signals that are associated with applied stress or strain and which vary in space and time. Many of these biophysical phenomena cannot be uncoupled in most in situ configurations, making it difficult to isolate and examine the roles of specific biophysical stimuli in regulating cell activity. Furthermore, the loading of an explant or artificial matrix in a controlled and isolated manner *in vitro* will not completely reproduce the *in vivo* environment of the chondrocytes, making it difficult to relate *in vitro* findings directly to physiologically relevant situations that are characteristic of daily living in an intact joint. Studies at the cellular level are likely to be the most useful for examining involvement of single pathways of signal transduction, or for isolating the effects of a single biophysical stimulus (e.g., osmotic or hydrostatic pressure, pH, deformation). Furthermore, studies of isolated cells will generally allow for direct stimulation of cells as well as rapid isolation of cells for analysis. As with explant studies, single-cell systems are further removed from the *in vivo* situation and present difficulties in extrapolating results to a physiologically relevant condition.

In Vivo Evidence of the Effects of Mechanical Stress on Cartilage

Studies of altered joint loading have provided strong evidence that "abnormal" loads lead to alterations in the composition, structure, metabolism, and mechanical properties of articular cartilage and other joint tissues. For example, disuse of the joint, achieved through casting or immobilization, results in a loss of proteoglycans, a decrease in tissue stiffness, and a decrease in cartilage thickness, which may be partly reversible with remobilization [e.g., (Akeson et al., 1987)]. Exercise may cause site-specific changes in proteoglycan content and cartilage stiffness, although these changes are not believed to be deleterious (Lammi et al., 1993) and may potentially have a beneficial effect in the normal joint (Lane, 1996). Altered joint loading due to instability or injury of the joint is now known to be a significant risk factor for the onset and progression of osteoarthritis (Buckwalter, 1995; Howell et al., 1992; Lohmander et al., 1999). These findings suggest that mechanical loading influences homeostasis in native tissues, but may also contribute to joint damage and degeneration under abnormal conditions. The thresholds of mechanical stimuli that define "abnormal" conditions, above or below which extracellular ma-

trix may be resorbed or remodeled, are still largely unknown thus motivating additional, more precisely controlled studies on the role of mechanical factors on cartilage.

In Vitro Studies of the Effects of Mechanical Loading on Cartilage Metabolism

Considerable research has been directed toward understanding the processes by which physical signals to the chondrocyte are converted to biochemical signals, although few previous studies have investigated these issues within the context of tissue engineering. Clarification of the specific signaling mechanisms in normal and engineered tissues would not only provide a better understanding of the processes that regulate the physiology of cartilage, but would also be expected to yield new insights on the success or failure of engineered repairs. In this respect, *in vitro* explant models of mechanical loading provide a model system in which the biomechanical and biochemical environments can be better controlled as compared with the *in vivo* situation. Explant models of cartilage loading have been utilized in a number of different loading configurations, including unconfined compression, indentation, tension, and osmotic and hydrostatic pressure. The general consensus of these studies is that static compression suppresses matrix biosynthesis, and cyclic and intermittent loading stimulate chondrocyte metabolism (e.g., (Gray et al., 1988; Guilak et al., 1994; Kim et al., 1995; Palmoski and Brandt, 1984; Sah et al., 1992; Torzilli and Grigiene, 1998). These responses have been reported over a wide range of loading magnitudes, and exhibit a stress-dose dependency [reviewed in (Guilak et al., 1997)]. Excessive loading (e.g., high magnitude, long duration) seem to have a deleterious effect, resulting in cell death, tissue disruption, and swelling (Farquhar et al., 1996; Guilak et al., 1994; Quinn et al., 1998). Other studies have shown that both static and intermittent mechanical compression influence the synthesis of nitric oxide, an important mediator of inflammation in the joint (Fermor et al., 2001).

The Mechanical Environment of the Chondrocyte in Articular Cartilage

The chondrocyte perceives its mechanical environment through biological, mechanical, and physicochemical interactions with the cartilage extracellular matrix. In addition to zonal variations in composition and structure, this matrix consists of several distinct regions based on proximity to the chondrocyte. These regions have been termed the pericellular, territorial, and interterritorial matrices (Hunziker, 1992). The bulk of the tissue is made up of interterritorial matrix, which consists primarily of water dissolved with small electrolytes (e.g., K^+, Na^+, Cl^-). The remaining solid matrix consists of collagen (mostly type II) and the proteoglycan aggrecan (Hardingham et al., 1992), with lesser amounts of other collagens, proteoglycans, proteins, and glycoproteins. Fibrillar collagen forms a dense, cross-linked network and provides primarily the tensile and shear properties of the tissue (Mow and Ratcliffe, 1997). Aggrecan contains glycosaminoglycan chains that consist of numerous carboxyl and sulfate groups that become negatively charged when dissolved in the interstitial fluid. The presence of these charges gives rise to repulsive forces and osmotic gradients so that a swelling pressure exists within the tissue (Lai et al., 1991; Maroudas, 1979). This swelling pressure influences the hydration state of the tissue, as well as the local physical environment of the chondrocytes in response to applied loading or deformation (Mow et al., 1999).

The pericellular matrix of chondrocytes consists of a distinct tissue region that contains large amounts of type VI collagen and a high concentration of proteoglycans. In addition to defining the physicochemical environment of the chondrocyte, the pericellular matrix interacts physically with the chondrocyte through cell-surface receptors (e.g., integrins, CD44, annexin V) that bind to matrix proteins such as fibronectin, hyaluronic acid, and various collagens in the pericellular region (Loeser, 1993; von der Mark and Mollenhauer, 1997). The chondrocyte together with this pericellular region and a surrounding capsule have been termed the "chondron" (Benninghoff, 1925; Poole et al., 1988; Smirzai, 1974). The structure, molecular anatomy, and metabolism of the chondron unit have been studied using both mechanical (Poole et al., 1991) and enzymatic (Lee et al., 1997) techniques for isolating them from the extracellular matrix. The functional significance of this distinct structural unit is as yet unknown. Because the pericellular matrix completely surrounds the chondrocyte, it is likely that any biophysical signals that the chondrocyte perceives are influenced by the structure and composition of this region. Indeed, there has been considerable speculation that the primary function of the chondron is biomechanical in nature (Greco et al., 1992; Poole et al., 1988; Smirzai, 1974).

Because of the intrinsic coupling between the mechanical and physicochemical properties of articular cartilage, it has been difficult to achieve a complete understanding of the mechanical signal transduction pathways of the chondrocytes. A fundamental aspect of this issue is an understanding of the mechanical environment of the chondrocytes within the articular cartilage extracellular matrix. For example, compressive loading of the cartilage extracellular matrix exposes the chondrocytes to spatially and time-varying stress, strain, fluid flow and pressure, osmotic pressure, and electric fields (Guilak, 2000; Mow et al., 1994, 1999). The relative contribution of each of these factors to the regulation of chondrocyte activity is an important consideration that is being studied by a number of investigators.

A fundamental step in determining the role of various factors in regulating chondrocyte activity is to characterize the mechanical environment within the tissue under physiological conditions of mechanical loading. This characterization would facilitate the reproduction of specific aspects of this environment in different model systems. For example, several novel microscopy techniques have provided important measurements of the in situ deformation behavior (i.e., shape and volume changes) of living chondrocytes or organelles in situ, showing that chondrocyte deformation is coordinated with that of the extracellular matrix (Broom and Myers, 1980; Buschmann et al., 1996; Guilak, 1995; Guilak et al., 1995; Wong et al., 1997). In other studies, similar techniques have investigated chondrocyte deformation within artificial matrices, generally consisting of hydrogels (Freeman et al., 1994; Knight et al., 1998 a,b; Lee et al., 2000). An important finding of these studies is that the mechanical environment of the chondrocyte within the native extracellular matrix may differ significantly from that in an artificial gel matrix. One principal difference is that chondrocytes undergo significant volume loss with compression of the native extracellular matrix, but little or no change in volume when compressed in agarose.

The signal(s) that the chondrocyte perceives may also involve other biophysical parameters related to the local stress-strain, fluid flow, or osmotic environment. As many of these parameters are difficult or impossible to measure directly (e.g., stress), theoretical models of the microenvironment of the chondrocyte are particularly valuable since they can provide quantitative predictions of the local mechanical environment. These models require information on the mechanical properties and constitutive behavior of the chondrocytes as well as those of the various tissue structures within articular cartilage. Theoretical predictions can be experimentally validated using information on the deformation behavior of the tissue at a cellular level. Several studies have sought to characterize the mechanical environment of the chondrocyte by combining experimentally determined measures of the mechanical properties and deformation behavior of chondrocytes with theoretical models of the cell within its pericellular and extracellular matrices (Bachrach et al., 1995; Baer and Setton, 2000; Guilak and Mow, 1992, 2000; Mow et al., 1994). The ultimate goal of this research has been to elucidate the sequence of biomechanical and biochemical events through which mechanical stress influences chondrocyte activity in both health and disease. Because of the complex geometry and constitutive behavior involved in this problem, numerical methods such as finite element analysis have generally been used to solve the theoretical models of cell-matrix interactions. These models can provide quantitative predictions of time-varying mechanical fields in the vicinity of the chondrocyte under dynamic loading of the cartilage matrix.

This approach was used to examine the effects of the relative mechanical properties of the chondrocyte and extracellular matrix, as well as those of cell shape and intercellular spacing, on the mechanical environment of the chondrocyte (Baer and Setton, 2000; Guilak and Mow, 1992, 2000). Results of these studies indicate that the mechanical environment of the chondrocyte is time-varying and spatially nonuniform. Further, the stress-strain environment of the chondrocytes depends on cell shape and the relative properties of the chondrocyte and the extracellular matrix. This model predicts that under physiologic magnitudes of matrix strain (Armstrong et al., 1979), the peak strains at the chondrocyte-matrix interface may be 50–100% higher than the nominal strains in the tissue, due to the difference in mechanical stiffness between the chondrocyte and the matrix. Nonlinear models have further suggested that cells may adopt an ellipsoidal geometry in the native extracellular matrix, in order to achieve a more uniform strain field within the cell when loaded (Baer et al., 2002). This finding has important implications in regard to studies that seek to duplicate the physiologic strain environment of the chondrocyte *in vitro* to examine mechanical signal transduction [e.g., (Dandrea and Vittur, 1997; De Witt et al., 1984; Donahue et al., 1995; Guilak et al., 1999; Lee et al., 1982)], to promote chondrogenesis (Elder et al., 2000; Freed et al., 1999; Mauck et al., 2000; Pazzano et al., 2000; Saris et al., 1999), or to investigate engineered cartilage repair (Ahsan and Sah, 1999; Chen and Sah, 1998; Li et al., 2000).

The finite element formulation of this theoretical model has also been used to examine the role of the pericellular matrix on the mechanical environment of the chondrocyte (Guilak and Mow, 2000). Chondrocytes were modeled with a surrounding pericellular matrix of varying size and properties, as measured experimentally by confocal microscopy and micropipette techniques. These theoretical models of chondrocyte deformation show good agreement with experimental microscopic measurements, but only if the theoretical model includes the geometry and mechanical properties of the pericellular matrix. This discrepancy between experimental data and theoretical predictions suggests that the pericellular matrix plays an important role in determining the mechanical stress environment and deformation behavior of the chondrocyte. This finding further supports the concept of a functional mechanical role for the chondron. For example, it has been hypothesized that the chondron contributes to a "protective" effect for the chondrocytes during loading (Poole et al., 1988), while others have suggested that the chondron serves as a mechanical transducer that influences chondrocyte-matrix interactions (Greco et al., 1992). Of particular interest is the finding that the lack of a pericellular matrix, as is typical during early time points following cell seeding on scaffolds, may have a significant influence on the mechanical environment of the cell. With time in culture, chondrocytes synthesize a new pericellular matrix (Hauselmann et al., 1996), which significantly affects their interaction with the extracellular matrix (Knight et al., 1998). These findings suggest that mechanical signals that the chondrocyte perceives are significantly altered within artificial matrices where the pericellular matrix may be newly generated or modified. The absence or modification of a pericellular matrix for chondrocytes in hydrogels and other tissue-engineered scaffolds can be expected to modify the metabolism of the contained cell population, and thus the biomechanical function of the engineered tissue replacements. A combination of theoretical analyses and detailed experimental measurements of the mechanical properties and metabolic activity of chondrocytes and chondrons, as described above, may be useful for providing new insights into effective strategies for preserving the biological function of the pericellular matrix in tissue-engineered constructs.

In addition, there is considerable evidence that the viscoelastic properties and deformation behavior of the chondrocyte plays an important role in its interactions with the extracellular matrix, particularly within matrices that possess properties similar to those of the chondrocytes (Baer and Setton, 2000; Guilak and Mow, 2000; LeRoux et al., 1999). From these studies, it is now believed that chondrocytes undergo significant changes in shape and volume under normal physiological conditions. Presumably, cellular deformation is one of many biophysical factors involved in the regulation of chondrocyte metabolism in response to mechanical stress. The mechanisms of intracellular signaling involved

in transducing cellular deformation are not fully understood but seem to involve several of the traditional messenger molecules such as calcium ion, inositol trisphophate, and cyclic AMP as well as other signaling pathways involving the cytoskeleton and nucleus (Guilak et al., 1997; Ingber, 1994). In the context of functional tissue engineering, it is clear that the interaction between cells and the extracellular matrix in an engineered tissue are dependent on a number of factors that are likely to be significantly altered relative to the native state (cell attachment, cell shape, extracellular matrix structure and properties, pericellular matrix structure and properties, etc.). The relative importance of each of these factors is unknown. More fundamentally, the importance of biomechanical signaling as regulator of cell activity *in vivo,* relative to the influence of biochemical mediators, remains to be determined.

Effects of Mechanical Stress on Engineered Tissues

The Stress-Strain Environment in Native Articular Cartilage

Articular cartilage is subjected to significant dynamic stresses during normal physiologic function. One goal of the functional tissue engineering initiative has been to improve our knowledge of the mechanical history that normal and repair tissues may encounter for different *in vivo* activities, in order to better define design parameters for the biomechanical function of repair tissues. These measurements can establish the patterns of activity and the limits of expected usage and may help to develop prescribed loading conditions for rehabilitation after repair. For articular cartilage, there are unique challenges with respect to the magnitude and rate of loading within the joint that make such measurements difficult. Subsequently, there is a dearth of information on the *in vivo* mechanical environment within the healthy joint. Peak stresses in the joint, measured against an endoprosthesis, have been shown to exceed 18 Mpa (Hodge et al., 1989) but stresses in a normal joint typically range from 5 to 10 MPa in animals and human joints (Brown and Shaw, 1983; Ronsky et al., 1995; von Eisenhart et al., 1999). In the presence of a cartilage defect, stresses may be altered significantly from their normal mechanical environment (Brown et al., 1991). Because of the difficulties involved in measuring loads and deformations of cartilage under realistic *in vivo* conditions, several investigators have used theoretical models of joint contact to predict these parameters (Ateshian and Wang, 1995; Blankevoort et al., 1991; Donzelli et al., 1999). Theoretical and experimental studies of cell-matrix interactions in cartilage, from the macroscopic to the microscopic levels (Guilak, 2000; Mow et al., 1994), allow prediction of biomechanical parameters that often cannot be measured. In combination with experimental measurements of cell-matrix interactions within the tissue (Guilak et al., 1995; Wong et al., 1996), theoretical models may allow researchers to investigate hypotheses on the influence of various mechanical factors on the biology or biomechanics of engineered cartilage repaired in situations where standard experimental measurements may not be feasible.

Biomechanical Failure of Engineered Tissues

A primary concern is whether the implant or repair tissue possesses the mechanical properties to withstand the physiologic loading conditions of the joint without undergoing overt failure, dislodgement, or fatigue. In many cases, the failure of engineered cartilage repair has been attributed, at least in part, to direct biomechanical factors (Breinan et al., 1997). Similarly, the failure of fresh meniscal allografts to protect the joint from degenerative changes appear to be due to poor biomechanical fixation and therefore extrusion of the grafts (Elliott et al., 1999). In this respect, it may be critical to control the *in vivo* mechanical environment so as not to exceed the capabilities of the repair. To date, such rehabilitation periods rarely have been used, particularly in animal models of cartilage repair. For example, a recent study showed frequent and overt delamination of periosteal flaps for chondrocyte implantation in the goat knee (Driesang and Hunziker, 2000),

but not following a simple rehabilitation period consisting of joint immobilization.

Biomechanical Influences on Cellular Activity in Engineered Grafts

In addition to material failure, it is well accepted that mechanical factors have the capability to influence the biologic activity of normal and repair tissues (Guilak et al., 1997; Helminen et al., 1987), and, therefore, may play an important role in the eventual success or failure of cellular grafts. In this regard, it would be important to better characterize the diverse array of physical signals that engineered cells may experience *in vivo* and their biologic response to such potential stimuli. This information may provide important insights into the long-term capabilities of engineered constructs to maintain the proper cellular phenotype. For example, the chemical, mechanical, and architectural properties of the scaffold with time will affect the physiological response of the cells. As a result, the mechanical influence on the cells will be related to the mechanical properties of the scaffold, the mechanical boundary conditions acting on the construct, and the interactions between the cells and their extracellular matrix (Guilak, 2000). In addition, the shape and morphology of the cells will be related to the cell/scaffold interactions, and cell shape alone may significantly influence the stress-strain environment within the tissue (Baer and Setton, 2000; Mow et al., 1994). All of these factors may contribute to the cell's ability to respond to both mechanical and biologic signals, and subsequently to synthesize and express extracellular matrix molecules.

As described earlier, much of the information currently available on chondrocyte response to mechanical stress has been realized from tissue explant culture experiments, which provide a more controlled mechanical and biochemical environment than the *in vivo* condition (Guilak et al., 1997). The great majority of such studies have been performed using chondrocytes embedded within their native extracellular matrix, although several experiments examining the response of chondrocytes in artificial matrices or engineered constructs suggest that cellular response to mechanical loading may be altered significantly in the absence of the native extracellular matrix. A comparison of the biological response of chondrocytes to compression within their native extracellular matrices to that within artificial matrices such as agarose reveals several important and distinct differences. For example, chondrocytes embedded within agarose show little metabolic response to compression until a newly synthesized matrix has accumulated within the agarose (Buschmann et al., 1995). This phenomenon has been attributed to alterations in the biomechanical and physicochemical microenvironment of the chondrocytes secondary to the increase in the local fixed-charge density of the matrix. Within such a model system, chondrocytes undergo significant deformation until a new pericellular matrix has been synthesized (Knight et al., 1998). Furthermore, dynamic compression of chondrocytes within the native extracellular matrix significant increases the production of nitric oxide, an important inflammatory mediator (Fermor et al., 2001); within an agarose matrix, similar regimens of compression suppress nitric oxide synthesis (Lee et al., 1998). As nitric oxide is an important regulator of both anabolic and catabolic activities in cartilage (Stefanovic-Racic et al., 1996), alteration of the "normal" response of the cells to mechanical stimuli may influence the balance between synthesis and degradation of the extracellular matrix. Furthermore, in the presence of mechanical stress, nitric oxide inhibits the production of prostaglandin E2 (Fermor et al., in press) and leukotriene B4 (Fermor et al., 2001), mediators that are believed to play a role in pain and inflammation in arthritis.

Other studies have revealed important characteristics of the response of repair cartilage to mechanical stress using novel *in vitro* models (Ahsan and Sah, 1999; Chen and Sah, 1998; Li et al., 2000). For example, static compression of chondrocytes transplanted onto articular cartilage significantly decreases proteoglycan synthesis rates (Chen and Sah, 1998) and cell proliferation (Li et al., 2000), and thus may affect subsequent integrative cartilage repair *in vivo*. Together, these findings suggest that various aspects of the repair process, including cellular proliferation, matrix biosynthesis, and tissue integration, may be influenced directly or indirectly by mechanical factors.

The Use of Biophysical Factors to Promote Cartilage Regeneration *In Vitro*

Mechanical stress is an important modulator of cell physiology, and there is significant evidence that controlled application of biophysical stimuli within "bioreactors" may be used to improve or accelerate tissue regeneration and repair *in vitro*. For example, early studies showed that cyclic mechanical stretch of skeletal myofibers increased the alignment of myotubes that assembled into organoids in culture (Vandenburgh, 1982). In other studies, mechanical stimulation has been shown to increase cellular alignment, proliferation, and matrix synthesis in many different cell types (Ives et al., 1986; Lee et al., 1982; Sumpio et al., 1988; Thoumine et al., 1995). Recent studies have shown improved success of tissue-engineered systems such as blood vessels and smooth muscle by preconditioning grafts with pulsatile fluid flow and pressure (Kim et al., 1999; Kim and Mooney, 2000; Niklason et al., 1999; Seliktar et al., 2000).

Specific to articular cartilage, there is growing evidence that mechanical stimuli can increase matrix deposition in tissue-engineered cartilage *in vitro*. Although the specific biophysical signals are not fully understood, a variety of different stimuli have been shown to increases matrix synthesis and accumulation. In particular, factors related to increased fluid within constructs seem to be particularly stimulatory (Freed et al., 1999; Pazzano et al., 2000). Increased fluid flow and perfusion are associated with increased shear stresses, nutrient transport, and metabolite exchange. Cyclic compression of chondrocytes in artificial matrices also increases matrix synthesis (Buschmann et al., 1995; Mauck et al., 2000) through a mechanism that potentially involves increases localized fluid flow. However, other stimuli such as hypogravity (Freed et al., 1997) also stimulate chondrogenesis, indicating a more complex relationship between mechanical stimuli and cell activity within bioreactors.

There is also increasing evidence that biomechanical stimuli can promote the differentiation of progenitor, or "stem" cells, into a chondrogenic phenotype. Dynamic fluid pressure has been shown to increase the proliferation of periosteal cells (Saris et al., 1999), potentially through a paracrine signaling mechanism between the cells in cambium and fibrous layers. In other studies, cyclic compressive loading of chick limb mesenchymal cells embedded in agarose was found to double the number of cartilage nodules and the rate of proteoglycan synthesis, whereas static compression had little effect on cellular differentiation (Elder et al., 2000).

These findings indicate a potentially important role for biomechanical and biophysical factors in stimulating cellular differentiation, gene expression, macromolecular synthesis, and matrix assembly in tissue engineering systems. The use of mechanical stimuli in place of or in combination with biochemical factors may provide a means to facilitate the development of functional tissues *in vitro*, prior to implantation in the body. A more thorough understanding of the response of native and engineered cartilage to mechanical stress, and the mechanisms involved in these responses, will hopefully improve the success of tissue-engineered cartilage repair.

Future Directions

It is now well accepted that mechanical factors play an important role in the health and maintenance of native articular cartilage. Mechanical signals, along with local and systemic biochemical factors, appear to contribute to the regulatory pathways used by chondrocytes to maintain a homeostatic balance of anabolic and catabolic activities. Once implanted within the joint, engineered tissues would be exposed to mechanical factors whose characteristics may differ from the native state. A better understanding of the *in vivo* mechanical environment, as well as the interaction of mechanical factors with cells in engineered tissues, may serve to improve the chances for success. It is important to note that other rapidly evolving technologies may have a significant impact on functional tissue engineering and cartilage repair, and these guidelines must be considered in light of the role of novel growth factors, new biomaterials, gene therapy, and other new technologies. In summary, several

important areas have been identified as targets for future studies in this area of research:

1. ***Understanding the sequence of biomechanical and biochemical events involved in the process of mechanical signal transduction in native tissues.*** A thorough understanding of these processes in native cartilage will provide important insight into the interactions of mechanical factors with repair tissues and constructs that are exposed to significant stresses *in vitro* or *in vivo*.

2. ***Determination of the mechanical environment of the cell in native and repair tissues.*** The relative importance of different mechanical factors in controlling cellular metabolism within engineered repair tissues is not known. A fundamental step in determining the relative importance of various aspects of the mechanical environment is the development of novel experimental and theoretical techniques to measure or predict the spatial and temporal variations in the stress-strain, fluid flow, fluid pressure, electric, and physicochemical fields at the cellular level within native and repair tissues. Such studies will require detailed information on the biomechanical properties and structure of artificial matrices and their interactions with cells.

3. ***Investigation of the effects of mechanical factors on tissue repair in vivo.*** Once implanted, tissue-engineered constructs will be subjected to significant loads and deformations *in vivo*. Both the mechanical and biological consequences of *in vivo* loading must be understood in order to improve the success of engineered repairs. Certain *in vitro* models of cartilage repair may be particularly useful in predicting potential outcomes *in vivo*.

4. ***The use of mechanical and biophysical factors to accelerate tissue regeneration in vitro.*** Mechanical and other biophysical factors are important modulators of cell physiology, and there is significant evidence that they may be controlled *in vitro* within bioreactors to improve or accelerate tissue regeneration or cellular differentiation prior to implantation *in vivo*. A more thorough understanding of chondrocyte response to mechanical and biophysical factors within native and artificial extracellular matrices will presumably improve the ability to develop functional tissue replacements *in vitro*. Further-

more, the use of nonchemical (i.e., biophysical) means of promoting the differentiation of progenitor cells may provide various advantages in tissue engineering.

Acknowledgments

We would like to thank Drs. David Butler, Steve Goldstein, and David Mooney for many important discussions on the topics outlined in this chapter. Supported in part by grants from the National Institutes of Health AG15768, AR43876, AR45644, AR47442, and the North Carolina Biotechnology Center.

References

Ahsan T, Sah RL. 1999. Biomechanics of integrative cartilage repair. *Osteoarthritis Cartilage* 7:29–40.

Akeson WH, Amiel D, Abel MF, Garfin SR, Woo SL. 1987. Effects of immobilization on joints. *Clin. Orthop.* 219:28–37.

Armstrong CG, Bahrani AS, Gardner DL. 1979. *In vitro* measurement of articular cartilage deformations in the intact human hip joint under load. *J. Bone Joint Surg. Am.* 61:744–755.

Ateshian GA, Wang H. 1995. A theoretical solution for the frictionless rolling contact of cylindrical biphasic articular cartilage layers. *J. Biomech.* 28:1341–1355.

Bachrach NM, Valhmu WB, Stazzone E, Ratcliffe A, Lai WM, Mow VC. 1995. Changes in proteoglycan synthesis of chondrocytes in articular cartilage are associated with the time-dependent changes in their mechanical environment. *J. Biomech.* 28:1561–1570.

Baer AE, Setton LA. 2000. The micromechanical environment of intervertebral disc cells: effect of matrix anisotropy and cell geometry predicted by a linear model. *J. Biomech. Eng.* 122:245–251.

Baer AE, Laursen T, Guilak F, Setton LA. 2002. The micromechanical environment of intervertebral disc cells determined by a finite deformation, anisotropic, and biphasic finite element model. *J. Biomech Eng.*, in press.

Benninghoff A. 1925. Form und bau der Gelenkknorpel in ihren Beziehungen Zur Funktion. In: *Zweiter Teil: der Aufbau des Gelenkknorpels in sienen Beziehungen zur Funktion.* Z. Zellforsch, ed. 2: pp. 783.

Blankevoort L, Kuiper JH, Huiskes R, Grootenboer HJ. 1991. Articular contact in a three-dimensional model of the knee. *J. Biomech.* 24:1019–1031.

Breinan HA, Minas T, Hsu HP, Nehrer S, Sledge CB, Spector M. 1997. Effect of cultured autologous chondrocytes on repair of chondral defects in a canine model. *J. Bone Joint Surg. Am.* 79:1439–1451.

Broom ND, Myers DB. 1980. A study of the structural response of wet hyaline cartilage to various loading situations. *Connect. Tissue Res.* 7:227–237.

Brown TD, Shaw DT. 1983. *In vitro* contact stress distributions in the natural human hip. *J. Biomech.* 16:373–384.

Brown TD, Pope DF, Hale JE, Buckwalter JA, Brand RA. 1991. Effects of osteochondral defect size on cartilage contact stress. *J. Orthop. Res.* 9:559–567.

Buckwalter JA. 1995. Osteoarthritis and articular cartilage use, disuse, and abuse: experimental studies. *J. Rheumatol. (Suppl.)* 43:13–15.

Buschmann MD, Gluzband YA, Grodzinsky AJ, Hunziker EB. 1995. Mechanical compression modulates matrix biosynthesis in chondrocyte/agarose culture. *J. Cell Sci.* 108:1497–1508.

Buschmann MD, Hunziker EB, Kim YJ, Grodzinsky AJ. 1996. Altered aggrecan synthesis correlates with cell and nucleus structure in statically compressed cartilage. *J. Cell Sci.* 109:499–508.

Butler DL, Goldstein SA, Guilak F. 2000. Functional tissue engineering: the role of biomechanics. *J. Biomech. Eng.* 122:570–575.

Chen AC, Sah RL. 1998. Effect of static compression on proteoglycan biosynthesis by chondrocytes transplanted to articular cartilage *in vitro*. *J. Orthop. Res.* 16:542–550.

Coutts RD, Woo SL, Amiel D, von Schroeder HP, Kwan MK. 1992. Rib perichondral autografts in full-thickness articular cartilage defects in rabbits. *Clin. Orthop.* 263–273.

Dandrea P, Vittur F. 1997. Propagation of intercellular Ca2+ waves in mechanically stimulated articular chondrocytes. *FEBS Lett.* 400:58–64.

De Witt MT, Handley CJ, Oakes BW, Lowther DA. 1984. In vitro response of chondrocytes to mechanical loading. The effect of short term mechanical tension. *Connect. Tissue Res.* 12:97–109.

Donahue HJ, Guilak F, Vander Molen M, McLeod KJ, Rubin CT, Grande DA, Brink PR. 1995. Chondrocytes isolated from mature articular cartilage retain the capacity to form functional gap junctions. *J. Bone Miner. Res.* 10:1359–1364.

Donzelli PS, Spilker RL, Ateshian GA, Mow VC. 1999. Contact analysis of biphasic transversely isotropic cartilage layers and correlations with tissue failure. *J. Biomech.* 32:1037–1047.

Driesang IM, Hunziker EB. 2000. Delamination rates of tissue flaps used in articular cartilage repair. *J. Orthop. Res.* 18:909–911.

Elder SH, Kimura JH, Soslowsky LJ, Lavagnino M, Goldstein SA. 2000. Effect of compressive loading on chondrocyte differentiation in agarose cultures of chick limb-bud cells. *J. Orthop. Res.* 18:78–86.

Elliott DM, Guilak F, Vail TP, Wang JY, Setton LA. 1999. Tensile properties of articular cartilage are altered by meniscectomy in a canine model of osteoarthritis. *J. Orthop. Res.* 17:503–508.

Farquhar T, Xia Y, Mann K, Bertram J, Burton-Wurster N, Jelinski L, Lust G. 1996. Swelling and fibronectin accumulation in articular cartilage explants after cyclical impact. *J. Orthop. Res.* 14:417–423.

Fermor B, Haribabu B, Brice Weinberg J, Pisetsky DS, Guilak F. 2001a. Mechanical stress and nitric oxide influence leukotriene production in cartilage. *Biochem. Biophys. Res. Commun.* 285:806–810.

Fermor B, Weinberg JB, Pisetsky DS, Misukonis MA, Banes AJ, Guilak F. 2001b. The effects of static and intermittent compression on nitric oxide production in articular cartilage explants. *J. Orthop. Res.* 19:729–737.

Fermor B, Weinberg JB, Pisetsky DS, Misukonis MA, Fink C, Guilak F. 2003. Induction of cyclooxygenase-2 by mechanical stress through a nitric oxide-regulated pathway. *Osteoarthritis Cartilage.* 10:792–798.

Freed LE, Langer R, Martin I, Pellis NR, Vunjak-Novakovic G. 1997. Tissue engineering of cartilage in space. *Proc. Natl. Acad. Sci. U.S.A.* 94:13885–13890.

Freed LE, Martin I, Vunjak-Novakovic G. 1999. Frontiers in tissue engineering. In vitro modulation of chondrogenesis. *Clin. Orthop.* 367:S46–58.

Freeman PM, Natarjan RN, Kimura JH, Andriacchi TP. 1994. Chondrocyte cells respond mechanically to compressive loads. *J. Orthop. Res.* 12:311–320.

Grande DA, Pitman MI, Peterson L, Menche D, Klein M. 1989. The repair of experimentally produced defects in rabbit articular cartilage by autologous chondrocyte transplantation. *J. Orthop. Res.* 7:208–218.

Gray ML, Pizzanelli AM, Grodzinsky AJ, Lee, RC. 1988. Mechanical and physiochemical determinants of the chondrocyte biosynthetic response. *J. Orthop. Res.* 6:777–792.

Greco F, Specchia N, Falciglia F, Toesca A, Nori S. 1992. Ultrastructural analysis of the adaptation of articular cartilage to mechanical stimulation. *Ital. J. Orthop. Traumatol.* 18:311–321.

Grodzinsky, AJ. 1983. Electromechanical and physicochemical properties of connective tissue. *Crit. Rev. Biomed. Eng.* 9:133–199.

Guilak F. 1995. Compression-induced changes in the shape and volume of the chondrocyte nucleus. *J. Biomech.* 28:1529–1542.

Guilak F. 2000. The deformation behavior and viscoelastic properties of chondrocytes in articular cartilage. *Biorheology* 37:27–44.

Guilak F, Mow VC. 1992. Determination of the mechanical response of the chondrocyte in situ using finite element modeling and confocal microscopy. *ASME Adv. Bioeng.* BED-20:21–23.

Guilak F, Mow VC. 2000. The mechanical environment of the chondrocyte: a biphasic finite element model of cell-matrix interactions in articular cartilage. *J. Biomech.* 33:1663–1673.

Guilak F, Meyer BC, Ratcliffe A, Mow VC. 1994. The effects of matrix compression on proteoglycan metabolism in articular cartilage explants. *Osteoarthritis Cartilage* 2:91–101.

Guilak F, Ratcliffe A, Mow VC. 1995. Chondrocyte deformation and local tissue strain in articular cartilage: a confocal microscopy study. *J. Orthop. Res.* 13:410–421.

Guilak F, Sah RL, Setton LA. 1997. Physical regulation of cartilage metabolism. In: *Basic Orthopaedic Biomechanics.* V.C. Mow and W.C. Hayes., eds. Lippincott-Raven, Philadelphia, pp. 179–207.

Guilak F, Zell RA, Erickson GR, Grande DA, Rubin CT, McLeod KJ, Donahue HJ. 1999. Mechanically induced calcium waves in articular chondrocytes are inhibited by gadolinium and amiloride. *J. Orthop. Res.* 17:421–429.

Guilak F, Butler DL, Goldstein SA. 2001. Functional tissue engineering—the role of biomechanics in articular cartilage repair. *Clin. Orthop.* 391:S295–S305.

Hardingham TE, Fosang AJ, Dudhia J. 1992. Aggrecan: the chondroitin sulfate/keratan sulfate proteoglycan from cartilage. In: *Articular Cartilage and Osteoarthritis.* K.E. Kuettner, R. Schleyerbach, J.G. Peyron and V.C. Hascall, eds. Raven Press, New York, pp. 5–20.

Hauselmann HJ, Masuda K, Hunziker EB, Neidhart M, Mok SS, Michel BA, Thonar EJ. 1996. Adult human chondrocytes cultured in alginate form a matrix similar to native human articular cartilage. *Am. J. Physiol.* 271:C742–752.

Helminen HJ, Jurvelin J, Kiviranta I, Paukkonen K, Saamanen AM, Tammi M. 1987. Joint loading effects on articular cartilage: a historical review. In: *Joint Loading: Biology and Health of Articular Structures.* HJ Helminen, I Kiviranta M Tammi et al., Eds. Wright and Sons, Bristol, pp. 1–46.

Hodge WA, Carlson KL, Fijan RS, Burgess RG, Riley PO, Harris WH, Mann. RW 1989. Contact pressures from an instrumented hip endoprosthesis. *J. Bone Joint Surg.* 71A:1378–1386.

Howell DS, Treadwell BV, Trippel SB. 1992. Etiopathogenesis of osteoarthritis. In: *Osteoarthri-*

tis, Diagnosis and Medical/Surgical Management. RW Moskowitz, DS Howell, VM Goldberg, HJ Mankin, eds. W.B. Saunders, Philadelphia, pp. 233–252.

Hunziker EB. 1992. Articular cartilage structure in humans and experimental animals. In: *Articular Cartilage and Osteoarthritis.* KE Kuettner, R Schleyerbach, JG Peyron, VC Hascall, eds. Raven Press, New York, pp. 183–199.

Ingber DE. 1994. Cellular tensegrity and mechanochemical transduction. In: *Cell Mechanics and Cellular Engineering.* VC Mow, F Guilak, R Tran-Son-Tay, RM Hochmuth, eds. Springer Verlag, New York, pp. 329–342.

Ives CL, Eskin SG, McIntire LV. 1986. Mechanical effects on endothelial cell morphology: *in vitro* assessment. *In Vitro Cell. Devel. Biol.* 22:500–507.

Kim BS, Mooney DJ. 2000. Scaffolds for engineering smooth muscle under cyclic mechanical strain conditions. *J. Biomech. Eng.* 122:210–215.

Kim BS, Nikolovski J, Bonadio J, Mooney DJ. 1999. Cyclic mechanical strain regulates the development of engineered smooth muscle tissue. *Nat. Biotechnol.* 17:979–983.

Kim YJ, Bonassar LJ, Grodzinsky AJ. 1995. The role of cartilage streaming potential, fluid flow and pressure in the stimulation of chondrocyte biosynthesis during dynamic compression. *J. Biomech.* 28:1055–1066.

Knight MM, Ghori SA, Lee DA, Bader DL. 1998a. Measurement of the deformation of isolated chondrocytes in agarose subjected to cyclic compression. *Med. Eng. Phys.* 20:684–688.

Knight MM, Lee DA, Bader DL. 1998b. The influence of elaborated pericellular matrix on the deformation of isolated articular chondrocytes cultured in agarose. *Biochim. Biophys. Acta* 1405:67–77.

Lai WM, Hou JS, Mow VC. 1991. A triphasic theory for the swelling and deformation behaviors of articular cartilage. *J. Biomech. Eng.* 113:245–258.

Lammi MJ, Hakkinen TP, Parkkinen JJ, Hyttinen M, Jortikka M, Helminen HJ, Tammi M. 1993. Effects of long-term running exercise on canine femoral head articular cartilage. *Agents Actions Suppl.* 39:95–99.

Lane NE. 1996. Physical activity at leisure and risk of osteoarthritis. *Ann. Rheum. Dis.* 55:682–684.

Lee DA, Frean SP, Lees P, Bader DL. 1998. Dynamic mechanical compression influences nitric oxide production by articular chondrocytes seeded in agarose. *Biochem. Biophys. Res. Commun.* 251:580–585.

Lee DA, Knight MM, Bolton JF, Idowu BD, Kayser MV, Bader DL. 2000. Chondrocyte deformation within compressed agarose constructs at the cellular and sub-cellular levels. *J. Biomech.* 33:81–95.

Lee GM, Poole CA, Kelley SS, Chang J, Caterson B. 1997. Isolated chondrons: a viable alternative for studies of chondrocyte metabolism *in vitro. Osteoarthritis Cartilage* 5:261–274.

Lee RC, Rich JB, Kelley KM, Weiman DS, Mathews MB 1982. A comparison of *in vitro* cellular responses to mechanical and electrical stimulation. *Am. Surg.* 48:567–574.

LeRoux MA, Guilak F, Setton LA. 1999. Compressive and shear properties of alginate gel: effects of sodium ions and alginate concentration. *J. Biomed. Mater. Res.* 47:46–53.

Li KW, Falcovitz YH, Nagrampa JP, Chen AC, Lottman LM, Shyy JY, Sah RL-2000. Mechanical compression modulates proliferation of transplanted chondrocytes. *J. Orthop. Res.* 18:374–382.

Loeser RF. 1993. Integrin-mediated attachment of articular chondrocytes to extracellular matrix proteins. *Arthritis Rheum.* 36:1103–1110.

Lohmander LS, Ionescu M, Jugessur H, Poole AR, 1999. Changes in joint cartilage aggrecan after knee injury and in osteoarthritis. *Arthritis Rheum.* 42:534–544.

Maroudas A. 1979. Physicochemical properties of articular cartilage. In: *Adult Articular Cartilage.* M. Freeman, ed. Pitman Medical, Tunbridge Wells, U.K., pp. 215–290.

Mauck RL, Soltz MA, Wang CC, Wong DD, Chao PH, Valhmu WB, Hung CT. 2000. Functional tissue engineering of articular cartilage through dynamic loading of chondrocyte-seeded agarose gels. *J. Biomech. Eng.* 122:252–260.

Moskowitz RW. 1992. Experimental models of osteoarthritis. In: *Osteoarthritis: Diagnosis and Medical/Surgical Management.* RW Moskowitz, DS Howell, VM Goldberg, HJ Mankin, eds. W.B. Saunders, Philadelphia, pp. 213–232.

Mow VC, Ratcliffe A. 1997. Structure and function of articular cartilage. In: *Basic Orthopaedic Biomechanics.* VC Mow, WC Hayes, eds. Lippincott-Raven, Philadelphia, pp. 113–177.

Mow VC, Ratcliffe A, Poole AR. 1992. Cartilage and diarthrodial joints as paradigms for hierarchical materials and structures. *Biomaterials* 13:67–97.

Mow VC, Bachrach N, Setton LA, Guilak F. 1994. Stress, strain, pressure, and flow fields in articular cartilage. In: *Cell Mechanics and Cellular Engineering.* VC Mow, F Guilak, R Tran-Son-Tay, R Hochmuth, eds. Springer Verlag, New York, pp. 345–379.

Mow VC, Sun DN, Guo XE, Hung CT, Lai WM. 1999a. Chondrocyte-extracellular matrix interactions during osmotic swelling. *ASME Bioengineering Conference* BED42:133–134.

Mow VC, Wang CC, Hung CT. 1999b. The extracellular matrix, interstitial fluid and ions as a mechanical signal transducer in articular cartilage. *Osteoarthritis Cartilage* 7:41–58.

Niklason LE, Gao J, Abbott WM, Hirschi KK, Houser S, Marini R, Langer R. 1999. Functional arteries grown *in vitro. Science* 284:489–493.

O'Driscoll SW, Keeley FW, Salter RB. 1986. The chondrogenic potential of free autogenous periosteal grafts for biological resurfacing of major full-thickness defects in joint surfaces under the influence of continuous passive motion. An experimental investigation in the rabbit. *J. Bone Joint Surg. Am.* 68:1017–1035.

Palmoski MJ, Brandt KD, 1984. Effects of static and cyclic compressive loading on articular cartilage plugs *in vitro. Arthritis Rheum.* 27:675–681.

Pazzano D, Mercier KA, Moran JM, Fong SS, DiBiasio DD, Rulfs JX, Kohles SS, Bonassar LJ. 2000. Comparison of chondrogensis in static and perfused bioreactor culture. *Biotechnology Progress* 16:893–896.

Poole CA, Flint MH, Beaumont BW. 1988. Chondrons extracted from canine tibial cartilage: preliminary report on their isolation and structure. *J. Orthop. Res.* 6:408–419.

Poole CA, Matsuoka, A, Schofield JR. 1991. Chondrons from articular cartilage. III. Morphologic changes in the cellular microenvironment of chondrons isolated from osteoarthritic cartilage. *Arthritis Rheum.* 34:22–35.

Pritzker KPH. 1994. Animal models for osteoarthritis: processes, problems and prospects. *Ann. Rheum. Dis.* 53:406–420.

Quinn TM, Grodzinsky AJ, Hunziker EB, Sandy JD. 1998. Effects of injurious compression on matrix turnover around individual cells in calf articular cartilage explants. *J. Orthop. Res.* 16:490–499.

Ronsky JL, Herzog W, Brown TD, Pedersen DR, Grood ES, Butler DL. 1995. *In vivo* quantification of the cat patellofemoral joint contact stresses and areas. *J. Biomech.* 28:977–983.

Sah RL, Grodzinsky AJ, Plaas AHK, Sandy JD. 1992. Effects of static and dynamic compression on matrix metabolism in cartilage explants. In: *Articular Cartilage and Osteoarthritis.* KE Kuettner, R Schleyerbach, JG Peyron, VC Hascall, eds. Raven Press, New York, pp. 373–392.

Sams AE, Nixon AJ. 1995. Chondrocyte-laden collagen scaffolds for resurfacing extensive articular cartilage defects. *Osteoarthritis & Cartilage* 3: 47–59.

Saris DB, Sanyal A, An KN, Fitzsimmons JS, O'Driscoll SW. 1999. Periosteum responds to dynamic fluid pressure by proliferating *in vitro. J. Orthop. Res.* 17:668–677.

Seliktar D, Black RA, Vito RP, Nerem RM. 2000. Dynamic mechanical conditioning of collagen-gel blood vessel constructs induces remodeling *in vitro. Ann. Biomed. Eng.* 28:351–362.

Smirzai JA. 1974. The concept of the chondron as a biomechanical unit. In: *Biopolymer und Biomechanik von Bindegewebssystemen.* F Hartmann, ed. Academic Press, Berlin, pp. 87.

Stefanovic-Racic M, Morales TI, Taskiran D, McIntyre LA, Evans CH. 1996. The role of nitric oxide in proteoglycan turnover by bovine articular cartilage organ cultures. *J. Immunol.* 156:1213–1220.

Stockwell RA. 1987. Structure and function of the chondrocyte under mechanical stress. In: *Joint Loading: Biology and Health of Articular Structures.* HJ Helminen, I Kiviranta, M Tammi et al., eds. Wright and Sons, Bristol, pp. 126–148.

Sumpio BE, Banes AJ, Buckley M, Johnson G, Jr. 1988. Alterations in aortic endothelial cell morphology and cytoskeletal protein synthesis during cyclic tensional deformation. *J. Vasc. Surg.* 7:130–138.

Temenoff JS, Mikos AG. 2000. Review: tissue engineering for regeneration of articular cartilage. *Biomaterials* 21:431–440.

Thoumine O, Ziegler T, Girard PR, Nerem RM. 1995. Elongation of confluent endothelial cells in culture: the importance of fields of force in the associated alterations of their cytoskeletal structure. *Exp. Cell Res.* 219:427–441.

Torzilli PA, Grigiene R. 1998. Continuous cyclic load reduces proteoglycan release from articular cartilage. *Osteoarthritis Cartilage* 6:260–268.

van Campen GPJ, van de Stadt RJ. 1987. Cartilage and chondrocytes responses to mechanical loading *in vitro.* In: *Joint Loading: Biology and Health of Articular Structures.* HJ Helminen, I Kiviranta, M Tammi et al., eds. Wright and Sons, Bristol, pp. 112–125.

Vandenburgh HH. 1982. Dynamic mechanical orientation of skeletal myofibers *in vitro. Dev. Biol.* 93:438–443.

von der Mark K, Mollenhauer J. 1997. Annexin V interactions with collagen. *Cell. Mol. Life Sci.* 53:539–545.

von Eisenhart R, Adam C, Steinlechner M, Muller-Gerbl M, Eckstein F. 1999. Quantitative determination of joint incongruity and pressure distribution during simulated gait and cartilage thickness in the human hip joint. *J. Orthop. Res.* 17:532–539.

Wakitani S, Goto T, Pineda SJ, Young RG, Mansour JM, Caplan AI, Goldberg VM. 1994. Mesenchymal cell-based repair of large, full-thickness defects of articular cartilage. *J. Bone Joint Surg. Am.* 76:579–592.

Wong M, Wuethrich P, Buschmann MD, Eggli P, Hunziker E. 1997. Chondrocyte biosynthesis corre-lates with local tissue strain in statically compressed adult articular cartilage. *J. Orthop. Res.* 15:189–196.

Wong M, Wuetrich P, Eggli P, Honziker EB 1996. Zone-specific cell biosynthetic activity in mature bovine articular cartilage: a new method using con-focal microscopic stereology and quantitative au-toradiography. *J. Orthop. Res.* 14:424–432.

22

Regulation of Cellular Response to Mechanical Signals by Matrix Design

Craig A. Simmons and David J. Mooncy

Introduction

Engineering replacement tissues and organs is a promising treatment strategy for the millions of patients who suffer tissue loss or end-stage organ failure annually, but cannot be treated by therapies such as transplantation (Langer and Vacanti, 1993). Although significant progress has been made over the past 20 years in the growth of tissues using, for instance, cells transplanted into three-dimensional biodegradable matrices, the mechanical properties of many engineered tissues remain inferior to those of native tissues (Cao et al., 1994; Carver and Heath, 1999; Kim et al., 1999a; Mauck et al., 2000; Niklason et al., 1999), presently limiting their clinical utility.

Appropriate mechanical properties are particularly important for tissues that support mechanical loads *in vivo*, including bone, cartilage, muscle, and vasculature. These tissues not only support mechanical loading, but are also responsive to mechanical forces during tissue development and maintenance. Because mechanical stimulation plays a critical role in natural tissue development, several investigators have attempted to improve the functionality of engineered tissues by mimicking the *in vivo* environment through application of external mechanical loads to tissues *in vitro*. This strategy has been moderately successful, resulting in improved strength and patency of small diameter blood vessels (Niklason et al., 1999), increased compressive modulus of articular cartilage (Carver and Heath, 1999; Mauck et al., 2000), and increased tensile modulus and strength of smooth muscle tissue (Kim et al., 1999a). In all of these cases, however, the properties of the engineered tissues were inferior to those of native tissues, suggesting that further study of the mechanisms by which mechanical forces regulate engineered tissue formation is warranted.

Tissue formation is regulated by several factors in the microenvironment of cells, including the adhesion of cells to their extracellular matrix (ECM) and to other cells, and chemical signals from growth factors, for example. In the absence of external mechanical stimuli, cell-matrix interactions regulate gene expression and ultimately, tissue development and function (Streuli, 1999). The ECM also serves as depot for growth factors that are released to the surrounding cells, thereby influencing their function and tissue formation (Streuli, 1999). Because the matrix plays a significant role in regulating tissue development, defining the roles of the matrix and its interaction with cells are critical to designing tissue engineering matrices that present the signals necessary for appropriate gene expression and functional tissue development.

In this review, we discuss how cell function is regulated by the properties of its matrix, with particular focus on the regulation of the cell's response to mechanical signals. In the first section, we consider the role integrins play in transducing external mechanical signals from the matrix to the cells and discuss how tissue engineering matrices may be designed to regulate the effects of mechanical forces on cell function. Our discussion is limited to mechanical effects caused by substrate or matrix deformation and does not consider effects due to other types of mechanical loads, such as hydrostatic pressure or fluid shear. In the second section, we discuss the role of the intrinsic mechanical properties of the matrix in regulating cell function. In particular, we examine how the stiffness of the matrix can dictate the fate and function of an adherent cell. Again, the implications for the design of tissue engineering scaffolds are discussed, with focus on novel hydrogels, materials that have wide utility in biomedical applications, including tissue engineering. In the third section, we consider the effect of external mechanical forces in regulating growth factor release from tissue engineering scaffolds and the implications for engineering tissues in mechanically dynamic environments. Finally, we conclude by summarizing and considering the challenges and directions for the future.

Role of Cell-Matrix Interactions in Cellular Response to External Mechanical Signals

Integrins and Cell Adhesion

The adhesion and interaction of cells with their ECM is mediated by cell-surface receptors. An important and well-studied class of these receptors is the integrins, a family of transmembrane heterodimeric glycoproteins that provide a physical linkage between the ECM, the cell surface, and the intracellular cytoskeleton (Hynes, 1992; Ingber, 1991; Juliano and Haskill, 1993). Integrins consist of α and β subunits, and specific combinations of these subunits interact with specific ECM proteins. The interactions are promiscuous, however, with an individual integrin often binding several distinct ECM protein ligands and individual ligands often being recognized by more than one integrin (Hynes, 1992). The ligand is typically a relatively short amino acid sequence on the ECM molecules. For instance, the integrin recognition site Arg-Gly-Asp (RGD) is found in fibronectin, vitronectin, and a number of other ECM proteins. Ligand occupation of activated integrin receptors can trigger a number of intracellular signals, often in concert with other receptors, such as those for growth factors (Giancotti and Ruoslahti, 1999; Howe et al., 1998; Hynes, 1992). The responses induced by integrin-mediated cell activation include adhesion, migration, proliferation, secretion, and morphological changes.

The functional significance of the multiplicity of integrins expressed by a cell and the apparent redundancy of multiple integrins binding the same ligand is that individual integrins may have discrete intracellular functions, transducing distinct signals from the extracellular matrix to the cell interior (Giancotti and Ruoslahti, 1999; Howe et al., 1998; Hynes, 1992). Within the context of tissue engineering, the functional consequence of specific integrin-ligand interactions is evident from cell behavior on different scaffold materials. For instance, vascular smooth muscle cells (SMCs) cultured on polyglycolic acid (PGA) scaffolds or two-dimensional PGA films exhibit higher rates of elastin production, but lower rates of collagen production, than SMCs on type I collagen sponges or films (Kim et al., 1999b). This effect is likely due to the specific ligands the SMCs use to adhere to the scaffold materials, since the proteins adsorbed to PGA scaffolds (primarily vitronectin) differ from those adsorbed to type I collagen sponges (both vitronectin and fibronectin) (Kim et al., 1999b). ECM production by chondrocytes is also dependent on the scaffold material, with enhanced proteoglycan synthesis and reduced collagen synthesis on PGA scaffolds compared with collagen sponges (Grande et al., 1997). As with SMCs, differences in ECM production by chondrocytes on different scaffolds may be due to the interactions of the cells with specific proteins adsorbed to the different materials, although this issue has not been addressed directly with the chondrocyte system.

Integrins and Mechanotransduction

Integrins also act as mechanotransducers, conveying mechanical signals from the ECM to within the cell to affect cell function and gene expression (Ingber, 1991; Shyy and Chien, 1997). This has been demonstrated directly by controlled twisting of magnetic microbeads bound to integrins on endothelial cells, which alters gene transcription only when the ligand binding site is occupied (Meyer et al., 2000), and indirectly by competitive inhibition of specific integrins, which blocks mechanically activated intracellular signaling pathways (MacKenna et al., 1998) and limits shear stress-induced vasodilation of coronary arterioles (Muller et al., 1997). The mechanisms by which integrins transduce mechanical signals is an active area of research. There is evidence that integrins can function as traditional receptors, mediating chemical signaling pathways that may intersect with growth factor receptor signaling pathways (Giancotti and Ruoslahti, 1999; Howe et al., 1998; Hynes, 1992; Schwartz and Ingber, 1994). Additionally, integrins transmit extracellular signals directly through the cellular cytoskeleton as evidenced by micromanipulation of microbeads bound to integrins on endothelial cells, which causes cytoskeletal filament reorganization and nuclear distortion, mediated by direct mechanical connections between integrins, cytoskeletal filaments, and nuclear scaffolds (Maniotis et al., 1997). Thus, mechanical signals can be transferred directly from integrins to distinct structures in the cell and nucleus, independent of chemical signaling pathways, although complementary interactions between mechanical- and chemical-mediated signaling pathways likely exist (Giancotti and Ruoslahti, 1999; Schwartz and Ingber, 1994).

Interestingly, and importantly for the design of tissue engineering matrices, the effects of external mechanical stimuli on cell function are significantly dependent on the specific integrins cells used to adhere to specific ECM proteins. For instance, cyclic straining of two-dimensional cultures of SMCs in serum-free media increases cell proliferation when the substrate is coated with fibronectin (Kim et al., 1999a; Wilson et al., 1995) or vitronectin (Wilson et al., 1995), but not laminin (Kim et al., 1999a; Wilson et al., 1995), elastin (Wilson et al., 1995), or type I collagen (Kim et al., 1999a). When serum-containing media is used, however, strain increases

cell proliferation on collagen-coated substrates. With serum present, fibronectin and vitronectin from the serum are able to adsorb onto the substrate (Nikolovski and Mooney, 2000; Kim et al., 1999a), apparently enabling the appropriate cell-matrix interactions necessary for mechanically regulated cell proliferation. An entirely different matrix dependency is observed when cell differentiation is examined. SMCs cultured on laminin increase their smooth muscle myosin expression (a marker for SMC differentiation and maturation) substantially in response to strain, whereas on fibronectin the differentiation response is muted (Reusch et al., 1996). Thus, specific matrix proteins appear to be responsible for signaling specific responses to strain. Furthermore, the strain-induced mitogenic response is inhibited by antibodies to β_3 and $\alpha_v\beta_5$, but not β_1, integrins (Wilson et al., 1995). Although SMCs utilize a variety of integrins to bind ECM proteins, the β_1 integrin is a key mediator of SMC adhesion, and hindering its action with a blocking antibody inhibits cell adhesion to type I collagen, fibronectin, and vitronectin (Nikolovski and Mooney, 2000). The observation that blocking β_1 integrin function does not inhibit the strain-induced mitogenic response in SMCs, but does inhibit adherence, suggests that specific integrins are responsible for transducing mechanical signals, and the cell-matrix interactions necessary for mechanotransduction may differ from those required for cell adhesion.

The dependence of strain-induced cell responses on cell-matrix interactions has also been demonstrated in three-dimensional tissue cultures using tissue engineering scaffolds. SMCs seeded on fibronectin-coated polyglycolide (PGA) scaffolds in serum-free media respond to cyclic strain with increased proliferation relative to those on nonstrained control scaffolds (Kim et al., 1999a) (Fig. 22.1A). On type I collagen sponges, however, there is no change in cell number with strain (Fig. 22.1A). Furthermore, the production rates of elastin and collagen by the SMCs increase with strain only with fibronectin-coated PGA scaffolds and not with collagen sponges (Fig. 22.1B,C). Thus, the mechanical responsiveness of the SMCs is influenced profoundly by the surface chemistry presented by

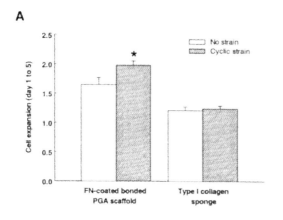

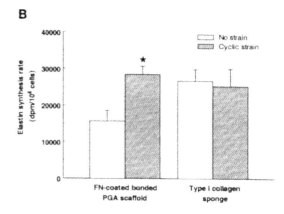

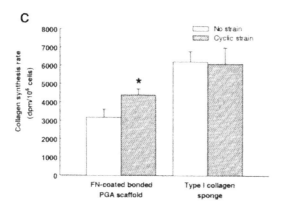

FIGURE 22.1. SMC response to mechanical stimulation in serum-free media. SMCs cultured on fibronection-coated bonded PGA scaffolds in serum-free media and subjected to cyclic strain (7% amplitude, 1 Hz) for four days show increased (*A*) proliferation, (*B*) elastin production, and (*C*) collagen production compared with those cultured under static conditions. On type I collagen sponges in serum-free media, however, SMCs do not respond to the application of cyclic strain. An asterisk indicates significant differences between strained and unstrained conditions ($p < 0.05$). After Kim et al. (1999a).

the matrix. In the presence of serum-containing media, however, cell number and elastin and collagen production increase in response to cyclic strain with *both* PGA and collagen scaffolds (Fig. 22.2A–C). As with two-dimensional collagen-coated substrates, fibronectin and vitronectin from the serum can adsorb to PGA scaffolds and collagen sponges (Fig. 22.2*D*), presumably thereby permitting the SMCs on both matrices to respond to the applied mechanical strain.

The long-term effects of cyclically straining SMC-seeded collagen sponges in serum-containing media for 20 weeks include higher cell density, increased elastin and type I collagen

mRNA levels, increased cellular alignment, and increased modulus and strength of the engineered tissue relative to control tissues (Kim et al., 1999a). These effects appear to be attributable directly to transduction of the cyclic deformation of the substrate through appropriate serum-derived ECM proteins and cell surface receptors. If secondary effects related to induced fluid flow (e.g., enhanced nutrient transport, fluid shear forces) had influenced the biological response, one would have expected no differences between the SMCs on collagen sponges subjected to cyclic strain and those in control (unstrained) cultures, since the external fluid flow was similar in each condition (Kim et al., 1999a).

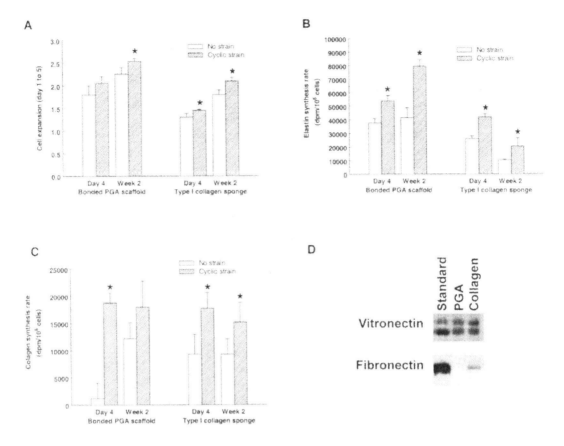

FIGURE 22.2. SMC response to mechanical stimulation in serum-containing media. In the presence of 10% vol/vol serum-containing media, SMCs cultured on both bonded PGA scaffolds and type I collagen sponges respond to cyclic strain (7% amplitude, 1 Hz) with increased (*A*) proliferation, (*B*) elastin production, and (*C*) collagen production compared with those cultured under static conditions. In the presence of serum-containing media, fibronectin and vitronectin absorb to PGA and collagen scaffolds, as shown in (*D*), a Western blot of adsorbed proteins stripped from serum-exposed PGA scaffolds and type I collagen sponges. The adsorption of fibronectin and vitronectin from the serum onto the matrices presumably permitted the SMCs to respond to the applied mechanical strain. It should be noted that fibronectin was seen as a faint band in PGA samples at higher exposure than shown here, confirming its presence. An asterisk indicates significant differences between strained and unstrained conditions ($p < 0.05$). After Kim et al. (1999a).

Implications for the Design of Tissue Engineering Matrices

The evidence that cells respond to mechanical forces in a matrix-dependent manner, with integrin-mediated mechanotransduction playing a major role, has important implications for the design of tissue engineering matrices. This issue will likely be particularly important for tissues in which mechanical stimulation is required for functional tissue formation. These data imply that mechanically regulated tissue formation is dependent not only on the characteristics of the applied external loads, but also on how the ECM modulates the response of cells to mechanical signals. Therefore, in order to present the cells with the appropriate signals to guide gene expression and tissue formation, one needs to design biomaterials that present the appropriate ligands required to elicit the desired response.

Incorporation of specific ligands into biomaterials, thereby making the material biologically recognizable or bioactive, has been the subject of extensive research. One strategy has been to incorporate adhesion-promoting oligopeptides into the surfaces of biomaterials (Hubbell, 1999). With this approach, greater control of cell adhesion can be achieved since adhesion is mediated by the specific peptides presented, rather than relying on nonspecific proteins adsorbed onto the material surface, as is the case with many traditional biomaterials. The peptides used are based on the primary structure of the receptor-binding domains of proteins such as fibronectin and laminin (Hubbell, 1999). Using short adhesion peptides, rather than the complete parent protein, has the advantage that nearly all the short peptides can be made available and active for binding (Massia and Hubbell, 1991). Furthermore, short peptides may provide higher levels of specificity and selectivity than complete parent proteins, which contain multiple adhesion-promoting domains cannot (Hubbell, 1999). This approach has been used with practical biomaterials for tissue engineering applications. For example, poly(lactic acid-co-lysine) (PLAL), a synthetic resorbable copolymer, improved cell spreading when modified with covalently attached RGD-containing peptides as compared with unmodified or negative control polymers (Cook et al., 1997). Similarly, three-dimensional agarose gels modified by covalent linking with the laminin-derived oligopeptide sequence CDPGYIGSR enhanced neurite outgrowth *in vitro* and *in vivo* compared with negative control gels (Borkenhagen et al., 1998).

In addition to modifying biomaterials to promote cell adhesion and related events, in some cases it may also be desirable to block nonspecific binding, thereby preventing undesired cellular responses. One practical biomaterial that has the potential in this regard is alginate. Alginates are naturally derived polysaccharides and have been used extensively as synthetic ECMs for cell immobilization, cell transplantation, and, most recently, tissue engineering (Jen et al., 1996; Rowley et al., 1999). Unlike many polymer systems, however, alginate and other hydrogels are hydrophilic and discourage protein adsorption and subsequent cell adhesion. Thus, alginate provides a "blank slate" that can be modified to present cells with the desired ligands, while inhibiting nonspecific interactions. This makes alginate an attractive model system with which to

specific cell-material interactions may be controlled precisely through chemical modifications.

Alginate hydrogels can be chemically modified by covalent coupling of cell adhesion ligands using aqueous carbodiimide chemistry, with reaction efficiencies greater than 80% (Rowley et al., 1999). Furthermore, the ligand density can be varied by several orders of magnitude. By presenting specific ligands and enabling specific cell-matrix interactions, this system has the potential to allow for control of gene expression of the cells within the matrices. This has been demonstrated with mouse skeletal myoblasts cultured on unmodified alginate surfaces and alginate surfaces coupled with GRGDY peptides (Rowley et al., 1999). On unmodified and negative control surfaces, no cell adhesion was observed. However, myoblasts on GRGDY-modified surfaces adhered, proliferated, and differentiated. The specificity of the interaction was demonstrated through inhibition of cell adhesion and spreading on the modified alginate by addition of soluble ligand to the medium. Furthermore, the rates of proliferation and differentiation were dependent on not only the presence of a binding ligand, but also the density of the ligand (Rowley and Mooney, 1999). A similar effect on differentiation was observed with pre-osteoblast (MC3T3) cells cultured in alginate beads, a model three-dimensional culture system. The cells in the RGD-modified alginate beads differentiated more rapidly than did those in the unmodified alginates, suggesting that the surface chemistry directed cell differentiation through integrin-mediated pathways (Alsberg et al., 2001)

The ability to control cell-matrix interactions in a specific and determined manner also makes alginate a promising system with which to study the role of cell-matrix interactions in transducing mechanical forces. As reviewed above, mechanical forces influence gene expression and tissue formation in an ECM-dependent manner. With traditional materials, however, protein adsorption and nonspecific cell interactions can inhibit the ability to direct mechanically regulated tissue formation precisely. Alginate and other biomaterials that do not interact specifically with mammalian cells provide a blank slate on which various ligands may be bound, thus providing an excellent model system to study how mechanical forces are transduced through specific integrin-mediated pathways. Furthermore, alginate is a practical, clinically relevant biomaterial, and the mechanistic studies on force transduction may have direct application to *in vivo* situations.

Intrinsic Mechanical Properties of the Matrix as a Design Parameter

Matrix Stiffness and Cellular Behavior

Although the chemical signals provided by tissue engineering matrices are clearly important, including in the transduction of external mechanical signals, the intrinsic mechanical properties of the matrix may play as important a role in directing gene expression and functional tissue formation. In particular, the stiffness of the material may regulate the ability of the matrix to resist cell-based tractional forces, thereby mediating changes in the force balance between the cell and matrix, and within the cell. Many cell types, when adhered to a substrate, are able to exert tractional forces that are large enough to contract a compliant substrate (Harris et al., 1980). The magnitudes of the tractions generated by fibroblasts have been estimated to be between 0.2–10 nN/μm^2 (Galbraith and Sheetz, 1997; Pelham and Wang, 1999), although variations exist between different cell types (Galbraith and Sheetz, 1997; Harris et al., 1980; Lee et al., 1994). On compliant substrates that cannot resist the tractional forces generated by the adherent cells, cells retract and round. In contrast, on rigid substrates that are stiff enough to resist the cell-generated tension, integrin-cytoskeleton linkages strengthen (Choquet et al., 1997), the cytoskeleton stiffens (Wang et al., 1993), and cells are able to spread. Thus, there is a force balance between the tension generated by the cytoskeleton of a cell and the resistance provided by the cell's adhesion to its matrix that determines the shape of the cell. Since cell shape correlates with several cellular behaviors, including growth (Chen et al., 1997; Folkman and Moscona, 1978; Ingber, 1990; Ingber and Folkman, 1989; Mooney et al., 1992, differentiation

(Mooney et al., 1992), apoptosis (Chen et al., 1997), motility (Pelham and Wang, 1997), and signaling [reviewed in (Chicurel et al., 1998; Huang and Ingber, 1999)], the fate of a cell may be regulated by the stiffness of the substrate to which it is adhered.

The role of substrate mechanics in regulating cell phenotype and function has been demonstrated in a number of model cell culture systems. In general, cells express tissue-specific functions and fail to proliferate when attached to compliant substrates, but tend to grow without differentiation on more rigid substrates (Ingber et al., 1994). For instance, epithelial cells on compliant substrates express the fully differentiated phenotype (Opas, 1989, 1994; Streuli and Bissell, 1990) and show elevated lamellipodial movement (Pelham and Wang, 1997), and fibroblasts on compliant substrates show reduced spreading and increased motility, compared with those on rigid substrates (Pelham and Wang, 1997). This ability of cells to respond to the stiffness of their substrate may play a critical role in regulating tissue patterning and remodeling (Huang and Ingber, 1999).

Regulation of cell phenotype by matrix mechanics has also been demonstrated using alginate, a practical material for tissue engineering applications. By modifying the ratio of the monomeric subunits of alginate, one can alter the extent of cross-linking and the mechanical properties of the hydrogel (further details are provided in the next section). Furthermore, as discussed in the previous section, alginate can be modified covalently with cell adhesion ligands (Rowley et al., 1999), and therefore both the mechanics and the surface chemistry of the matrix can be altered. Skeletal myoblasts cultured on RGD-coupled alginates with high compressive modulus show increased proliferation and differentiation compared with those on medium and low stiffness gels (Rowley and Mooney, 1999). Importantly, the matrix stiffness effect is coupled to the peptide density, with enhanced proliferation and differentiation on high modulus gels with high ligand densities. On low modulus gels, however, increasing the ligand density does not promote differentiation. Therefore, skeletal myoblast differentiation with this system requires a combination of high substrate stiffness and high ligand density. These conditions differ from those found to promote differentiation in other cell types. For example, epithelial cells differentiate only on compliant substrates (Opas, 1989, 1994; Streuli and Bissell, 1990), endothelial cells differentiate at intermediate adhesion ligand densities (Ingber and Folkman, 1989), and hepatoctyes lose differentiated functions with increasing ligand density (Mooney et al., 1992). Together, these findings suggest that the effect of the matrix properties on cell function may be cell type dependent, with specific ranges of responsiveness for different cell types. For instance, the response of different cell types to substrates of varying stiffnesses may be dictated by the tractional force the cells are able to generate, the magnitude of which is cell-type dependent (Galbraith and Sheetz, 1997; Harris et al., 1980; Lee et al., 1994).

Implications for the Design of Tissue Engineering Matrices

The ability of the stiffness of the adhesion matrix to control cell phenotype has important implications for the design of tissue engineering matrices. Controlling material stiffness may allow one to regulate several cell responses, in essence tailoring the material properties to elicit the desired cell response. This novel approach to biomaterial design requires processing methods be developed to alter the mechanical properties of tissue engineering matrices in a controlled manner.

As discussed previously, the favorable characteristics of alginate, including the ability to control its interaction with cells, make it a promising biomaterial for tissue engineering applications. Controlling the mechanical properties of alginate hydrogels precisely is difficult, however. Typically, alginate hydrogels are formed by ionic cross-linking with divalent cations. The mechanical properties of ionically cross-linked alginates are dependent on the nature of the alginate, specifically the ratio of the mannuronic acid (M) and guluronic acid (G) monomeric subunits that comprise the linear polysaccharide. The range of properties available from ionically cross-linked alginate hydrogels is limited, however. For instance, decreasing the M:G ratio from 70:30 to 40:60 increases the compressive modulus only

five-fold from 12 kPa to 62 kPa (Rowley and Mooney, 1999). Furthermore, the mechanical properties change in an uncontrollable manner due to ion exchange between the binding divalent cations and monovalent cations in the surrounding fluid (Shoichet et al., 1996).

These limitations of alginate have motivated efforts to developed biomaterials that maintain the favorable characteristics of alginate, but permit more precise control of the mechanical properties. One promising approach is to covalently cross-link alginate with various macromolecules. Poly(ethylene glycol) (PEG)-diamine is one molecule that has been investigated for this application' (Eiselt et al., 1999; Lee et al., 2000b). Covalent cross-linking of alginate with PEG-diamines produces hydrogels with widely varying mechanical properties, and the moduli and swelling properties of the hydrogels can be controlled by varying the chain length of the cross-linking molecule, the cross-linking density, or the weight fraction of the cross-linking molecule in the hydrogel (Eiselt et al., 1999). Further tailoring of the mechanical properties of alginate hydrogels can be achieved with other cross-linking molecules, such as adipic dihydrazide and lysine (Lee et al., 2000b). By using different types of cross-linking molecules and by controlling the cross-linking density, the stiffness and swelling properties can be tightly regulated in an independent manner (Table 22.1). Specifically, the shear modulus of the hydrogel is controlled primarily by the cross-linking density, with moderate dependence on the type of cross-linking molecule, and the swelling behavior is dependent on the hydrophilic/hydrophobic nature of the cross-linking molecule.

Another approach to biomaterial design is to synthesize new polymers that are derived from alginate, thereby maintaining the favorable characteristics (such as the gentle gelling behavior and biocompatibility) while introducing more control over the mechanical properties. This strategy has been used to synthesize poly(aldehyde guluronate) (PAG) hydrogels from sodium poly(guluronate), the portion of the alginate molecule responsible for its gelling behavior (Bouhadir et al., 1999). Oxidation of sodium poly(guluronate) with sodium periodate yields PAG, which can then be cross-linked both covalently with adipic dihydrazide and ionically with calcium, for in-

TABLE 22.1. Characteristics of alginate hydrogels cross-linked with various cross-linking molecules. After Lee et al. (2000b).

Cross-linking molecule	c_{max}[b] (mol %)	G_{max}[c] (kPa)	Q_{max}[d]
Ca^{2+}	96	24.4 ± 0.6	N.A.
Adipic dihydrazide	35	36.9 ± 3.7	43.3 ± 3.4
Lysine	35	31.1 ± 1.8	78.1 ± 3.5
PEG-diamine 1000[a]	15	36.5 ± 3.2	78.3 ± 3.0
PEG-diamine 3400[a]	5	29.8 ± 2.1	197.7 ± 14.7

[a]Number indicates the molecular weight of the PEG.
[b]Concentration of cross-linking molecule at which the hydrogel showed maximum shear modulus.
[c]Maximum shear modulus.
[d]Degree of swelling of the hydrogel in water at the maximum shear modulus.

stance. A wide range of mechanical properties and degradation rates can be achieved by varying the cross-linker concentration, the PAG concentration, the cation concentration, and the degree of PAG oxidation (Bouhadir et al., 1999; Lee et al., 2000a). Additionally, cell adhesion peptides can be coupled to PAG using carbodiimide chemistry or reductive amination, thereby permitting defined cell interactions.

Polyacrylamide is another material that has been used as a model system to investigate the effect of altering substrate stiffness on cell function (Pelham and Wang, 1997, 1999). By varying the relative concentrations of acrylamide and bis-acrylamide, the flexibility of the polyacrylimide substrate can be controlled in a systematic and reproducible manner. Furthermore, polyacrylimide does not interact specifically with cells, and therefore cell-matrix interactions occur only between the cells and any ECM ligands that are coated onto the material. Although useful as a model system, polyacrylimide has not been demonstrated to be a practical biomaterial for tissue engineering applications.

Controlled Growth Factor Release by External Mechanical Stimuli

The ability of the ECM to regulate cell function is not limited to adhesive and mechanical-based interactions between the cells and matrix. Natural ECMs also serve as depots for growth factors that are released to the tissue microenvironment

providing chemical signals to affect cell function and guide tissue formation (Streuli, 1999). Similarly, synthetic and naturally derived polymers have been used in tissue engineering as reservoirs and delivery vehicles for soluble signaling proteins, such as vascular endothelial growth factor (VEGF) (Murphy et al., 2000; Sheridan et al., 2000), and bone morphogenic protein (BMP) (Ripamonti and Reddi, 1997), and DNA (Bonadio et al., 1999, 2000; Richardson et al., 2001; Shea et al., 1999). Although there have been many attempts to develop biomaterials for controlled drug delivery in response to external stimuli such as ultrasound, and electric or magnetic fields (Langer, 1998), little attention has been paid to the effect of external mechanical stimuli on the release of growth factors from polymeric matrices.

Since many natural and engineered tissues experience mechanical stimulation during their development and maintenance, it is important to consider how matrix deformation and increased intramatrix pressure due to mechanical loading may affect growth factor release from the matrix. To that end, the properties of the matrix, including its abilities to undergo repeated deformation and to reversibly bind growth factors, will be critical considerations in the design of tissue engineering matrices that can respond to mechanical stimuli and regulate growth factor release.

Growth Factor Release from Mechanically Loaded Hydrogels

Alginate hydrogels can be used as growth factor carriers and are capable of undergoing repeated

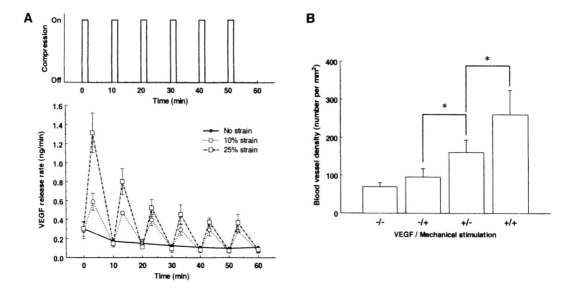

FIGURE 22.3. Growth factor release from mechanically stimulated matrices. (*A*) Compressive loading (2 minutes compression followed by 8 minutes relaxation) of alginate hydrogels containing VEGF causes increased release rates, in a strain amplitude-dependent manner. (*B*) In NOD mice with femoral ligations, blood vessel density was significantly higher with mechanically stimulated alginate gels loaded with VEGF (+/+) than with VEGF-loaded gels that were not mechanically stimulated (+/−). VEGF alone (+/−) increased blood vessel density compared with control gels without VEGF under static conditions (−/−) and mechanically stimulated gels without VEGF (−/+). An asterisk indicates significant differences between experimental conditions ($p < 0.05$). After Lee et al. (2000c).

deformation, making them an excellent model system to investigate mechanically regulated growth factor release. Using this system, the release kinetics of model drug molecules have been investigated *in vitro* under conditions of dynamic mechanical loading (Lee et al., 2000c). Compressive mechanical loading had no effect on the release behavior of trypan blue, a model molecule that does not bind to the hydrogel and therefore is depleted rapidly from the matrix. However, with VEGF, which binds polysaccharides reversibly, compressive loading increased the release rate compared with unloaded gels, and the increase could be modulated by the amplitude of the compression (Fig. 22.3*A*). As a result, the total amount of VEGF released from mechanically stimulated alginate gels was up to double that released from unloaded gels, confirming that external mechanical loading can control the release of growth factors from matrices. Importantly for the design of tissue engi-

neering matrices, the effect of mechanical loading on the growth factor release profile is dependent on the properties of the matrix, including its ability to bind the factor to be delivered and possibly its mechanical properties.

The potential for this mechanism to regulate tissue formation has been demonstrated *in vivo* using alginate hydrogels containing VEGF (Lee et al., 2000c). In severe combined immunodeficient (SCID) mice, increased granulation tissue thickness and vascularization was observed around hydrogels that were mechanically stimulated, as compared with nonstimulated hydrogels and hydrogels without growth factor. Similarly, in nonobese diabetic (NOD) mice with femoral artery ligations, alginate hydrogels containing VEGF enhanced collateral blood vessel formation 14 days postimplantation when mechanically stimulated, as compared with nonstimulated VEGF-containing hydrogels or hydrogels without growth factor (Fig. 22.3*B*).

Implications for the Design of Tissue Engineering Matrices

These results suggest a number of implications for tissue engineering and matrix design. For instance, by designing matrices with mechanical and drug-interaction properties that are tailored for the drug to be delivered and the mechanical environment experienced by the matrix, the delivery dose and rate may be regulated in a precise manner. Furthermore, these findings also indicate a general mechanism by which mechanical stimuli may influence natural tissue development and cell function: Mechanically regulated release of growth factor from natural ECMs may be an important and previously unrecognized means by which mechanical stimuli influence tissue adaptation.

Summary and Future Directions

Despite the promise offered by tissue engineering and the significant advances made in the field, many engineered tissues still do not provide the functionality necessary for clinical applications. Gene expression, tissue development, and, ultimately, tissue function are determined by the chemical and mechanical cues presented to cells within the tissue. Therefore, restoring function to engineered tissues requires a better understanding of how the cellular microenvironment influences tissue formation and function, and the development of strategies to control the microenvironment to guide specific programs of gene expression.

The ECM plays a vital role in regulating the response of cells to signals from their microenvironment through, as reviewed in this chapter, mediation of cell adhesion, regulation of the cellular force balance, and release of growth factors. This is critically important for tissue engineering, since the ability of the matrix to regulate the cellular microenvironment can be exploited by designing matrices that present the appropriate signals necessary to direct cell function to engineer functional tissues. Several design strategies were reviewed in this chapter, including incorporation of specific cell adhesion ligands, manipulation of the intrinsic mechanical properties of the matrix, and control of growth factor release. These approaches will be particularly important to engineer tissues that require mechanical stimulation for proper development, since the cellular response to mechanical signals is regulated by the properties of the matrix.

Future Directions

The design strategies presented in this chapter are promising and in many cases, represent novel approaches to engineering functional tissues. There are several challenges that remain to be addressed, however, before functional, clinically successful engineered tissues can be achieved using these approaches and others. Summarized below are several directions for future research that will address some of the challenges confronting tissue engineers today.

Development of Novel Polymeric Matrices

One clear future direction is further development of polymeric matrices that can control the local microenvironment precisely. Alginate and alginate-derived polymers are promising materials in this regard, since they inhibit nonspecific adhesion and can be modified to present specific ligands and to have specific intrinsic mechanical properties. However, it is likely that alginate-based materials are not ideal for all tissue engineering applications and additional synthetic ECMs that can be customized in a similar manner need to be developed. To that end, it is probable that specific tissues will require unique combinations of adhesion ligands, matrix mechanical properties, and growth factor release kinetics, and therefore there is a critical need to determine the tissue-specific matrix characteristics necessary to guide tissue regeneration. A significant challenge will be to develop "smart" matrices that respond to feedback provided by the cells involved in tissue development, thereby modifying the signals provided by the matrix to the cells as required by the regeneration process. Such a system could be used to regulate the switch from a proliferation program to a differentiation program or from a tissue modeling phase to a tissue remodeling phase, for example. Clearly, the matrix degradation properties will play an important role in this regard.

Matrix-Regulated Mechanotransduction

Designing matrices that regulate the local cellular microenvironment is further complicated when one considers external mechanical stimuli. As discussed in this review, the matrix properties regulate the cell response to external mechanical loading. An immediate aim, therefore, is to determine how various matrix design parameters influence the cell response to mechanical loading. For instance, using the alginate system, one could determine the combined effects of ligand type, ligand density, and material stiffness on mechanotransduction. Again, the matrix properties likely influence cell response to mechanical stimulation in a cell-type dependent manner, and therefore the optimal matrix properties need to be determined on a tissue- and cell-specific basis.

Mechanically Regulated Growth Factor Delivery

The regulation of growth factor delivery from mechanically stimulated tissue engineering matrices is an intriguing concept with a number of potential applications for tissue engineering and drug delivery. The release kinetics under mechanical stimulation are dependent on the intrinsic properties of the matrix and future investigations should consider how delivery profiles can be controlled precisely by manipulating the matrix properties. Again, the desired delivery profiles will be dependent on the intended application, and therefore unique matrix designs will be required to engineer specific tissues.

Cellular Mechanotransduction

Another approach to identifying important matrix design principles is basic science studies on mechanotransduction at the cellular level. An improved understanding of how cells sense and transduce mechanical signals from the matrix to alter cell behavior could potentially have profound implications on how we design tissue engineering matrices, particularly for mechanically stimulated tissues. A significant challenge will be to translate the basic science findings to determine and implement practical matrix design strategies. To that end, cell mechanics studies that utilize biomaterials that are practical for tissue engineering applications may ease the transition.

Mechanistic Studies Using Tissue Engineering Matrices as Model Systems

Finally, synthetic matrices designed for tissue engineering applications also have utility as model systems to investigate the mechanisms of natural biological processes. For instance, the release of growth factors from mechanically stimulated tissue engineering matrices may mimic what occurs with natural ECMs that are stressed mechanically. Therefore, this system may be used to investigate a previously unrecognized mechanism by which tissues may respond to mechanical loading. Additionally, the ability to control a variety of matrix properties, including ligand type, density, and matrix mechanics, allows one to investigate the mechanisms by which cells respond to specific environmental cues. Accordingly, advances in the design of matrix materials that regulate cell function will have important implications not only for improving engineered tissue functionality, but also for our basic understanding of how signals from the environment regulate cell function in natural and pathological tissue development.

References

Alsberg E, Anderson KW, Albeiruti A, Franceschi RT, Mooney DJ. 2001. Cell-interactive alginate hydrogels for bone tissue engineering. *J. Dent. Res.* 80:2025–2029.

Bonadio J. 2000. Tissue engineering via local gene delivery: update and future prospects for enhancing the technology. *Adv. Drug Delivery Rev.* 44:185–194.

Bonadio J, Smiley E, Patil P, Goldstein S. 1999. Localized, direct plasmid gene delivery *in vivo:* prolonged therapy results in reproducible tissue regeneration. *Nat. Med.* 5:753–759.

Borkenhagen M, Clémence J-F, Sigrist H, Aebischer P. 1998. Three-dimensional extracellular matrix engineering in the nervous system. *J. Biomed. Mater. Res.* 40:392–400.

Bouhadir KH, Hausman DS, Mooney DJ. 1999. Synthesis of cross-linked poly(aldehyde guluronate) hydrogels. *Polymer* 40:3575–3584.

Cao Y, Vacanti JP, Ma X, Paige KT, Upton J, Chowanski Z, Schloo B, Langer R, Vacanti CA. 1994. Generation of neo-tendon using synthetic polymers seeded with tenocytes. *Transplant. Proc.* 26:3390–3392.

Carver SE, Heath CA. 1999. Influence of intermittent pressure, fluid flow, and mixing on the regenerative properties of articular cartilage. *Biotechnol. Bioeng.* 65:274–281.

Chen CS, Mrksich M, Huang S, Whitesides GM, Ingber DE. 1997. Geometric control of cell life and death. *Science* 276:1425–1428.

Chicurel ME, Chen CS, Ingber DE. 1998. Cellular control lies in the balance of forces. *Curr. Opin. Cell Biol.* 10:232–239.

Choquet D, Felsenfeld DP, Sheetz MP. 1997. Extracellular matrix rigidity causes strengthening of integrin-cytoskeleton linkages. *Cell* 88:39–48.

Cook AD, Hrkach JS, Gao NN, Johnson IM, Pajvani UB, Cannizzaro SM, Langer R. 1997. Characterization and development of RGD-peptide-modified poly(lactic acid-co-lysine) as an interactive, resorbable biomaterial. *J. Biomed. Mater. Res.* 35:513–523.

Eiselt P, Lee KY, Mooney DJ. 1999. Rigidity of two-component hydrogels prepared from alginate and poly(ethylene glycol)-diamines. *Macromolecules* 32:5561–5566.

Folkman J, Moscona A. 1978. Role of cell shape in growth control. *Nature* 273:345–349.

Galbraith CG, Sheetz MP. 1997. A micromechanical device provides a new bend on fibroblast traction forces. *Proc. Natl. Acad. Sci. U.S.A.* 94:9114–9118.

Giancotti FG, Ruoslahti E. 1999. Integrin signaling. *Science* 285:1028–1032.

Grande DA, Halberstadt C, Naughton G, Schwarz R, Manji R. 1997. Evaluation of matrix scaffolds for tissue engineering of articular cartilage grafts. *J. Biomed. Mater. Res.* 34:211–220.

Harris AK, Wild P, Stopak D. 1980. Silicone rubeer substrata: a new wrinkle in the study of cell locomotion. *Science* 208:177–179.

Howe A, Aplin AE, Alahari SK, Juliano RL. 1998. Integrin signaling and cell growth control. *Curr. Opin. Cell Biol.* 10:220–231.

Huang S, Ingber DE. 1999. The structural and mechanical complexity of cell-growth control. *Nat. Cell Biol.* 1:E131–E138.

Hubbell JA. 1999. Bioactive biomaterials. *Curr. Opin. Biotechnol.* 10:123–129.

Hynes RO. 1992. Integrins: versatility, modulation, and signaling in cell adhesion. *Cell* 69:11–25.

Ingber DE. 1990. Fibronectin controls capillary endothelial cell growth by modulating cell shape. *Proc. Natl. Acad. Sci. U.S.A.* 87:3579–3583.

Ingber D. 1991. Integrins as mechanochemical transducers. *Curr. Opin. Cell Biol.* 3:841–848.

Ingber DE, Folkman J. 1989. Mechanochemical switching between growth and differentiation during fibroblast growth factor-stimulated angiogenesis *in vitro:* role of extracellular matrix. *J. Cell Biol.* 109:317–330.

Ingber DE, Dike L, Hansen L, Karp S, Liley H, Maniotis A, McNamee H, Mooney DJ, Plopper G, Sims J, Wang N. 1994. Cellular tensegrity: exploring how mechanical changes in the cytoskeleton regulate cell growth, migration, and tissue pattern during morphogenesis. *Int. Rev. Cytol.* 150:173–224.

Jen AC, Wake C, Mikos AG. 1996. Review: hydrogels for cell immobilization. *Biotechnol. Bioeng.* 50:357–364.

Juliano RL, Haskill S. 1993. Signal transduction from the extracellular matrix. *J. Cell Biol.* 120:577–585.

Kim B-S, Nikolovski J, Bonadio J, Mooney DJ. 1999a. Cyclic mechanical strain regulates the development of engineered smooth muscle tissue. *Nat. Biotechnol.* 17:979–983.

Kim B-S, Nikolovski J, Bonadio J, Smiley E, Mooney, DJ. 1999b. Engineered smooth muscle tissues: regulating cell phenotype with the scaffold. *Exp. Cell Res.* 251:318–328.

Langer R. 1998. Drug delivery and targeting. *Nature* 392 (Supp):5–10.

Langer R, Vacanti JP. 1993. Tissue engineering. *Science* 260:920–926.

Lee J, Leonard M, Oliver T, Ishihara A, Jacobson K. 1994. Traction forces generated by locomoting keratocytes. *J. Cell Biol.* 127:1957–1964.

Lee KY, Bouhadir KH, Mooney DJ. 2000a. Degradation behavior of covalenty cross-linked poly(aldehyde guluronate) hydrogels. *Macromolecules* 33:97–101.

Lee KY, Rowley JA, Eislet P, Moy EM, Bouhadir KH, Mooney DJ. 2000b. Controlling mechanical and swelling properties of alginate hydrogels independently by cross-linker type and cross-linking density. *Macromolecules* 33:4291–4294.

Lee KY, Peters MC, Anderson KW, Mooney DJ. 2000c. Controlled growth factor release from synthetic extracellular matrices. *Nature* 408:998–1000.

MacKenna DA, Dolfi F, Vuori K, Ruoslahti E. 1998. Extracellular signal-regulated kinase and c-Jun

NH$_2$-terminal kinase activation by mechanical stretch is integrin-dependent and matrix-specific in rat cardiac fibroblasts. *J. Clin. Invest.* 101:301–310.

Maniotis AJ, Chen CS, Ingber DE. 1997. Demonstration of mechanical connections between integrins, cytoskeletal filaments, and nucleoplasm that stabilize nuclear structure. *Proc. Natl. Acad. Sci. U.S.A.* 94:849–854.

Massia SP, Hubbell JA. 1991. An RGD spacing of 440 nm is sufficient for integrin $\alpha_5\beta_3$-mediated fibroblast spreading and 140 nm for focal contact and stress fiber formation. *J. Cell Biol.* 114:1089–1100.

Mauck RL, Soltz MA, Wang CCB, Wong DD, Chao P-HG, Valhmu WB, Hung CT, Ateshian GA. 2000. Functional tissue engineering of articular cartilage through dynamic loading of chondrocyte-seeded agarose gels. *J. Biomech. Eng.* 122:252–260.

Meyer CJ, Alenghat FJ, Rim P, Fong JH-J, Fabry B, Ingber DE. 2000. Mechanical control of cyclic AMP signalling and gene transcription through integrins. *Nat. Cell Biol.* 2:666–668.

Mooney DJ, Hansen L, Vacanti J, Langer R, Farmer S, Ingber D. 1992. Switching from differentiation to growth in hepatocytes: control by extracellular matrix. *J. Cell. Phys.* 151:497–505.

Muller JM, Chilian WM, Davis MJ. 1997. Integrin signaling transduces shear stress-dependent vasodilation of coronary arterioles. *Circ. Res.* 80:320–326.

Murphy WL, Peters MC, Kohn DH, Mooney DJ. 2000. Sustained relase of vascular endothelial growth factor from mineralized poly(lactide-co-glycolide) scaffolds for tissue engineering. *Biomaterials* 21:2521–7.

Niklason LE, Gao J, Abbott WM, Hirschi KK, Houser S, Marini R, Langer R. 1999. Functional arteries grown *in vitro*. *Science* 284:489–493.

Nikolovski J, Mooney DJ. 2000. Smooth muscle cell adhesion to tissue engineering scaffolds. *Biomaterials* 21:2025–2032.

Opas M. 1989. Expression of the differentiated phenotype by epithelial cells *in vitro* is regulated by both biochemistry and mechanics of the substratum. *Devel. Biol.* 131:281–293.

Opas M. 1994. Substratum mechanics and cell differentiation. *Int. Rev. Cytol.* 150:119–137.

Pelham Jr RJ, Wang Y-L. 1997. Cell locomotion and focal adhesions are regulated by substrate flexibility. *Proc. Natl. Acad. Sci. U.S.A.* 94:13661–13665.

Pelham Jr RJ, Wang Y-L. 1999. High reolution detection of mechanical forces exerted by locomoting fi-broblasts on the substrate. *Mol. Biol. Cell* 10:935–945.

Reusch P, Wagdy H, Reusch R, Wilson E, Ives HE. 1996. Mechanical strain increases smooth muscle and decreases nonmuscle myosin expression in rat vascular smooth muscle cells. *Circ. Res.* 79:1046–1053.

Richardson TP, Murphy WL, Mooney DJ. 2001. Polymeric delivery of proteins and plasmid DNA for tissue engineering and gene therapy. *Crit. Rev. Eukaryot. Gene Expr.* 11:47–58.

Ripamonti U, Reddi AH. 1997. Tissue engineering, morphogenesis, and regeneration of the periodontal tissues by bone morphogenetic proteins. *Crit. Rev. Oral. Biol. Med.* 8:154–163.

Rowley JA, Madlambaya G, Mooney DJ. 1999. Alginate hydrogels as synthetic extracellular matrix materials. *Biomaterials* 20:45–53.

Rowley JA, Mooney DJ. 1999. Alginate synthetic ECMs: Controlling myoblast function with chemistry and mechanics. *Society for Biomaterials 25th Annual Meeting Transactions,* 290.

Schwartz MA, Ingber DE. 1994. Integrating with integrins. *Mol. Biol. Cell* 5:389–393.

Shea LD, Smiley E, Bonadio J, Mooney DJ. 1999. DNA delivery from polymer matrices for tissue engineering. *Nat. Biotechnol.* 17:551–554.

Sheridan MH, Shea LD, Peters MC, Mooney DJ. 2000. Bioabsorbable polymer scaffolds for tissue engineering capable of sustained growth factor delivery. *J. Controlled Release* 64:91–102.

Shoichet MS, Li RH, White ML, Winn SR. 1996. Stability of hydrogels used in cell encapsulation: An *in vitro* comparison of alginate and agarose. *Biotechnol. Bioeng.* 50:374–381.

Shyy JY-J, Chien S. 1997. Role of integrins in cellular responses to mechanical stress and adhesion. *Curr. Opin. Cell Biol.* 9:707–713.

Streuli C. 1999. Extracellular matrix remodelling and cellular differentiation. *Curr. Opin. Cell Biol.* 11:634–640.

Streuli CH, Bissell MJ. 1990. Expression of extracellular matrix components is regulated by substratum. *J. Cell Biol.* 110:1405–1415.

Wang N, Butler JP, Ingber DE. 1993. Mechanotransduction across the cell surface and through the cytoskeleton. *Science* 260:1124–1127.

Wilson E, Sudhir K, Ives HE. 1995. Mechanical strain of rat vascular smooth muscle cells is sensed by specific extracellular matrix/integrin interactions. *J. Clin. Invest.* 96:2364–2372.

23

Artificial Soft Tissue Fabrication from Cell-Contracted Biopolymers

Robert T. Tranquillo and Brett C. Isenberg

Introduction

Artificial tissues created by combining cultured cells and polymer scaffolds have the potential to revolutionize health care by providing a supply of tissues on demand (Langer and Vacanti, 1993). While the use of synthetic biodegradable polymer scaffolds has received considerable attention (Atala et al., 1997), an attractive alternative is the tissue-equivalent (TE) a reconstituted type I collagen gel compacted by entrapped tissue cells [traction by cells on surrounding collagen fibrils causes compaction of the entangled fibrillar network and exudation of the interstitial culture medium (Barocas et al., 1995; Bell et al., 1979; Ehrlich et al., 1989; Grinnell and Lamke, 1984; Guidry and Grinnell, 1985, 1986; Harris et al., 1981; Moon and Tranquillo, 1993; Stopak and Harris, 1982; Tranquillo et al., 1992)]. Investigators have fabricated a variety of soft tissues using TE [e.g. skin (Auger et al., 2000; Bell et al., 1981; Coulomb et al., 1998; Eaglstein and Falanga, 1998; Michel et al., 1993; Wilkins et al., 1994), cartilage (de Chalain et al., 1999; Toolan et al., 1996; Schreiber and Ratcliffe, 2000; Weiser et al., 1999), vessel (Bader et al., 2000; Hirai and Matsuda, 1995; Kobashi and Matsuda, 1999; L'Heureux et al., 1993; Weinberg and Bell, 1986; Ye et al., 2000b; Ye et al., 2000a), ligament/tendon (Awad et al., 2000; Butler and Awad, 1999; Huang et al., 1993; Woo

et al., 1999), skeletal muscle (Shansky et al., 1997), myocardium (Eschenhagen et al., 1997), and at least one has Food and Drug Administration (FDA) approval, the skin-equivalent product, Apligraf™, by Organogenesis, Inc. In addition to having FDA approval for therapeutic use like certain synthetic biodegradable polymers,[1] type I collagen also shares the desirable feature of being permissive to host remodeling, that is, being resorbed over time by the action of collagenase and other enzymes. It confers several advantages over synthetic polymers investigated to date: it is a natural cell substrate being the major structural protein in most soft tissues, being conducive to the critical events of cell spreading and binding of many extracellular matrix (ECM) components such as fibronectin; cellularity is achieved directly (because the cells are present in the solution of monomeric collagen when fibrillogenesis is initiated); an appropriate applied mechanical constraint to the compaction yields alignment of collagen fibrils and cells character-

---

[1]As with synthetic polymers (e.g., PLA-PGA), only certain forms of type I collagen are FDA approved for use in clinical devices, such as the injectable form of pepsin-extracted type I collagen from bovine dermis, marketed as Zyderm™ by Collagen Corp. with PMA for long-term implantation.

istic of the tissue (Barocas and Tranquillo, 1997a; Barocas et al., 1998; Grinnell and Lamke, 1984; Huang et al., 1993; Klebe, 1989; Kolodney and Elson, 1993; L'Heureux et al., 1993; Lopez Valle et al., 1992).

It is this last feature, the development of strong alignment that results from providing a mechanical constraint to compaction, which makes the use of TEs[2] most attractive. This follows from two axioms: (1) that native tissue function, particularly mechanical function, depends on structure (cell and fiber alignment) as much as it depends on composition, and (2) that artificial tissue should serve as a functional regeneration template. There is considerable evidence for the first axiom across a wide range of load bearing soft connective tissues, including orthopedic tissues (e.g., ligaments and tendons) and cardiovascular tissues (e.g. arterial media and valve leaflet). In the second axiom, "functional regeneration template" means that in addition to providing necessary function, the artificial tissue should provide alignment cues (template) as it is replaced by cell-derived ECM (regeneration).

The first part of this chapter will review studies that have been aimed at developing a fundamental understanding of the interrelationship between compaction and alignment during the initial compaction phase of tissue-equivalent fabrication. This necessarily includes a summary of topics ranging from collagen gel rheology to cell contact guidance (the tendency for a cell to align with aligned fibrils). The second part will present mechanical issues that arise after the initial compaction phase, which are important to the comprehensive design rationale for artificial tissues that is ultimately desired.

---

[2]While "tissue-equivalent" is only used in the literature to refer to tissue cells cultured within a collagen gel, there are many similarities, whether or not tissue cells are present, between collagen gel and fibrin gel, the latter being of interest because of its potential for stimulating ECM synthesis as occurs in wound healing. Thus, studies using fibrin gel are considered as well, and usage of "gel" or "tissue-equivalent" without explicit reference to collagen or fibrin applies to either.

Initial Compaction Phase of Tissue-Equivalent Fabrication

The initial compaction phase has been relatively well studied as compared with the subsequent phases of TE fabrication. Since this phase involves cells exerting traction on a collagen or fibrin gel, understanding the compaction and associated alignment requires a knowledge of the microstructural and rheological properties of collagen and fibrin gels. Following a summary of these properties, studies aimed at characterizing and modeling the initial compaction and alignment are reviewed.

Collagen gel is typically reconstituted from a solution of pepsin-digested or acid-extracted type I collagen by restoring physiological pH and temperature, which induces self-assembly (end-to-end and lateral) of collagen molecules. Fibrin gel is typically prepared by combining fibrinogen and thrombin solutions containing calcium ions; thrombin-mediated cleavage of fibrinopeptides from fibrinogen leads to self-assembly in a similar manner. As seen for the case of collagen gel in Figure 23.1, a highly hydrated network (only 0.1–0.5 wt% protein) of long, highly entangled fibrils results, with the appearances being very similar for both gels in scanning and transmission electron microscopy (Allen et al., 1984; Muller et al., 1984). The effective "pore size" is about 1 μm, which along with the fibril diameter (0.05–0.5 μm) is small compared with the cell body dimension (50–100 μm) but comparable to that of tip of a pseudopod. Tractional structuring causes consolidation of collagen fibrils around each cell (discussed further in the next section), so even before compaction leads to a significant reduction in pore size throughout the gel, pore size is rapidly reduced around the cells. This may not be true in fibrin gels if fibrinolysis is not inhibited.

The similarity between collagen and fibrin gels extends to their mechanical behavior, the manner in which they respond to an applied force, which is determined by the interaction of their two component phases: the fibrillar network and the interstitial solution, typically tissue culture medium. The network effectively resists shear and extension but has little com-

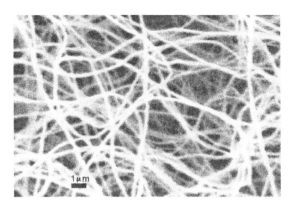

FIGURE 23.1. Scanning electron micrograph of a collagen gel. See text for description.

pressive stiffness because fibrils buckle easily. Significant resistance to interstitial flow of the solution through the network (inversely related to the network permeability) can, however, lead to high solution pressures that impart compressive stiffness to the gel. While this is not of significance for TEs due to the very small force exerted by the cells on the network via cell traction, it is relevant when characterizing the rheology of the network, which is critical to interpreting the compaction of TEs. Such a study has recently been accomplished for collagen gel in the case of an applied confined compression.

The results of this study indicate that qualitatively the network flows like a viscoelastic fluid (Knapp et al., 1997), as it does in response to an applied shear (Barocas et al., 1995), although the values of the shear modulus and the viscosity depend on whether the deformation is compressive or shearing. (These results were obtained in an acellular collagen gel so only apply to tissue-equivalents during their initial incubation before significant cell-induced compaction of the collagen and cell alteration of the collagen network by secretion or proteolysis.) An important consequence of fluidlike behavior is that a stress in the network is completely relaxed with sufficient time. This stress relaxation property proves to be useful in elucidating the nature of the contact guidance signal in a gel with aligned fibrils (see the section titled, "Confined Compression of Tissue-Equivalents; Direct Evidence for Strain-Based Contact Guidance"). The network of a fibrin gel similarly exhibits viscoelas-

tic fluid behavior in shear (Bale et al., 1985) and compression (D. Knapp and R. Tranquillo, unpublished work), which is not surprising given the similar protein concentration and fibrillar network microstructure.

Cell Traction and Compaction of Tissue-Equivalents

Traction exerted by tissue cells on the fibrillar network of the gel is manifested as macroscopic gel compaction. Pioneering work in the documentation of traction *in vitro*, and its morphogenetic implications *in vivo* was performed by Harris and co-workers. They documented the extensive consolidation of collagen fibrils that occurred over time around cells dispersed in collagen gel, termed tractional structuring (Harris et al., 1981; Stopak and Harris, 1982). The macroscopic manifestation of this phenomenon was documented earlier by Bell and co-workers in their seminal fibroblast-populated collagen lattice (FPCL) assay (Bell et al., 1979). They reported that fibroblasts cultured in a small floating disk of collagen gel dramatically compacted the gel. (This will be termed *free compaction,* to distinguish it from cases where a mechanical constraint to compaction is imposed, such as the periphery of the disk being anchored.) Notably, no observations were reported about any cell alignment in the FPCL. Our detailed analysis of freely compacting spheres of collagen and fibrin gel populated uniformly with fibroblasts (the spherical analogue of the FPCL assay) shows that cell orientation is random, that is, there is no macroscopic alignment (Bromberek et al., in press).

Grinnell and co-workers conducted a series of investigations to elucidate the nature of tractional structuring during initial compaction (Grinnell and Lamke, 1984; Guidry and Grinnell, 1985, 1986). Several important observations and conclusions were made: fibrils in the gel interior are rearranged even when the cells reside only at the gel surface, and disrupting the network connectivity inhibits compaction, both implying the transmission of traction force through a connected fibrillar network; only 5% of the network is degraded even though the gel

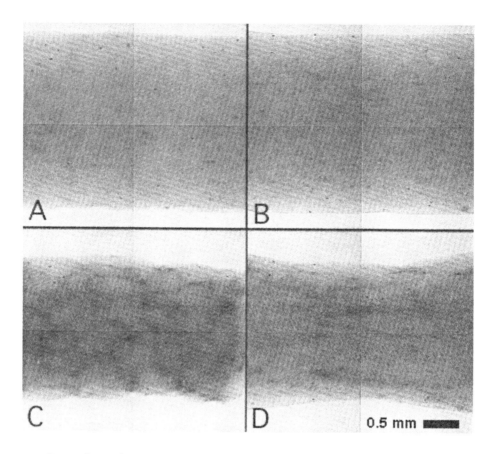

FIGURE 23.2. Comparison of compaction and alignment in free and constrained compaction. (*A*) Initially isotropic, uncompacted free-floating collagen gel. (*B*) Initially isotropic, uncompacted collagen gel that is attached to platens at both ends (not visible). (*C*) Compacted free-floating collagen gel, showing initial isotropic orientation is retained. (*D*) Compacted collagen gel attached at both ends (i.e., compaction is constrained in the axial direction), showing fibril alignment develops in the axial direction.

volume may decrease by 85% or more, implying compaction involves primarily a consolidation of existing fibrils rather than their degradation and replacement; few covalent modifications of the collagen occur; a partial reexpansion of compacted gel occurs after treatment of the cells with cytochalasin D or removal of the cells with detergent; and cell-free gel compacted under centrifugal force exhibits a partial reexpansion upon removal of the force similar to a fibroblast-compacted gel. Based on these observations, a two-step mechanism was hypothesized for the mechanical stabilization of collagen fibrils during gel compaction: cells pull collagen fibrils into proximity via traction-exerting pseudopods, and over a longer time scale, the fibrils become non-

covalently cross-linked, independent of cell-secreted factors.

There have been numerous observations of "spontaneous" cell alignment in TEs that involve a mechanical constraint to compaction:

- along the long axis of a rectangular slab constrained along its four sides (Klebe, 1989)
- along the axis of a thin rectangular slab constrained at opposite sides (Kolodney and Elson, 1993)
- along the axis of a cylindrical rod constrained at its two ends (Fig. 23.2) (Knapp et al., 1999)
- in the plane of a disk constrained around its periphery (Lopez Valle et al., 1992)

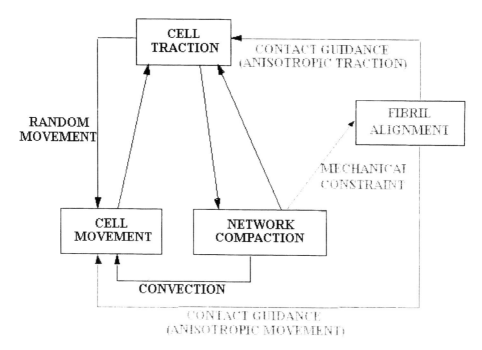

FIGURE 23.3. Self-organizing biomechanical feedback mechanisms in tissue-equivalents. Contact guidance generates biomechanical feedback that affects subsequent gel compaction due to cell traction and associated fibril alignment.

- parallel to the base of a hemisphere constrained at its base (Tuan et al., 1996), even if cells were initially seeded only on the upper or lower surface of the hemisphere (Grinnell and Lamke, 1984)
- circumferentially in a sphere when cells were initially excluded from the core, but radially when cells were initially only included in the core (Bromberek et al., 2002); in these cases the acellular regions of the gel provide the mechanical constraint
- circumferentially in a tube compacting around a mandrel that constrained radial but not axial compaction, but axially in the case of an adherent mandrel that constrained axial as well as radial compaction (L'Heureux et al., 1993)

These observations can be explained and thereby unified by the anisotropic biphasic theory (ABT) of tissue-equivalent mechanics (Barocas and Tranquillo, 1997a). It assumes that cells align, exert traction, and migrate preferentially in the direction that surrounding fibrils are aligned (i.e., exhibit contact guidance in response to aligned

fibrils) and that deformations of the fibrillar network (induced by cell traction in these observations) that create an anisotropic strain cause fibril alignment. Specifically, fibrils align along an axis of extension and, being the more relevant case for tissue-equivalent fabrication, align normal to an axis of compression. The proposed interplay between cell traction, fibrillar network deformation, fibril alignment, and cell contact guidance is depicted in Figure 23.3. Predictions consistent with all of the above observations are presented in (Barocas and Tranquillo, 1997a) [i.e., for those cases that are either spherically symmetric or axi-symmetric, which is required for the numerical analysis of the model equations detailed in (Barocas and Tranquillo, 1997b)]. An extensive comparison of ABT predictions and data for media-equivalent fabrication are presented in (Barocas et al., 1998).

The success of the ABT in explaining observations of evolving alignment in TEs quantitatively as well as qualitatively in many cases (Barocas and Tranquillo, 1997a, 1998), suggests the validity of its assumptions, including the strain-based origin of the contact guidance field and the nature of the contact guidance response described above. Independent and direct support for these key assumptions is provided by observation of the effects of an applied (rather than cell induced) deformation of the network, as discussed in the next section. As with all theories, this one is limited in its validity, and can only be applied to the relatively short time before the cells complicate the gel mechanics by extensive compaction and by alteration of the extracellular matrix composition. Experiments can be carried out at lower cell concentrations to make these complications of lesser significance, or the theory could be extended to include them. Since higher cell concentrations are generally of interest in obtaining TEs with improved functional properties, extending the theory is called for but presents new challenges, including nonlinear modeling of viscoelasticity, permeability, and fibril reorientation, and improved modeling of traction inhibition that occurs as compaction proceeds. Another important development will be a full three-dimensional optimizing code for mold design, which is critical for fabricating tissue equivalents that have nonsimple geometries, such as a valve equivalent.

Another key to understanding the relationship between cell traction, contact guidance, and fibril alignment in TEs is knowledge of the direct molecular interactions between tissue cells and the surrounding matrix, and how they influence evolving TE structure and function. To this end, Ogle and Mooradian (1999) have investigated the role of cell surface integrins in the compaction and initial strengthening of TEs, particularly type I collagen-based media-equivalents (MEs). By selectively inhibiting various integrin subunits, the authors determined that blocking the $\alpha2\beta1$ integrin inhibited gel compaction during the initial 24 hours of incubation and reduced the tensile modulus of the constructs after 72 hours as compared with controls. These results indicate a direct link between the mechanism by which cells adhere to the surrounding fibrils and their ability to restructure the network. However, the details of the signaling mechanisms involved have yet to be resolved.

Confined Compression of Tissue-Equivalents: Direct Evidence for Strain-Based Contact Guidance

That contact guidance in an aligned gel is correlated with fibril alignment is well established, although the mechanism by which cells sense and respond to aligned fibrils is still an open question, as discussed in the next section. However, this correlation does not necessarily imply that the cells exhibit contact guidance in direct response to aligned fibrils, since aligned fibrils associated with an anisotropic network strain will also generally be associated with an anisotropic network stress. Exactly how cells might directly sense an anisotropic stress is also an open question, and a moot point if the network is perfectly elastic, since stress is then proportional to strain and therefore mathematically equivalent, although it is far from irrelevant with respect to the cellular mechanism. However, the initial state of the gels is, as described previously, that is, the network is viscoelastic, so that stress also depends on the strain rate and the equivalence of stress and strain that applies in the case of an elastic gel does not apply. Moreover, the network exhibits fluidlike behavior in which all stress is completely relaxed with sufficient time, unlike solidlike behavior. This stress relaxation property allows an experiment to be conducted to directly verify the key assumptions about the nature of the contact guidance field and the accompanying cellular response, namely confined compression of a tissue equivalent.

In this experiment (Girton and Tranquillo, 2001), a TE formed within a rectangular chamber having an impermeable surface at one end and a flat porous piston, which is permeable to the interstitial fluid but not the network. The assumed fibril reorientation in response to a confined compression is normal to the compression axis. Two of the chamber walls are made of glass so that the network strain (based on displacement of marker beads), the direction and degree of fibril alignment (as measured by the angle of extinction and birefringence, respectively), and cell orientation can be measured using microscopy and image analysis.

The key results are as follows: the direction of fibril alignment is perpendicular to the direction of compression (consistent with the ABT assumption), to a degree that increases with strain; cells align in the direction of fibril alignment, to a degree that increases with strain; and the degrees of both fibril alignment and cell alignment are approximately the same 24 hours after the compression is applied as they are immediately after it is first applied. Taken together, these results verify the assumptions that cells align preferentially in the direction that surrounding fibrils are aligned and that deformations of the fibrillar network that create an anisotropic strain cause fibril alignment. The related assumption that cells migrate preferentially in the direction that surrounding fibrils are aligned is verified by the data presented in Dickinson et al. (1994); the assumption that cells exert traction preferentially in the direction that surrounding fibrils are aligned is difficult to verify directly, but follows from the fact that cells preferentially align in that direction and preferentially exert traction along their long axis as concluded from the observed pattern of birefringence around elongated cells in gels, emanating radially outward from both ends of the cells.

Moreover, the observation that cell alignment is sustained at 24 hours, at which time the stress should be virtually relaxed in the gel since the longest measured relaxation time is 6 hours, implies that the cells do in fact respond directly to the anisotropic strain (i.e., aligned fibrils) and not to any associated anisotropic stress. This is also supported by the observation of contact guidance in gels aligned with a magnetic field and using low cell concentrations (Guido and Tranquillo, 1993), where little stress should exist. Of course, this does not preclude the possibility that cells can directly sense and respond to an anisotropic network stress in other situations, such as in an oscillatory strain field generated by external means, which has obvious cardiovascular relevance.

Insight into the Mechanism of Contact Guidance

Beyond this discrimination between anisotropic strain (aligned fibrils) vs. anisotropic stress as the macroscopic signal, little progress has been made in distinguishing between the possibilities for the dominant cellular mechanism underlying contact guidance among those proposed by Dunn (1982):

1. focal adhesions are confined to fibrils, which since aligned (chemical anisotropy) orient the adhesions, hence locomotion,
2. the cell distorts differently when migrating in different directions and when oriented with fibrils distorts less, due to the structural anisotropy, hence favoring locomotion oriented along fibrils,
3. as elaborated by Haston et al. (1983), that because exertion of cell traction on the substratum is necessary for locomotion, pseudopods pulling in the direction of maximum elastic modulus (i.e., along fibrils) are more efficient since the displacement of fibrils toward the cell is much less than for pseudopods pulling in the direction of minimal modulus of elasticity (i.e., across fibrils), the result of this elastic anisotropy (more generally, a viscoelastic anisotropy) again being locomotion oriented along the fibrils.

The main reason for the lack of progress in distinguishing between these possibilities is the intrinsic difficulty in changing just one property of an aligned network (chemical, structural, mechanical anisotropy) without changing the others, not to mention the lack of reproducibility of gel properties, let alone the network alignment.

Realizing Target Mechanical Properties

A major issue in cardiovascular and orthopedic tissue engineering is realizing target mechanical properties. It does not suffice to have a TE that mimics the native alignment in order to guide remodeling if it will experience catastrophic failure upon implantation because of insufficient strength, or lead to failure because of creep or compliance mismatch. There are two major approaches to realizing target mechanical properties of TE: exogenously cross-linking the aligned network fibrils, and inducing the entrapped cells to compositionally remodel the aligned network into an extracellular matrix that better resembles native tissue in terms of composition and presumably mechanical properties.

We have reported that glycation can be exploited to increase the circumferential tensile stiffness and ultimate tensile strength of MEs and increase their resistance to collagenolytic degradation, all without loss of cell viability (Girton et al., 1999). The glycated MEs were fabricated by entrapping high passage adult rat aorta SMC in collagen gel made from pepsin-digested bovine dermal collagen, and incubated for up to 10 wk in complete medium with 30 mM ribose added. We have reported more recently on experiments showing that ME compaction due to traction exerted by the SMC with consequent alignment of collagen fibrils was necessary to realize the glycation-mediated stiffening and strengthening, but that synthesis of extracellular matrix constituents by these cells likely contributed little, even when 50 μg/ml ascorbate, which is reported to stimulate collagen synthesis (Phillips et al., 1992), was added to the medium. These glycated MEs exhibited a compliance similar to arteries, but possessed less tensile strength and much less burst strength. MEs fabricated with low rather than high passage adult rat aorta SMC possessed almost 10 times greater tensile strength, suggesting that alternative SMC sources and biopolymer gels may yield sufficient strength by compositional remodeling prior to implantation in addition to the structural remodeling (i.e., circumferential alignment) already obtained.

It is known that the proliferation and ECM synthesis of cells entrapped in collagen are inhibited relative to conventional cell culture. Consequently, we have investigated fibrin as an alternative biopolymer to type I collagen for TE fabrication. Fibrin is the major structural protein of a blood clot and plays a vital role in the subsequent wound healing response. It forms a fibrillar network that can be compacted and aligned by entrapped tissue cells forming a TE similar to the fibrillar collagens. Studies of fibroblasts entrapped in fibrin gel have suggested an increased amount of collagen synthesis as compared with fibroblasts entrapped within collagen gel, and there is evidence of collagen fibrils

and other ECM accumulating in the fibrin network (Clark et al., 1995; Tuan et al., 1996). We have confirmed this directly (Grassl et al., in press), and we have endeavored to optimize incubation conditions to maximize cell-mediated compositional remodeling of fibrin gel to produce TEs that are more mechanically robust and tissue-like (Neidert et al., 2002).

TEs formed by entrapping human foreskin fibroblasts in fibrin gel and incubated for 7 weeks with 10% fetal bovine serum (FBS) supplemented with optimal concentrations of transforming growth factor (TGF)-β or insulin showed three times as much collagen produced compared with controls incubated with only 10% FBS. TEs incubated with just TGF-β, however, were both stiffer and stronger than incubated with just insulin, suggesting some mechanism in addition to collagen synthesis for modulating TE mechanical properties. Furthermore, an additive effect was apparent in the modulus and ultimate tensile strength for TEs incubated with both TGF-β and insulin.

Initial experiments with rat aorta smooth muscle cells revealed rapid and complete fibrinolysis that did not occur with the fibroblasts. Subsequent experiments performed in complete medium (10% FBS) supplemented with the plasmin inhibitor ϵ-aminocaproic acid revealed that, as with fibroblasts, smooth muscle cells synthesized and incorporated more collagen into the network when entrapped in fibrin gel as compared to collagen gel. In order to examine the potential effects of fibrinolysis on fibrin-based TEs fabricated with fibroblasts, we examined the effect of TE incubation with plasmin. The addition of plasmin to medium supplemented with the combination of 10% FBS, TGF-β, and insulin increased both the collagen content and modulus of the TE, suggesting a beneficial effect of fibrin degradation products or some other proteolytic product. In fact, if medium supplemented with optimal concentrations of TGF-β, insulin, and plasmin in addition to serum was changed three times weekly, the amount of collagen produced by the fibroblasts in fibrin constructs was 100 times greater than the amount produced in medium containing only serum and changed weekly (Neidert et al., 2002). This translated into constructs with a protein composition of at least 33% collagen by weight possessing an ultimate tensile strength approaching native load bearing tissues.

While these results for fibrin-based TEs are very promising, many questions remain. In terms of functional tissue engineering, a key question is will fibrin induce ECM production in a manner that allows all target mechanical properties to be realized? This will hinge on the structure (i.e., alignment) and quality (i.e., cross-linking) of the ECM produced by cells in the aligned fibrin network.

Since many of the target applications for TEs are load-bearing tissues, such as the artery and the heart valve, the importance of these loads in the development and maintenance of such tissues cannot be overlooked. A number of investigators have reported on the beneficial effects of physiological mechanical loading on the mechanical properties of tissue constructs (Kim et al., 1999; Niklason et al., 1999; Seliktar et al., 2000). Nerem and co-workers reported that cyclic mechanical stretching of collagen-based MEs increased their modulus (108%), ultimate tensile strength (240%), and fibril alignment after eight days (Seliktar et al., 2000). This is correlated to increased levels of active MMP-2, implicating a beneficial role for network degradation in its structural remodeling. Other investigators have reported increased ECM synthesis in response to cyclic mechanical stretching of SMCs and fibroblasts cultured on planar substrata (Carver et al., 1991; Chiquet et al., 1996; Li et al., 1998; Riser et al., 1998; Sumpio et al., 1988), which may contribute to beneficial compositional remodeling in longer-term cyclic loading of TEs. In addition to enhanced mechanical properties, cyclic loading appears to shift SMCs toward the contractile phenotype (Birukov et al., 1995; Birukov et al., 1998; Kanda et al., 1993), which is highly advantageous considering that a major role of arterial SMCs is to control the diameter of the arterial wall.

Future Directions

There are many directions that need to be pursued in terms of functional tissue engineering based on TE fabrication. Some have already been indicated in the previous sections, such as extending the ABT to phenomena that occur at large strains/compactions. This would be complemented by development of a micromechanical-macromechanical model (Agoram and Barocas, 2001). In turn, this will require quantitative characterization of tractional structuring of the surrounding fibrillar networks by cells (Fig. 23.4) along with macroscopic compaction and alignment data. A major challenge will be extending these mechanical models to the post-compaction compositional remodeling (tissue growth) phase. This will require quantitative composition-structure-mechanical property data as a function of incubation time and conditions. Mechanical testing with alignment imaging will be valuable in this regard (Fig. 23.5). Complicated interrelated compositional changes involving biopolymer degradation, ECM production, and cell proliferation must be considered, which in turn depend on a myriad of factors, including biopolymer type, cell type/source, medium composition, mechanical stress state, autocrine factors, and cell phenotype. Superimposed on this complexity are effects associated with cell-cell interactions during co-culture fabrication (e.g., SMC-EC), and strategies based on hybrid or layered constructs (Bell et al., 1978; L'Heureux et al., 1993, 1998). Models proposed for tissue growth and remodeling (Fung et al., 1993; Humphrey, 1999) will have to be refined accordingly and perhaps other modeling approaches developed.

Artificial soft tissue fabrication from cell-compacted biopolymers derives from the interactions between cells and a biopolymer network, and major advances with this approach will depend on major advances related to the cell type/source and biopolymer network properties. Concerning the cells, the use of genetically altered adult cells and adult stem cells offers much potential, particularly in regard to the major hurdle of realizing elastin fiber production to confer recoverable deformation. Concerning biopolymer network properties, a major advance might result from synthetic fibrillar matrices which would be permissive to cell induced compaction and alignment as with biopolymer gels, but be more conducive to cell synthesis and assembly of extracellular matrix.

The measurement of the mechanical properties of highly compacted TEs has largely been

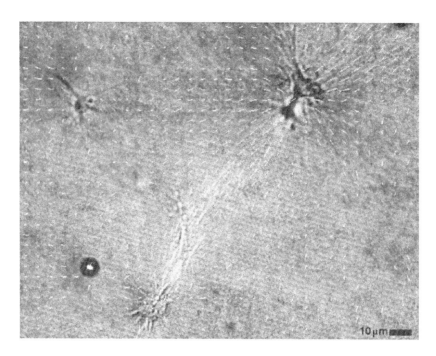

FIGURE 23.4. Polarized light image of tractional restructuring. A birefringence map showing strong alignment of the fibrils that developed between nearby cells as a consequence of tractional structuring. The direction and length of each segment reflect the local direction and strength of alignment as determined from retardation using the method described in (Tower et al., in press).

FIGURE 23.5. Simultaneous alignment imaging and tensile testing. Polarized light imaging during uniaxial mechanical testing of a media-equivalent reveals that the alignment is suite nonuniform. Such imaging can be performed at video rate and subsequently correlated to the applied load and displacement. The sample in this image is being subjected to 50% strain.

limited to quasi-static uniaxial tensile testing. However, TEs are clearly complicated viscoelastic composite materials, just as the tissues whose properties we seek to emulate. In order for a TE to be viable, it must match, or at least approach, the mechanical properties of the corresponding native tissue under a wide range of loading conditions, which are often dynamic and involve several different types of loading simultaneously. Therefore, a more comprehensive characterization of the mechanical properties of TEs is required in order to determine if a TE is capable of appropriately responding to physiologically relevant loading and gain insight into how to address current shortcomings. Moreover, an investigation of the failure mechanisms of TEs will give valuable information regarding the molecular basis for TE mechanics and improved fabrication, and determining the fatigue properties of TEs will be critical to understanding their long-term success as viable tissue replacements.

Ultimately, *in vivo* experience and feedback based on the nature of tissue remodeling and functional changes that occur upon implantation to guide artificial tissue fabrication will be required.

Acknowledgments

The research performed in my laboratory on these subjects over the years by Victor Barocas, Stacey Dixon, Tim Girton, Erin Grassl, Brett Isenberg, Dave Knapp, Evie Lee, Mike Neidert, David Shreiber, and Ted Tower, and the collaboration with Jack Lewis and Ted Oegema are very gratefully acknowledged, as is funding from NHLBI 1R01 HL60495 (R.T.T.) and the NSF MRSEC-Artificial Tissues Program (DMR-9809364).

References

Agoram B, Barocas VH. 2001. Coupled macroscopic and microscopic scale modeling of fibrillar tissues and tissue equivalents. J Biomech Eng 123(4):362–9.

Allen TD, Schor SL, Schor AM. 1984. An ultrastructural review of collagen gels, a model system for cell-matrix, cell-basement membrane and cell-cell interactions. *Scan. Electron Microsc.* (Pt 1):375–390.

Atala A, Mooney DJ, Vacanti JP, Langer RS, eds. 1997. Synthetic biodegradable polymer scaffolds. Birkhauser.

Auger FA, Pouliot R, Tremblay, N, Guignard R, Noel P, Juhasz J, Germain L, Goulet F. 2000. Multistep production of bioengineered skin substitutes: sequential modulation of culture conditions. *In Vitro Cell. Dev. Biol. Anim.* 36:96–103.

Awad HA, Butler DL, Harris MT, Ibrahim RE, Wu Y, Young RG, Kadiyala S, Boivin GP. 2000. In vitro characterization of mesenchymal stem cell-seeded collagen scaffolds for tendon repair: effects of initial seeding density on contraction kinetics. *J. Biomed. Mater. Res.* 51:233–240.

Bader A, Steinhoff G, Strobl K, Schilling T, Brandes G, Mertsching H, Tsikas D, Froelich J, Haverich A. 2000. Engineering of human vascular aortic tissue based on a xenogeneic starter matrix. *Transplantation* 70:7–14.

Bale MD, Muller MF, Ferry JD. 1985. Rheological studies of creep and creep recovery of unligated fibrin clots: comparison of clots prepared with thrombin and ancrod. *Biopolymers* 24:461–482.

Barocas VH, Tranquillo RT. 1997a. An anisotropic biphasic theory of tissue-equivalent mechanics: the interplay among cell traction, fibrillar network deformation, fibril alignment, and cell contact guidance. *J. Biomech. Eng.* 119:137–145.

Barocas VH, Tranquillo RT. 1997b. A finite element solution for the anisotropic biphasic theory of tissue-equivalent mechanics: the effect of contact guidance on isometric cell traction measurement. *J. Biomech. Eng.* 119:261–269.

Barocas VH, Moon AG, Tranquillo RT. 1995. The fibroblast-populated collagen microsphere assay of cell traction force—Part 2: Measurement of the cell traction parameter. *J. Biomech. Eng.* 117:161–170.

Barocas VH, Girton, TS, Tranquillo RT. 1998. Engineered alignment in media-equivalents: Magnetic prealignment and mandrel compaction. *J. Biomech. Eng.* 120:660–666.

Bell E, Ivarsson B, Merrill C. 1978. Production of a tissue-like structure by contraction of collagen lattices by human fibroblasts of different proliferative potential *in vitro*. *Proc. Nat. Acad. Sci.* U.S.A. 76:1274–1278.

Bell E, Ivarsson B, Merrill C. 1979. Production of a tissue-like structure by contraction of collagen lattices by human fibroblasts of different proliferative potential *in vitro*. *Proc. Nat. Acad. Sci.* U.S.A. 76:1274–1278.

Bell E, Ehrlich HP, Buttle DJ, Nakatsuji T. 1981. Living tissue formed *in vitro* and accepted as skin-equivalent tissue of full thickness. *Science* 211:1052–1054.

Birukov KG, Shirinsky, VP, Stepanova OV, Tkachuk VA, Hahn AWA, Resnick TJ, Smirnov VN. 1995. Stretch affects phenotype and proliferation of vascular smooth muscle cells. *Mol. Cell. Biochem.* 144:131–139.

Birukov KG, Bardy N, Lehoux S, Merval R, Shirinsky VP, Tedgui A. 1998. Intraluminal pressure is essential for the maintenance of smooth muscle caldesmon and filamin content in aortic organ culture. *Arterioscl. Thromb. Vasc. Biol.* 18:922–927.

Bromberek BA, Enever PAJ, Caldwell MD, Tranquillo RT. 2002. Macrophages influence a competition of contact guidance and chemotaxis for fibroblast alignment in a fibrin gel coculture assay. *Exp. Cell Res.* 275:230–42.

Butler DL, Awad HA. 1999. Perspectives on cell and collagen composites for tendon repair. *Clin. Orthop.* (367 Suppl.) S324–332.

Carver W, Nagpal ML, Nachtigal M, Borg TK, Terracio L. 1991. Collagen expression in mechanically stimulated cardiac fibroblasts. *Circ. Res.* 69:116–122.

Chiquet M, Matthisson M, Koch M, Tannheimer M, Chiquet-Ehrismann R. 1996. Regulation of extracellular matrix synthesis by mechanical stress. *Biochem. Cell Biol.* 74:737–744.

Clark RA, Nielsen LD, Welch MP, McPherson JM. 1995. Collagen matrices attenuate the collagen-synthetic response of cultured fibroblasts to TGF-beta. *J. Cell Sci.* 108:1251–1261.

Coulomb B, Friteau L, Baruch J, Guilbaud J, Chretien-Marquet B, Glicenstein J, Lebreton-Decoster C, Bell E, Dubertret L. 1998. Advantage of the presence of living dermal fibroblasts within *in vitro* reconstructed skin for grafting in humans. *Plast. Reconstr. Surg.* 101:1891–1903.

de Chalain T, Phillips JH, Hinek A. 1999. Bioengineering of elastic cartilage with aggregated porcine and human auricular chondrocytes and hydrogels containing alginate, collagen, and kappa-elastin. *J. Biomed. Mater. Res.* 44:280–288.

Dickinson RB, Guido S, Tranquillo RT. 1994. Biased cell migration of fibroblasts exhibiting contact guidance in oriented collagen gels. *Ann. Biomed. Eng.* 22:342–356.

Dunn GA. 1982. Contact guidance of cultured tissue cells: a survey of potentially relevant properties of the substratum. In: *Cell Behaviour.* R. Bellairs, A. Curtis, G. Dunn, eds. Cambridge University Press, Cambridge pp. 247–280.

Eaglstein WH, Falanga V. 1998. Tissue engineering and the development of Apligraf, a human skin equivalent. *Cutis* 62:1–8.

Ehrlich HP, Buttle DJ, Bernanke DH. 1989. Physiological variables affecting collagen lattice contraction by human dermal fibroblasts. *Exp. Mol. Pathol.* 50:220–229.

Eschenhagen T, Fink C, Remmers U, Scholz H, Wattchow J, Weil J, Zimmermann W, Dohmen HH, Schafer H, Bishopric N, Wakatsuki T, Elson EL. 1997. Three-dimensional reconstitution of embryonic cardiomyocytes in a collagen matrix: a new heart muscle model system. *FASEB J.* 11:683–694.

Fung YC, Liu SQ, Zhou JB. 1993. Remodeling of the constitutive equation while a blood vessel remodels itself under stress. *J. Biomech. Eng.* 115:453–459.

Girton TS, Oegema TR, Tranquillo RT. 1999. Exploiting glycation to stiffen and strengthen tissue-equivalents for tissue engineering. *J. Biomed. Mater. Res.* 46:87–92.

Girton TS, Tranquillo RT. 2001. Confined compression of a tissue-equivalent: collagen fibril and cell alignment in response to anisotropic strain. *J. Biomech. Eng.* (accepted).

Grassl ED, Oegema TR, Tranquillo RT. 2002. Fibrin as an alternative biopolymer to type I collagen for fabrication of a media-equivalent. *J Biomed Mat Res* 60(4):607–612.

Grinnell F, Lamke CR. 1984. Reorganization of hydrated collagen lattices by human skin fibroblasts. *J. Cell Sci.* 66:51–63.

Guido S, Tranquillo RT. 1993. A methodology for the systematic and quantitative study of cell contact guidance in oriented collagen gels: correlation of fibroblast orientation and gel birefringence. *J. Cell Sci.* 105:317–331.

Guidry C, Grinnell F. 1985. Studies on the mechanism of hydrated collagen gel reorganization by human skin fibroblasts. *J. Cell Sci.* 79:67–81.

Guidry C, Grinnell F. 1986. Contraction of hydrated collagen gels by fibroblasts: evidence for two mechanisms by which collagen fibrils are stabilized. *Collagen Rel. Res.* 6:515–529.

Harris AK, Stopak D, Wild P. 1981. Fibroblast traction as a mechanism for collagen morphogenesis. *Nature* 290:249–251.

Haston WS, Shields JM, Wilkinson PC. 1983. The orientation of fibroblasts and neutrophils on elastic substrata. *Exp. Cell Res.* 146:117–126.

Hirai J, Matsuda T. 1995. Self-organized, tubular hybrid vascular tissue composed of vascular cells and collagen for low-pressure-loaded venous system. *Cell Transplant.* 4:597–608.

Huang D, Chang TR, Aggarwal A, Lee RC, Ehrlich HP. 1993. Mechanisms and dynamics of mechanical strengthening in ligament-equivalent fibroblast-populated collagen matrices. *Ann. Biomed. Eng.* 21:289–305.

Humphrey JD. 1999. Remodeling of a collagenous tissue at fixed lengths. *J. Biomech. Eng.* 121:591–597.

Kanda K, Matsuda T, Oka T. 1993. Mechanical stress induced cellular orientation and phenotypic modulation of 3-D cultured smooth muscle cells. *ASAIO J.* 39:M686–M690.

Kim B-S, Nikolovski J, Bonadio J, Mooney DJ. 1999. Cyclic mechanical strain regulates the development of engineered smooth muscle tissue. *Nat. Biotechnol.* 17:979–983.

Klebe RJ, Caldwell H, Milam S. 1989. Cells transmit spatial information by orienting collagen fibers. *Matrix* 9:451–458.

Knapp DM, Barocas VH, Moon AG, Yoo K, Petzold LR, Tranquillo RT. 1997. Rheology of reconstituted type I collagen gel in confined compression. *J. Rheol.* 41:971–993.

Knapp DM, Barocas VB, Tower TT, Tranquillo RT. 1999. Estimation of cell traction and migration in an isometric cell traction assay. *AIChE J* 45:2628–2640.

Kobashi T, Matsuda T. 1999. Fabrication of branched hybrid vascular prostheses. *Tissue Eng.* 5:515–524.

Kolodney MS, Elson EL. 1993. Correlation of myosin light chain phosphorylation with isometric contraction of fibroblasts. *J. Biol. Chem.* 268:23850–23855.

L'Heureux N, Germain L, Labbe R, Auger FA. 1993. *in vitro* construction of a human blood vessel from cultured vascular cells: a morphologic study. *J. Vasc. Surg.* 17:499–509.

L'Heureux N, Paquet S, Labbe R, Germain L, Auger FA. 1998. A completely biological tissue-engineered human blood vessel. *FASEB J.* 12:47–56.

Langer R, Vacanti JP. 1993. Tissue engineering. *Science* 260:920–926.

Li Q, Muragaki Y, Hatamura I, Ueno H, Ooshima A. 1998. Stretch-induced collagen synthesis in cultured smooth muscle cells from rabit aortic media and a possible involvement of angiotensin II and transforming growth factor-β. *J. Vasc. Res.* 35:93–103.

Lopez Valle CA, Auger FA, Rompre R, Bouvard V, Germain L. 1992. Peripheral anchorage of dermal equivalents. *Br. J. Dermatol.* 127:365–371.

Michel M, Germain L, Auger FA. 1993. Anchored skin equivalent cultured *in vitro:* a new tool for percutaneous absorption studies. *In Vitro Cell. Dev. Biol.* 29A:834–837.

Moon AG, Tranquillo RT. 1993. The fibroblast-populated collagen microsphere assay of cell traction force—Part 1. Continuum Model. *AIChE J* 39:163–177.

Muller MF, Ris H, Ferry JD. 1984. Electron microscopy of fine fibrin clots and fine and coarse fibrin films. Observations of fibers in cross-section and in deformed states. *J. Mol. Biol.* 174:369–384.

Neidert MR, Lee ES, Tower TT, Oegema TR, Tranquillo RT. 2002. Enhanced fibrin remodeling *in vitro* for improved tissue-equivalents. Biomaterials 23(17):3717–31.

Niklason LE, Gao J, Abbot WM, Hirschi KK, Houser S, Marini R, Langer R. 1999. Functional arteries grown *in vitro. Science* 284:489–493.

Ogle BM, Mooradian DL. 1999. The role of vascular smooth muscle cell integrins in the compaction and mechanical strengthening of a tissue-engineered blood vessel. *Tissue Eng.* 5:387–402.

Phillips CL, Tajima S, Pinnel SR. 1992. Ascorbic acid and transforming growth factor-β1 increase collagen biosynthesis via different mechanisms: coordinate regulation of Proα1(I) and Proα1(III) collagens. *Arch. Biochem. Biophys.* 295:397–403.

Riser BL, Cortes P, Yee J, Sharba AK, Asano K, Rodriguez-Barbero A, Narins RG. 1998. Mechanical strain- and high glucose-induced alterations in mesangial cell collagen metabolism: role of TGF-β. *J. Am. Soc. Nepphrol.* 9:827–836.

Schreiber RE, Ratcliffe A. 2000. Tissue engineering of cartilage. *Methods Mol. Biol.* 139:301–309.

Seliktar D, Black RA, Vito RP, Nerem RM. 2000. Dynamic mechanical conditioning of collagen-gel blood vessel constructs induces remodeling *in vitro. Ann. Biomed. Eng.* 28:351–362.

Shansky J, Del Tatto M, Chromiak J, Vandenburgh H. 1997. A simplified method for tissue engineering skeletal muscle organoids *in vitro* [letter]. *In Vitro Cell. Dev. Biol. Anim.* 33:659–661.

Stopak D, Harris AK. 1982. Connective tissue morphogenesis by fibroblast traction. I. Tissue culture observations. *Dev. Biol.* 90:383–398.

Sumpio BE, Banes AJ, Link WG, Johnson G. 1988. Enhanced collagen production by smooth muscle cells during repetitive mechanical stretching. *Arch. Surg.* 123:1233–1236.

Toolan BC, Frenkel SR, Pachence JM, Yalowitz L, Alexander H. 1996. Effects of growth-factor-enhanced culture on a chondrocyte-collagen implant for cartilage repair. *J. Biomed. Mater. Res.* 31:273–280.

Tower TT, Neidert MR, Tranquillo RT. accepted. Concurrent alignment imaging and mechanical testing of tissues.

Tranquillo RT, Durrani MA, Moon AG. 1992. Tissue engineering science: consequences of cell traction force. *Cytotechnology* 10:225–250.

Tuan TL, Song A, Chang S, Younai S, Nimni ME. 1996. *in vitro* fibroplasia: matrix contraction, cell growth, and collagen production of fibroblasts cultured in fibrin gels. *Exp. Cell Res.* 223:127–134.

Weinberg CB, Bell E. 1986. A blood vessel model constructed from collagen and cultured vascular cells. *Science* 231:397–400.

Weiser L, Bhargava M, Attia E, Torzilli PA. 1999. Effect of serum and platelet-derived growth factor on chondrocytes grown in collagen gels. *Tissue Eng.* 5:533–544.

Wilkins LM, Watson SR, Prosky SJ, Meunier SF, Parenteau NL. 1994. Development of a bilayered living skin construct for clinical applications. *Biotechnol. Bioeng.* 43:747–756.

Woo SL, Hildebrand K, Watanabe N, Fenwick JA, Papageorgiou CD, Wang JH. 1999. Tissue engineering of ligament and tendon healing. *Clin. Orthop. Rel. Res.* (367 Suppl) S312–323.

Ye Q, Zund G, Benedikt P, Jockenhoevel S, Hoerstrup SP, Sakyama S, Hubbell JA, Turina M. 2000a. Fibrin gel as a three dimensional matrix in cardiovascular tissue engineering. *Eur. J. Cardiothorac. Surg.* 17:587–591.

Ye Q, Zund G, Jockenhoevel S, Hoerstrup SP, Schoeberlein A, Grunenfelder J, Turina M. 2000b. Tissue engineering in cardiovascular surgery: new approach to develop completely human autologous tissue. *Eur. J. Cardiothorac. Surg.* 17:449–454.

24

Cytomechanics: Signaling to Mechanical Load in Connective Tissue Cells and Its Role in Tissue Engineering

Albert J. Banes, Michelle Wall, Joanne Garvin, and Joanne Archambault

Introduction

Engineering replacement tissues has been a theoretical reality since the inception of synthetic dermis and cell growth in organized collagen gels (Bell et al., 1979; Yannas et al., 1982). However, though the formulation and fabrication methods for this embodiment of engineered dermis have been known for some time, commercializing the concept has been problematic. Nevertheless, scientists, clinicians and businessmen alike, involved in the tissue engineering endeavor, understand that theory will match reality eventually.

At the tissue level, striated skeletal muscle, bone, cartilage, tendon, ligament, meniscus, and intervertebral disc are subjected to mechanical deformation. Repetitive loading stimulates muscle cell hypertrophy, whereas no loading results in tissue atrophy (Vandenburgh, 1987). Likewise, bone responds to weight bearing by mineralizing and to immobilization by demineralizing (Lanyon, 1987; Wolff, 1892). Cartilage responds to cyclic compression by synthesizing matrix and to static loading by degrading matrix (Buschmann et al., 1995; Gray et al., 1988). Tendons and ligaments also respond positively to mechanical stimuli by expressing matrix and strengthening, and negatively to immobilization, by forming adhesions, losing range of motion and weakening (Banes et al., 1999; Hyman and Rodeo, 2000). However, connective tissues can also be weakened by application of excessive mechanical loading, especially repetitive motion of the same type (Archambault et al., 1995; Backman et al., 1990; Carpenter et al., 1999). These examples substantiate the thought that cells respond to load stimuli both anabolically and catabolically. At the cellular level, mechanisms defining how cells detect and respond to load have been under intense investigation over the past 10 years. Techniques that permit controlled application of tension, compression, and fluid flow to cells *in vitro* have accelerated the field of cytomechanics (Archambault et al., 2002; Banes et al., 1985; Buschmann et al., 1995; Dennis et al., 2000; Graff et al., 2000; Vandenburgh 1988). Investigators have utilized strains that have been measured *in vivo* or have been calculated (Burton et al., 1999; Lee et al., 1994, Oliver et al., 1994). Values in percentage elongation range from 0.0001 to 0.0035 to for bone cells, 0.1 to 0.15 for cartilage cells and meniscus, 0.01 to 0.09 for tendon and ligament, and up to 100% for muscle (where 1% elongation = 0.01 strain, where ϵ or strain = $\delta l/l$ or change in length divided by original length) (Archambault et al., in press; Binderman et al., 1988; Boitano et al., 1992, 1998; Bouman et al., 2002). In addition, the force that cells can exert on their substrate has been measured as 3–20×10^{-3} dyne (Burton et al., 1999; Lee et al., 1994; Oliver et al., 1994).

Application of Mechanical Load to Tissue Engineered Cells and Constructs

Mechanical load has been applied to cells and tissue constructs with a variety of devices to mechanically condition cells and matrix (Figs. 24.1–24.5) (Yu et al., 2000). One application of mechanical load to engineered tissues is simply to increase nutrification of the cell-populated constructs by utilizing fluid flow to provide more growth medium and remove metabolites (examples, engineered vascular tissue, cartilage, liver, kidney constructs). Cells grown in hollow fiber devices have utilized this principle for 20 years. More recent tissue engineering applications include the culture of liver cells in such modified bioreactors (Sussman and Kelly, 1995). Other applications involve culture of dermal or epidermal cells on a poly-lactic-polyglycolic scaffold increased nutrification as well as mechanical load due to increasing head pressure in the system. Fabrication and testing of biologic and biosynthetic linear embodiments of connective tissues (tendons and ligaments) have been achieved in tension-providing devices (Banes et al., 1999, 2001; Garvin and Banes, 2002; Hannafin et al., 1995; Figs. 24.3 and 24.4). The study of how cells respond to mechanical forces has taken on a new perspective as the field of cytomechanics meets that of tissue engineering: mechanical loading can positively modify cell responses to achieve stronger, more *in vivo*–like matrices and cell phenotypes.

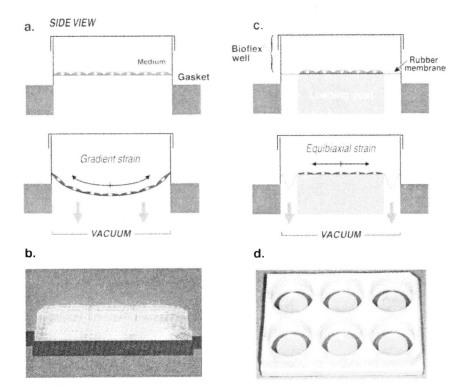

FIGURE 24.1. Diagram of a single well of a flexible bottomed culture plate (*b*) capable of strain application to cultured cells using a microprocessor-controlled, strain providing device (Flexercell Strain Unit™ Tension+). This vacuum-operated device controls pressure (vacuum) to the substratum of flexible bottom culture plates resulting in gradient strain that is maximal at the periphery and minimal at the center of the well (*a*). (*c*) A flexible bottom membrane deformed in equibiaxial strain by deforming the membrane across a planar faced cylinder. Vacuum draws the periphery of the membrane downward deforming the membrane across the planar faced loading post resulting in equibiaxial strain. The flexible bottom culture plate (BioFlex culture plate) is shown in *d* with a loading post beneath each well. Uniaxial strain may be applied to the membrane as well using an arctangle-shaped loading post (see Figure 24.4 *c*, lower left).

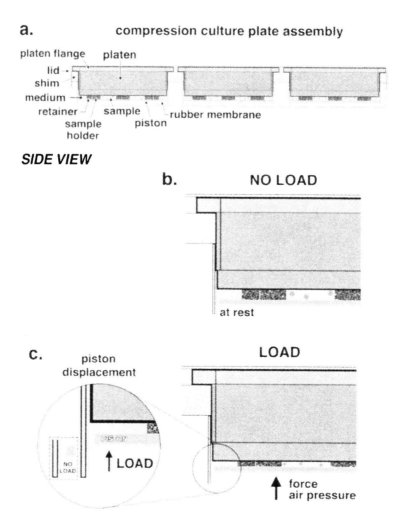

FIGURE 24.2. Side view of a diagram of a culture plate designed to provide compression to a sample when positive pressure is applied from the bottom with Flexercell Strain Unit™ Compression + (*a,* three wells of a six-well, BioPress™ culture plate). In this embodiment, the flexible membrane is a sterile barrier stiffened with a disc to act as a piston (*b*). Positive pressure moves the piston upward and can compress a cartilage sample placed between the disc and an upper platen (*c*). In this way, cyclic pressure provides intermittent force to the cartilage sample simulating compressive forces experienced by cartilage *in vivo.*

How Cells Detect Deformation

Organelles such as the Golgi tendon organ, the Vater-Pacini corpuscle, and the motor plate ending are responsible for reception of strain and transmission of signals through innervation to the brain (Jozsa and Kannus, 1997). Two models for detection and response to mechanical deformation on the cellular and molecular levels include touch reception in *Caenorhabditis elegans* and sound detection in the mammalian ear (Gu

et al., 1996; Howard et al., 1988). These model systems are examples of "outside-in" signaling to external forces (Banes et al., 1995; Clark and Brugge, 1995; Duncan and Turner, 1995). In a nematode touch reception model, sodium ion channels comprised of degenerin proteins are located at the front and rear of the worm. These channels are responsible for detecting movement in the reverse direction when touched (Garcia-Anoveras and Corey, 1996). In sound reception, hair cells in the inner ear have a link protein con-

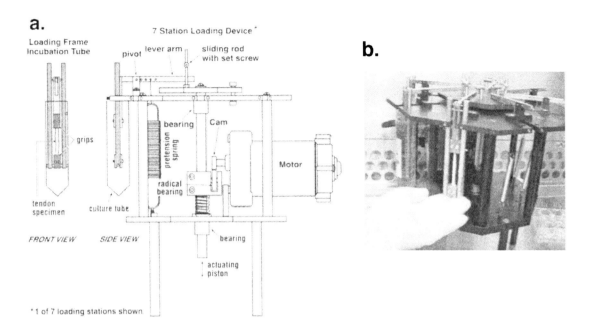

FIGURE 24.3. (*a*) Diagram of a displacement controlled uniaxial loading device for linear pieces of connective tissue. A tendon tissue sample clamped in the jaws of the loading device subjected to uniaxial loading by regulating the timing of the motor through a cam and lever (*b*). This device is displacement controlled but can be modified for load control.

nected to an ion channel that is activated by sound (Howard et al., 1988). Greater activation results in the channel remaining in the open state longer and thus more signal is sent to the brain where it is interpreted as sound. Therefore, mechano-gated ion channels are fundamental in the detection of deformation in touch and hearing. Mechano-gated ion channels are important in connective tissues as well (Duncan and Turner, 1995; Elfervig et al., 2001a; Hamill and Martinac, 2001; Hamill and McBride, 1996; Sachs, 1988; Sukharev et al., 1994). Some pathways may be dominant for mechanical load activation. However, it is likely that deformation stimuli share corronalities with ligand-activated systems such as the receptor protein tyrosine kinase (RPTK) or Jun activated kinase/JAK/STAT pathways contributing to pathway activation, fail-safe redundancy, signal amplification/dampening, and overall regulation and diversity in the systems. A load stimulus increases system strain from a basal state. As a cell responds to a load stimulus orienting to the strain field and polymerizing actin, the intrinsic strain field in the

cell syncytium, increases in intensity (Buckley et al., 1988). Strain on individual cell-substratum (focal adhesions) or cell-cell contacts [entercellular adhesion molecules (ICAMs), gap junctions, other] may induce conformational changes that result in activation of signaling pathways (Riveline et al., 2001). Signaling involves activation events that open and close membrane channels and are dependent on kinases/phosphatases to activate/deactivate focal adhesion proteins, cytoplasmic filaments or receptor or nonreceptor protein tyrosine/serine/threonine kinases which activate specific pathways and elicit transcriptional and translational events (Fig. 24.6) (Banes et al., 1995, 2001). Figure 24.6 shows how individual cells may detect load deformation by the latter mechanism (Banes et al., 2001). Key among these mechanisms are stretch-activated ion channels, other ion channels, integrins, and cell adhesion molecules. In many cases, a connection from a membrane spanning receptor, such as an integrin to the cytoskeleton and nucleus may transmit a signal (Ingber, 1993, 1997). In other cases, deformation of the plasma membrane

Tissue Train Culture System
Technique for Forming and Mechanically Loading Cells and Gel

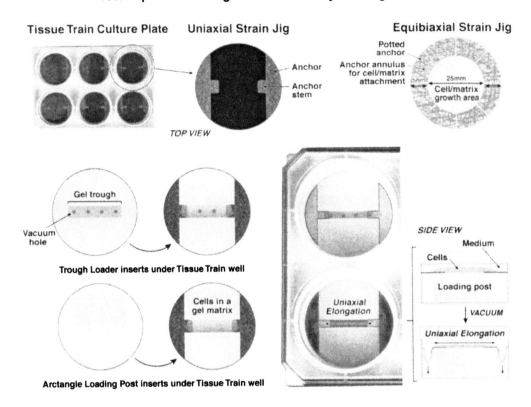

FIGURE 24.4. Culture plate capable of providing uniaxial or equibiaxial strain to three-dimensional cultures (*a*, Tissue Train™ culture device). This device uses a flexible bottom Tissue Train™ culture plate containing bonded nylon mesh as anchors to which adhere the cells in a collagen gel (*a*, right). A three-dimensional cell-populated gel is formed by plating the mixture into a loading trough (*c*) formed by deforming the membrane downward into the trough placed beneath the flexible membrane. After the mixture gels, the vacuum is released and the construct is a linear material attached at east and west poles to the nylon mesh anchors (*c*, upper and lower right). To apply uniaxial strain, an arctangle-shaped loading post (*c*, lower left, a loading post that is rectangular with curved short ends) is placed beneath the membrane such that the sides beneath the anchors are not bounded by the loading post. When vacuum is applied, the membrane deforms downward applying uniaxial strain to the construct (*d*). This mode is designed for use in tendon, ligament, and other constructs that use uniaxial strain. To apply equibiaxial strain, a segmented annular anchor is bonded to the periphery of the Tissue Train culture plate well (*b*). Cells and collagen gel are formed and a circular loading post is placed beneath the well. When vacuum is applied, the construct is subjected to equibiaxial substrate strain. This device is useful for loading, cornea constructs, heart, skin and other such engineered tissues.

itself may result in activation of G proteins (Hamill and Martinac, 2001).

Ion Channels

Mechanically gated ion channels (MGC) are connected to the cytoskeleton and can pass ions at a rate of 10^5 to 10^6 ions/sec (Fig. 24.7) (Guharry and Sachs, 1984; Hamill and Macbride, 1996; Hamill and Martinac, 2001; Sachs, 1988). Nominally closed ion channels are physically linked to the cytoskeleton and are activated by phosphorylation of their cytoplasmic tails to achieve an open state. Calcium channels may play a key role in an initial response to cell deformation in osteoblasts since

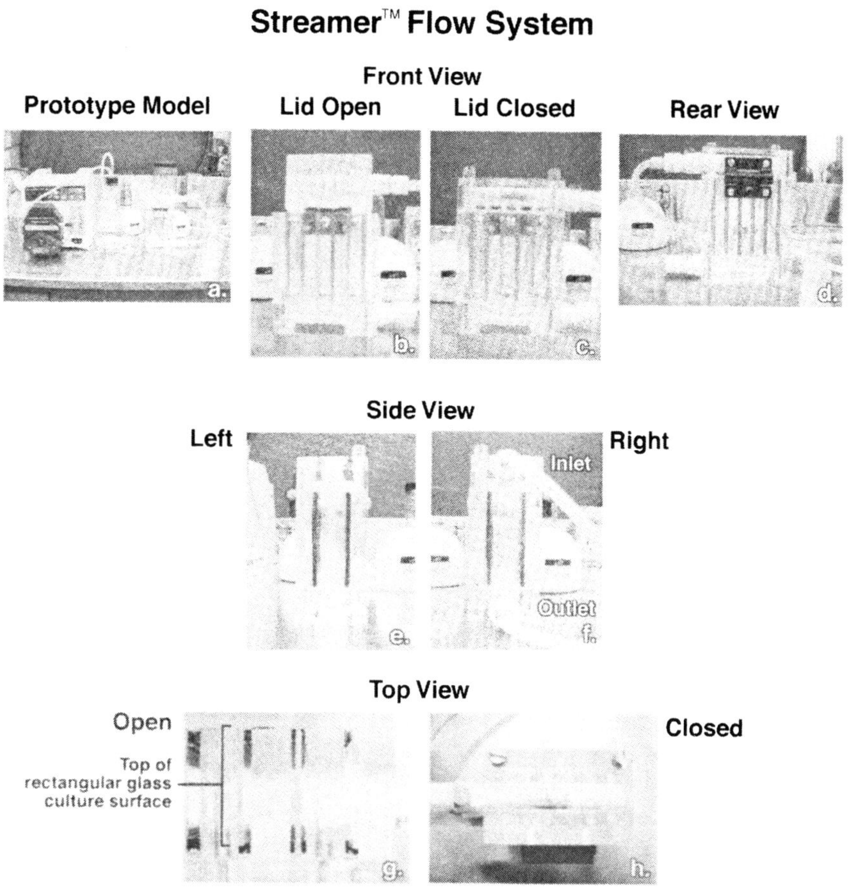

Streamer™ Flow System

Front View

Prototype Model | Lid Open | Lid Closed | Rear View

Side View

Left | Right

Inlet

Outlet

Top View

Open | Closed

Top of rectangular glass culture surface

FIGURE 24.5. Flow device to apply fluid flow to cultured cells (upper left and front and rear views, Streamer™ fluid flow device). The device utilizes rectangular growth surfaces fabricated from glass or rigid plastic as the growth surface (Culture Slips™ 25 × 75 mm or 100 × 75 mm. The Culture Slips™ insert vertically into each of six channels in the body of the device (side and top views). Flow rate is controlled by a microprocessor to achieve steady laminar, oscillating, and flow reversals. The flow system can be modified to utilize tubes or blood vessels.

the open state increases in response to cyclic stretch. Furthermore, an increase in intracellular calcium concentration, $[Ca^{2+}]_{ic}$, is induced by substrate deformation or fluid flow, a response that requires extracellular calcium (Archambault et al., 2001; Elfervig et al., 2001a; Hung et al., 1995, 1997; Kenamond et al., 1998; Yu et al., 2001). Calcium channels are sensitive to ion concentrations inside and outside the cell. Once activated, these channels can change in m sec, the $[Ca^{2+}]_{ic}$ from a basal level of 50 nM to a maximum concentration of over 1000 nM (Banes et al., 1995; Boitano et al., 1992). Cell deformation by plasma membrane indentation, tension, or shear stress has been shown to induce a rapid internal calcium transient

(Boitano et al., 1998; Olesin et al., 1988). Immediate and downstream responses to deformation can be inhibited or retarded in osteoblasts and tendon cells by stretch-activated channel blockers such as gadolinium or calcium channel blockers like verapamil and nifedipine (Vadiakas and Banes, 1992; You et al., 2001).

Figure 24.7a shows a stretch-activated ion channel in a plasma membrane in the closed position (Duncan and Turner, 1995; Hamill and Martinac, 2001; Sachs, 1988). As a positive pressure is applied to the plasma membrane effectively stretching the lipid bilayer, the channel increases its probability of being in the open state and allows extracellular ions to pass to the interior as in

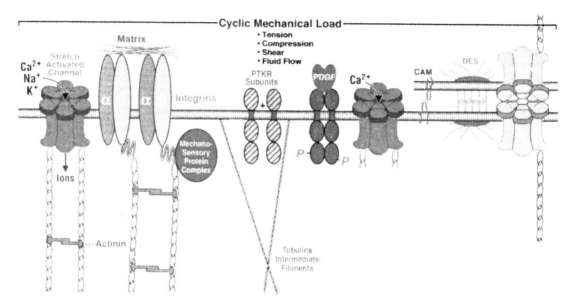

FIGURE 24.6. Multiple pathways through which cells detect mechanical deformation as tension, compression, shear, or fluid flow. Proven mechanisms include stretch-activated or inactivated ion channels, activation of transmembrane spanning integrins that link matrix to the cytoskeleton with members of a mechanosensory complex, with src, focal adhesion kinase (FAK), paxillin and other binding partners, tyrosine kinase receptors and other kinase receptors, ligand-activated receptors such as the platelet-derived growth factor receptor or IL-1β receptor whose effects synergize with load, calcium and other ion channels, cell adhesion molecules, desmosomes, and gap junctions.

a stretch-activated cation (SA-CAT) channel. These channels move ions at 10^5 to 10^6 ions/sec and are ubiquitous. Channels can also be inactivated by stretch, an example of load signal dampening (Hamill and Martinac, 2001; Sachs, 1988). Figure 24.7*b* shows a voltage-activated, nominally closed, Na^+, K^+, or Cl^- ion channel that is opened when a voltage change occurs across a bilayer by a local depolarization. Classically, the conformation of these channels is the nominally closed state. Activation occurs by a depolarizing event followed by conduit blockage as the C-terminal protein segment flips into the channel pore. Figure 24.7*c* shows an example of a ligand-gated channel that is nominally closed but may be opened by binding of a specific ligand. Figure 24.7*d* shows an example of an adenosine triphosphate (ATP) transport mechanism. One example is the CFTR, or cystic fibrosis transport regulator, that is involved in volume regulation and may facilitate ATP transport from the cell (Schweib-

ert, 1999). Recently, it has been demonstrated that ATP is an important second messenger secreted by load-stimulated chondrocytes, annulus cells, and tendon cells (Elfervig et al. 2001; Graff et al., 2000; Tsuzaki, unpublished).

Gap Junctions and Hemichannels

Figure 24.7*e* shows two examples of how connexins that form gap junction channels may operate. Two hemichannels, each in the plasma membranes of adjacent cells, interact with each other in the extracellular space to form a functional channel (Bruzzone et al.,1996). Clusters of hundreds of these channels associate in one region of the cell-cell contact region forming an immunochemically detectable gap junction. Cells that are connected in syncytia, such as tendon internal fibroblasts, osteoblasts, annulus cells of the intervertebral disc, and other connective tissue cells, signal load stimuli to each

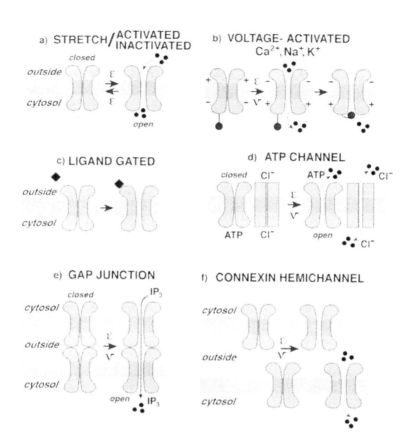

FIGURE 24.7. Embodiments of ion channels: (*a*) stretch-activated that move from the nominally closed to the nominally open state, or stretch-inactivated and move from the nominally open to the nominally closed state; (*b*), voltage-activated and open when the plasma membrane is depolarized; (*c*), ligand gated and open when a ligand binds to the channel; (*d*), nominally closed but cotransport ATP and Cl^- ions when activated by strain or a voltage change; (*e*), a gap junction channel that may be opened by a voltage change or strain to pass inositol trisphosphate (IP^3) or other small molecules; (*f*), a connexin hemichannel, or half a gap junction channel, that may be opened by strain or a voltage change to pass ions or even ATP.

other through gap junctions (Banes et al., 1994; Sood et al., 1999; Kenamond et al 1997). Functional gap junctions are important in tendon cells to transduce mechanical signals into Ca^{2+} and IP_3 intracellular second messengers *in vitro* and in tendons ex vivo to stimulate DNA and collagen synthesis (Banes et al., 1994, 1999; Kenamond et al., 1997, 1998). Figure 24.7*f* shows a single connexin hemichannel that does not appose another hemichannel. It is hypothesized that these hemichannels may pass ATP or other ions from the cell to the exterior (Cotrina et al., 1998).

Linkage of Matrix to Integrins to the Cytoskeleton

In cultured cells, integrins clustered in the plane of the membrane form a focal adhesion (Burridge et al., 1996; Schoenwaelder and Burridge, 1999). Integrins provide the transmembrane spanning portion of a direct mechanical linkage between matrix and the cell interior to transduce mechanical stimuli (Fig. 24.8). Focal adhesions are important structures in anchorage-dependent cultured cells because they mediate substrate attachment and act in one form of

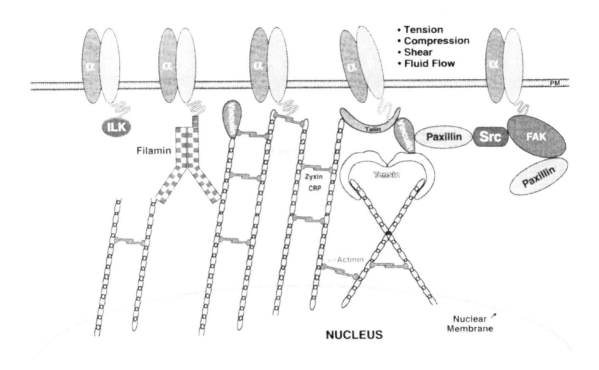

FIGURE 24.8. How transmembrane-spanning integrin subunits that unite extracellular matrix with actin-binding proteins such as a-actinin cross bridges, talin, filimin or focal adhesion kinase (FAK) can directly transmit strain in an outside-in signaling pathway where the stimulus may be tension, compression, shear or fluid flow. Deformation of the matrix-bound integrin activates talin, integrin-linked kinase (ILK), vinculin, tensin, paxillin and src. Activation of these proteins initiates cascades that may lead to further pathway activation stimulating cell-cell signaling, mediator release, mitogenesis, matrix synthesis or degradation or other reactions.

transmembrane signal transduction. Most connective tissue cells enter the apoptotic state and die if they become rounded for a prolonged period. Therefore, rapid and continued attachment to a matrix is vital. However, although integrins link matrix to cells *in vivo*, evidence of focal adhesions in tissues is lacking. During cell spreading, α and β integrin subunits bind matrix molecules at their extracellular face while the β subunit cytoplasmic tail binds α-actinin, filamin, or talin (Burridge et al., 1996; Schoenwaelder and Burridge, 1999). A second group of focal adhesion proteins that is involved in integrin-mediated signal transduction includes focal adhesion kinase (FAK), integrin-linked kinase (ILK), and c-src (pp60src) (Dedhar, 2000; Schwartz et al., 1995). The phosphorylation pathways that are downstream of these kinases have not been fully explored to date, although FAK plays a critical role in anchorage-dependent cell survival: FAK activation via integrin engagement suppresses apoptosis in a number of cell systems (Frisch and Ruoslahti, 1996). A third group of focal adhesion proteins, vinculin, paxillin and zyxin, function as molecular scaffolds that connect signaling pathways to focal adhesions and the actin cytoskeleton. Vinculin binds to both actin and paxillin (Johnson and Criag, 1995; Turner et al., 1990; Wood et al., 1994). Paxillin also binds to FAK (Hilldebrand et al., 1995; Turner and Miller, 1994). Paxillin integrates the actin cytoskeleton and an essential signaling pathway that impacts cell survival. Similarly, zyxin binds to both α-actinin (Crawford et al., 1992; Drees et al., 1999; Reinhard et al., 1999) and to members of the C-reactive protein (CRP) family of proteins (Sadler et al., 1992) that have been implicated in cellular differentiation (Stronach et al., 1999; Zhao et al., 1999). In summary, focal adhesions are multifunctional organelles that provide both physical connections between the exterior and internal compartments of the cell, and connect to signaling pathways that transduce mechanical stimuli to the nucleus (Banes et al., 1995, 2001).

Extracellularly, integrins bind to short amino acid sequences such as the RGD (Arginine, Glycine, Aspartic Acid) sequences in collagens and fibronectins, the YIGSR (Tyrosine, Isoleucine, Glycine, Serine, Arginine) sequence in laminins, and like sequences in proteoglycans, to the cytoskeletal network (Burridge and Chrzanowska-Wodnicka, 1996; Clark and Brugge, 1995). Integrins transduce mechanical stimuli by connecting extracellular matrix components to cytoskeletal elements and intracellular signaling pathways. In fact, mechanical stimuli have been shown to activate both src and FAK, and to result in increased phosphorylation of paxillin in tendon cells (Banes et al., 1995a,b; Brigman et al., 1996, 1997). In addition, mechanical stimuli cause physical remodeling and changes in the molecular composition of focal adhesions (Drees et al., 1999; Olesin et al., 1994; Qi et al., 2002). These new data support the view that focal adhesions are exquisitely sensitive to mechanical changes outside of the cell, and suggest that the cell can respond rapidly and dramatically to these changes by altering the very structures that detect mechanical load.

However, it should be noted that focal adhesions, per se, are observed almost exclusively in cultured cells. These structures appear to be exaggerated by growth of cells on a rigid planar surface. It is clear that integrins perform a similar function in linking matrix to the cytoskeleton *in vivo*. Examples of integrin-rich adhesive structures in organized tissues include the myotendinous junctions of skeletal muscle and the intercalated discs of cardiac muscle. These structures contain many of the same proteins found in the focal adhesions of cultured cells and serve the same purpose in transmitting tension at these sites (Tidball, 1986).

Signal transduction mechanisms discussed thus far have involved outside-in signaling. However, the matrix-integrin-cytoskeleton linkage also can exert force from the inside outward, onto the matrix. Cells can apply measurable forces to flexible substrates resulting in formation of wrinkles in the rubber, usually 90° to the long axis of the cell, perpendicular to the direction of movement (Burton et al, 1999; Lee et al., 1994; Oliver et al, 1994). Wang and Ingber showed that RGD ligand-coated magnetic beads could be oriented in a magnetic field, bound tightly to the cell surface in culture then resist rotation in a second magnetic field oriented 90° to the direction of the first field (Wang and Ingber, 1995). These data confirmed that integrins are functionally united to the cytoskeleton via matrix.

Cell-Cell Contacts

Proteins such as cadherins, catenins, and desmosomal proteins, that bind cells together, are intrinsically important in resisting normal forces generated in tissues (Fig. 24.9). A precedent example is desmoglein, a desmosomal protein, that when mutated, results in blister formation in the epidermis of patients with pemphigous vulgaris. Recent data have indicated that in MC3T3-E1 osteoblasts, two-dimensional substrate tension upregulates gene expression of desmocollin 3, an extracellular protein that links adjacent cells at a desmosome (Qi et al., 2002). Laminar flow up-regulates this gene but pulsating or flow reversals may inhibit or retard its expression.

Mechanical Load-Generated Second Messengers

A precedent report on rapid responses to mechanical load in cultured cells indicated that a static strain applied to osteoblasts *in vitro* stimulated secretion of cAMP, PGE_2, and Ca^{2+} (Binderman et al., 1988). Subsequently, it was reported that strain could stimulate secretion of NO (Nitric Oxide) and ATP in chondrocytes under cyclic compressive load (Fermor et al., 2001; Graff et al., 2000) while substrate strain and fluid shear increased intracellular calcium concentration in tenocytes, osteoblasts, chondrocytes, and annulus cells (Banes et al., 1994; Elfervig et al., 2001a; Guilak et al., 1999; Kenamond et al., 1997, 1998; Sood et al., 1999; You et

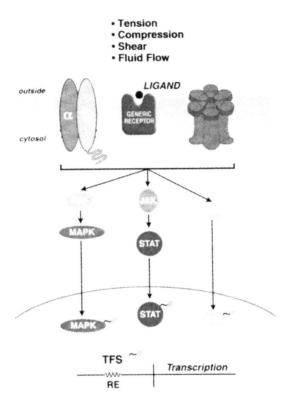

- Tension
- Compression
- Shear
- Fluid Flow

FIGURE 24.9. How outside-in mechanical stimuli can activate specific pathways involved in regulation of mitogenesis (MAPK), stress responses (JAK/STAT), or apoptosis (JNK). Tension, compression shear, or fluid flow can stimulate matrix-integrin-cytoskeletal connections, ligands activate specific receptors and ion channels can open or close in response to mechanical stimuli. Specific pathways can be activated depending on the type, magnitude, frequency, duration, and interval timing of the mechanical stimulus. Second messenger secretion may be affected in an acute response, or in an adaptive response, transcription factors can be activated that drive specific gene expression that regulates the cell phenotype.

al., 2000). Figure 24.10 shows a general model of second messengers that are active in response to mechanical load. In general, second messengers that are involved in ligand-activated pathways are also involved in mechano-stimulated pathways. Studies in which inhibitors have been used to test a role for a given pathway in a load response implicate stretch-activated ion channels in a gadolinium-sensitive step or some other ion channel in a connective tissue mechano-response

(Elfervig et al., 2000; Gu et al., 1996; Guilak et al., 1999; You et al., 2000).

Mechanical Load Activated Pathways

Load-activated pathways, such as kinase and phosphatase pathways, have been demonstrated in tendon cells especially phosphorylation of src, FAK and paxillin, proteins associated with focal adhesions and a putative mechanosensory complex (Banes et al., 1995b, 2001; Brigman et al., 1996, 1997). Other pathways have been reported that are involved in actin polymerization (Heidemann, 1999), and cell contraction such as rac and rho (Ridley, 1995), MAPK (Mitogan Activated Protein Kinase), ERK-(Extracellularly Regulated Kinase) integrin-stimulated pathways (MacKenna et al., 1998), SAPK, (Stress Activated Protein Kinase), JAK (Janus Kinase)/ STAT (Signal Transoucer and Activator of Transcription)/JNK (C-Jun-Terminal Kinase) in stress responses and potential apoptosis (Arnoczsky et al., 2001). Mechanical stimuli may act synergistically with mechanical load and stimulate connective tissue pathways such as Ca^{2+}-based signal transduction, synergy with growth factors to drive mitogenesis or increase responsiveness to destructive cytokines such as interleukin(IL)-1β (Banes et al., 1994, 1995a; Elfervig et al., 2001a; Xu et al., 2000). DNA synthesis in growth-arrested tendon cells is stimulated synergistically by picomolar amounts of platelet-derived growth factor (PDGF) or insulin-like growth factor-I (IGF)-I and cyclic substrate strain. Stimulation caused transit through the cell cycle due to cyclin D1 expression stimulated in G1 phase (Hu et al., 1994). Annulus cells pretreated with IL-1β then subjected to fluid flow had a heightened and prolonged increase in $[Ca^{2+}]_{ic}$, indicating an interaction in the pathways for a cytokine response and fluid flow (Elfervig et al., 2001a). Stretch increased the growth rate and collagen synthesis in nucleus pulposus cells (Matsumoto et al., 1999).

ATP acts as an autocrine/paracrine ligand acting at purinoceptors in the plasma membrane (Dubyak and el-Moatassim, 1993). ATP binding

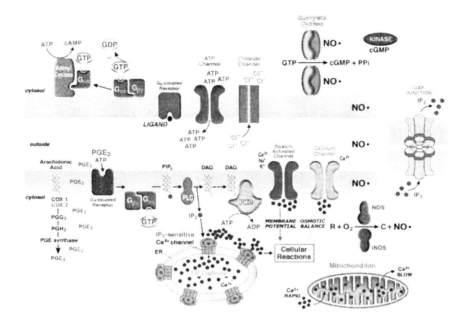

FIGURE 24.10. A generalized compendium of responses to load once the cell has detected a load signal. A first line of response is usually a shift in ion channel open state, either with a stretch-activated channel or a calcium, sodium, potassium, or chloride channel. These events are rapid, occurring in milliseconds, resulting in movement of 10^5 to 10^6 ions per second. Many cells respond to a deformation with a massive influx of extracellular calcium ions and signaling from cell to cell with IP_3, through gap junctions. This influx may occur with both L-type calcium channels and stretch-activated channels. IP_3 is generated at the plasma membrane from PIP2, phosphoinositol phosphate, by action of phospholipase C. IP_3 acts at a Ca^{2+} channel in the rough endoplasmic reticulum (ER cisterna) to liberate stores of intracellular calcium, available for intracellular reactions. Diacyl-glycerol is also liberated and acts with protein kinase C to phosphorylate proteins, including ion channel proteins, regulating channel activation. The mitochondrion is involved in a rapid uptake of intracellular Ca^{2+} and slow release of this ion. These events may affect membrane polarization and osmotic balance. Other rapid events may include activation of adenyl cyclase, a G protein-dependent enzyme, that produces CAMP, a second messenger. Likewise, prostanoids, such as prostaglandin E_2, especially synthesized from inducible cyclooxygenase 2 (COX 2), is released and reacts with its G protein-coupled receptor and may activate metalloproteinase expression. ATP released from the chloride-linked channels acts at purinoreceptors in another G-protein-coupled reaction. ATP may act as a load-modulating signal. Arginine is converted to nitric oxide (NO) by nitric oxide synthase (NOS) or the inducible enzyme (iNOS). NO diffuses through the cytoplasm and plasma membrane to adjacent cells to activate guanyl cyclase to convert GTP to cGMP, making it available for kinase reactions. Lastly, load acts synergistically with some ligand-activated receptors to drive certain pathways such as mitogenesis or synthesis of matrix metalloproteinases.

to P2Y2 receptors activates G proteins and increased $[Ca^{2+}]_{ic}$ through phospholipase C (PLC) action on phosphoinositol phosphate $(PIP)_2$ generating inositol tri-phosphate (IP_3) and intracellular calcium release from internal stores in chondrocytes, tendon cells and annulus cells (Cotrina et al., 1998; Elfervig et al., 2001b; Graff et al., 2000). P2U receptor activation is not G-protein dependent.

Cells Respond Positively and Negatively to Mechanical Load Using Diverse and Redundant Pathways

Most of the literature addressing responses to mechanical load demonstrate that cells respond in a positive way to tension, compression or fluid flow. Motion is vital to proper function of all tissues and cells. Immobilization or skeletal unloading leads to muscle atrophy, bone demineralization, loss of biomechanical strength, and weakness. On the cellular level, removal of contacts with the substrate in anchorage dependent cells results in cell rounding and eventual cell death (Duncan and Jurner, 1995). Recently, Archambault and co-workers showed that tendon cells respond to tension and fluid flow by expressing matrix metalloproteinases (MMP) (Archambault 2002). MMP 1 and 3 genes are sensitive to cytokines such as IL-1β and can be stimulated synergistically with load. However, the enzymes require activation before matrix can be degraded (Bowman et al., 2002).

Y.C. Fung showed that all tissues have an intrinsic strain level (Fung, 1981). Tissues depend on mechanical cues for homeostasis and to model and remodel (Banes et al., 1995, 2001; Frost, 1988). The controversies and questions in the cytomechanics and tissue engineering fields at present, with respect to connective tissue cell responses to load involve the following: (1) the optimum magnitude, frequency, duration and type of load applied to the cell in two- or three-dimensional environments, and (2) understanding the hierarchy of load signals to achieve a positive outcome. It has been reported that cells, particularly osteoblasts, may be more sensitive to fluid flow than to substrate strain (Duncan and Turner, 1995; You et al., 2000, 2001). However, one must test a number of responses and gene read-outs before a general statement can be made with certainty. On the cellular level, responses to mechanical load are now well documented and the challenge is to apply strain to cells and tissues in as near a physiologic mode as is possible and measure tissue-specific outcomes.

One aspect of mechanical load effects on cells that has been understudied is that of pulse dampening rather than stimulation by load (Banes et al., 1995, 2001). Cells must desensitize to load signals after some point. We interpret the minimum effective strain (MES) concept for bone put forth in the mechanostat theory to indicate that cells take an early cue from mechanical stimuli to activate a pathway, for instance, toward mineralization in bone or matrix synthesis in several connective tissues

(Frost, 1987). Generally, we feel that mechanical signals are involved in regulating a setpoint for tissue homeostasis.

Future questions to address include investigating commonalities and differences in cell deformation responses from tension, compression, shear, and fluid flow. Are there markers for specific gene expression and protein profiles that are load responsive? Is there a dominant load pathway? If so, can this pathway be utilized in a tissue engineering application to build functional connective tissue constructs or improve healing when applied as a physical therapy regimen? Are there load-specific drugs that can positively modify connective tissue development, homeostasis and healing? These are some of the challenges to be met in uniting the fields of cytomechanics and tissue engineering.

Acknowledgments

The authors would like to thank Ms. Betty Horton in the UNC Medical Illustrations Department for the detailed figures in this manuscript. Supported inpart by NIH grants AR38121 and AR45833

References

Applegate DO, Applegate M, Baumgartner M, Bennett JW, Danssaert J, Hardin R, Laiterman L, Schramm L, Tolbert WR. 1998. Apparatus for the growth and packaging of three dimensional tissue cultures. US patent # 5,843,766, assignee: Advanced Tissue Sciences, Inc, La Jolla, CA.

Archambault JM, Wiley JP, Bray RC. 1995. Exercise loading of tendons and the development of overuse injuries. *Sports Med.* 20:77–89.

Archambault J, Elfervig-Wall M, Tsuzaki M, Herzog W, Banes AJ. 2002a Rabbit tendon cells produce MMP-3 in response to fluid flow without significant calcium transients. *J Biomech.* 35:303–309, 2002.

Archambault J, Tsuzaki M, Herzog W, Banes AJ. 2002b. Stretch and interleukin-1b induce matrix metalloproteinase in rabbit tendon cells. *J. Orthop Res.* 20:36–39, 2002.

Arnoszcky S, Tiun T, Schuler P, Morse R, Lavaguno M, Gardner K. 2001. Upregulation of stress-activated protein kinases (SAPK) in response to increased cytosolic calcium levels due to cyclic strain: a potential cellular mechanism for repetitive stress injuries in tendons. *Trans. Orthop. Res. Soc.* 47:20.

Backman C, Boquist L, Friden J, Lorentzon R, Toolanen G. 1990. Chronic Achilles paratenonitis with tendinosis: an experimental model in the rabbit. *J. Orthop. Res.* 8:541–547.

Banes AJ, Gilbert J, Taylor D, Monbureau O. 1985. A new vacuum-operated stress-providing instrument that applies static or variable duration cyclic tension or compression to cells *in vitro. J Cell Sci.* 75:35–42.

Banes AJ, Sanderson M, Boitano S, Hu P, Brigman B, Tsuzaki M, Fischer T, Lawrence WT. 1994. Mechanical load +\− growth factors induce [Ca^{2+}]$_{ic}$ release, cyclin D1 expression and DNA synthesis in avian tendon cells. In: *Cell Mechanics and Cellular Engineering.* VC Mow, F Guilak, R Tran Son Tay, R Hochmuth, eds Springer Verlag, New York pp 210–232.

Banes AJ, Tsuzaki M, Hu P, Brigman B, Brown T, Almekinders L, Lawrence WT, Fischer T. 1995a. Cyclic mechanical load and growth factors stimulate DNA synthesis in avian tendon cells. *J. Biomech.* 28:1505–1513.

Banes A, Tsuzaki M, Yamamoto J, Fischer T, Brigman B, Brown T, Miller L. 1995b. Mechanoreception at the cellular level: the detection, interpretation, and diversity of responses to mechanical signals. *Biochem. Cell. Biol.* 73:349–365.

Banes AJ, Weinhold P, Yang X, Tsuzaki M, Bynum D, Bottlang M, Brown T. 1999. Gap junctions regulate responses of tendon cells ex vivo to mechanical loading. *Clin. Orthop.* S356–S370.

Banes AJ, Lee G, Graff R, Otey C, Archambault J, Tsuzaki M, Elfervig M, Qi J. 2001. Mechanical forces and signaling in connective tissue cells: Cellular mechanisms of detection, transduction and responses to mechanical deformation. *Curr. Opin. Orthop.* 12:389–396.

Banes AJ, Garvin J, Wall M, Qi J. 2001. BATS Bioartificial tissues: dynamic 3D cell culture as a model for tissue engineering, factor/drug response. Biomechanics Society, San Diego, 2001.

Bell E, Ivarsson B, Merrill C. 1974. Production of a tissue-like structure by contraction of collagen lattices by human fibroblasts of different proliferative potential *in vitro. Proc. Natl. Acad. Sci. USA.* 76:1274.

Binderman I, Zor U, Kaye AM, Shimshoni Z, Harell A, Somjen D. 1988. The transduction of mechanical force into biochemical events in bone cells may involve activation of phospholipase A2. *Calcif. Tissue Intl.* 42:261–266.

Boitano S, Dirksen ER, Sanderson MJ. 1992. Intercellular propagation of calcium waves mediated by inositol trisphosphate. *Science* 258:292–295.

Boitano S, Dirksen E, Evans W. 1998. Sequence-specific antibodies to connexins block intercellular calcium signaling through gap junctions. *Cell Calcium* 23:1–9.

Bowman K, Herzog W, Tsuzaki M, Guyton G, Spencer K, Archambault A, Banes AJ. 2002. Il-*1B* induced MMPs degrade collagen in rabbit Achilles tendons. *Trans. Orthop. Res. Soc.* 48:60.

Brigman B, Tsuzaki M, Schaller M, Fischer T, Brown T, Horesovsky G, Miller L, Benjamin M, Ralphs J, McNeilly C, Banes A. A mechanosensory protein complex: Paxillin, c-SRC and FAK associate and activate in response to mechanical load. *Transactions of the 43rd Annual Meeting of the ORS,* Vol. 22 (1), 1997, p. 711.

Brigman B, Fischer T, Tsuzaki M, Brown T, Miller L, Banes A. 1996. Mechanical load activates tendon cell pp60src phosphorylation: Role in a proposed mechanosensory complex. *Trans. Orthop. Res. Soc.* 21:1.

Bruzzone R, White T, Paul D. 1996. Connections with connexins: the molecular basis of direct intercellular signaling. *J. Biochem.* 238:1–27.

Buckley MJ, Banes AJ, Levin LG, Sumpio BE, Jordan R, Sato M. 1988. Osteoblasts increase their rate of division in response to cyclic mechanical strain *in vitro. J Bone Miner* 4:225–236.

Burridge K, Chrzanowska-Wodnicka M. 1996. Focal adhesions, contractility and signaling. *Annu. Rev. Cell. Dev. Biol.* 12:463–519.

Burton K, Paark JH, Taylor DL. 1999. Keratinocytes generate traction forces in two phases. *Mol. Biol. Cell* 10:3745–3769.

Buschmann MD, Hunziker EB, Kim YJ, Grodzinsky AJ. 1995. Coordinated changes in biosynthesis and morphology of cells and nuclei in articular cartilage under compression. *Trans. Orthop. Res. Soc.* 20:289.

Carpenter JE, Thomopoulos S, Soslowsky LJ. 1999. Animal models of tendon and ligament injuries for tissue engineering applications. *Clin. Orthop* 367: S296–S311.

Carvalho R, Scott E. Yen E. 1995. The effects of mechanical stimulation on the distribution of β_1 integrin and expression of β_1-integrin mRNA in te-85 human osteosarcoma cells. *Arch. Oral Biol.* 40: 257–264.

Clark E, Brugge J. 1995. Integrins and signal transduction pathways: the road taken. *Science* 268: 233–239.

Cotrina ML, Lin JHC, Alves-Rodrigues, Liu S, Li J, Azmi-Ghadmi H, Kang J, Naus CCG, Nedergaard M. 1998. Connexins regulate calcium signaling by controlling ATP release. *Proc. Natl. Acad. Sci. U.S.A.* 95:15735–15740.

Crawford AW, Michelsen JW, Beckerle MC. 1992. An interaction between zyxin and alpha-actinin. *J. Cell. Biol.* 116:1381–1193.

Dedhar S. 2000. Cyclic stretching stimulates subsequent cell attachment and spreading to matrix cell-substrate interactions and signaling through ILK. *Curr. Opin. Cell. Biol.* 12:250–256.

Dennis RG, Kosnick P. 2000. System and method for emulating an *in vitro* environment of a muscle tissue specimen. United States Patent patent number 6, Sept. 5, 2000, 114,164.

Drees BE, Andrews KM, Beckerle MC. 1999. Molecular dissection of zyxin function reveals its involvement in cell motility. *J. Cell. Biol.* 147:1549–1560.

Dubyak GR, el-Moatassim C. 1993. Signal transduction via P2-purinergic receptors for extracellular ATP and other nucleotides. J Phys Am 265: C577–C606.

Duncan RL, Turner CH. 1995. Mechanotransduction and the functional response of bone to mechanical strain. *Calcif Tissue Int.* 57:344–358.

Elfervig M, Minchew J, Francke E, Tsuzaki M, Banes A. 2001a. IL-1β sensitizes intervertebral disc annulus cells to fluid-induced shear stress. *J. Cell Biochem.* 82:290–298.

Elfervig MK, Graff RD, Lee GM, Kelley SS, Sood A, Banes AJ. 2001b. ATP induces Ca^{2+} signaling in human chondrons cultured in three-dimensional agarose films. Osteo Cart 9:518–526.

Elfervig M, Francke E, Archambault J, Herzog W, Tsuzaki M, Bynum D, Brown TD, Banes AJ. 2000. Fluid-induced shear stress activates human tendon cells to signal through multiple Ca^{2+} dependent pathways. 46th Trans. Orthop. Res. Soc. 25:179.

Fermor B, Weinberg J, Pisetsky D, Misukonis M, Banes A, Guilak F. 2001. The effects of static and dynamic compression on nitric oxide production in articular cartilage explants. *J. Orthop. Res.* 19:72–80.

Frisch SM, Ruoslahti E. 1997. Integrins and anoikis. *Curr. Opin. Cell. Biol.* 9:701–706.

Frisch SM, Vuori K, Ruoslahti E, Chan-Hui PY. 1996. Control of adhesion-dependent cell survival by focal adhesion kinase. *J. Cell. Biol.* 134:793–799.

Frost HM. 1987. The mechanostat: a proposed pathogenic mechanism of osteoporoses and the bone mass effects of mechanical and nonmechanical agents. *Bone Miner. Res.* 2:73–85.

Fung YC. 1981. *Biomechanics. Mechanical Properties of Living Tissues.* Springer Verlag, New York.

Garcia-Anoveros J, Corey D. 1996. Mechanosensation: touch at the molecular level. *Curr. Biol.* 65:541–543.

Garvin J, Banes AJ. 2002. Novel culture system for the development of a bioartificial tendon. *Trans. Orthop. Res. Soc.* 48:31.

Graff RD, Lazarowski ER, Banes AJ, Lee AJ. 2000. ATP release by mechanically loaded porcine chondrons in pellet culture. Arthritis Rheum 43:1571–1579.

Gray ML, Pizzanelli AM, Grodzinsky AJ, Lee RC. 1988. Mechanical and physicochemical determinants of the chondrocyte biosynthetic response. *J. Orthop. Res.* 6:777–792.

Gu G, Caldwell G, Chalfie M. 1996. Genetic interactions affecting touch sensitivity in caenorhabditis elegans. *Proc Natl Acad Sci* U.S.A. 93:6577–6582.

Guharay F, Sachs F. 1984. Stretch-activated signal ion channel currents in tissue-cultured embryonic chick skeletal muscle. *J. Physiol.* 352:685–701.

Guilak F. 1995. Compression-induced changes in the shape and volume of the chondrocyte nucleus. *J. Biomech.* 28:1529–1541.

Guilak F, Zell R, Erickson G, Grande D, Rubin C, McLeod K, Donahue H. 1999. Mechanically induced calcium waves in articular chondrocytes are inhibited by gadolinium and amiloride. *J. Orthop. Res.* 17:421–429.

Elfervig M, Francke E, Archambault J, Herzog W, Tsuzaki M, Bynum D, Brown TD, Banes AJ. 2000. Fluid-induced shear stress activates human tendon cells to signal through multiple Ca^{2+} dependent pathways. 46th Trans. Orthop. Res. Soc. 25:179.

Hamill O, Martinac B. 2001. Molecular basis of mechanotransduction in living cells. *Phys. Rev.* 81:685–739.

Hamill OP, McBride DW. 1996. The pharmacology of mechanogated membrane ion channels. *Pharm. Rev.* 48:231–252.

Hannafin JA, Arnoczsky SA, Hoonjan A, Torzilli PA. 1995. Effect of stress deprivation and cyclic tensile loading on the material and morphologic properties of canine flexor digitorum profundus tendon: an *in vitro* study. *J. Orthop. Res.* 13:907–914.

Heidemann S, Kaech S, Buxbaum R, Matus A. 1999. Direct observations of the mechanical behaviors of the cytoskeleton in living fibroblasts. *J. Cell. Biol.* 145:109–122.

Hildebrand JD, Schaller MD, Parsons JT. 1995. Paxillin, a tyrosine phosphorylated focal adhesion-associated protein binds to the carboxyl terminal domain of focal adhesion kinase. *Mol. Biol. Cell.* 6:637–647.

Howard J, Roberst WM, Hudspeth AJ. 1988. Mechanoelectrical transduction by hair cells. *Annu. Rev. Biophys. Chem.* 17:99–124.

Hu P, Xiao H, Brigman B, Lawrence WT, Banes AJ. 1994. G1 D cyclins are differentially regulated in tendon epitenon and internal fibroblasts. *Trans. Orthop. Res. Soc.* 19(2):639.

Hung CT, Pollack SR, Reilly TM, Brighton CT. 1995. Real-time calcium response of cultured bone cells to fluid flow. *Clin Orthop* 313:256–269.

Hung C, Allen F, Pollack S, Attia E, Hannafins J, Torzcilli P. 1997. Intracellular calcium response of ACL and MCL ligament fibroblasts to fluid-induced shear stress. *Cell Signal* 9:587–594.

Hyman J, Rodeo SA. 2000. Injury and repair of tendons and ligaments. *Phys. Med. Rehab. Clin. North. Am.* 11:267–288.

Ilic D, Almeida EA, Schlaepfer DD, Dazin P, Aizawa S, Damsky CH. 1998. Extracellular matrix survival signals transduced by focal adhesion kinase suppress p53-mediated apoptosis. *J. Cell. Biol.* 143:547–560.

Ingber DR. 1993. Cellular tensegrity: defining new rules of biological design that govern the cytoskeleton. *J. Cell. Sci.* 104:613–627.

Ingber DE. 1997. Tensegrity: the architectural basis of cellular mechanotransduction. *Annu. Rev. Physiol.* 59:575–599.

Johnson RP, Craig SW. 1995. F-actin binding site masked by the intramolecular association of vinculin head and tail domains. *Nature* 373:261–264.

Jozsa L, Kannus P. 1997. Innervation and mechanoreceptors of tendons. In: *Human Tendons: Anatomy, Physiology and Pathology.* Jozsa L, Kannus P, eds. Champaign, IL, Human Kinetics, 87–90.

Kenamond C, Weinhold P, Bynum D, Tsuzaki M, Benjamin M, Ralphs J, McNeilly C, Banes AJ. 1997. Human tendon cells express connexin-43 and propagate a calcium wave in response to mechanical stimulation. *Transactions of the 43rd Annual Meeting of the Orthopaedic Research Society,* Vol. 22(1), p.179–180.

Kenamond C, Boitano S, Francke E, Sood A, Yang X, Faber J, Bynum D, Banes A. 1998. Cyclic strain activates tendon cells and primes them for a second mechano-stimulus to increase intracellular calcium. *Trans of the 44th Annual Mtg of the ORS,* Vol 23, Sect 1, p. 92.

Kerr R, Wang XT, Pike AV. 2000. Fatigue quality of mammalian tendons. *J. Exp. Biol.* 203:1317–1327.

Lanyon L. 1987. Functional strain in bone tissue as an objective and controlling stimulus for adaptive bone remodeling. *J. Biomech.* 20:1083–1093.

Lee J, Leonard M, Oliver T, Ishihara A, Jacobson K. 1994. Traction forces generated by locomoting keratinocytes. *J. Cell. Biol.* 127:1957–1964.

MacKenna DE, Dolfi F, Vuori K, Ruoslahti E. 1998. Extracellular signal-regulated kinase and c-jun NH2-terminal kinase activation by mechanical stretch is integrin-dependent and matrix-specific in rat cardiac fibroblasts. *J. Clin. Invest.* 101: 301–310.

Matsumoto T, Kawakami M, Kuribayashi K, Takenaka T, Tamaki T. 1999. Cyclic mechanical stretch stress increases the growth rate and collagen synthesis of nucleus pulposus cells *in vitro. Spine* 24:315–319.

Olesin SP, Clapham DE, Davies PF. 1988. Haemodynamic shear stress activates a K^+ current in vascular endothelial cells. *Nature* 331:168–170.

Oliver T, Lee J, Jacobson K. 1994. Forces exerted by locomoting cells. *Cell Biol.* 5:139–147.

Qi J, Samuhel K, Jones J, Banes AJ. 2002. Expression of desmocollin 3, a cell adhesion protein, is upregulated by stretch and laminar flow but suppressed by oscillating flow and flow reversal. *Trans. Orthop. Res. Soc.* 27:[CD].

Reinhard M, Zumbrunn J, Jaquemar D, Kuhn M, Walter U, Trueb B. 1999. An alpha-actinin binding site of zyxin is essential for subcellular zyxin localization and alpha-actinin recruitment. *J. Biol. Chem.* 274:13410–13418.

Ridley AJ. 1995. Rho-related proteins: actin cytoskeleton and cell cycle. *Curr. Opin. Genet. Dev.* 5:24–30.

Riveline D, Zamir E, Balaban NQ, Schwarz US, Ishizaki T, Narumiya S, Kam Z, Geiger B, Bershadsky A. 2001. Focal contacts as mechanosensors: externally applied local mechanical force induces growth of focal contacts by an mDia1-dependent and ROCK-independent mechanism. *J. Cell. Biol.* 153:1175–1185.

Sachs F. 1988. Mechanical transduction in biological systems. *Crit. Rev. Biomed. Eng.* 16:141–169.

Sadler I, Crawford AW, Michelsen JW, Beckerle MC. 1992. Zyxin and cCRP: two interactive LIM domain proteins associated with the cytoskeleton. *J. Cell. Biol.* 119:1573–1587.

Schoenwaelder SM, Burridge K. 1999. Bidirectional signaling between the cytoskeleton and integrins. *Curr. Opin. Cell. Biol.* 11:274–286.

Schwartz MA, Schaller MD, Ginsberg MH. 1995. Integrins: emerging paradigms of signal transduction. *Annu. Rev. Cell. Dev. Biol.* 11:549–599.

Schweibetrt EM. 1999. ABC transporter-facilitated ATP conductive transport. *Am. Phys. Soc.* 276:1–8.

Sood A, Bynum D, Boitano S, Weinhold P, Tsuzaki M, Brown T, Banes A. 1999. Gap junction blockade inhibits Ca^{2+} signaling *in vitro* and mechanical load-induced mitogenesis and collagen synthesis in avian tendons ex vivo. *45th Trans. Orthop. Res. Soc.* Vol 24, (2), p1083.

Stronach BE, Renfranz PJ, Lilly B, Beckerle MC. 1999. Muscle LIM proteins are associated with muscle sarcomeres and require dMEF2 for their expression during Drosophila myogenesis. *Mol. Biol. Cell.* 10:2329–2342.

Sukharev S, Blount P, Martinac B, Blattner F, Kung C. 1994. A large-conductance mechanosensitive channel in E. coli encoded by mscL alone. *Nature* 368:265–268.

Sussman NL, Kelly JH. 1995. The artificial liver. *Sci. Med.* 2:68–77.

Tidball JG. 1986. Energy stored and dissipated in skeletal muscle basement membranes during sinusoidal oscillations. *Biophys. J.* 50:1127–1138.

Turner CE, Miller JT. 1994. Primary sequence of paxillin contains putative SH2 and SH3 domain binding motifs and multiple LIM domains: identification of a vinculin and pp125Fak-binding region. *J. Cell. Sci.* 107:1583–1591.

Turner CE, Glenney JR Jr, Burridge K. 1990. Paxillin: a new vinculin-binding protein present in focal adhesions. *J. Cell. Biol.* 111:1059–1068.

Vadiakas G, AJ Banes. 1992. Verapamil decreases cyclic load-induced calcium incorporation in ROS 17/2.8 osteosarcoma cell cultures. *Matrix* 12: 439–447.

Vandenburgh H. 1987. Motion into mass: How does tension stimulate muscle growth? *Med. Sci. Sports Exerc.* 19:S142–S149.

Vandenburgh H. 1988. A computerized mechanical cell stimulator for tissue culture: effects of skeletal muscle organogenesis. *in vitro Cell Dev. Biol.* 24:609–618.

Wang N, Ingber DE. 1995. Probing transmembrane mechanical coupling and cytomechanics using magnetic twisting cytometry. *Biochem. Cell. Biol.* 73:327–335.

Wolff J. 1892. The Law of Bone Remodeling (trans. Maquet P and Furlong R). Springer—Verlag, New York. 1–26.

Wood CK, Turner CE, Jackson P, Critchley DR. 1994. Characterisation of the paxillin-binding site and the C-terminal focal adhesion targeting sequence in vinculin. *J. Cell. Sci.* 107:709–717.

Xu Z, Buckley M, Evans C, Agarwal S. 2000. Cyclic tensile strain acts as an antagonist of IL-1β actions in chondrocytes: *J. Immunol.* 165:453–460.

Yannas IV, Burke JF, Orgill DP, Skrabut EM. 1982. Wound tissue can utilize a polymeric template to synthesize a functional extension of skin. *Science* 215:174–176.

You J, Reilly GC, Zhen X, Yellowley CE, Chen Q, Donahue HJ, Jacobs C. 2001. Osteopontin gene regulation by oscillatory fluid flow via intracellular calcium mobilization and activation of mitogen-activated protein kinase in MC3T3-E1 osteoblasts. *J. Biol. Chem.* 276:13365–13371.

You J, Yellowley CE, Donahue HJ, Zhang Y, Chen Q, Jacobs CR. 2000. Substrate deformation levels associated with routine physical activity are less stimulatory to bone cells relative to loading-induced oscillatory fluid flow. *J. Biomech. Eng.* 122:387–393.

Yu J, Banes AJ, Otey C. 2000. Mechanical conditioning of connective tissue cells. *Amer Soc. Cell Biol.* 11:226A

Zhao MK, Wang Y, Murphy K, Yi J, Beckerle MC, Gilmore TD. 1999. LIM domain-containing protein trip6 can act as a coactivator for the v-Rel transcription factor. *Gene Exp* 8:207–217.

Part VI

Bioreactors and the Role of Biophysical Stimuli in Tissue Engineering

25

The Role of Biomechanics in Analysis of Cardiovascular Diseases: Regulation of the Fluid Shear Response by Inflammatory Mediators

Geert W. Schmid-Schönbein and Shunichi Fukuda

Introduction

Mechanics is an essential tool to develop a quantitative analysis of the circulation. While biomechanics of the circulation is still in development, many details of blood flow have already been studied in both large and small blood vessels. Biomechanics permits analysis with realistic and testable assumptions about the organ microvasculature, the mechanical properties of the blood vessel, and the rheological properties of the blood (Fung, 1997). Biomechanical models have facilitated the independent testing of theoretical predictions, a process that has served to considerably increase the quantitative understanding of many aspects in the circulation (Lee, 2000; Schmid-Schönbein, 1999).

It is important to apply biomechanical analysis to the complex events that occur in circulatory failure, such as ischemia, reperfusion injury, thrombosis, and shock. Biomechanics has already been an effective tool in selected aspects that occur during circulatory complication. But a more extensive analysis of cardiovascular disease requires new directions in biomechanics. At the center is the so-called *inflammatory process,* long recognized in medicine as a cellular and immunological defense and repair mechanism (Zweifach et al., 1974) and more recently also as a poten-

tially pathogenic process (Granger and Schmid-Schönbein, 1995). Thus an analysis of the inflammatory process is of importance not only in regards to our understanding of cardiovascular disease but also in regards to tissue engineering in which wound healing plays a central role.

The Inflammatory Process

An increasing body of evidence has accumulated to suggest that most cardiovascular diseases and potentially many other diseases are accompanied by an inflammatory reaction (Entman et al., 1991; Granger and Schmid-Schönbein, 1995; Ho and Bray, 1999; Oparil and Oberman, 1999; Ray et al., 1998; Ross, 1999; Thomas, 1974). Inflammatory reactions consist of several events and are readily detected in the microcirculation. Inflammation starts with a relatively innocent elevation of endothelial permeability (Lush and Kvietys, 2000; McDonald et al., 1999; Moore et al., 1998). It is followed by expression of membrane adhesion molecules on leukocytes, platelets, and endothelial cells, formation of oxygen free radicals, synthesis of inflammatory mediators, such as platelet activating factor, growth factors, and proteases, as well as suppression of anti-inflammatory mediators (Granger and Kubes, 1994; Gute et al., 1998). There is mast cell activation and degranulation. Segments of the capillary network become obstructed; leukocytes accumulate in microvessels, they adhere to the endothelium, to each other, and to patellets; and they migrate into the interstitium (Del Zoppo et al., 1991; Engler et al., 1983; Harris et al., 1994; Ritter et al., 1995). Many cell functions become compromised and eventually are followed by apoptotic and necrotic cell death (Suematsu et al., 1992, 1994; Tung et al., 1997). To analyze cardiovascular diseases, the inflammatory process is one of the most important phenomena to understand, from its trigger mechanisms to the long-term pathophysiological consequences. Leukocyte rolling and attachment to the endothelium has been analyzed in considerable detail (Damiano et al., 1996; Dong et al., 1999; Rodgers et al., 2000; Zhao et al., 1995) and there is a large biological literature.

In this chapter we will focus on one particular aspect of inflammation—cell activation. Cell activation is a requirement to trigger inflammation and therefore we are interested in this process as a vehicle to explore possible pathways for prevention. Traditionally, inflammation has been thought of as a consequence of the action of stimulatory mediators and many different biochemical candidates have been proposed (Mazzoni and Schmid-Schönbein, 1996). More recently, it has also become evident that in addition to humoral mediators, mechanical stresses serve to control the inflammatory reaction. The endothelium, leukocytes, and platelets respond to fluid shear. The endothelial cell has been studied in greatest detail (Barakat, 1999; Davies, 1995; Papadaki et al., 1999; Topper and Gimbrone, 1999). Physiological levels of fluid shear stress serve to maintain low levels of inflammation while disturbance of the fluid shear leads to synthesis of proinflammatory genes and a proinflammatory reaction (Gimbrone et al., 1997; Traub and Berk, 1998). In the following, we will examine the fluid shear response in the case of circulating leukocytes, one of the key mediators of inflammation in the circulation.

These observations in a tissue culture have already been expanded to tissue engineered vascular grafts (Golledge, 1997). For example, preservation of the fluid shear field in a vascular prosthesis as close as possible to a normal physiological values serves to minimize leukocyte attachment (Liu et al., 1999) as well as blood flow-related focal intimal hyperplasia (Liu, 1998).

Fluid Shear Inactivation in Circulating Leukocytes

Pseudopod formation is a requirement for amoeboid migration and phagocytosis of leukocytes. Pseudopods are formed by redistribution of actin and local polymerization with local extension of the cytoplasm. The process is regulated by a class of proteins that control the actin fiber length, the branching of actin fibers and their binding to each other (Lee et al., 2001; Zigmond, 2000). Pseudopods are relatively rigid structures that exhibit reduced viscoelastic creep compared with the main cell body (Schmid-Schönbein and Skalak, 1984). Circulating leukocytes have few pseudopods (Schmid-Schönbein et al., 1980). In the undeformed state, leukocytes have a spherical shape with numerous membrane folds, and there are no large membrane projections that can be designated as pseudopods (we typically classify projections greater than 1 μm as pseudopods). Leukocytes with pseudopods in the circulation have high probability to be trapped in the capillary network (Harris and Skalak, 1993; Ritter et al., 1995; Schmid-Schönbein, 1987; Worthen et al., 1989) and therefore are quickly removed from the active circulation. Pseudopod formation and retraction is a relatively slow process, which is of the order of magnitude of minutes, a period of time that is long compared with the transit time through capillaries, which is in seconds.

Leukocytes use pseudopods during amoeboid motion to migrate from the bone marrow across the endothelium into the lumen of microvessels (Campbell, 1972; Chamberlain and Lichtman, 1978; De Bruyn et al., 1971; Möhle et al., 1997). In peripheral microvessels they adhere to the endothelium and then migrate in reversed direction back across the microvascular wall out into the interstitium (Farr and De Bruyn, 1975; Hammersen and Hammersen, 1987; Marchesi and Florey, 1961; Ohashi et al., 1996). In both direc-

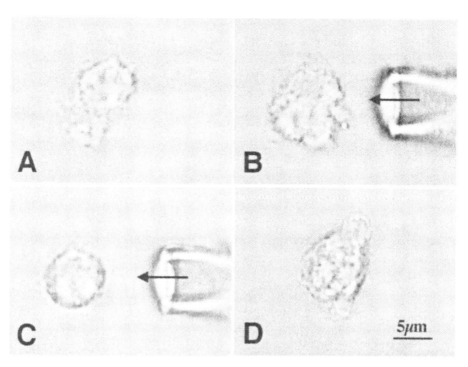

FIGURE 25.1. Micrographs of a human leukocyte observed on an inverted light microscope with 100 x objective magnification (numerical aperture 1.4) during application of fluid shear stress from a micropipette (right side in panels B and C). Note initial pseudopod formation in (A), progressive pseudopod retraction at 60 sec (B) and 120 sec after (C) application of fluid shear from the micropipette, as well as new formation of pseudopods already 60 sec (D) after discontinuation of a fluid shear. The magnitude of the fluid shear stress is of the order of 1 dyn/cm2 and is nonuniform over the membrane of the cell.

tions of migration the cells require pseudopod formation to achieve a displacement. This situation raises the question, what mechanism may exist that prevents pseudopod formation in the circulation? Furthermore, if a mechanism exists that serves to prevent pseudopod formation in leukocytes in the active circulation, how is it possible that leukocytes will eventually adhere to the endothelium and project pseudopods in spite of normal physiological shear rates, as is frequently observed in inflammation?

To examine whether fluid shear may play a role in this process, we exposed individual leukocytes to a set of well-defined fluid stresses, as illustrated in Figure 25.1. Leukocytes were collected by sedimentation of heparinized blood from the supernatant and placed on a coverslip into a cell chamber on a high-resolution microscope. Fluid motion of a mixture of buffer and plasma over the surface of the cells was gener-

ated from a micropipette placed adjacent to the leukocyte. The fluid shear stress on the surface of the cell was computed in the following fashion. Since the buffer/plasma mixture is incompressible and has a Newtonian viscosity and since the Reynolds number is about 10^{-5} the fluid velocity is governed by the Stokes approximation of the equation of fluid motion. We solved these equations numerically, assuming no slip condition on the surface of the cell and the glass surface. There is a parabolic velocity profile out of the micropipette tip with maximum velocity and constant pressure at a distance in the fluid far away from the leukocyte.

For typical experimental values of cell and pipette diameter (about 8 μm and 2 μm, respectively) and position relative to the cell (about 10 μm), the fluid shear stress in these experiments is nonuniform over the cell membrane. It reaches its maximum close to the micropipette tip at about 1.6 dyn/cm^2 (Moazzam et al., 1997). At the cell membrane positioned away from the pipette, the fluid shear stress assumes values close to zero. The normal stress due to the flow out of the pipette has values that are comparable in magnitude to the fluid shear stress. The values serve as order of magnitude estimates, the exact values of the fluid shear stress need to be computed in the future considering the details of the surface membrane folds or pseudopods.

The application of fluid shear stress leads to a rapid retraction of pseudopods by the leukocytes (Moazzam et al., 1997). The retraction rate is relatively insensitive with respect to the magnitude of the shear stress in a physiological range. Removal of the shear stress leads to projection of the pseudopods. The process is reproducible. If the fluid shear stress is applied over a prolonged period of time (several minutes), the leukocyte will swell and dramatically reduce the cytoplasmic mechanical stiffness (Moazzam et al., 1997). The fluid shear response can be inhibited by membrane K^+ ion channel blocker or by chelation of calcium.

Reduction of physiological fluid stress *in vivo* in vessels of the microcirculation by occlusion leads to projection of pseudopods on many leukocytes. In contrast, restoration of fluid shear stress serves to stimulate retraction of pseudopods and eventually permit detachment from the endothelium. Fluid shear inactivation on leukocytes is reproducible in the microcirculation. But this observation raises also the important question: what mechanisms exist that serve to prevent the fluid shear response? Without blockade of the fluid shear inactivation, leukocyte cannot spread on the endothelium in a flowing blood vessel and migrate into the adjacent tissue, as in the case of inflammation.

The Regulation of Fluid Shear Stress Response by Inflammatory Mediators

Leukocytes in a passive spherical state without pseudopods cannot spread on the endothelium in order to establish firm attachment and to migrate into the tissue. Thus, a normal inflammatory or immune reaction cannot proceed in a perfused blood vessel unless there exist mechanisms that serves to inhibit the fluid shear response. There must exist a mechanism by which cell spreading may occur on the endothelium even under conditions of normal blood flow and shear stress. To examine this issue in-vitro, we applied during fluid shear some selected inflammatory mediators, such as platelet activating factor (PAF), the chemotactic peptide formyl-methionl-leucyl-phenylalanine (FMLP) or the cytokine tumor necrosis factor-α. Such mediators serve to suppress the shear stress response of leukocytes in a dose-dependent manner (Fukuda et al., 2000). There are differences between inflammatory stimulators. For example, at a concentration of 10^{-7} M FMLP, about 30% of fresh human leukocytes no longer respond to shear stress, while still 100% of the cells respond at the same concentration of PAF. At the concentration of 10^{-5} M FMLP, on average 60% of the leukocytes no longer respond to shear, and 40% in the presence of the same concentration of PAF. Leukocytes that exhibit a reduced response to shear stress tend to spread on a substrate (Fig. 25.2). Fully spread cells exhibit little shear stress response.

Attenuation of leukocyte fluid shear inactivation can also be observed *in vivo* (Fukuda et al., 2000). To demonstrate this effect, postcapillary venules in the rat mesentery microcirculation were occluded for 3 minutes with a micropipette during which time the shear stress is near zero. The pseudopod formation was observed during flow occlusion and after flow restoration. Nonstimulated leukocytes project pseudopods during vessel obstruction in every direction, retract their pseudopods immediately after restoration of fluid stress, and are washed away from the endothelium once they approach their passive spherical shape. In contrast, in the presence of 10^{-8} M PAF, the majority of pseudopods were projected over the endothelium during flow obstruction, and after flow restoration about 40% of the cells still continued to adhere on the endothelium without pseudopod retraction. The fluid shear stresses in the microvessels of the mesentery are of the order of 1–10 dyn/cm.

What could be the mechanism that permits an inhibition of the fluid shear response in these

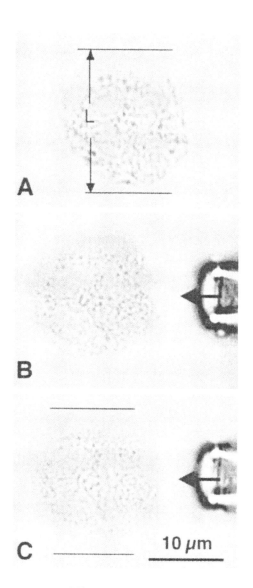

with depleted cGMP levels no longer respond to shear stress and they also spread out as seen after stimulation with FMLP or PAF at higher concentrations. In contrast, cGMP analogues (8-bromoguanosine 3':5'-cyclic monophosphate and dibutyryl cGMP) serve to enhance in a dose-dependent manner the fluid shear stress response even after it has been attenuated by inflammatory mediators.

Furthermore, since NO serves to synthesize cGMP, we expect to see an affect of the NO on the leukocyte shear stress response. The shear stress response in endothelial NO synthase ($-/-$) mice, in which NO level in blood is decreased, is reduced compared with that in wild-type mice. Alternatively, NO donors (Fukuda et al., 2000) serve to increase the fluid shear response. Inflammatory mediators and cGMP have the opposite effect on the shear stress response and on pseudopod formation during inflammation. Under the blood flow and fluid shear stress, cGMP increases the sensitivity of leukocytes to fluid shear stress and promotes pseudopod retraction, while inflammatory mediators reduce the shear stress response by induction of pseudopod projection. In inflammation, both biochemical mediators and cGMP may regulate the shear stress response of leukocytes. The balance between these two factors may serve to control pseudopod formation, cell spreading, and migration in the microcirculation. Thus, inflammation, NO, and the fluid shear response appear to be closely inter-linked quantities (Frangos et al., 1996; Granger and Kubes, 1996; Kubes and McCafferty, 2000).

FIGURE 25.2. Micrographs of fresh human neutrophils after deletion of the second messenger cGMP with methylene blue (10-4 M). After such treatment, the neutrophils are spread and with few pseudopods (A). In contrast to leukocytes without depletion of cGMP, they exhibit no significant retraction of the cell cytoplasm after 1 min (B) or 2 min (C) application of fluid shear stress.

circulating cells. We have identified a second messenger, cyclic guanosine monophosphate (cGMP) that appears to play a key role in the fluid shear response. The interesting aspect about cGMP is that its synthesis is controlled by nitric oxide (NO) (Fukuda et al., 2000). Cells

CD18 Distribution under Fluid Shear Stress

Several reports have demonstrated that membrane adhesion molecules on endothelial cells, such as intercellular adhesion molecule (ICAM)-1, vascular cell adhesion molecule (VCAM)-1, and E-selectin, are regulated by physiological fluid shear stresses at the transcriptional level (Ando and Kamiya, 1996; Chappell et al., 1998; Mohan et al., 1999; Morigi et al., 1995). McIntire and his colleagues have discussed the importance of these adhesion molecules in organ rejection and atherosclerosis (Jones et al., 1996).

Regulation of adhesion molecules on leukocytes by fluid shear stress poses some specific issues. Leukocyte rolling on the endothelium mediated by L-selectin has been reported to require "critical threshold" fluid shear stresses, which may prevent excessive leukocyte-endothelium interaction under low shear stress (Finger et al., 1996; Taylor et al., 1996). The levels of the critical threshold shear stress range between 0.6–2.0 dyn/cm^2 (Finger et al., 1996) and 4–7 dyn/cm^2 (Taylor et al., 1996).

The CD18 (β_2 integrin) expression on leukocytes suspended in whole blood is down-regulated by fluid shear stress of about 5.0 dyn/cm^2 in a cone and plate shear device (Fukuda et al., 2000). Fluid shear stress down-regulates CD18 expression, even in the presence of inflammatory mediators, such as PAF and FMLP. Using laser confocal microscopy after immunolabeling with a fluorescent antibody against CD18, we also observed that the CD18 expression on adherent human leukocyte is reduced in response to fluid shear stress (unpublished results). In the absence of fluid shear stress, the CD18 distribution on leukocytes migrating on a glass surface is relatively uniform. In the presence of shear stress, down-regulation of CD18 is more pronounced in regions on the cell with higher shear stress (1.5 dyn/cm^2 at the upstream position of the cell) compared with regions of the cell exposed to lower shear stress. In addition to CD18 down-regulation in the cell as a whole, CD18 molecules on the cell surface were redistributed from higher shear stress areas to lower shear areas within 2 minutes after shear application. We could find no evidence for cytoplasmic internalization of CD18 during this surface redistribution under fluid shear.

CD18 down-regulation and redistribution in response to fluid shear stress occur simultaneously with pseudopod retraction. CD18 signaling depends upon integrin-cytoskeleton interaction, and the actin network is involved in regulation of CD18 signaling during granulocyte locomotion (Hellberg et al., 1999). The CD18 down-regulation by shedding may be linked to the local actin depolymerization and loss of anchoring to the F-actin fibers.

In contrast to the regulation of L-selectin, physiological levels of fluid shear stress reduce CD18 expression on leukocytes. Fluid shear stress may serve to suppress inflammatory responses on circulating leukocytes by reduction of CD18 expression. Once leukocytes adhere to the endothelium in inflammation, the fluid shear stress in the contact region with the substrate is lower than that on the membrane that is exposed to the blood flow, leading to redistribution of CD18 molecules from the noncontact region to the contact region. Increased surface expression of CD18 is needed not only for leukocyte adhesion to the endothelium, but also for adherence-dependent cell locomotion (Hughes et al., 1992; Schleiffenbaum et al., 1989; Vedder and Harlan, 1988).

More recent evidence suggests, furthermore, that the fluid shear stress response depends on the attachment via integrins. Attachment via β_2 integrins serves to maintain the fluid shear inactivation. In contrast, attachment via β_1 integrins leads to spreading of the cells and suppression of shear inactivation (Marschel, 1999). Mechanotransduction processes in response to mechanical stresses share many common features with processes in cell adhesion, such as an increase in tyrosine phosphorylation of proteins in the focal adhesion sites (Shyy and Chien, 1997). Recent findings suggest that integrins may function as mechanotransducers in endothelial cells. Receptor tyrosine kinases and integrins may serve as mechanosensors to transduce mechanical stimuli into chemical signals via their association with adapter protein Shc (Chen et al., 1999). But it is also possible to demonstrate with small-diameter pipettes that the fluid shear response acts locally on different parts of the membrane, independent of a particular choice of attachments (Fukuda, unpublished results).

The Significance for Tissue Engineering

The fluid shear response is a fundamental control mechanism in living tissues and in the circulation. It serves to control cells in a fashion that is adapted to the mechanical environment in which individual cells need to function. The mechanism is present in many different cell types. It is remarkable that the fluid shear response is insensitive with respect to normal stress and only requires a small shear stress, far below the level to achieve a large passive deformation of the cells. Thus it is well adapted to the stresses encountered in the circulation. No individual molecule thus far has been identified to play a singular role in the fluid shear response. Instead, fluid shear may affect a wider range of membrane kinetics, the bilipid membrane layer thermal motion and the kinetics of underlying receptor signaling proteins. The signal is transmitted to all segments of the cell cytoplasm, especially the actin matrix. Fluid shear controls the shape that a cell assumes in a tissue, the membrane adhesion mechanisms and the genes that are expressed.

The fluid shear response is modified in inflammation, possibly due to the direct action of inflammatory biochemical mediators. As a consequence, endothelial cells become activated; leukocytes start to accumulate in tissue regions and initiate an inflammatory cascade. This is one of the fundamental complications in inflammation and it is the starting point for eventual organ dysfunction. If inflammation is limited to local regions of wound healing or immune protection against invading microorganisms, the process is life preserving. But if inflammatory reactions spread into the general circulation or reach innocent bystander organs, the consequences are cardiovascular complications with potentially severe consequences (Mitsuoka et al., 2000).

While fluid shear stress is one of the key control mechanisms, it may not be the only stress component that influences the inflammation. In vascular grafts, circumferential and axial stress components also influence the inflammatory response. Liu has reviewed the field (Liu, 1999). In organ transplantations the minimization of leukocyte activation and infiltration serves to improve survival and reduce organ rejection (Ozer et al., 1999; Robinson et al., 1998; Shirasawa et al., 2000).

Conclusion

Biomechanics at the level of individual cells, the microcirculation and the tissue is an effective tool to analyze important diseases such as stroke, myocardial infarction, or physiological shock. The biomechanics of inflammation needs to take center stage in the analysis of disease, organ rejection, or tissue engineering. There is a rich collection of problems that require analysis and finding solutions will be a rewarding exercise. The role of fluid shear stress in control of cells in the cardiovascular system is a major aspect of these problems.

Acknowledgment

This research was supported in part by NIH Grants HL10881 and HL43026.

References

Ando J, Kamiya A. 1996. Flow-dependent regulation of gene expression in vascular endothelial cells *Jpn. Heart J.* 37:19–32.

Barakat AI. 1999. Responsiveness of vascular endothelium to shear stress: potential role of ion channels and cellular cytoskeleton (review) *Int. J. Mol. Med.* 4:323–332.

Campbell FR. 1972. Ultrastructural studies of transmural migration of blood cells in the bone marrow of rats, mice and guinea pigs *Am. J. Anat.* 135:521–535.

Chamberlain JK, Lichtman MA. 1978. Marrow egress: specificity of the site of penetration into the sinus. *Blood* 52:959–968.

Chappell DC, Varner SE, Nerem RM, Medford RM, Alexander RW. 1998. Oscillatory shear stress stimulates adhesion molecule expression in cultured human endothelium *Circ. Res.* 82:532–539.

Chen KD, Li YS, Kim M, Li S, Yuan S, Chien S, Shyy JY. 1999. Mechanotransduction in response to shear stress. Roles of receptor tyrosine kinases, integrins, and Shc. *J. of Biol. Chem.* 274:18393–18400.

Damiano ER, Westheider J, Tözeren A, Ley K. 1996. Variation in the velocity, deformation, and adhesion energy density of leukocytes rolling within venules. *Circ. Res.* 79:1122–1130.

Davies P. 1995. Flow-mediated endothelial mechanotransduction. *Physiol. Rev.* 75:519–560.

De Bruyn PP, Michelson S, Thomas TB. 1971. The migration of blood cells of the bone marrow through the sinusoidal wall. *J. Morphol.* 133:417–437.

Del Zoppo GJ, Schmid-Schönbein GW, Mori E, Copeland BR, Chang C-M. 1991. Polymorphonuclear leukocytes occlude capillaries following middle cerebral artery occlusion and reperfusion. *Stroke* 22:1276–1283.

Dong C, Cao J, Struble EJ, Lipowsky HH. 1999. Mechanics of leukocyte deformation and adhesion to endothelium in shear flow. *Ann. Biomed. Eng.* 27:298–312.

Engler RL, Schmid-Schönbein GW, Pavelec RS. 1983. Leukocyte capillary plugging in myocardial ischemia and reperfusion in the dog. *Am. J. Pathol.* 111:98–111.

Entman ML, Michael L, Rossen RD, Dreyer WJ, Anderson DC, Taylor AA, Smith CW. 1991. Inflammation in the course of early myocardial ischemia. *FASEB J.* 5:2529–2537.

Farr AG, De Bruyn PP. 1975. The mode of lymphocyte migration through postcapillary venule endothelium in lymph node. *Am. J. Anat.* 143:59–92.

Finger EB, Puri KD, Alon R, Lawrence MB, von Andrian UH, Springer TA. 1996. Adhesion through L-selectin requires a threshold hydrodynamic shear. *Nature* 379:266–269.

Frangos JA, Huang TY, Clark CB. 1996. Steady shear and step changes in shear stimulate endothelium via independent mechanisms—superposition of transient and sustained nitric oxide production. *Biochem. Biophys. Res. Comm.* 224:660–665.

Fukuda S, Yasu T, Predescu DN, Schmid-Schönbein GW. 2000. Mechanisms for regulation of fluid shear stress response in circulating leukocytes. *Circ. Res.* 86:E13–18.

Fung YC. 1997. *Biomechanics. Circulation.* Springer-Verlag, New York.

Gimbrone MA, Jr, Nagel T, Topper JN. 1997. Biomechanical activation: an emerging paradigm in endothelial adhesion biology. *J. Clin. Invest.* 99:1809–1813.

Golledge J. 1997. Vein grafts: haemodynamic forces on the endothelium—a review. *Eur. J. Vasc. Endovasc. Surg.* 14:333–343.

Granger DN, Kubes P. 1994. The microcirculation and inflammation: modulation of leukocyte-endothelial cell adhesion. *J. Leukoc. Biol.* 55:662–675.

Granger DN, Kubes P. 1996. Nitric oxide as antiinflammatory agent. *Methods Enzymol.* 269:434–442.

Granger ND, Schmid-Schönbein GW. 1995. *Physiology and Pathophysiology of Leukocyte Adhesion.* Oxford University Press, New York.

Gute DC, Ishida T, Yarimizu K, Korthuis RJ. 1998. Inflammatory responses to ischemia and reperfusion in skeletal muscle. *Mol. Cell. Biochem.* 179:169–187.

Hammersen F, Hammersen E. 1987. The ultrastructure of endothelial gap-formation and leukocyte emigration. *Prog. Appl. Microcirc.* 12:1–34.

Harris AG, Skalak TC. 1993. Leukocyte cytoskeletal structure is a determinant of capillary plugging and flow resistance in skeletal muscle. *Am. J. Physiol.* 265:H1670–H1675.

Harris AG, Skalak TC, Hatchell DL. 1994. Leukocyte-capillary plugging and network resistance are increased in skeletal muscle of rats with streptozotocin-induced hyperglycemia. *Int. J. Microcirc. Clin. Exp.* 14:159–166.

Hellberg C, Ydrenius L, Axelsson L, Andersson T. 1999. Disruption of beta(2)-integrin-cytoskeleton coupling abolishes the signaling capacity of these integrins on granulocytes. *Biochem. Biophys. Res. Comm.* 265:164–169.

Ho E, Bray TM. 1999. Antioxidants, NFkappaB activation, and diabetogenesis. *Proc. Soc. Exp. Biol. Med.* 222:205–213.

Hughes BJ, Hollers JC, Crockett-Torabi E, Smith CW. 1992. Recruitment of CD11b/CD18 to the neutrophil surface and adherence-dependent cell locomotion. *J. Clin. Invest.* 90:1687–1696.

Jones DA, Smith CW, McIntire LV. 1996. Leucocyte adhesion under flow conditions: principles important in tissue engineering. *Biomaterials* 17:337–347.

Kubes P, McCafferty DM. 2000. Nitric oxide and intestinal inflammation. *Am. J. Med.* 109:150–158.

Lee E, Pang K, Knecht D. 2001. The regulation of actin polymerization and cross-linking in Dictyostelium. *Biochim. Biophys. Acta* 1525:217–227.

Lee JS. 2000. 1998. Distinguished lecture: biomechanics of the microcirculation, an integrative and therapeutic perspective. *Ann. Biomed. Eng.* 28:1–13.

Liu SQ. 1998. Prevention of focal intimal hyperplasia in rat vein grafts by using a tissue engineering approach. *Atherosclerosis* 140:365–377.

Liu SQ. 1999. Biomechanical basis of vascular tissue engineering. *Crit. Rev. Biomed. Eng.* 27:75–148.

Liu SQ, Moore MM, Glucksberg MR, Mockros LF, Grotberg JB, Mok AP. 1999. Partial prevention of monocyte and granulocyte activation in experimental vein grafts by using a biomechanical engineering approach. *J. Biomech.* 32:1165–1175.

Lush CW, Kvietys PR. 2000. Microvascular dysfunction in sepsis. *Microcirculation* 7:83–101.

Marchesi VT, Florey HW. 1961. Electron micrographic observations on the emigration of leukocytes. *Q. J. Exp. Physiol.* 45:343–361.

Marschel P. 1999. "*The Influence of Integrins on the Leukocyte Response to Fluid Shear Stress.*" Master's thesis, University of California San Diego.

Mazzoni MC, Schmid-Schönbein GW. 1996. Mechanisms and consequences of cell activation in the microcirculation. *Cardiovasc. Res.* 32:709–719.

McDonald DM, Thurston G, Baluk P. 1999. Endothelial gaps as sites for plasma leakage in inflammation. *Microcirculation* 6:7–22.

Mitsuoka H, Kistler EB, Schmid-Schönbein GW. 2000. Generation of *in vitro* activating factors in the ischemic intestine by pancreatic enzymes. *Proc. Natl. Acad. Sci. U.S.A.* 97:1772–1777.

Moazzam F, DeLano FA, Zweifach BW, Schmid-Schönbein GW. 1997. The leukocyte response to fluid stress [see comments]. *Proc. Natl. Acad. Sci. U.S.A.* 94:5338–5343.

Mohan S, Mohan N, Valente AJ, Sprague EA. 1999. Regulation of low shear flow-induced HAEC VCAM-1 expression and monocyte adhesion. *Am. J. Physiol.* 276:C1100–C1107.

Möhle R, Moore MA, Nachman RL, Rafii S. 1997. Transendothelial migration of CD34+ and mature hematopoietic cells: an *in vitro* study using a human bone marrow endothelial cell line. *Blood* 89:72–80.

Moore TM, Chetham PM, Kelly, JJ, Stevens T. 1998. Signal transduction and regulation of lung endothelial cell permeability. Interaction between calcium and cAMP. *Am. J. Physiol.* 275:L203–L222.

Morigi M, Zoja C, Figliuzzi M, Foppolo M, Micheletti G, Bontempelli M, Saronni M, Remuzzi G, Remuzzi A. 1995. Fluid shear stress modulates surface expression of adhesion molecules by endothelial cells. *Blood* 85:1696–1703.

Ohashi KL, Tung DK-L, Wilson JM, Zweifach BW, Schmid-Schönbein GW. 1996. Transvascular and interstitial migration of neutrophils in rat mesentery. *Microcirculation* 3:199–210.

Oparil S, Oberman A. 1999. Nontraditional cardiovascular risk factors. *Am. J. Med. Sci.* 317:193–207.

Ozer K, Adanali G, Zins J, Siemionow M. 1999. In vivo microscopic assessment of cremasteric microcirculation during hindlimb allograft rejection in rats. *Plast. Reconstr. Surg.* 103:1949–1956.

Papadaki M, Eskin SG, Ruef J, Runge MS, McIntire LV. 1999. Fluid shear stress as a regulator of gene expression in vascular cells: possible correlations with diabetic abnormalities. *Diabetes Res. Clin. Pract.* 45:89–99.

Ray WJ, Ashall F, Goate AM. 1998. Molecular pathogenesis of sporadic and familial forms of Alzheimer's disease. *Mol. Med. Today* 4:151–157.

Ritter LS, Wilson DS, Williams SK, Copeland JG, McDonagh PF. 1995. Early in reperfusion following myocardial ischemia, leukocyte activation is necessary for venular adhesion but not capillary retention. *Microcirculation* 2:315–327.

Robinson LA, Tu L, Steeber DA, Preis O, Platt JL, Tedder TF. 1998. The role of adhesion molecules in human leukocyte attachment to porcine vascular endothelium: implications for xenotransplantation. *J. Immunol.* 161:6931–6938.

Rodgers SD, Camphausen RT, Hammer DA. 2000. Sialyl Lewis(x)-mediated, PSGL-1-independent rolling adhesion on P-selectin. *Biophys. J.* 79:694–706.

Ross R. 1999. Atherosclerosis—an inflammatory disease [see comments]. *N. Eng. J. Med.* 340:115–126.

Schleiffenbaum B, Moser R, Patarroyo M, Fehr J. 1989. The cell surface glycoprotein Mac-1 (CD11b/CD18) mediates neutrophil adhesion and modulates degranulation independently of its quantitative cell surface expression. *J. Immunol.* 142:3537–3545.

Schmid-Schönbein GW. 1999. Biomechanics of microcirculatory blood perfusion. *Annu. Rev. Bioeng.* 1:73–102.

Schmid-Schönbein GW. 1987. Mechanisms of granulocyte-capillary-plugging. *Prog. Appl. Microcirc.* 12:223–230.

Schmid-Schönbein GW, Skalak R. 1984. Continuum mechanical model of leukocytes during protopod formation. *J. Biomech. Eng.* 106:10–18.

Schmid-Schönbein GW, Shih YY, Chien S. 1980. Morphometry of human leukocytes. *Blood* 56:866–875.

Shirasawa B, Hamano K, Ueda M, Ito H, Kobayashi T, Fujimura Y, Kojima A, Esato K. 2000. Contribution of proliferating leukocytes to phenotypic change in smooth muscle cells during the development of coronary arteriosclerosis in transplanted hearts. *Eur. Surg. Res.* 32:30–38.

Shyy JY, Chien S. 1997. Role of integrins in cellular responses to mechanical stress and adhesion. *Curr. Opin. Cell Biol.* 9:707–713.

Suematsu M, DeLano FA, Poole D, Engler RL, Miyasaka M, Zweifach BW, Schmid-Schönbein GW. 1994. Spatial and temporal correlation between leukocyte behavior and cell injury in postischemic rat skeletal muscle microcirculation. *Lab. Invest.* 70:684–695.

Suematsu M, Suzuki H, Ishii H, Kato S, Yanagisawa T, Asako H, Suzuki M, Tsuchiya, M. 1992. Early midzonal oxidative stress preceding cell death in hypoperfused rat liver. *Gastroenterology* 103:994–1001.

Taylor AD, Neelamegham S, Hellums JD, Smith CW, Simon SI. 1996. Molecular dynamics of the transition from L-selectin- to beta 2-integrin-dependent neutrophil adhesion under defined hydrodynamic shear. *Biophys. J.* 71:3488–3500.

Thomas L. 1974. Inflammation as a disease mechanism. In: *The Inflammatory Process.* Zweifach BW,

Grant L, McCluskey RT, eds Academic Press, New York.

Topper JN, Gimbrone MA, Jr. 1999. Blood flow and vascular gene expression: fluid shear stress as a modulator of endothelial phenotype. *Mol. Med. Today* 5:40–46.

Traub O, Berk BC. 1998. Laminar shear stress: mechanisms by which endothelial cells transduce an atheroprotective force. *Arterioscler. Thromb. Vasc. Biol.* 18:677–685.

Tung DK-L, Bjursten LM, Zweifach BW, Schmid-Schönbein, GW. 1997. Leukocyte contribution to parenchymal cell death in an experimental model of inflammation *J. Leukoc Biol.* 62:163–175.

Vedder NB, Harlan JM. 1988. Increased surface expression of CD11b/CD18 (Mac-1) is not required for stimulated neutrophil adherence to cultured endothelium. *J. Clin. Invest.* 81:676–682.

Worthen GS, Schwab III B, Elson EL, Downey GP. 1989. Cellular mechanics of stimulated neutrophils: stiffening of cells induces retention in pores *in vitro* and lung capillaries *in vivo. Science* 245:183–186.

Zhao Y, Chien S, Skalak R. 1995. A stochastic model of leukocyte rolling. *Biophys. J.* 69: 1309–1320.

Zigmond SH. 2000. How WASP regulates actin polymerization. *J. Cell Biol.* 150:F117–20.

Zweifach BW, Grant L, McCuskey RT. 1974. *The Inflammatory Process.* Academic Press, New York.

26

A Full Spectrum of Functional Tissue-Engineered Blood Vessels: From Macroscopic to Microscopic

François A. Auger, Guillaume Grenier, Marielle Rémy-Zolghadri, and Lucie Germain

Introduction

Tissue engineering has created several original and new avenues of investigation in biology (Auger et al., 2000). This new domain of research in biotechnology was introduced in the 1980s as a life-saving procedure for burn patients. The successful engraftment of autologous living epidermis was the first proof of concept of this powerful approach. From the efforts in this field, two schools of thought emerged. A first one is the seeding of cells into various gels or scaffolds in which the cells secrete and/or reorganize the surrounding extracellular matrix (ECM), and a second one, the coaxing of cells onto the secretion of an abundant autologous ECM, thus creating their own environment in the absence of any exogenous material. This latter methodology, which we called the "self-assembly approach," takes advantage of the ability of cells to recreate *in vitro* tissue-like structures when appropriately cultured (Auger et al., 2000). The conditions entail particular media composition and adapted mechanical straining of these three-dimensional structures.

Our own experience with the culture of autologous epidermal sheets gave us some insight in the property of cells to recreate such *in vitro* tissue-like structures. This expertise led us to develop tissue-engineered structures on the basis of the following two concepts: the living substitutes that we created have no artificial biomaterial, and the ECM is either a biological one repopulated by the cells or an ECM neosynthesized by the cells themselves. Such living substitutes have distinct advantages because of their cellular composition that confer to them superior physiological characteristics when implanted into the human body, that is, their ability to renew themselves over time and their healing property if they are damaged. Moreover, the presence of autologous cells in the living reconstructed tissue should facilitate its interactions with the surrounding host environment.

Here, we describe our own experience in the reconstruction of a full spectrum of blood vessels by tissue engineering: macroscopic and microscopic. We applied the self-assembly approach with some impressive results to the reconstruction of a small-diameter blood vessel and the use of a cell-seeded scaffold leading to the formation of capillary-like structures in a full-thickness skin. The following highlights the major points for the generation of these organs.

In Vitro Production of a Tissue-Engineered Blood Vessel

Capillaries are necessary for the nutrition of tissues, whereas larger blood vessels are essential for the survival of whole organs. Therefore, vascular surgery has strived to replace diseased blood vessels in order to restore the blood flow and to save limbs. Atherosclerosis is a systemic disease affecting blood vessels (Berceli et al., 1991; Geer et al., 1961) and leading to their partial or complete obstruction. The smaller arteries are even more affected since the low blood flow also favors the production of blood clots in these diseased vessels.

To solve the problems of blood vessel replacement, the grafting of commercially available vascular synthetic prostheses is successful in the case of large arteries (Brewster and Rutherford, 1995) but of limited success when small-diameter blood vessels (<5 mm) are concerned (limb salvage or coronary arterial grafts, for example) (Langer and Vacanti, 1993). Indeed, the thrombogenic properties of the biomaterials associated with the low blood flow in these arteries are responsible in many cases for the formation of blood clots and obstruction of the conduct (O'Donnell et al., 1984; Sayers et al., 1998; Stephen et al., 1977). The alternative approaches using allogeneic or xenogeneic implants have given few encouraging results because of the high percentage of prostheses degradation over time in long-term implantation (Allaire et al., 1994; Charara et al., 1989; Dardik, 1989; Stanke et al., 1998). Another approach attempted to combine the advantages of the synthetic grafts with the biological characteristics of an active endothelial intima (Herring et al., 1978; Zilla et al., 1993). Many investigators have then tried to seed endothelial cells (Graham et al., 1980; Herring et al., 1985; Meerbaum et al., 1992; Ortenwall et al., 1990; Park et al., 1990; Rémy et al., 1994; Walker et al., 1987; Zilla et al., 1993), mesothelial cells (Bellon et al., 1997; Verhagen et al., 1998) or bone marrow cells (Noishiki et al., 1996, 1998) upon synthetic grafts. These attempts have been hampered by the weak attachment of cells to these various surfaces. The cells were dislodged by the arterial flow and consequently their physiological function also disappeared (Gourevitch et al., 1988; Miyata et al., 1991; Nerem et al., 1998; Vohra et al., 1992). However, this field is still very active, with various methods directed toward better anchoring of the endothelial cells (Bath et al., 1998; Bujan et

al., 1996; Clowes, 1996; Falk et al., 1998; Flugelman et al., 1992; Zilla et al., 1993) and consequently enhanced clinical results (Bordenave et al., 1999; Deutsch et al., 1999; Meinhart et al., 1997).

The attempt to produce a reconstructed blood vessel by tissue-engineering methods appeared in 1986 with the first model published by Weinberg and Bell (1986). The method devised by these researchers was based on collagen gels seeded with bovine vascular cells. Such a technique was the basis for subsequent research conducted by our and other teams (L'Heureux et al., 1993; Tranquillo et al., 1996; Ziegler et al., 1995). But the resulting structures were not resistant enough to sustain normal blood pressure (L'Heureux, 1993; Weinberg and Bell, 1986) and these prostheses had to be reinforced with a synthetic mesh (Hirai and Matsuda, 1996; Weinberg and Bell, 1986) making them hybrid artificial substitutes (composed of living cells in association with a synthetic support) with all the untoward properties associated with such constructions.

Since the previous approaches did not seem to be conducive to appropriate clinical results, we worked to develop living tissue-engineered blood vessels (TEBVs) that would not contain synthetic material. For the replacement of small-diameter arteries, these constructs would then be ideal substitutes as long as they could withstand physiological blood pressure (L'Heureux et al., 1998). We published in 1998, the first *in vitro* production of such a tissue-engineered blood vessel (Fig. 26.1A) (L'Heureux et al., 1998). Each layer of the vascular wall was reconstructed: the intima (composed of endothelial cells), the media (composed of smooth muscle cells) and the adventitia (composed of fibroblasts). Macroscopically, the TEBV appeared as a homogeneous tubular tissue strikingly resembling a human artery.

Concept for the Production of a TEBV

The cells used for the preparation of the TEBV were isolated from human umbilical cord vein using an enzymatic method for endothelial cells (Jaffe et al., 1973) and the method of explants

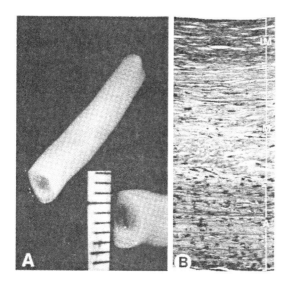

FIGURE 26.1. Macroscopic (*A*) and microscopic (*B*) views of the mature TEBV. (*A*) When removed from the tubular mandrel, the TEBV is self-supporting with an open lumen (3 mm internal diameter). (*B*) Paraffin cross-section of the TEBV wall stained with Masson's trichrome. Collagen fibers are stained in gray and cells nuclei in dark. Inner membrane (IM) = 125 μm, media (M) = 320 μm and adventitia (A) = 235 μm. Note the endothelium covering the luminal surface of the TEBV. Reprinted by permission from L'Heureux et al. (1998).

published by Ross (1971) for smooth muscle cells isolation. The fibroblasts were provided by the enzymatic treatment of a small biopsy of human skin (Germain et al., 1993). In order to obtain an abundant ECM production, fibroblasts and smooth muscle cells were cultured in media supplemented with ascorbic acid until they self-assembled into sheet that could be detached from the culture support and then be wrapped around a tubular mandrel. Thus, the steps necessary to produce a complete blood vessel are the following: smooth muscle cell sheet is rolled over an acellular inner membrane (dehydrated tubular tissue formed with a 5-week cultured fibroblast sheet). After a week of maturation, a fibroblast sheet is wrapped around the media layer to produce, after 7 weeks of maturation, the adventitia. Finally, the endothelial cells are injected in the lumen to be seeded on the inner membrane and form a confluent endothelium (Fig. 26.1*B*) (L'Heureux et al., 1998).

Under these conditions of maturation, the smooth muscle cello (SMCs) and fibroblasts produce and organize their own ECM composed of collagen and glycosaminoglycans.

The advantage of this tissue engineering method for blood vessel reconstruction is that no ECM or synthetic material, which could be associated with inflammatory reactions after grafting, are added. The starting materials are only human cells and culture medium. The cell isolation process should favor both a high yield and also the clean separation of each cell type from the others. Thus, specific procedures could be designed for this purpose. In culture, cells must be observed regularly with a phase-contrast microscope and characterized with specific markers (L'Heureux et al., 1993; Mc-Keehan et al., 1990; Owens, 1995) to identify the cells and to assess the purity of the cultures. The culture conditions (e.g., addition of irradiated fibroblasts in keratinocyte cultures), culture medium, and its additives must be carefully determined (Green et al., 1979; L'Heureux et al., 1993). The serum used could lead to differentiation or changes in the cell phenotype, as it was demonstrated for human dermal fibroblasts (Moulin et al., 1997) so that its choice is of crucial importance.

Functional Characteristics of the TEBV

Although completely biologic, this reconstructed human blood vessel is highly resistant with a burst strength of over 2500 mm Hg (Fig. 26.2*A*). This is significantly higher than that of human saphenous veins, which are currently considered the best grafts for lower limb vascular reconstruction (Abbott and Vignati, 1995; Veith et al., 1986). This resistance is provided mainly by the adventitia (at a level of 2238 ± 251 mm Hg at 7 week of culture) and increases with the maturation time of the tissue in culture (L'Heureux et al., 1998). This improved resistance, compared with other previously designed biological tissue-engineered blood vessels (Hirai and Matsuda, 1996; Weinberg and Bell, 1986), which had to be reinforced with synthetic material to resist the

A.

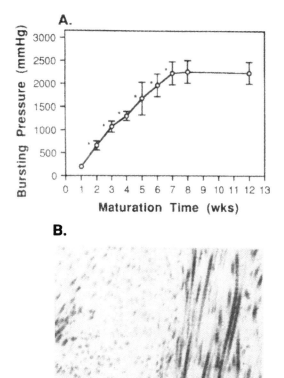

B.

FIGURE 26.2. (*A*) Burst strenght of the adventitia over time of maturation *in vitro*. *Significantly different than the precedent point (*$p < 0.001$, **$p < 0.005$, ¥$p < 0.05$) with the Student's *t* test ($n = 8$–13). (*B*) Adventitial ECM ultrastructure observed by transmission electron microscopy. Uranyl acetate and lead citrate staining (scale bar = 500 nm). Reprinted by permission from L'Heureux et al. (1998).

size tropoelastin, the soluble precursor of elastin, without formation of elastic fibers (Fleischmajer et al., 1991). However, in some complex three-dimensional scaffolds, tropoelastin is transformed into elastin by lysyl-oxydase and deposited on the microfibrillar network to form a complete elastic tissue (Duplan-Perrat et al., 2000). Such a phenomenon is likely occuring in our tissue-engineered blood vessel because of particular three-dimensional culture conditions.

It is well known that monolayered cultures of human smooth muscle cells are accompanied by the loss of expression of differentiated proteins (Thyberg et al., 1990). We have observed the re-expression of such proteins like desmin in a significant proportion of human smooth muscle cells present in the media (L'Heureux et al., 1998), as well as markers of the contractile phenotype of smooth muscle cells such as myosin (unpublished results). Therefore, the formation of tissue-like structures, in contrast to the culture in monolayer, provides conditions for adequate cell redifferentiation. Moreover, the similarity of the macroscopic aspect with the native tissue makes possible the use of a classic system to perform pharmacological studies (Stoclet et al., 1996). The response of the reconstructed media to vasoactive agonists was measured with an isolated organ bath system and our results have shown an active vasocontractility (L'Heureux et al., 2001; Stoclet et al., 1996).

The endothelial cells seeded on the inner membrane form a confluent and functional endothelium (Fig. 26.3*B*). Indeed, endothelial cells express von Willebrand factor and incorporate acetylated low-density lipoproteins (L'Heureux et al., 1993, 1998). They inhibit platelet adhesion (Fig. 26.3*B*), in contrast to the acellular inner membranes, which promote platelet adhesion and aggregation (Fig. 26.3*A*) (L'Heureux et al., 1998). Therefore, the endothelium provides an antithrombogenic surface (Cines et al., 1998; L'Heureux et al., 1998) which characteristic is essential for a successful engraftment.

We have transplanted a human reconstructed blood vessel in dogs in order to demonstrate its properties regarding implantation and flow patency. The TEBV demonstrated a good suturability by the use of conventional surgical tech-

physiological pressure, is likely due to the impressive organization of the ECM under our culture conditions (Fig. 26.2*B*). Another contributing factor is probably the controlled stable gelatinase activity in our three-dimensional tissue reconstructions in the long term (data not shown). This high resistance was obtained without pulsatile conditions in the culture, in contrast to the model of Niklason et al. (1999). Another remarkable aspect of our reconstructed blood vessel is the presence of elastin fibers, an important component of the ECM (L'Heureux et al., 1998) absent in other reconstructed models (Weinberg and Bell, 1986). Fibroblasts and smooth muscle cells *in vitro* are able to synthe-

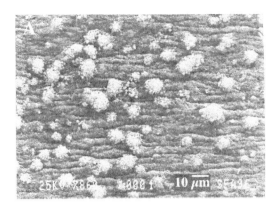

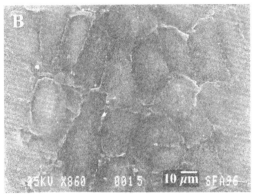

FIGURE 26.3. Inhibition of platelet adhesion by the endothelium. Scanning electron micrographs of unendothelialized inner membrane (IM) (*A*) promoted platelet adhesion and activation whereas endothelialized IM (*B*) almost completely inhibited the process. Reprinted by permission from L'Heureux et al. (1998).

sated by the benefit of producing autologous implants for patients in various future clinical applications. Presently, experiences are being conducted to ensure TEBV long-term stability after implantation. Furthermore, appropriate culture condition under pulsatile flow should decrease the maturation time of these TEBVs.

In Vitro Neovascularization of Tissue-Engineered Skin

The first clinical application of tissue engineering has been the grafting of epidermal sheets expanded *in vitro* to extensively burned patients (O'Connor et al., 1981). Indeed, the possibility of isolating and culturing epidermal cells from a small biopsy and expanding them *in vitro* to many thousands of square centimeters of epidermis (Rheinwald and Green, 1975) has paved the way for clinical applications (Auger, 1988; Berthod et al., 1997a; Boyce, 1996; Damour et al., 1997; Gallico et al., 1984; Green and Barrandon, 1988). Although the epidermal sheets are thin, comprising only three to five cell layers at the time of transplantation, they reform a differentiated epidermis within a few weeks after grafting under the influence of the *in vivo* environment. In this way, the crucial barrier function of the skin can be reestablished.

The dermal portion of the skin is responsible for its mechanical properties. Different approaches have been designed to reconstruct the dermis: the inclusion of fibroblasts in a collagen gel, their culture in various scaffold such as collagen sponges containing or not glycosaminoglycans, or in a nylon or polyglycolic acid mesh (Auger et al., 1998; Bell et al., 1979; Berthod et al., 1997b; Bouvard et al., 1992; Boyce et al., 1990, 1995; Contard et al., 1993; Cooper et al., 1991; Damour et al., 1994; Hansbrough et al., 1989). The living fibroblasts then organize their surrounding ECM composed of collagens and glycoproteins (Berthod et al., 1996, 1997b; Sahuc et al., 1996). For all the various types of reconstructed dermis, the addition of keratinocytes on its surface will lead to the production of reconstructed skin (Auger et al., 1995; Bell et al. 1981; Berthod et al., 1997a, 1997b;

niques and a mechanical stability by the absence of early tearing or dilatation after 1 week of femoral interposition grafting (L'Heureux et al., 1998). In these experiments, the 50% patency rate 1 week after grafting was very encouraging. Indeed, endothelial cells were deliberately not seeded in this xenogeneic situation since they would have promoted hyperacute rejection (Lu et al., 1994).

Thus, the TEBV that we have produced offers exciting perspectives in clinical as well as in pharmacological applications. Presently, its production *in vitro* takes approximately 3 months to obtain a completely mature blood vessel that possesses a supraphysiological resistance. This relatively long delay could be largely compen-

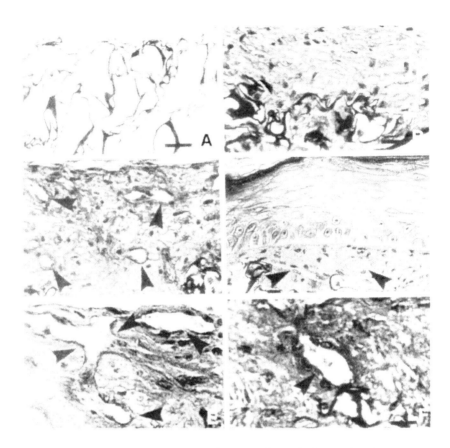

FIGURE 26.4. Histological staining of *in vitro* tissue-engineered skin equivalents. Biopolymers were seeded with human umbilical vein endothelial cells (HUVEC) (*A*), fibroblasts alone (*B*), HUVEC and fibroblasts (*C, E, F*), and HUVEC, fibroblasts, and keratinocytes (endothelialized skin equivalent or ESE) (*D*) and cultured 31 days. Sections were stained with Masson's trichrome. Note that fibroblasts were numerous and filled the pores with newly synthesized ECM either when cultured alone (*B*) or with HUVEC (*C, E, F*). In contrast, when HUVEC were alone in the biopolymer, endothelial cells were scarce (*A*). In coculture of fibroblasts and HUVEC, the formation of capillary-like tubular structures in the newly synthesized ECM was observed (*E, F*). In the ESE, keratinocytes produced a differentiated epidermis. Moreover, tubular structures (arrowhead) were present in the dermal portion of the ESE (*D*). Bars indicate 60 μm (*A*), 40 μm (*E*), and 10 μm (*F*). Reprinted by permission from Black et al. (1998).

Boyce et al., 1990; Contard et al., 1993; Michel et al., 1999; Slivka et al., 1991). These reconstructed skin substitutes exhibit many characteristics similar to the normal skin (Duplan-Perrat et al., 2000).

Most of the methods of dermis reconstruction by tissue engineering allow adjustment of the dermal thickness *in vitro*. However, the slow revascularization in these reconstructed tissues will delay the nutrition of the epidermis after grafting. Indeed, the survival of transplanted native skin depends on the imbibition phenome-

non for the first few days after grafting. It seems that the skin substitutes are not conducive to a rapid neovascularization, which entails a significant negative effect on epidermal layer. But the inosculation process by which blood vessels already present in the graft rapidly connect to the vessels of the granulation tissue reestablishes a good nutrition of the epidermis, as has been shown in full-thickness skin grafting (Young et al., 1996). Until recently the absence of vascularization in reconstructed skin limited the possibility of grafting an appropriate dermis since the

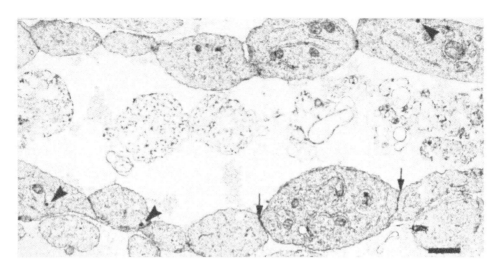

FIGURE 26.5. Transmission electron microscopy of a capillary-like structure observed in a 31-day-old endothelialized skin equivalent (ESE) ESE. An endothelial capillary-like tube formed by at least 13 endothelial cells was observed along its longitudinal axis, delimiting an internal lumen. Weibel-Palade bodies were also seen (*arrowheads*). Note that lumen was filled with cellular debris, but did not contains ECM components. The intercellular junctions between individual endothelial cells can be seen (*arrows*). Bar indicates 1 μm. Reprinted by permission from Black et al. (1998).

time necessary for its revascularization could reduce the chance of survival of the epidermis.

In order to circumvent this important clinical stumbling block, we have seeded endothelial cells together with fibroblasts within a collagen-glycosaminoglycan (GAG) sponge to produce the first reconstructed skin containing capillary-like structures (Black et al., 1998). One must note that the seeding of endothelial cells alone in the sponge does not lead to the formation of tubular structures, showing that other factors are necessary to obtain a three-dimensional reconstruction (Fig. 26.4*A*). However, when endothelial cells are seeded together with fibroblasts within the sponge, capillary-like structures, that is, endothelial cells surrounding a lumen, are observed in the reconstructed tissue (Fig. 26.4 C, E, F). Under these culture conditions, fibroblasts produce a dense ECM and reexpress type XIV collagen, which is usually lost in monolayer cultures (Berthod et al., 1997b; Trueb and Trueb, 1992). It is likely that the organization of this ECM favors the neovascularization. Capillary formation is observed in the absence (reconstructed endothelialized dermis) (Fig. 26.4*C, E,*

F) and in the presence (reconstructed endothelialized skin) (Fig. 26.4*D*) of keratinocytes contrary to ECM component like elastic fibers, which need the presence of keratinocytes to be organized.

The tubular structures reformed in the reconstructed endothelialized skin are delimited by endothelial cells which express the von Willebrand factor and which are surrounded by ECM components characteristic of capillary basement membrane: laminin and type IV collagen.

The ultrastructural study of these reconstructed endothelialized skin by electron microscopy demonstrates the typical ultrastructure associated with capillaries: closed tubes formed by endothelial cells, which contain Weibel-Palade bodies and are attached together by intercellular junctions. These cells lie on a discontinuous basement membrane. A dense ECM is observed around the reconstructed capillaries but not in the lumen. Longitudinal sections as well as cross-sections of reconstructed capillaries were observed (Fig. 26.5).

This improved reconstructed skin containing capillary-like structures is a step forward in the production of optimal living skin substitutes for the treatment of burned patients. The addition of capillaries to the reconstructed dermis *in vitro* should accelerate the nutrition of the epidermis after grafting because of inosculation of the reconstructed capillaries by those of the underlying granulation tissue. Therefore, this could allow the production and successful grafting of a more complex reconstructed skin substitute.

In clinical situations, it would be advantageous to promote angiogenesis to improve wound healing, or to block it to restrict the growth of tumors by starvation related to the blocking of angiogenesis (Bergers et al., 1999). The reconstructed endothelialized skin provides a new tool not only for the understanding of the capillary development but also for the testing of drugs affecting the angiogenic process.

Applications of Reconstructed Tissues for Gene Therapy

Tissue engineering offers new tools for gene therapy. Indeed, the cells could be safely transfected during their culture *in vitro* without using viruses that induce immunogenic response (Clowes et al., 1997a, 1997b; Pickering et al., 1994; Veelken et al., 1994; Xu et al., 1993). After evaluation of the gene expression, modified cells could be incorporated in reconstructed tissues. Endothelial cells, with their immediate contact with blood, are an attractive target for gene transfer, since the gene product could be secreted directly into the circulation (Clowes et al., 1997a). Since the hyperproliferation of smooth muscle cells is a problem in atherosclerosis, these cells could be targeted to limit their proliferation potential before being included in the reconstructed blood vessels that would be grafted to replace those affected by this disease.

Skin is also an attractive tissue for gene therapy (Ghazizadeh et al., 1998; Greenhalgh et al., 1994; Khavari and Krueger, 1997; Krueger et al., 1994). Skin biopsies are easy to harvest and genes can be introduced into skin cells (Jiang et al., 1991; Morgan et al., 1987). Although the epidermis is not vascularized, the products secreted by transfected keratinocytes do reach the

circulation (Teumer et al., 1990). Moreover, the epidermis is a rapidly renewing tissue, thus containing stem cells. Their high regeneration potential and long life span make stem cells ideal candidates for transfection and long-term gene expression. Several teams, including ours, are working on the characterization of skin stem cells and their isolation and culture *in vitro* (Bickenbach, 1981; Bickenbach and Chism, 1998; Jones and Watt, 1993; Jones et al., 1995; Michel et al., 1996; Morris and Potten, 1994, 1999; Morris et al., 1985). Another advantage of the skin resides in its accessibility. The reconstructed tissue could easily be removed if necessary in the event of problems with the gene therapy.

Conclusion

The self-assembly method for organ and tissue reconstruction is a truly different approach in the field of tissue engineering of soft tissues (L'Heureux et al., 1998; Michel et al., 1999). This methodology demonstrated that the association of cells in well-defined conditions (spatial as well as nutritional) was able to induce the reorganization of a complete tissue by various cells and to induce many of the physiological cellular properties lost in traditional bidimensional culture conditions (L'Heureux, 1998). The integration of various tissues and organs created by such a method should be more rapid and without the resorption phase that characterizes the scaffold approach. Indeed, any synthetic or biosynthetic material has to be degraded by the implanted patient. This has led to a chronic inflammatory reaction in many instances. The self-assembly approach is more akin to the physiological phenomenon that occurs in the womb during organogenesis. This may prove to be a distinct advantage for certain type of grafts and is then conform to the principles of the functional tissue engineering discipline. Thus, the advances in tissue engineering will help to provide a therapeutic alternative to solve the problem of storage of organ for human transplantation and provide human *in vitro* models for various testing from physiology to pharmacotoxicology.

References

Abbott WM, Vignati JJ. 1995. Prosthetic grafts: when are they a reasonable alternative? *Semin. Vasc. Surg.* 8:236–245.

Allaire E, Guettier C, Bruneval P, Plissonnier D, Michel JB. 1994. Cell-free arterial grafts: morphologic characteristics of aortic isografts, allografts, and xenografts in rats. *J. Vasc. Surg.* 19:446–456.

Auger FA. 1988. The role of cultured autologous human epithelium in large burn wound treatment. *Transplantation/Implantation Today* 5:21–26.

Auger FA, Lopez Valle CA, Guignard R, Tremblay N, Noël B, Goulet F, Germain L. 1995. Skin equivalent produced with human collagen. *In Vitro Cell. Dev. Biol. Anim.* 31:432–439.

Auger FA, Rouabhia M, Goulet F, Berthod F, Moulin V, Germain L. 1998. Tissue-engineering human skin substitutes developed from collagen-populated hydrated gels: clinical and fundamental applications. *Med. Biol. Eng. Comput.* 36:801–812.

Auger FA, Rémy-Zolghadri M, Grenier G, Germain L. 2000. The self-assembly approach for organ reconstruction by tissue engineering. *E-Biomed:* a regenerative medicine 1:75–85.

Bath VD, Truskey GA, Reichert WM. 1998. Using avidin-mediated binding to enhance initial endothelial cell attachment and spreading. *J. Biomed. Mater. Res.* 40:57–65.

Bell E, Ivarsson B, Merrill C. 1979. Production of a tissue-like structure by contraction of collagen lattices by human fibroblasts of different proliferative potential *in vitro. Proc. Natl. Acad. Sci. U.S.A.* 76:1274–1278.

Bell E, Ehrlich HP, Buttle DJ, Nakatsuji T. 1981. Living tissue formed *in vitro* and accepted as skin-equivalent tissue of full thickness. *Science* 211:1052–1054.

Bellon JM, Garcia-Honduvilla N, Escudero C, Gimeno MJ, Contreras L, de Haro J, Bujan J. 1997. Mesothelial versus endothelial cell seeding: evaluation of cell adherence to a fibroblastic matrix using ^{111}In-oxine. *Eur. J. Vasc. Endovasc. Surg.* 13:142–148.

Berceli SA, Borovetz HS, Sheppeck RA, Moosa HH, Warty VS, Armany MA, Herman IM. 1991. Mechanisms of vein graft atherosclerosis: LDL metabolism and endothelial actin reorganization. *J. Vasc. Surg.* 13:336–347.

Bergers G, Javaherian K, Lo KM, Folkman J, Hanahan D. 1999. Effects of angiogenesis inhibitors on multistage carcinogenesis in mice. *Science* 284:808–812.

Berthod F, Sahuc F, Hayek D, Damour O, Collombel C. 1996. Deposition of collagen fibril bundles by long-term culture of fibroblasts in a collagen sponge. *J. Biomed. Mater. Res.* 32:87–93.

Berthod F, Damour O. 1997a. *In vitro* reconstructed skin models for wound coverage in deep burns. *Br. J. Dermatol.* 136:809–816.

Berthod F, Germain L, Guignard R, Lethias C, Garrone R, Damour O, van der Rest, M, Auger FA. 1997b. Differential expression of collagens XII and XIV in human skin and in reconstructed skin. *J. Invest. Dermatol.* 108:737–742.

Bickenbach JR. 1981. Identification and behavior of label-retaining cells in oral mucosa and skin. *J. Dent. Res.* 60:1611–1620.

Bickenbach JR, Chism E. 1998. Selection and extended growth of murine epidermal stem cells in culture. *Exp. Cell. Res.* 244:184–195.

Black AF, Berthod F, L'Heureux N, Germain L, Auger FA. 1998. *In vitro* reconstruction of capillary-like network in a tissue-engineered skin equivalent. *FASEB J.* 12:1331–1340.

Bordenave L, Rémy-Zolghadri M, Fernandez P, Bareille R, Midy D. 1999. Clinical performance of vascular grafts lined with endothelial cells. *Endothelium* 6:267–275.

Bouvard V, Germain L, Rompré P, Roy B, Auger F A. 1992. Influence of dermal equivalent maturation on the development of a cultured skin equivalent. *Biochem. Cell. Biol.* 70:34–42.

Boyce S T, Michel S, Reichert U, Shroot B, and Schmidt R. 1990. Reconstructed skin from cultured human keratinocytes and fibroblasts on a collagen-glycosaminoglycan biopolymer substrate. *Skin Pharmacol.* 3:136–143.

Boyce S T, Supp A P, Harriger M D, Greenhalgh D G, Warden G D. 1995. Topical nutriments promote engraftment and inhibit wound contraction of cultured skin substitutes in athymic mice. *J. Invest. Dermatol.* 104:345–349.

Boyce S T. 1996. Cultured skin substitutes: a review. *Tissue Eng.* 2:255–266.

Brewster D C, Rutherford R B. 1995. Prosthetic grafts in vascular surgery. In: Vascular Surgery, RB Rutherford, ed. WB Saunders, Philadelphia. pp. 492–521.

Bujan J, Garcia-Honduvilla N, Contreras L, Gimeno M J, Escudero C, Bellon J M, San-Roman J. 1996. Coating PTFE vascular prostheses with a fibroblastic matrix improves cell retention when subjected to blood flow. *J. Biomed. Mater. Res.* 39:32–39.

Charara J, Beaudoin G, Fortin C, Guidoin R, Roy P E, Marble A, Schmitter R, Paynter R. 1989. *In vivo* biostability of four types of arterial grafts with impervious walls: their haemodynamic and pathological characteristics. *J. Biomed. Eng.* 11:416–428.

Cines D B, Pollak E S, Buck C A, Loscalzo J, Zimmerman G A, McEver R P, Pober J S, Wick T M, Konkle B A, Schwartz B S, Barnathan E S, McCrae K R, Hug B A, Schmidt A M, Stern D M 1998. Endothelial cells in physiology and in the pathophysiology of vascular disorders. *Blood* 91:3527–3561.

Clowes A W 1996. Improving the interface between biomaterials and the blood: the gene therapy approach. *Circulation* 93:1319–1320.

Clowes A W. 1997a. Vascular gene therapy in the 21st century. *Thromb. Haemost.* 78:605–610.

Clowes A W. 1997b. Vascular gene transfer using smooth muscle cells. *Ann. N.Y. Acad. Sci.* 811: 293–297.

Contard P, Bartel R L, Jacobs L II, Perlish J S, MacDonald E D II, Handler L, Cone D, Fleischmajer R. 1993. Culturing keratinocytes and fibroblasts in a three-dimensional mesh results in epidermal differentiation and formation of a basal lamina-anchoring zone. *J. Invest. Dermatol.* 100:35–39.

Cooper, M L, Hansbrough J F, Spielvogel R L, Cohen R, Bartel R L, Naughton G. 1991. *In vitro* optimization of a living dermal substitute employing cultured human fibroblasts on a biodegradable polyglycolic acid or polyglactin mesh. *Biomaterials* 12:243–248.

Damour O, Gueugniaud P Y, Berthin-Maghit P, Rousselle P, Berthod F, Sahuc F, Collombel C. 1994. A dermal substrate made of collagen-GAG-chitosan for deep burn coverage: first clinical uses. *Clin. Mater.* 15:273–276.

Damour O, Braye F, Foyatier J L, Fabreguette A, Rousselle P, Vissac S, Petit P. 1997. Cultured autologous epidermis for massive burn wounds: 15 years of practice. In: *Skin Substitute Production by Tissue Engineering: Clinical and Fundamental Applications.* Landes, Austin, Rouabhia M, ed. pp. 23–45.

Dardik H. 1989. Modified human umbilical vein allograft. In: *Vascular Surgery,* RB Rutherford, ed WB Sauders Company, Philadelphia, pp. 474–480.

Deutsch M, Meinhart J, Fischlein T, Preiss P, Zilla P. 1999. Clinical autologous *in vitro* endothelialization of infrainguinal ePTFE grafts in 100 patients: a 9-year experience. *Surgery* 126:847–855.

Duplan-Perrat F, Damour O, Montrocher C, Peyrol S, Grenier G, Jacob MP, Braye F. 2000. Keratinocytes influence the maturation and organization of the elastin network in a skin equivalent. *J. Invest. Dermatol.* 114:365–370.

Falk J, Townsend LE, Vogel LM, Boyer M, Olt S, Wease GL, Trevor KT, Seymour M, Glover JL, Bendick PJ. 1998. Improved adherence of genetically modified endothelial cells to small-diameter expanded polytetrafluoroethylene grafts in a canine model. *J. Vasc. Surg.* 27:902–908.

Fleischmajer R, Contard P, Schwart E, MacDonald ED II, Jacobs L II, Sakai LY. 1991. Elastin-associated microfibrils (10 nm) in a three-dimensional fibroblast culture. *J. Invest. Dermatol.* 97:638–643.

Flugelman MY, Virmani R, Leon MB, Bowman RL, Dichek DA. 1992. Genetically engineered endothelial cells remain adherent and viable after stent deployment and exposure to flow *in vitro. Circ. Res.* 70:348–354.

Gallico GG III, O'Connor NE, Compton CC, Kehinde O, Green H. 1984. Permanent coverage of large burn wounds with autologous cultured human epithelium. *N. Engl. J. Med.* 331: 448–451.

Geer JC, McGill HC, Strong JP. 1961. The fine structure of human atherosclerotic lesions. *Am. J. Pathol.* 38:263–275.

Germain L, Rouabhia M, Guignard R, Carrier L, Bouvard V, Auger FA. 1993. Improvement of human keratinocyte isolation and culture using thermolysin. *Burns* 2:99–104.

Ghazizadeh S, Kolodka TM, Taichman LB. 1998. The skin as a vehicle for gene therapy. In: *Principles of Molecular Medicine.* JL Jameson, ed. Humana Press, Totowa, NJ. pp. 775–779.

Graham LM, Vinter DW, Ford JW, Kahn RH, Burkel WE, Stanley JC. 1980. Endothelial cell seeding of prosthetic vascular grafts: early experimental studies with cultured autologous canine endothelium. *Arch. Surg.* 115:929–933.

Gourevitch D, Jones CE, Crocker J, Goldman M. 1988. Endothelial cell adhesion to vascular prosthetic surfaces. *Biomaterials* 9:97–100.

Green H, Barrandon Y. 1988. Cultured epidermal cells and their use in the generation of epidermis. *NIPS* 3:54–56.

Green H, Kehinde O, Thomas J. 1979. Growth of cultured human epidermal cells into multiple epithelia suitable for grafting. *Proc. Natl. Acad. Sci. U.S.A.* 76:5665–5668.

Greenhalgh DA, Rothnagel JA, Roop DR. 1994. Epidermis: an attractive target tissue for gene therapy. *J. Invest. Dermatol.* 103:63S–69S.

Hansbrough JF, Boyce ST, Cooper ML, Foreman TJ. 1989. Burn wound closure with cultured autologous keratinocytes and fibroblasts attached to a collagen-glycosaminoglycan substrate. *JAMA* 262:2125–2130.

Herring MB, Gardner A, Glover J. 1978. A single-staged technique for seeding vascular grafts with autogenous endothelium. *Surgery* 84:498–504.

Herring MB, Baughman S, Glover J. 1985. Endothelium develops on seeded human arterial prosthesis: a brief clinical note. *J. Vasc. Surg.* 2:727–730.

Hirai J, Matsuda T. 1996. Venous reconstruction using hybrid vascular tissue composed of vascular cells and collagen tissue regeneration process. *Cell Transplant.* 5:93–105.

Jaffe EA, Nachman RL, Becker CG, Minick CR. 1973. Culture of human endothelial cells derived from umbilical veins. Identification by morphologic and immunologic criteria. *J. Clin. Invest.* 52:2745–2756.

Jiang CK, Connolly D, Blumenberg M. 1991. Comparison of methods for transfection of human epidermal keratinocytes. *J. Invest. Dermatol.* 97:969–973.

Jones PH, Watt FM. 1993. Separation of human epidermal stem cells from transit amplifying cells on the basis of differences in integrin function and expression. *Cell* 73:713–724.

Jones PH, Harper S, Watt FM. 1995. Stem cell patterning and fate in human epidermis. *Cell* 80:83–93.

Khavari PA, Krueger GG. 1997. Cutaneous gene therapy. *Dermatol. Clin.* 15:27–35.

Krueger GG, Morgan JR, Jorgensen CM, Schmidt L, Li HL, Kwan MK, Boyce ST, Wiley HS, Kaplan J, Petersen MJ. 1994. Genetically modified skin to treat disease: potential and limitations. *J. Invest. Dermatol.* 103:76S–84S.

Langer R, Vacanti JP. 1993. Tissue engineering. *Science* 260:920–926.

L'Heureux N, Germain L, Labbé R, Auger FA. 1993. *In vitro* construction of human vessel from cultured vascular cells: a morphologic study. *J. Vasc. Surg.* 17:499–509.

L'Heureux N, Pâquet S, Labbé R, Germain L, Auger FA. 1998. A completely biological tissue-engineered human blood vessel. *FASEB J.* 12:47–56.

L'Heureux N, Stoclet JC, Auger FA, Lagaud GJL, Germain L, Andriantsitohaina R. 2001. Human tissue-engineered vascular media: a new model for pharmacological studies of contractile responses. *FASEB J.* 15:515–524.

Lu CY, Khair-el-Din TA, Davidson IA, Butler TM, Brasky KM, Vazquez MA, Sicher SC. 1994. Xenotransplantation. *FASEB J.* 8:1122–1130.

McKeehan WL, Barnes D, Reid L, Stanbridge E, Murakami H, Sato GH. 1990. Frontiers in mammalian cell culture. *In Vitro Cell. Dev. Biol.* 26:9–23.

Meinhart J, Deutsch M, Zilla P. 1997. Eight years of clinical endothelial cell transplantation. Closing the gap between prosthetic grafts and vein grafts. *ASAIO J.* 43:M515–M521.

Meerbaum SO, Sharp WV, Schmidt SP. 1992. Lower extremity revascularization with polytetrafluo-roethylene grafts seeded with microvascular endothelial cells. In: *International Society of Applied Cardiovascular Biology.* Zilla P, Fasol R, Callow A, eds. Basel, Karger AG, 2:107–119.

Michel M, Torok N, Godbout MJ, Lussier M, Gaudreau P, Royal A, Germain L. 1996. Keratin 19 as a biochemical marker of skin stem cells *in vivo* and *in vitro*: keratin 19 expressing cells are differentially localized in function of anatomic sites, and their number varies with donor age and culture stage. *J. Cell Sci.* 109:1017–1028.

Michel M, L'Heureux N, Pouliot R, Xu W, Auger FA, Germain L. 1999. Characterization of a new tissue-engineered human skin equivalent with hair. *In Vitro Cell Dev. Biol. Anim.* 35:318–326.

Miyata T, Conte MS, Trudell LA, Mason D, Whittemore AD, Birinyi LK. 1991. Delayed exposure to pulsatile shear stress improves retention of human saphenous vein endothelial cells on seeded ePTFE grafts. *J. Vasc. Res.* 50:485–493.

Morgan JR, Barrandon Y, Green H, Mulligan RC. 1987. Expression of an exogenous growth hormone gene by transplantable human epidermal cells. *Science* 237:1476–1479.

Morris RJ, Potten CS. 1994. Slowly cycling (label-retaining) epidermal cells behave like clonogenic stem cells *in vitro*. *Cell Prolif.* 27:279–289.

Morris RJ, Potten CS. 1999. Highly persistent label-retaining cells in the hair follicles of mice and their fate following induction of anagen. *J. Invest. Dermatol.* 112:470–475.

Morris RJ, Fisher SM, Slaga TJ. 1985. Evidence that the centrally and peripherally located cells in the murine epidermal proliferative unit are two distinct cell populations. *J. Invest. Dermatol.* 84:277–281.

Moulin V, Auger FA, O'Connor-McCourt M, Germain L. 1997. Fetal and postnatal sera differentially modulate human dermal fibroblast phenotypic and functional features *in vitro*. *J. Cell. Physiol.* 171:1–10.

Nerem RM, Alexander RW, Chappell DC, Medford RM, Varner SE, Taylor WR. 1998. The study of the influence of flow on vascular endothelial biology. *Am. J. Med. Sci.* 316:169–175.

Niklason LE, Gao J, Abbott WM, Hirschi KK, Houser S, Marini R, Langer R. 1999. Functional arteries grown *in vitro*. *Science* 284:489–493.

Noishiki Y, Tomizawa Y, Yamane T, Matsumoto A. 1996. Autocrine angiogenic vascular prosthesis with bone marrow transplantation. *Nat. Med.* 2:90–93.

Noishiki Y, Yamane Y, Okoshi T, Tomizawa Y, Satoh S. 1998. Choice, isolation and preparation of cells for bioartificial vascular grafts. *Artif. Organs.* 22:50–62.

O'Connor NE, Mulliken JB, Banks-Schlegel S, Kehinde O, Green H. 1981. Grafting of burns with cultured epithelium prepared from autologous epidermal cells. *Lancet* 1:75–78.

O'Donnell TF, Mackey W, McCullough JL, Maxwell SL, Farber SP, Deterling RA, Callow AD. 1984. Correlation of operative findings with angiographic and noninvasive hemodynamic factors associated with failure of polytetrafluoroethylene grafts. *J. Vasc. Surg.* 1:136–148.

Ortenwall P, Wadenvik H, Kutti J, Risberg B. 1990. Endothelial cell seeding reduces thrombogenicity of Dacron grafts in humans. *J. Vasc. Surg.* 11:403–410.

Owens GK. 1995. Regulation of differentiation of vascular smooth muscle cells. *Physiol. Rev.* 75:487–517.

Park PK, Jarrell BE, Williams SK, Carter TL, Rose DG, Martinez-Hernandez A, Carabasi RA. 1990. Thrombus-free, human endothelial surface in the midregion of a Dacron vascular graft in the splanchnic venous circuit. Observations after nine months of implantation. *J. Vasc. Surg.* 11:468–475.

Pickering JG, Jekanowski J, Weir L, Takeshita S, Losordo DW, Isner JM. 1994. Liposome-mediated gene transfer into human vascular smooth muscle cells. *Circulation* 89:13–21.

Rheinwald JG, Green H. 1975. Serial cultivation of strains of human epidermal keratinocytes: the formation of keratinizing colonies from single cells. *Cell* 6:331–344.

Rémy M, Bordenave L, Bareille R, Gorodkov A, Rouais F, Baquey C. 1994. Endothelial cell compatibility testing of various prosthetic surfaces. *J. Mater. Sci. Mater. Med.* 5:808–812.

Ross R. 1971. The smooth muscle cell. II. Growth of smooth muscle in culture and formation of elastic fibers. *J. Cell. Biol.* 50:172–186.

Sahuc F, Nakazawa K, Berthod F, Damour O, Collombel C. 1996. Mesenchymal-epithelial interactions regulate gene expression of type VII collagen and kalinin in keratinocytes and dermal-epidermal junction formation in a skin equivalent model. *Wound Repair Regen.* 4:93–102.

Sayers RD, Raptis S, Berce M, Miller JH. 1998. Long-term results of femorotibial bypass with vein or polytetrafluoroethylene. *Br. J. Surg.* 85:934–938.

Stanke F, Riebel D, Carmine S, Cracowski JL, Caron, F, Magne JL, Egelhoffer H, Bessard G, Devillier P. 1998. Functional assessment of human femoral arteries after cryopreservation. *J. Vasc. Surg.* 28:273–283.

Slivka SR, Landeen L, Zeigler F, Zimber M, Bartel RL. 1993. Characterization, barrier function, and drug metabolism of an *in vitro* skin model. *J. Invest. Dermatol.* 100:40–46.

Stephen M, Loewenthal J, Little JM, May J, Sheil AG. 1977. Autogenous veins and velour Dacron in femoropopliteal arterial bypass. *Surgery* 81:314–318.

Stoclet JC, Andriantsitohaina R, L'Heureux N, Martinez C, Germain L, Auger FA. 1996. Use of human vessels and human smooth muscle cells in pharmacology. *Cell Biol. Toxicol.* 12:223–225.

Teumer J, Lindahl A, Green H. 1990. Human growth hormone in the blood of athymic mice grafted with cultures of hormone-secreting human keratinocytes. *FASEB J.* 4:3245–3250.

Thyberg J, Hedin U, Sjolund M, Palmberg L, Bottger BA. 1990. Regulation of differentiated properties and proliferation of arterial smooth muscle cells. *Arteriosclerosis* 10:966–990.

Tranquillo RT, Girton TS, Bromberek BA, Triebes TG, Mooradian DL. 1996. Magnetically orientated tissue-equivalent tubes: application to a circumferentially orientated media-equivalent. *Biomaterials* 17:349–357.

Trueb J, Trueb B. 1992. Type XIV collagen is a variant of undulin. *Eur. J. Biochem.* 207:549–557.

Veelken H, Jesuiter H, Mackensen A, Kulmburg P, Schultze J, Rosenthal F, Mertelsmann R, Lindemann A. 1994. Primary fibroblasts from human adults as target cells for ex vivo transfection and gene therapy. *Hum. Gene Ther.* 5:1203–1210.

Veith FJ, Gupta SK, Ascer E, White-Flores S, Samson RH, Scher LA, Towne JB, Bernhard VM, Bonier P, Flinn WR, Astelford P, Yao, JST, and Bergan JJ. 1986. Six year prospective multicenter randomized comparison of autologous saphenous vein and expanded polytetrafluoroethylene grafts in infrainguinal arterial reconstructions. *J. Vasc. Surg.* 3:104–114.

Verhagen HJM, Blankensteijn DJ, de Groot PG, Heijnen-Snyder GJ, Pronk A, Vroom TM, Muller HJ, Nicolay K, van Vroonhoven TJ, Sixma JJ, and Eikelboom BC. 1998. *In vivo* experiments with mesothelial cell seeded ePTFE vascular grafts. *Eur. J. Vasc. Endovasc. Surg.* 15:489–496.

Vohra R, Thomson GJ, Carr HM, Sharma H, Walker MG. 1992. The response of rapidly formed adult human endothelial cell monolayers to shear stress of flow: a comparison of fibronectin-coated Teflon and gelatin-impregnated Dacron grafts. *Surgery* 111:210–220.

Walker MG, Thomson GJL, Shaw JW. 1987. Endothelial cell seeded versus non-seeded ePTFE grafts in patients with severe peripheral vascular disease. In: *Endothelialization of Vascular Grafts*. Zilla P, Fasol R, Deutsch M, eds. Basel S, Karger A.G., pp. 245–248.

Weinberg CB, Bell E. 1986. A blood vessel model constructed from collagen and cultured vascular cells. *Science* 231:397–400.

Xu XM, Ohashi K, Sanduja SK, Ruan KH, Wang LH, Wu KK. 1993. Enhanced prostacyclin synthesis in endothelial cells by retrovirus-mediated transfer of prostaglandin H synthase cDNA. *J. Clin. Invest.* 91:1843–1849.

Young DM, Greulich KM, Weier HG. 1996. Species-specific in situ hybridization with fluorochrome-labeled DNA probes to study vascularization of human skin grafts on athymic mice. *J. Burn Care Rehabil.* 17:305–310.

Ziegler T, Alexander RW, Nerem RM. 1995. An endothelial cell-smooth muscle cell co-culture model for use in the investigation of flow effects on vascular biology. *Ann. Biomed. Eng.* 23:216–225.

Zilla P, von Oppell U, Deutsch M. 1993. The endothelium: a key for the future. *J. Card. Surg.* 8:32–60.

27

Engineering Functional Cartilage and Cardiac Tissue: *In vitro* Culture Parameters

Lisa E. Freed, Maria A. Rupnick, Dirk Schaefer and Gordana Vunjak-Novakovic

Introduction

The musculoskeletal and the cardiovascular systems perform structural and mechanical functions vital for health and survival. The incidence of diseases including osteoarthritis and heart failure is increasing and, in the absence of curative interventions, has resulted in substantial human suffering and medical expense.

Articular cartilage is an avascular tissue containing only one cell type, the chondrocyte, which generates and maintains an extracellular matrix (ECM) consisting of a fibrous network of collagen type II and glycosaminoglycan (GAG)-rich proteoglycan (Buckwalter and Mankin, 1997a). The main function of articular cartilage is to allow for maximal joint mobility while transferring compressive and shear forces. Once damaged, cartilage has a limited capacity for self-repair (Buckwalter and Mankin, 1997b). Osteoarthritis, the hallmark of which is progressive cartilage degeneration, results in chronic pain and disability and affects an estimated 40 million Americans (Oddis, 1996). Therapeutic interventions, such as analgesics, physical therapy, and surgery have had mixed results (Buckwalter and Mankin, 1997b; Lim et al., 1996).

The myocardium (cardiac muscle) is a highly differentiated muscular organ composed of cardiac myocytes, fibroblasts, and macrophages with a dense supporting vasculature and collagenous ECM (Brilla et al., 1995; MacKenna et al., 1994). The myocytes form a three-dimensional syncytium that enables propagation of electrical signals across specialized intracellular junctions to produce coordinated mechanical contractions that pump blood forward. Once damaged, the heart is unable to regenerate. Heart failure affects over five million Americans (Rich, 1997), and is the final common pathway of cardiovascular diseases, which are the leading cause of morbidity and mortality in developed countries (Dominguez et al., 1999). Currently, the only definitive treatment for end stage heart failure is cardiac transplantation, but limited organ availability has led to prolonged waiting periods that are often not survivable (Evans, 2000).

We believe that these problems may be alleviated, at least in part, by engineering tissue constructs based on isolated cells and polymeric scaffolds that, when implanted *in vivo*, can help restore the function of a damaged tissue (Fig. 27.1). A cell-based method in which autologous chondrocytes are amplified *in vitro* and then transplanted into localized areas of damaged cartilage is currently used clinically (Brittberg et al., 1994) with mixed results (Messner and Gillquist, 1996; Newman, 1998). However, the efficacy of this procedure is limited by leakage of the cells out of the defect (Breinan et al., 1998), and it is unsuitable in cases where osteoarthritis involves the entire articular surface There is currently no clinically used cell-based method to repair damaged myocardium. Therefore, there is a

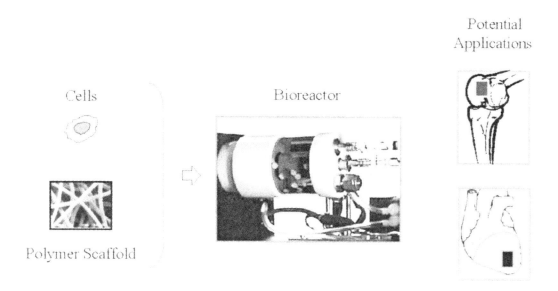

FIGURE 27.1. Model system. Isolated cells are cultured on polymer scaffolds in bioreactors to engineer functional tissues. Engineered cartilage and cardiac tissue can potentially be used to repair damaged articular cartilage or myocardium (Freed and Vunjak-Novakovic, 2002).

clear and present need to further our understanding of these systems and devise new, disease modifying interventions.

Functional Engineered Cartilage

Relatively few *in vivo* studies involving engineered cartilage include a functional evaluation of the repair tissue. Wakitani et al. (1994, 1998) repaired osteochondral defects in skeletally immature rabbits using auto- and allografts made of cells and collagen gel. In one study, autologous progenitor cells obtained from bone marrow or periosteum were expanded in monolayers, embedded in a gel of bovine type I collagen, and implanted within 2 hours into $6 \times 3 \times 3$ mm defects in the medial femoral condyles of 3–4-month-old rabbits (Wakitani et al., 1994). The 6-month repair tissues were about twice as stiff in defects implanted with engineered cartilage as in defects left empty, but only about half as stiff as native articular cartilage, as estimated from the relative compliance obtained by indentation testing. The repair tissue remodeled into a cartilaginous surface and new subchondral bone.

The same investigators used allogenic articular chondrocytes in conjunction with methods described above for the repair of similarly sized condylar and femoropatellar groove defects in young rabbits (Wakitani et al., 1998). Six-month repair tissues were again about twice as stiff in defects implanted with engineered cartilage as in defects left empty, but did not remodel even after a period of almost 1 year such that the investigators expressed doubts about the longterm mechanical stability of the repair tissue.

Chu et al. (1997) repaired osteochondral defects in skeletally mature rabbits using allografts made of cells and polylactic acid (PLA). Perichondrocytes obtained from the costal cartilage of adult rabbits were expanded in monolayers, seeded on PLA sponges, and implanted within 2 hours into 3.7 mm diameter $\times$ 5 mm deep defects in the medial femoral condyles of 9–12-month-old rabbits. Equilibrium moduli of the repair tissue, measured in confined compression, were approximately 280 kPa after periods ranging from 6 weeks to 1 year. One-year repair tissues displayed subchondral bone regeneration in approximately half of the cases and contained approximately half as much GAG per unit dry weight as native articular cartilage. The authors speculated that donor cell immunogenicity and PLA degradation may have adversely affected tissue repair.

Schaefer et al. (2002) repaired osteochondral defects in skeletally mature rabbits using composites based on engineered cartilage. Articular chondrocytes obtained from 3-month-old rabbits were expanded in monolayers, dynamically seeded on polyglycolic acid (PGA) mesh in bioreactors, cultured for 4–6 weeks *in vitro,* sutured to an osteoconductive support, and implanted into 7 × 5 × 5 mm femoropatellar groove defects in 8-month-old rabbits. Following implantation of engineered cartilage composites, 6-month repair tissues had Young's moduli of approximately 790 kPa that were comparable to values previously measured by indentation testing of similar cartilage from similarly aged rabbits (Hoch et al., 1983), and were significantly stiffer than untreated defects (Fig. 27.2a). Defects implanted with engineered cartilage were repaired with uniformly thick cartilage with characteristic architectural features and new subchondral bone, whereas defects implanted with cell-free scaffolds or left untreated were filled with irregularly shaped and histologically inferior fibrocartilage (Fig. 27.2b,c). We concluded that engineered cartilage based on cultured articular chondrocytes provided a mechanically functional template capable of undergoing orderly remodeling *in vivo* into osteochondral tissue with physiologically stiff cartilage.

Buschmann et al. (1992, 1995) studied the effects of mechanical stimulation on *in vitro* chondrogenesis. Articular chondrocytes obtained from bovine calves were embedded in agarose gel and cultured exposed to static, dynamic, or no mechanical compression (static strain amplitudes up to 50%, dynamic strain amplitudes of 6%, and frequencies of 0.01–1 Hz) for up to 7 weeks. Engineered cartilage had equilibrium moduli of approximately 75 kPa, as assessed in confined compression. The ECM appeared spatially discontinuous, due to the continued presence of the agarose, and had wet weight fractions of DNA and GAG that were approximately 25% those of native calf cartilage. Dynamic compression increased and static compression decreased ECM synthesis rates, as

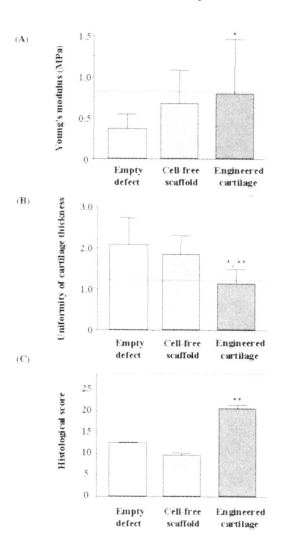

FIGURE 27.2. Application of functional engineered cartilage to osteochondral defect repair. Engineered cartilage based on articular chondrocytes and PGA mesh was cultivated for 4–6 weeks *in vitro* and then implanted as a composite with an osteoconductive support into 7 × 5 × 5 mm osteochondral defects in femoropatellar grooves of adult rabbits (Schaefer et al., 2002). Six month repair tissues in defects left empty, implanted with cell-free scaffolds or implanted with composites based on engineered cartilage were compared with respect to: (*A*) Young's moduli, assessed by indentation testing; (*B*) uniformity of cartilage thickness (i.e., the ratio of thickness of the repair cartilage at the defect center to its thickness overall); and (*C*) overall histological score. Dotted lines represent average values obtained for native rabbit articular cartilage. Data represent average ± standard deviation of 4 to 7 independent samples.* Statistically different from empty defects;** statistically different from cell-free scaffolds by ANOVA ($p < 0.05$).

determined from sulfate and proline incorporation, in a manner that depended on the amount of matrix which in turn depended on *in vitro* culture duration. In a related study, engineered cartilage based on calf chondrocytes cultured for 3 weeks in chitosan or agarose gel had equilibrium stiffnesses of 11 or 28 kPa, respectively, as assessed in unconfined compression (Hoemann et al., 2001). Cartilaginous ECM accumulated in both groups, with progressive degradation of the chitosan but not the agarose. In addition, the chitosan adhered to osteochondral defects, implying that this gel could serve as a chondrocyte delivery vehicle.

Lee and Bader (1997) studied the effects of short-term dynamic compression on chondrocytes cultured in agarose. Articular chondrocytes obtained from 1.5-year-old steer were embedded in agarose and cultured exposed to no, static, or dynamic compression (strain amplitudes of 15% and frequencies of 0.3–3 Hz) for 2 days. Dynamic compression generally increased the rate of GAG synthesis, decreased the rate of proline incorporation, and increased the rate of thymidine incorporation, and each parameter was influenced in a distinct manner by the dynamic compression regimen.

Grande et al. (1997) engineered cartilage using different polymer scaffolds and culture systems. Articular chondrocytes were obtained from bovine calves, expanded in monolayers, seeded onto meshes made of PGA, polylactic-co-glycolic acid (PLGA), or bovine type I collagen, and cultured for 5 weeks either in perfused Teflon® bags or static dishes. PGA or PLGA meshes were associated with increased GAG synthesis rates, whereas collagen mesh was associated with increased protein synthesis rates. ECM synthesis rates were generally higher in perfused than static cultures.

Freed et al. (1997, 1998) and Vunjak-Novakovic et al. (1999) studied engineered cartilage cultured in rotating bioreactors for periods of up to 7 months. Articular chondrocytes obtained from bovine calves were dynamically seeded at high density on PGA mesh in spinner flasks and then cultured in rotating bioreactors. After 6 weeks *in vitro,* the engineered cartilage had an ECM that was spatially continuous over its entire cross-sectional area (7 mm diameter × 5 mm thick) (Freed et al., 1998). Time histories of GAG and protein synthesis rates and of the fractions of components in the engineered cartilage are shown in Figure 27.3*a* and *b.* ECM synthesis rates were initially high and comparable to native cartilage, then decreased by approximately 40% over 6 weeks. Fractions of GAG and collagen type II increased at rates comparable to the rate of PGA degradation. At lower cell seeding densities, the nonwoven PGA mesh unravelled and constructs lost their structural integrity over 2 weeks (Vunjak-Novakovic et al., 1998). These findings imply that for a given set of culture conditions, ECM must be deposited at a rate comparable to that at which the scaffold's structural integrity is lost.

Equilibrium moduli of engineered cartilage cultured for 6 weeks and 7 months *in vitro* were respectively approximately 172 and 932 kPa, the latter of which was similar to native articular cartilage, as measured in confined compression (Fig. 27.3*c*) (Freed et al., 1997; Vunjak-Novakovic et al., 1999). Engineered cartilage cultured for 7 months had a GAG fraction 30% higher than that of native cartilage (Fig. 27.3*d*), whereas its total collagen fraction and dynamic stiffness were 34% and 46% as high, respectively. We concluded that the deposition of ECM components might precede their functional organization and postulated that the observed structural and functional deficiencies could be due to the absence of specific biochemical and/or mechanical stimuli during *in vitro* culture.

Obradovic et al. (1999) demonstrated the need for efficient gas exchange during the *in vitro* cultivation of engineered cartilage. Articular chondrocytes obtained from bovine calves were cultured for 5 weeks in rotating bioreactors with or without gas exchange. Culture medium oxygen tension (pO_2) and pH were respectively 87 mm Hg and 7.0, or 43 mm Hg and 6.7, in the presence or absence of gas exchange. Gas exchange significantly increased synthesis rates of GAG and collagen, as assessed by the rate of sulfate incorporation and the fraction of hydroxyproline in newly synthesized protein (Fig. 27.4*a*). Engineered constructs grown in the presence of efficient gas exchange were thick and had a high fraction of uniformly distributed GAG, whereas those grown without gas ex-

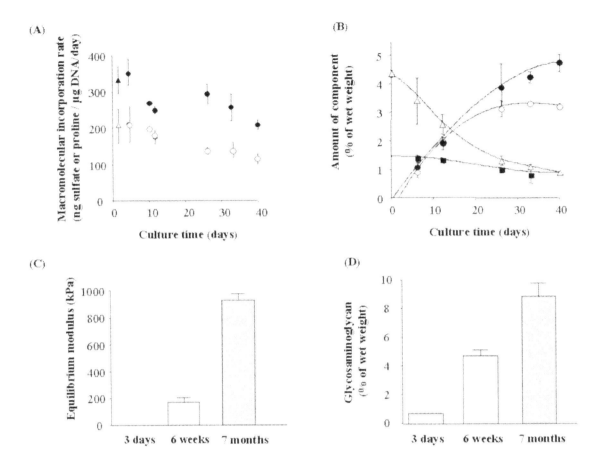

FIGURE 27.3. Effects of culture time on engineered cartilage. Engineered cartilage based on bovine calf articular chondrocytes and PGA mesh and cultured for up to 7 months in rotating bioreactors was assessed with respect to: (*A*) rates of incorporation of sulfate (*closed symbols*) and proline (*open symbols*). Engineered cartilage (*circular symbols*) was also compared to native bovine calf cartilage (*triangular symbols*); (*B*) fractional amounts of glycosaminoglycans (GAG, *closed circles*), collagen type II (*open circles*), cells (*squares*), and PGA (*triangles*); (*C*) equilibrium modulus, assessed in radially confined compression (3 day constructs were too fragile to measure); and (*D*) GAG fraction (Freed et al., 1997, 1998; Vunjak-Novakovic et al., 1999). *Dotted lines* represent average values obtained for native bovine calf articular cartilage. Data represent average ± standard deviation of 3 to 6 independent samples.

change were thinner and had markedly lower GAG fractions that decreased with increasing depth from the tissue surfaces (Fig. 27.4*b*). The finding that GAG synthesis rate depended on pO_2 was used to develop a mathematical model that predicted GAG concentrations in engineered cartilage as a function of position and culture time (Obradovic et al., 2000). We concluded that hypoxic conditions comparable to those found in articular cartilage *in vivo* were suboptimal for *in vitro* chondrogenesis.

Vunjak-Novakovic et al. (1999) studied the effects of bioreactor hydrodynamics on the structural and functional properties of engineered cartilage. Articular chondrocytes obtained from bovine calves were cultured for 6 weeks in static flasks, mixed flasks or rotating bioreactors, which respectively corresponded to static, turbulent, or laminar flow conditions (Freed and Vunjak-Novakovic, 2000, 2002). The corresponding equilibrium moduli were approximately 53, 51, and 172 kPa, as measured in con-

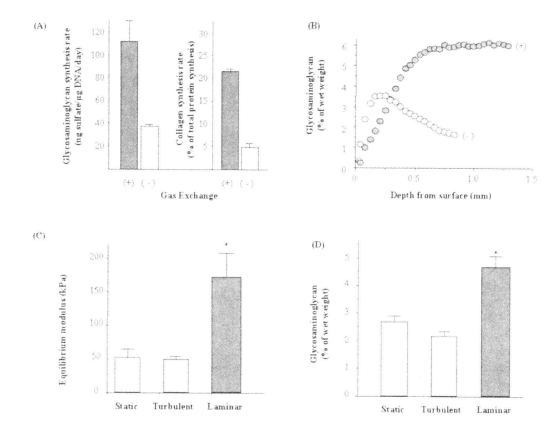

FIGURE 27.4. Effects of culture conditions on engineered cartilage. Engineered cartilage based on bovine calf articular chondrocytes and PGA mesh and cultured for 5–6 weeks in rotating bioreactors was assessed with respect to: (*A*) synthesis rate of glycosaminoglycans (GAG) or collagen and (*B*) spatial distribution of GAG in constructs cultivated in bioreactors with (+) or without (−) gas exchange membranes (Obradovic et al., 1999); (*C*) equilibrium modulus and (*D*) GAG fraction of constructs cultivated under static, turbulent, or laminar flow conditions, in static cultures, spinner flasks, and rotating bioreactors, respectively (Vunjak-Novakovic et al., 1999). GAG and protein synthesis rates (*A*) were respectively assessed from macromolecular incorporation rates of sulfate and proline; collagen synthesis rate was determined as the fraction of new proteins containing hydroxyproline (Freed et al., 1998). GAG distribution (*B*) was determined by image analysis of histological sections (Martin et al., 1999a). Equilibrium modulus (*C*) was assessed in radially confined compression (Vunjak-Novakovic et al., 1999). Data represent average ± standard deviation of 3 to 6 independent samples.* Statistically different from other groups by ANOVA ($p < 0.05$).

fined compression (Fig. 27.4*c*). GAG fractions were also highest for engineered cartilage grown in rotating bioreactors (Fig. 27.4*d*). We concluded that hydrodynamic factors such as mixing and flow modulated the structure and function of engineered cartilage, and that laminar flow conditions at the tissue surfaces enhanced *in vitro* chondrogenesis.

Carver et al. (1999) studied the effects of hydrostatic pressure on the development of engineered cartilage. Articular chondrocytes ob-

tained from foal cartilage were seeded on PGA mesh and cultured in a perfused bioreactor such that constructs were exposed to intermittent hydrostatic pressure (3 MPa in 60 cycles of 5 sec on/15 sec off over 20 minutes, 6 times per day). Pressure-induced increases in GAG content correlated with increases in compressive stiffness estimated using a dynamic mechanical analyzer.

Mauck et al. (2000) studied the effects of mechanical compression on engineered cartilage. Articular chondrocytes obtained from bovine calf cartilage were embedded in agarose and cultured statically or exposed to intermittent dynamic compression (strain amplitudes of 10% and frequencies of 1 Hz in three consecutive cycles of 1 hour on/1 hour off once per day, 5 days per week). Engineered constructs cultured for 1 month with intermittent dynamic compression had an equilibrium modulus of approximately 100 kPa, which was significantly higher than that of constructs cultured statically. Dynamic compression also significantly improved construct biochemical composition.

Functional Engineered Cardiac Tissue

Li et al. (1999) recently studied the *in vivo* survival and function of implanted engineered cardiac tissue. Cardiac cells were obtained from fetal rat hearts, seeded onto foams of denatured collagen (Gelfoam®), cultured for 1 week *in vitro,* and implanted into adult rats of the same inbred strain as the donor cells either subcutaneously or sutured to a cryogenically generated scar on the epicardial surface of the left ventricular wall. Subcutaneously implanted engineered tissue contracted spontaneously at 1 week, as assessed by echocardiography, and consisted of interconnected myocytic cells and new blood vessels at 5 weeks, as assessed histologically. Epicardially grafted engineered tissue was adherent to the host scar tissue and consisted of proliferating cells, as assessed by bromodeoxyuridine incorporation. Myocardial function, assessed by left ventricular pressure measurements, was not improved by treatment with engineered cardiac tissue as compared to cell-free Gelfoam® or no treatment. In complementary *in vitro* studies, engineered cardiac tissue contracted spontaneously for periods of up to 2 months.

Eschenhagen and Colleagues (1997) engineered cardiac tissue *in vitro* using cells and collagen gel. Cardiac cells were obtained from embryonic chicks, enriched for myocytes, mixed with rat tail collagen type I, cast between two Velcro®-coated tubes separated by spacers and cultured under tension for 11 days. At timed intervals, constructs were placed in a test chamber,

the spacers removed, and contractile properties measured in response to (i) tensile strains (up to 20%), (ii) electrical stimulation (10 ms pulses at frequencies of 0.8–2.5 Hz), and (iii) pharmacological agents. Engineered cardiac tissue had dimensions of $15 \times 6 \times 0.18$ mm, with the majority of the cells distributed at the surfaces, and incorporated radiolabeled thymidine at rates that decreased with increasing culture time over 6 days. Engineered cardiac tissue contracted spontaneously at approximately 72 beats per minute (bpm), exhibited characteristic positive force-length and negative force-frequency behaviors, and were responsive to pharmacological agents. Gene transfer into the component cardiac cells was also demonstrated. The authors concluded that engineered cardiac tissue could provide a model system for controlled *in vitro* studies of cardiac cell-polymer interactions, cardiac tissue development and function, and genetic manipulations. The authors further suggested that, as compared to monolayers, the three-dimensionality of engineered cardiac tissues reduced the undesirable proliferation of nonmyocytic cells.

Zimmermann et al. (2000) engineered cardiac tissue starting from neonatal rat heart cells in conjunction with methods similar to those described above for avian cells except that the cells were embedded in a mixture of collagen and Matrigel®. Engineered cardiac tissue contracted spontaneously at rates of up to 174 bpm that generally increased with increasing cell density. Contractile force increased with increasing fractions of Matrigel® and with culture duration up to day 18, then remained stable until day 26. The authors concluded that the functional properties of engineered cardiac tissues based on neonatal rat heart cells depended on initial cell density, soluble growth factors provided by the Matrigel®, and *in vitro* culture duration.

Fink et al. (2000) engineered cardiac tissue starting from mammalian and avian heart cells using methods similar to those described above except that the constructs were exposed to dynamic stretch (tensile strains of 0–20% at a frequency of 1.5 Hz) between culture days 4 and 10. Engineered cardiac tissues based on neonatal rat heart cells were superior to those based on avian cells with respect to spontaneous contraction rate (approximately 142 and 91 bpm, respectively) and responsiveness to stretch and pharmacological agents. Dynamic stretch generally improved the contractility and pharmacological responsiveness of engineered cardiac tissue and was associated with increases in cell diameter and protein content, expression of sarcomeric α-actin, longitudinal cell orientation, sarcomere density and length, and mitochondrial density. The authors concluded that dynamic stretch induced cellular hypertrophy which improved the structure and contractile function of engineered cardiac tissue.

Akins et al. (1997, 1999) engineered cardiac tissue using two different substrates in rotating bioreactors. Cardiac cells were obtained from neonatal rat hearts, co-inoculated with either fibronectin-coated microcarrier beads or collagen fibers into rotating vessels, and cultured for 6 days. Cardiac constructs based on microcarrier beads and control monolayers of cardiac cells contracted spontaneously at approximately 50 and 190 bpm, respectively. Constructs based on microcarrier beads were similar to control monolayers with respect to the fraction of myocytes, amount of DNA per unit protein, specific activities of various metabolic enzymes, expression levels of myosin heavy chain and filamentous actin, and pharmacological responsiveness. Structural features of differentiated cardiac tissue were observed for cardiac cells cultured on both microcarrier beads and collagen fibers.

Freed, Vunjak-Novakovic, and collaborators engineered cardiac tissue using cells from neonatal rat or embryonic chick hearts, PGA mesh, and various bioreactors including rotating vessels. Engineered cardiac tissue based on neonatal rat heart cells and cultured for 2–3 weeks contracted spontaneously and synchronously at rates of approximately 30–130 bpm (Freed and Vunjak-Novakovic, 1997). Bursac et al. (1999) studied the functional properties of engineered cardiac tissue using a linear array of stimulating and recording electrodes. Cardiac cells obtained from neonatal rats were enriched for myocytes, dynamically seeded on PGA mesh and cultured in mixed flasks for 1 week. Engineered cardiac constructs had an electrophysiologically functional, 0.05–0.07 mm thick tissue-like region at the surfaces that propagated electrical impulses over distances of up to 5 mm and contracted synchronously at rates that could be controlled over a range of stimulation frequencies (approximately 80–270 bpm). We concluded that engineered cardiac tissue could provide a model system for controlled *in vitro* studies of cardiac tissue development and electrophysiological function.

Carrier et al. (1999) studied the effects of bioreactor hydrodynamics on the structure and metabolic activity of engineered cardiac tissue. Cardiac cells obtained from embryonic chick or neonatal rat hearts were dynamically seeded on PGA mesh and cultured for 2 weeks in three different environments. The relative rates of lactate production and glucose consumption for engineered cardiac tissue cultured statically, in mixed flasks, and in rotating vessels were respectively 2.0, 1.6, and 1.0 mol/mol, implying anaerobic, partially aerobic, and aerobic cell metabolism. Cell metabolic activity and tissue structure were markedly inferior when cultures were carried out under static rather than mixed conditions. We concluded that hydrodynamic factors affected engineered cardiac tissue in a manner consistent with that observed for engineered cartilage, and that aerobic conditions were needed to engineer cardiac tissue.

Papadaki et al. (2001) studied the effects of scaffold coating and bioreactor hydrodynamics on the electrophysiological and molecular properties of engineered cardiac tissue. Cardiac cells obtained from neonatal rats were enriched for myocytes, seeded on PGA mesh that was either laminin-coated or not, and cultured in either mixed flasks or rotating bioreactors for 1 week. Engineered cardiac tissue contained a functional, peripheral region 0.12–0.16 mm thick consisting of multiple cell layers. The use of laminin-coated PGA mesh was associated with significantly improved conduction velocity and expression of muscle-specific creatine kinase (CK-MM), as compared to uncoated PGA (Fig. 27.5*a,b*). The laminar flow conditions in rotating bioreactors were associated with significantly improved maximum capture rate and expression level of the gap junctional protein connexin 43, as compared to the turbulent conditions in mixed flasks (Fig. 27.5*c,d*). Engineered cardiac tissue made using laminin-coated PGA and cul-

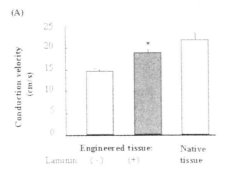

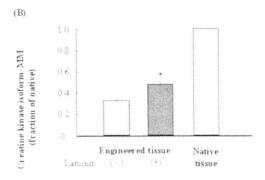

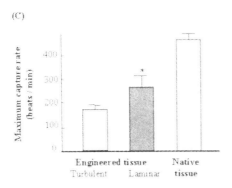

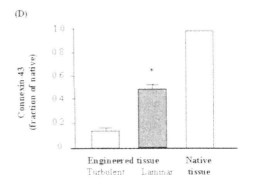

FIGURE 27.5. Effects of culture conditions on engineered cardiac tissue. (*A, B*) Engineered cardiac tissue based on neonatal rat heart cells and PGA mesh without (−) or with (+) laminin coating and cultured for 1 week in bioreactors were compared with respect to: (*A*) conduction velocity, assessed using a linear electrode array, (*B*) expression of creatine kinase (CK-MM), determined by Western blot, for constructs cultured in rotating bioreactors, (*C*) maximum capture rate (i.e., the maximum rate at which each stimulus was followed by a response), and (*D*) expression of connexin-43, determined by Western blot, for constructs based on uncoated PGA cultured under either turbulent or laminar conditions in spinner flasks or rotating bioreactors, respectively (Papadaki et al., 2001). Data represent average ± standard error of 4 to 10 independent samples. *Statistically different from corresponding group of engineered tissue by ANOVA ($p < 0.05$).

tured in rotating bioreactors had conduction velocities and maximum capture rates that were, respectively, 86% and 65% as high as those in native neonatal rat heart. We concluded that functional and molecular properties of engineered cardiac tissues were interrelated and depended on model system parameters including the scaffold coating and bioreactor vessel.

In Vitro Factors

The approach to tissue engineering shown in Figure 27.1 involves seeding of a high density of uniformly distributed cells on a three-dimensional polymeric scaffold, and cultivating the resulting construct under conditions that permit functional tissue development. The criteria for cell seeding and tissue cultivation translate into a set of design requirements, some of which can be generalized, whereas others depend on the specific tissue type and application. Engineered tissues should ideally possess a certain minimal size, thickness, and mechanical integrity to allow for handling and permit construct survival under physiological conditions (e.g., in an articular joint for engineered cartilage or the myocardial wall for engineered cardiac tissue). One of the main paradigms of functional tissue engineering is that once implanted, an engineered tissue should: (i) continue to develop and integrate with adjacent host tissues until normal architecture has been restored, and (ii) provide some minimal level of function immediately postimplantation that should improve progressively until normal function has been restored. Specific functional requirements depend on the application (i.e., engineered cartilage should withstand and transmit loads, and engineered cardiac tissue should contract in a coordinated manner). *In vitro* factors related to the cells, scaffold, and the conditions and duration of cultivation can modulate the functionality of engineered cartilage and cardiac tissue as follows.

Cell and Scaffold-Related Factors

The cells used thus far to engineer functional cartilage and cardiac tissues have varied with respect to donor age (embryonic, neonatal, immature, or adult), differentiation state (precursor or phenotypically mature), *in vitro* processing method (expanded or enriched for a specific cell type), and *in vitro* potential for proliferation and ECM biosynthesis, and have provided a test bed for genetic manipulations (e.g. Madry et al., 2002; Zimmermann et al., 2000).

In the case of engineered cartilage, *in vitro* studies were mainly done using bovine calf chondrocytes (e.g., Buschmann et al., 1992; Freed et al., 1998), since these cells are readily available in large quantities, whereas *in vivo* studies were done using allogeneic articular chondrocytes (Schaefer et al., 2002; Wakitani et al., 1998), allogeneic perichondrocytes (Chu et al., 1997), and autologous periosteal cells or precursor cells from the bone marrow (Wakitani et al., 1994). Differentiated chondrocytes and their precursors each offer advantages for *in vivo* cartilage repair as follows. Chondrocytes can maintain their differentiation potential if expanded and cultured under appropriate conditions (Benya and Shaffer, 1982; Martin et al., 1999b, 2001b) and express their differentiated phenotype for prolonged periods of time (7–8 months) in three-dimensional cultures (Freed et al., 1997; Hauselmann et al., 1994). Allogeneic engineered cartilage did not provoke a significant immune

response *in vivo* (Noguchi et al., 1994; Schaefer et al., 2002; Schreiber et al., 1999), which might be explained by the masking of cell surface antigens by cartilaginous ECM (Langer and Gross, 1974). In contrast to chondrocytes, precursor cells from the bone marrow are easy to harvest, remain metabolically active in older donors (Haynesworth et al., 1998), and can recapitulate some aspects of embryonic skeletal tissue development (Caplan et al., 1997).

In the case of engineered cardiac tissue, *in vitro* and *in vivo* studies were done using heart cells obtained from embryonic chicks and fetal or neonatal rats (e.g., Carrier et al., 1999; Li et al., 1999; Zimmermann et al., 2000). *In vivo* studies avoided immunorejection by using donor heart cells from inbred rats (Li et al., 1999). In general, the younger the cell donor, the higher the proliferative capacity and metabolic activity of the harvested cells. Whereas all cell types used to engineer cartilage were able to proliferate *in vitro*, only embryonic and fetal heart cells proliferated *in vitro* (Fink et al., 2000; Li et al., 1999). Engineered cardiac tissue based on rat cells was superior to that based on chick cells with respect to the presence of spontaneous contractility (Carrier et al., 1999), and contractile force, frequency, and responsiveness to pharmacological stimuli (Fink et al., 2000; Zimmermann et al., 2000).

Tissue engineering scaffolds that have been investigated to date vary with respect to material chemistry (e.g., collagen, agarose, or synthetic polymers with or without coatings), general structure (e.g., gels, fibrous meshes, porous sponges), detailed structure (e.g., porosity, pore size distribution, orientation and connectivity of the polymer phase), mechanical properties (e.g., elasticity), and degradation (degradability, degradation rate). At present, selecting an existing scaffold or designing a new one is an empirical process for several reasons. First, it is not yet possible to predict how a given cell type will attach, proliferate, and differentiate when cultured on a given scaffold. Second, it can be technically difficult to fabricate two scaffolds that differ with respect to only one of the above characteristics and therefore to carry out systematically controlled studies (Pei et al., 2002). Finally, differences in experimental design often make it difficult to distinguish the effects of specific scaffold properties and to compare the results of studies reported in the literature.

In the case of engineered cartilage, a variety of different scaffolds have been successfully used, including gels of agarose (Buschmann et al., 1995; Lee and Bader, 1997; Mauck et al., 2000), collagen (Wakitani et al., 1994, 1998), fibrin (Ameer et al., 2002; Homminga et al., 1993), and chitosan (Hoemann et al., 2001); meshes of collagen (Grande et al., 1997) and PGA (Carver and Heath, 1999; Freed et al., 1997, 1998; Grande et al., 1997; Obradovic et al., 1999; Schaefer et al., 2002; Vunjak-Novakovic et al., 1999), sponges of PLA (Chu et al., 1997; Freed et al., 1993), and meshes and sponges of benzylated hyaluronic acid (Brun et al., 1999; Pei et al., 2002; Solchaga et al., 1999). In the case of engineered cardiac tissue, successfully used scaffolds include gels of collagen with or without Matrigel® (Eschenhagen et al., 1997; Fink et al., 2000; Zimmermann et al., 2000), meshes of PGA with or without laminin-coating (Bursac et al., 1999; Carrier et al., 1999; Freed et al., 1997; Papadaki et al., 2001), sponges of denatured or native collagen (e.g., Gelfoam®, Ultrafoam®) with or without Matrigel® (Li et al., 1999; Radisic, 2003), fibers of collagen and polystyrene beads (Akins et al., 1999).

Cell-free scaffolds used alone (Freed et al., 1994; Grande et al., 1998) and in conjuction with growth factors such as bone morphogenic protein (Sellers et al., 2000) or transforming growth factor beta (Athanasiou et al., 1997) improved the repair of critically sized osteochondral defects compared with natural healing. However, engineered cartilage was superior to cell-free scaffolds or no treatment with respect to structural and functional properties of the repair tissue (Wakitani et al., 1994, 1998; Schaefer et al., 2002) (Fig. 27.2). Functional myocardial repair has not yet been achieved by implanting engineered cardiac tissue, and cell-free scaffolds can cause complications including thrombogenicity, material failure, and inability to either contract or grow (Li et al., 1999).

High cell density was associated with improved chondrogenesis (Freed et al., 1998; Vunjak-Novakovic et al., 1998) and myogenesis (Carrier et al., 1999; Radisic, 2003; Zimmermann et al., 2000) in engineered tissues. High cell densities in engineered cartilage increased ECM synthesis and deposition by chondrocytes (Freed et al., 1998) and induced chondrogenesis by bone marrow derived progenitor cells (Butnariu-Ephrat et al., 1996). Likewise, high cell densities in engineered cardiac tissue improved contractile force and frequency (Zimmermann et al., 2000).

Cell metabolic activity influenced the growth and *in vivo* integration of engineered tissues. Poor integration between engineered cartilage and the surrounding host cartilage is one of major problems limiting the success of articular cartilage repair (e.g., Buckwalter and Mankin, 1997b; Newman, 1998). *In vitro,* the presence of viable cells was associated with the integration of engineered and native cartilage with native cartilage (Obradovic et al., 2001), and required for the integration of two explanted pieces of native cartilage (Reindel et al., 1995). *In vivo,* graft integration was reportedly improved by partial enzymatic digestion of the adjacent native cartilage (Caplan et al., 1997; Hunziker and Kapfinger, 1998) and treatment with bone morphogenic protein (Sellers et al., 2000).

Scaffolds should be composed of biocompatible, biodegradable materials to minimize their immunogenicity *in vivo*. PGA, PLA, and collagen are currently used in products approved by the Food and Drug Administration (e.g., Dexon®, Vicryl®, Gelfoam®, Ultrafoam®), whereas agarose and polystyrene microcarriers do not degrade and are therefore mainly useful as *in vitro* research tools. The rate of scaffold degradation should ideally match the rate of ECM deposition by a specific cell type and under a specific set of *in vitro* culture conditions. For example, PGA nonwoven mesh, which loses its structural stability over approximately 2 weeks *in vitro*, was able to support *in vitro* chondrogenesis starting from bovine calf articular chondrocytes (Fig. 27.3b) but not from bone marrow progenitor cells from the same donor, probably due to lower rates of ECM biosynthesis by the latter cell type (Martin et al., 2001a).

In Vitro Culture Conditions

Hydrodynamic factors (e.g., static or mixed; laminar or turbulent culture conditions), biochemical factors (e.g., oxygen, growth factors), and physical stimuli (e.g., compression, stretch) can modulate the function of engineered cartilage and cardiac tissue. The use of bioreactors generally increased the size and improved the structure and function of engineered tissues, due to combination of enhanced mass transfer and direct hydrodynamic stimulation. Mass transfer generally occurs by molecular diffusion and convective flow that results from concentration, osmotic, and pressure gradients. Mass transfer requirements vary from tissue to tissue, depending on the cellularity, metabolic activity, and developmental stage, and determine its degree of vascularization. For example, cartilage, which is avascular, contains a relatively low density of cells, whereas the heart, which is highly vascular, contains a high density of metabolically active cells.

Oxygen plays a particularly important role in engineering functional tissues. GAG synthesis and deposition were significantly improved by culturing engineered cartilage at a pO_2 of 87 rather than 43 mm Hg (Obradovic et al., 1999) (Fig. 27.4 *a,b*). GAG synthesis rate in engineered cartilage exhibited a first-order dependence on local pO_2, as shown by mathematical modeling (Obradovic et al., 2000). For comparison, the pO_2 in native adult cartilage ranges from approximately 50 mm Hg at the articular surface to less than 7 mm Hg in the deep zone (Brighton and Heppenstall, 1971), although higher values are likely to be present in immature cartilage due to the presence of blood vessels. The finding that higher than physiological pO_2 increased GAG content in engineered cartilage was consistent with results reported for explant cultures of cartilage and periosteum (O'Driscoll et al., 1997; Ysart and Mason, 1994), and is significant because GAG content correlated positively with improved mechanical properties (Vunjak-Novakovic et al., 1999).

Hydrodynamic factors can be exploited in functional tissue engineering. In particular, convective mixing improved the kinetic rate, yield, and spatial uniformity of cell seeding on three-dimensional polymer scaffolds (Vunjak-Novakovic et al., 1998), and improved the structure and composition of engineered tissues by enhancing mass transport within the culture medium and at the tissue surfaces (Carrier et al.,

1999; Gooch et al., 2001; Vunjak-Novakovic et al., 1996). In rotating bioreactors, convective mixing in conjunction with an internal gas exchange membrane maintained a pO_2 of 85–90 mm Hg in the bulk culture medium and promoted oxygen transfer to the surfaces of engineered cartilage, which grew to a thickness of approximately 5 mm (Freed et al., 1997, 1998; Vunjak-Novakovic et al., 1999). Diffusional transport of oxygen within the engineered cartilage did not appear to limit its growth. In contrast, engineered cardiac constructs consisted of a peripheral tissue-like layer that was only approximately 0.1 mm thick (Carrier et al., 2002). This finding could be attributed to diffusional limitations within the engineered cardiac tissue, as supported by calculations based on literature values for the metabolic activity of cardiac myocytes that showed that the pO_2 would decrease to zero at a depth of approximately 0.1 mm (Carrier et al., 2002). The engineering of a thick layer of functional cardiac tissue is therefore likely to require the perfusion of oxygen-rich culture medium directly through the growing construct and/or its neovascularization *in vitro*.

Physical forces can also be exploited in functional tissue engineering, presumably via the same processes by which they determine the architecture of native tissues (e.g., in bone (Thompson, 1977)). *In vitro*, dynamic but not static compression increased ECM biosynthesis rates in cartilage explants (Bonassar et al., 2000, 2001; Grodzinsky et al., 2000; Sah et al., 1989) and improved the GAG content and mechanical function of engineered cartilage (Buschmann et al., 1992, 1995; Carver and Heath, 1999; Lee and Bader, 1997; Mauck et al., 2000). Improvements were reported for a variety of cells, scaffolds, and culture systems and for physical stimuli that varied in nature (mechanical compression, hydrostatic pressure), amplitude, frequency, duration, and regime (various cycles of stimulation followed by rest, applied continuously or intermittently). Likewise, both dynamic and passive stretch improved structural properties, contractility, and pharmacological responsiveness of engineered cardiac tissues (Eschenhagen et al., 1997; Zimmermann et al., 2000). The hydrodynamic stresses acting at the surfaces of constructs cultured in rotating bioreactors (approximately 1 dyn/cm²) (Freed and Vunjak-Novakovic, 1995), although different in nature and several orders of magnitude lower than the forces associated with joint loading or cardiac contraction, appeared to enhance *in vitro* chondrogenesis (Freed et al., 1998; Vunjak-Novakovic et al., 1999) (Fig. 27.4*c,d*) and myogenesis (Carrier et al., 1999; Papadaki et al., 2001) (Fig. 27.5*c,d*) according to as yet unknown mechanisms.

In Vitro Culture Duration and *In Vivo* Remodeling

The structural and functional properties of engineered cartilage and cardiac tissue can be improved to some degree by increasing the duration of *in vitro* culture (Fink et al., 2000; Freed et al., 1997, 1998) (Fig. 27.3*b*). Engineered cartilage cultured for 7 months had very high equilibrium moduli comparable to native cartilage and GAG fractions (Fig. 27.3*c,d*), but dynamic stiffnesses and collagen fractions remained subnormal (Freed et al., 1997). Engineered cardiac tissue cultured for 1 month exerted contractile forces that were less than 1 mN (Fink et al., 2000), several orders of magnitude lower than native cardiac muscle. It is likely that functional deficiencies present in engineered cartilage and cardiac tissue grown *in vitro* were due to the absence of specific biochemical and/or physical factors normally present *in vivo*.

In vivo, engineered cartilage remodeled into physiologically stiff cartilage and new subchondral bone (e.g., Schaefer et al., 2002; Wakitani et al., 1994) according to a process that is thought to be triggered by osteogenic factors derived from the bone and its marrow and chondroprotective chemical and physical factors present at the joint surface (e.g., Caplan et al., 1997; Sellers et al., 2000). Notably, chondrocytes in engineered cartilage aligned into characteristic columns following *in vivo* exposure to physiological loading (Freed et al., 1994; Schaefer et al., 2002), whereas no such cell alignment has yet been reported for *in vitro* grown engineered cartilage. The optimal duration of *in vitro* cultivation of engineered cartilage prior to its implantation *in vivo* has not yet been determined. In some studies, implantation was done within 2

hours of cell seeding, requiring functional tissue development to occur *in vivo* (Chu et al., 1997; Wakitani et al., 1994, 1998). Another study found no difference in the histological quality of the repair tissues following the implantation of engineered cartilage cultured for either 2 or 4 weeks *in vitro* (Schreiber et al., 1999). We implanted engineered cartilage that had been cultured for 4 to 6 weeks *in vitro* based on previous studies showing that similar tissue constructs contained metabolically active cells and high concentrations of cartilaginous ECM, and were mechanically functional (Schaefer et al., 2002). In particular, 6 week constructs had wet weight fractions of GAG and collagen that were respectively two thirds and one third those of native calf cartilage, and equilibrium moduli that were one third that of native calf cartilage (Freed et al., 1998; Vunjak-Novakovic et al., 1999). However, another study demonstrated a trade-off between the initial stiffness of engineered cartilage and its integration potential (mechanically nonfunctional constructs cultured for less than 1 week integrated better than mechanically functional constructs cultured for 5 weeks (Obradovic et al., 2001), implying the need for further studies aimed at optimizing *in vitro* culture duration.

Summary and Future Directions

Articular cartilage and cardiac muscle are two representative examples of engineered tissues that differ in general but share the lack of intrinsic capacity for self repair. Both tissues can be engineered by culturing cells on three-dimensional polymer scaffolds in bioreactors under conditions that promote the development of functional tissue structures. This chapter reviewed recent studies in which functional engineered cartilage and cardiac tissues were grown *in vitro* and then either implanted *in vivo* (for osteochondral and myocardial repair, respectively) or used for *in vitro* research (for controlled studies of tissue development and function).

We will define a set of general requirements for functional engineered cartilage and cardiac muscle as follows. At the time of implantation, an engineered tissue should be sufficiently thick and structurally stable to allow for handling and survival under physiological conditions. The structural integrity and functionality of the engineered tissue should then improve progressively *in vivo* until normal tissue architecture and function have been restored. The cells within the engineered tissue should be metabolically active, in order to potentiate integration of the graft with surrounding host tissues. Tissue- and application-specific functional requirements include the ability to withstand compressive and shear loading in the case of engineered cartilage, and the ability to propagate electrical signals and contract in a coordinated manner in the case of engineered cardiac muscle.

Engineered cartilage can already meet some of the above criteria. Several studies reported that osteochondral defects repaired with engineered cartilage were approximately as stiff as native articular cartilage 6 months postimplantation, and stiffer than the repair tissue in untreated defects (Chu et al., 1997; Schaefer et al., 2002; Wakitani et al., 1994, 1998). On the other hand, these and other studies have also reported high variability in mechanical testing data and low correlation between repair tissue structure and function, and suggested that repair tissue that appeared functional at 6 months may fail mechanically after a longer period of time (Chu et al., 1997; Schaefer et al., 2002; Wakitani et al., 1998; Wei and Messner, 1997).

In contrast, engineered cardiac tissue does not yet meet most of the above criteria, and is an estimated two to three orders of magnitude lower in thickness and contractile force generation than native cardiac tissue. Engineered cardiac tissue, although already useful as a model system for controlled *in vitro* studies (e.g., Bursac et al., 1999; Eschenhagen et al., 1997; Fink et al., 2000; Papadaki et al., 2001), therefore requires further scale-up before it can be useful *in vivo*. Damaged myocardial tissue was not yet functionally improved following implantation of engineered cardiac tissue (Li et al., 1999).

Additional research needs to include the exploration of new cell sources (in particular those allowing the generation of nonimmunogenic implants), customization of the scaffold (chemistry, geometry, degradation rate, and mechanical properties), and optimization of bioreactor design (e.g., to provide perfusion for cardiac muscle and physical stimuli such as compression and stretch for cartilage and cardiac muscle, respectively). In addition, coordinated *in vitro* and *in vivo* studies are needed to better define the design requirements for engineered tissues and to correlate their structural and functional features with *in vitro* culture parameters.

Aknowledgments

This work was supported by NASA Grant NCC8-174. The authors thank R. Langer for helpful advice, F. Guilak, K. Athanasiou, and A. Grodzinsky for useful discussions regarding cartilage biomechanics, and S. Kangiser for expert help with manuscript preparation.

References

Akins RE, Schroedl NA, Gonda SR, Hartzell CR. 1997. Neonatal rat heart cells cultured in simulated microgravity. *In Vitro Cell. Dev. Biol.* Anim. 33: 337–343.

Akins RE, Boyce RA, Madonna ML, Schroedl NA, Gonda SR, McLaughlin TA, Hartzell CR. 1999. Cardiac organogenesis *in vitro:* reestablishment of three-dimensional tissue architecture by dissociated neonatal rat ventricular cells. *Tissue Eng.* 5:103–118.

Ameer GA, Mahmood TA, Langer R. 2002. A biodegradable composite scaffold for cell transplantation. *J. Orthop. Res.*, in press. 20:16–19.

Athanasiou K, Korvick D, Schenck R. 1997. Biodegradable implants for the treatment of osteochondral defects in a goat model. *Tissue Eng.* 3:363–373.

Benya PD, Shaffer JD. 1982. Dedifferentiated chondrocytes reexpress the differentiated collagen phenotype when cultured in agarose gels. *Cell* 30:215–224.

Bonassar LJ, Grodzinsky AJ, Srinivasan A, Davila SG, Trippel SB. 2000. Mechanical and physicochemical regulation of the action of insulin-like growth factor-I on articular cartilage. *Arch. Biochem. Biophys.* 379:57–63.

Bonassar LJ, Grodzinsky AJ, Frank EH, Davila SG, Bhaktav NR, Trippel SB. 2001. The effect of dynamic compression on the response of articular cartilage to insulin-like growth factor-I. *J. Orthop. Res.* 19:11–17.

Breinan HA, Minas T, Barone L, Tubo R, Hsu HP, Shortkroff S, Nehrer S, Sledge CB, Spector M.

1998. Histological evaluation of the course of healing of canine articular cartilage defects treated with cultured autologous chondrocytes. *Tissue Eng.* 4:101–114.

Brighton CT, Heppenstall RB. 1971. Oxygen tension in zones of the epiphyseal plate, the metaphysis and diaphysis. *J. Bone Joint Surg.* 53A:719–728.

Brilla CG, Maisch B, Rupp H, Sunck R, Zhou G, Weber KT. 1995. Pharmacological modulation of cardiac fibroblast function. *Herz* 20:127–135.

Brittberg M, Lindahl A, Nilsson A, Ohlsson C, Isaksson O, Peterson L. 1994. Treatment of deep cartilage defects in the knee with autologous chondrocyte transplantation. *N. Engl. J. Med.* 331:889–895.

Brun P, Abatangelo G, Radice M, Zacchi V, Guidolin D, Daga Gordini D, Cortivo R. 1999. Chondrocyte aggregation and reorganization into three-dimensional scaffolds. *J. Biomed. Mater. Res.* 46:337–346.

Buckwalter JA, Mankin HJ. 1997a. Articular cartilage, part I: tissue design and chondrocyte-matrix interactions. *J. Bone Joint Surg.* 79A:600–611.

Buckwalter JA, Mankin HJ. 1997b. Articular cartilage, part II: degeneration and osteoarthrosis, repair, regeneration, and transplantation. *J. Bone Joint Surg.* 79A:612–632.

Bursac N, Papadaki M, Cohen RJ, Schoen FJ, Eisenberg SR, Carrier R, Vunjak-Novakovic G, Freed LE. 1999. Cardiac muscle tissue engineering: toward an *in vitro* model for electrophysiological studies. *Am. J. Physiol.* 277:H433–H444.

Buschmann MD, Gluzband YA, Grodzinsky AJ, Kimura JH, Hunziker EB. 1992. Chondrocytes in agarose culture synthesize a mechanically functional extracellular matrix. *J. Orthop. Res.* 10:745–752.

Buschmann MD, Gluzband YA, Grodzinsky AJ, Hunziker EB. 1995. Mechanical compression modulates matrix biosynthesis in chondrocyte/agarose culture. *J. Cell Sci.* 108:1497–1508.

Butnariu-Ephrat M, Robinson D, Mendes DG, Halperin N, Nevo Z. 1996. Resurfacing of goat articular cartilage from chondrocytes derived from bone marrow. *Clin. Orthop.* 330:234–243.

Caplan AI, Elyaderani M, Mochizuki Y, Wakitani S, Goldberg VM. 1997. Principles of cartilage repair and regeneration. *Clin. Orthop.* 342:254–269.

Carrier RL, Papadaki M, Rupnick M, Schoen FJ, Bursac N, Langer R, Freed LE, Vunjak-Novakovic G. 1999. Cardiac tissue engineering: cell seeding, cultivation parameters and tissue construct characterization. *Biotechnol. Bioeng.* 64:580–589.

Carrier RL, Freed LE, Rupnick M, Langer R, Schoen FJ, Vunjak-Novakovic G. 2002. Engineered cardiac

muscle: perfusion improves tissue architecture. *Tissue Eng.* 8:175–188.

Carver SE, Heath CA. 1999. Semi-continuous perfusion system for delivering intermittent physiological pressure to regenerating cartilage. *Tissue Eng.* 5:1–11.

Chu C, Dounchis JS, Yoshioka M, Sah RL, Coutts RD, Amiel D. 1997. Osteochondral repair using perichondrial cells: a 1 year study in rabbits. *Clin. Orthop.* 340:220–229.

Dominguez L, Parriaello G, Amato P, Licata G. 1999. Trends of congestive heart failure: epidemiology contrast with clinical trial results. *Cardiologia* 44:801–808.

Eschenhagen T, Fink C, Remmers U, Scholz H, Wattchow J, Woil J, Zimmermann W, Dohmen HH, Schafer H, Bishopric N, Wakatsuki T, Elson E. 1997. Three-dimensional reconstitution of embryonic cardiomyocytes in a collagen matrix: a new heart model system. *FASEB J.* 11:683–694.

Evans RW. 2000. Economic impact of mechanical cardiac assistance. *Prog. Cardiovasc. Dis.* 43:81–94.

Fink C, Ergun S, Kralisch D, Remmers U, Weil J, Eschenhagen T. 2000. Chronic stretch of engineered heart tissue induces hypertrophy and functional improvement. *FASEB J.* 14:669–679.

Freed LE, Vunjak-Novakovic G. 1995. Cultivation of cell-polymer constructs in simulated microgravity. *Biotechnol. Bioeng.* 46:306–313.

Freed LE, Vunjak-Novakovic G. 1997. Microgravity tissue engineering. *In Vitro Cell. Dev. Biol. Anim.* 33:381–385.

Freed LE, Vunjak-Novakovic G. 2002. Culture environments: cell-polymer-bioreactor systems. In: *Methods of Tissue Engineering.* A Atala, RP Lanza, eds. Academic Press, San Diego, pp. 97–111.

Freed LE, Marquis JC, Nohria A, Emmanual J, Mikos AG, Langer R. 1993. Neocartilage formation *in vitro* and *in vivo* using cells cultured on synthetic biodegradable polymers. *J. Biomed. Mater. Res.* 27:11–23.

Freed LE, Grande DA, Lingbin Z, Emmanual J, Marquis JC, Langer R. 1994. Joint resurfacing using allograft chondrocytes and synthetic biodegradable polymer scaffolds. *J. Biomed. Mater. Res.* 28:891–899.

Freed LE, Langer R, Martin I, Pellis N, Vunjak-Novakovic G. 1997. Tissue engineering of cartilage in space. *Proc. Natl. Acad. Sci. U.S.A.* 94:13885–13890.

Freed LE, Hollander AP, Martin I, Barry JR, Langer R, Vunjak-Novakovic G. 1998. Chondrogenesis in a cell-polymer-bioreactor system. *Exp. Cell Res.* 240:58–65.

Freed LE, Vunjak-Novakovic G. 2000. Tissue engineering bioreactors. In: *Principles of Tissue Engi-*

neering. RP Lanza, R Langer, J Vacanti, eds. Academic Press, San Diego, pp. 143. 143–156.

Gooch KJ, Blunk T, Courter DL, Sieminski AL, Bursac PM, Vunjak-Novakovic G, Freed LE. 2001. IGF-I and mechanical environment interact to modulate engineered cartilage development. *Biochem. Biophys. Res. Com.* 286:909–915.

Grande DA, Halberstadt C, Naughton G, Schwartz R, Manji R. 1997. Evaluation of matrix scaffolds for tissue engineering of articular cartilage grafts. *J. Biomed. Mater. Res.* 34:211–220.

Grande DA, Athanasiou K, Schwartz R. 1998. Matrix engineering for repair of articular cartilage defects. *Trans. Orthop. Res. Soc.* 23:801.

Grodzinsky AJ, Levenston ME, Jin M, Frank EH. 2000. Cartilage tissue remodeling in response to mechanical forces. *Annu. Rev. Biomed. Eng.* 2:691–713.

Hauselmann HJ, Fernandes RJ, Mok SS, Schmid TM, Block JA, Aydelotte MB, Kuettner KE, Thonar EJM. 1994. Phenotypic stability of bovine articular chondrocytes after long-term culture in alginate beads. *J. Cell Sci.* 107:17–27.

Haynesworth SE, Reuben D, Caplan AI. 1998. Cell-based tissue engineering therapies: the influence of whole body physiology. *Adv. Drug Deliver. Rev.* 33:3–14.

Hoch DH, Grodzinsky AJ, Koob TJ, Albert ML, Eyre DR. 1983. Early changes in material properties of rabbit articular cartilage after menisacectomy. *J. Orthop. Res.* 1:2–12.

Hoemann CD, Sun J, Legare A, McKee MD, Ranger P, Buschmann MD. 2001. A thermosensitive polysaccharide gel for cell delivery in cartilage repair. *Trans. Orthop. Res. Soc.* 26:626.

Homminga GN, Buma P, Koot HW, Van der Kraan PM, van den Berg WB. 1993. Chondrocyte behaviour in fibrin glue *in vitro*. *Acta Orthop. Scand.* 64:441–445.

Hunziker EB, Kapfinger E. 1998. Removal of proteoglycans from the surface of defects in articular cartilage transiently enhances coverage by repair cells. *J. Bone Joint Surg.* 80B:144–150.

Langer F, Gross AE. 1974. Immunogenicity of allograft articular cartilage. *J. Bone Joint Surg.* 56-A:297–304.

Lee DA, Bader DL. 1997. Compressive strains at physiological frequencies influence the metabolism of chondrocytes seeded in agarose. *J. Orthop. Res.* 15:181–188.

Li RK, Jia ZQ, Weisel RD, Mickle DAG, Choi A, Yau TM. 1999. Survival and function of bioengineered cardiac grafts. *Circulation* 100:II63–II69.

Lim K, Shahid M, Sharif M. 1996. Recent advances in osteoarthritis. *Singapore Med. J.* 37:189–193.

MacKenna DA, Omens JH, McCulloch AD, Covell JW. 1994. Contribution of collagen matrix to passive left ventricular mechanics in isolated rat heart. *Am. J. Physiol.* 266:H1007–H1018.

Madry, H, R Padera, J Seidel, R Langer, LE Freed, SB Trippel, and G Vunjak-Novakovic. 2002. Gene transfer of a human insulin-like growth factor I cDNA enhances tissue engineering of cartilage. Hum. Gene Ther. 13: 1621–1630.

Martin I, Obradovic B, Freed LE, Vunjak-Novakovic G. 1999a. A method for quantitative analysis of glycosaminoglycan distribution in cultured natural and engineered cartilage. *Ann. Biomed. Eng.* 27:656–662.

Martin I, Vunjak-Novakovic G, Yang J, Langer R, Freed LE. 1999b. Mammalian chondrocytes expanded in the presence of fibroblast growth factor-2 maintain the ability to differentiate and regenerate three-dimensional cartilaginous tissue. *Exp. Cell Res.* 253:681–688.

Martin I, Shastri VP, Padera RF, Yang J, Mackay AJ, Langer R, Vunjak-Novakovic G, Freed LE. 2001a. Selective differentiation of mammalian bone marrow stromal cells cultured on three-dimensional polymer foams. *J. Biomed. Mater. Res.* 55:229-235.

Martin I, Suetterlin R, Baschong W, Heberer M, Vunjak-Novakovic G, Freed LE. 2001b. Enhanced cartilage tissue engineering by sequential exposure of chondrocytes to FGF-2 during 2D expansion and BMP-2 during 3D cultivation. *J. Cell. Biochem.* 83:121–128.

Mauck RL, Soltz MA, Wang CCB, Wong DD, Chao PG, Valhmu WB, Hung CT, Ateshian GA. 2000. Functional tissue engineering of articular cartilage through dynamic loading of chondrocyte-seeded agarose gels. *J. Biomech. Eng.* 122:252–260.

Messner K, Gillquist J. 1996. Cartilage repair: a critical review. *Acta Orthop. Scand.* 67:523.

Newman AP. 1998. Articular cartilage repair. *Am. J. Sports Med.* 26:309–324.

Noguchi T, Oka M, Fujino M, Neo M, Yamamuro T. 1994. Repair of osteochondral defects with grafts of cultured chondrocytes: comparison of allografts and isografts. *Clin. Orthop.* 302:251–258.

Obradovic B, Carrier RL, Vunjak-Novakovic G, Freed LE. 1999. Gas exchange is essential for bioreactor cultivation of tissue engineered cartilage. *Biotechnol. Bioeng.* 63:197–205.

Obradovic B, Meldon JH, Freed LE, Vunjak-Novakovic G. 2000. Glycosaminoglycan deposition in engineered cartilage: experiments and mathematical model. *AIChE J.* 46:1860–1871.

Obradovic B, Martin I, Padera RF, Treppo S, Freed LE, Vunjak-Novakovic G. 2001. Integration of engineered cartilage. *J. Orthop. Res.* 19:1089–1097.

Oddis CV. 1996. New perspectives on osteoarthritis. *Am. J. Med.* 100:10S–15S.

O'Driscoll SW, Fitsimmons JS, Commisso CN. 1997. Role of oxygen tension during cartilage formation by periosteum. *J. Orthop. Res.* 15:682–687.

Papadaki M, Bursac N, Langer R, Merok J, Vunjak-Novakovic G, Freed LE. 2001. Tissue engineering of functional cardiac muscle: molecular, structural and electrophysiological studies. *Am. J. Physiol.* 280:H168–H178.

Pei M, LA Solchaga, J Seidel, L Zeng, G Vunjak-Novakovic, AI Caplan, LE Freed. *2002.* Bioreactors mediate the effectiveness of tissue engineering scaffolds. FASEB J. (August 7, 2002) 10.1096/fj.02-0083fje

Radisic M., Euloth M., Yang L., Langer R., Freed L.E., Vunjak-Novakovic G. High density seeding of myocyte cells for tissue engineering. *Biotechnology and Bioengineering* 82:403–414 (2003).

Reindel ES, Ayroso AM, Chen AC, Chun DM, Schinagl RM, Sah RL. 1995. Integrative repair of articular cartilage *in vitro:* adhesive strength of the interface region. *J. Orthop. Res.* 13:751–760.

Rich M. 1997. Epidemiology, pathophysiology, and etiology of congestive heart failure in older adults. *J. Am. Geriatr. Soc.* 45:968–974.

Sah RLY, Kim YJ, Doong JYH, Grodzinsky AJ, Plaas AHK, Sandy JD. 1989. Biosynthetic response of cartilage explants to dynamic compression. *J. Orthop. Res.* 7:619–636.

Schaefer D, I Martin, G Jundt, J Seidel, M Heberer, AJ Grodzinsky, I Bergin, G Vunjak-Novakovic, LE Freed. 2002. Tissue engineered composites for the repair of large oesteochondral defects. Arthritis Rheum. 46: 2524–2534.

Schreiber RE, Ilten-Kirby BM, Dunkelman NS, Symons KT, Rekettye LM, Willoughby J, Ratcliffe A. 1999. Repair of osteochondral defects with allogeneic tissue-engineered cartilage implants. *Clin. Orthop.* 367S:S382–S395.

Sellers RS, Zhang R, Glasson SS, Kim HD, Peluso D, D'Augusta DA, Beckwith K, Morris EA. 2000. Repair of articular cartilage defects one year after treatment with recombinant human bone morphogenetic protein-2 (rhBMP-2). *J. Bone Joint Surg.* 82:151–160.

Solchaga LA, Dennis JE, Goldberg VM, Caplan AI. 1999. Hyaluronic acid-based polymers as cell carriers for tissue-engineered repair of bone and cartilage. *J. Orthop. Res.* 17:205–213.

Thompson DW. 1977. *On Growth and Form.* Cambridge University Press, New York.

Vunjak-Novakovic G, Freed LE, Biron RJ, Langer R. 1996. Effects of mixing on the composition and morphology of tissue-engineered cartilage. *AIChE J.* 42:850–860.

Vunjak-Novakovic G, Obradovic B, Bursac P, Martin I, Langer R, Freed LE. 1998. Dynamic cell seeding of polymer scaffolds for cartilage tissue engineering. *Biotechnol. Prog.* 14:193–202.

Vunjak-Novakovic G, Martin I, Obradovic B, Treppo S, Grodzinsky AJ, Langer R, Freed LE. 1999. Bioreactor cultivation conditions modulate the composition and mechanical properties of tissue engineered cartilage. *J. Orthop. Res.* 17:130–138.

Wakitani S, Goto T, Pineda SJ, Young RG, Mansour JM, Caplan AI, Goldberg VM. 1994. Mesenchymal cell-based repair of large, full-thickness defects of articular cartilage. *J. Bone Joint Surg.* 76A:579–592.

Wakitani S, Goto T, Young RG, Mansour JM, Goldberg VM, Caplan AI. 1998. Repair of large full-thickness articular cartilage defects with allograft articular chondrocytes embedded in a collagen gel. *Tissue Eng.* 4:429–444.

Wei X, Messner K. 1997. Maturation-dependent repair of untreated osteochondral defects in the rabbit knee. *J. Biomed. Mater. Res.* 34:63–72.

Ysart GE, Mason RM. 1994. Responses of articular cartilage explant cultures to different oxygen tensions. *Biochim. Biophys. Acta* 1221:15–20.

Zimmermann WH, Fink C, Kralish D, Remmers U, Weil J, Eschenhagen T. 2000. Three-dimensional engineered heart tissue from neonatal rat cardiac myocytes. *Biotechnol. Bioeng.* 68:106–114.

28

Tissue Engineering Skeletal Muscle

Paul E. Kosnik, Robert G. Dennis, and Herman H. Vandenburgh

Introduction

The hierarchical structure of skeletal muscle allows for the generation of directed force and power, permitting locomotion and the activities of daily living. Skeletal muscle aging, disease, and injury result in functional deficits and an increased dependence upon others and decreased quality of life. The ultimate goal of functional skeletal muscle tissue engineering is to reduce or eliminate functional deficits and maintain quality of living. This goal may be accomplished by at least two independent pathways. In situ tissue engineering may be possible by the targeted introduction of myogenic stem cells (i.e., satellite cells) and their fusion into existing or new postmitotic muscle fibers, leading to enhanced functional capability. Ex vivo skeletal muscle tissue engineering entails the creation of functional tissue from these same myogenic stem cells in the tissue culture environment, and their reimplantation into an approximate site *in vivo*. The in situ approach has been reviewed (Partridge, 1991) and will not be further discussed. Ex vivo tissue engineering is a more daunting task and will be the subject of this chapter.

While the goal of using ex vivo engineered skeletal muscle for functional work *in vivo* is still years away from clinical use, near term applications of the rapidly evolving tissue engineered skeletal muscle technology are possible. These in-clude their use as implantable protein delivery devices (Vandenburgh et al., 1996), high content drug screening for skeletal muscle wasting disorders (Vandenburgh et al., 1999), as mechanical ex vivo actuators, and for pharmacogenomic screening for testing patient's reactions to drugs (Fig. 28.1). This chapter is a review of the current status of tissue engineering skeletal muscle ex vivo, the existing technological challenges, and future directions to overcome these challenges to developing functional implants. Detailed methods for tissue engineering skeletal muscle are not included since they are covered in other reviews (Vandenburgh et al., 1998b) or from the primary sources referenced in this chapter. For a review of the structure and function of skeletal muscle as it relates to functional tissue engineering, see Chapter 4 by Faulkner and Dennis in this text.

Skeletal muscle generates force and power when activated, transfers force to the skeleton through tendons, is infused by a complex vasculature, and is innervated by motor and sensory nerves. Skeletal muscle is therefore composed of not only contractile skeletal muscle tissue, but connective, vascular, and nervous tissues as well. Eventually, functional tissue engineered skeletal muscle will need to consist of skeletal muscle tissue integrated with connective tissue to transmit power to interact with the environment, nervous tissue to elicit contractions physiologically and provide force and displacement feedback, and

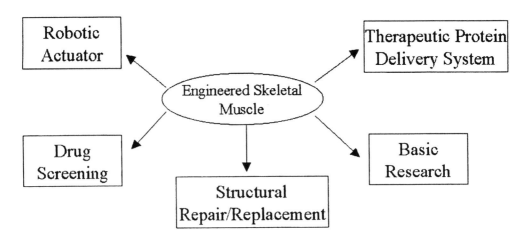

FIGURE 28.1. Uses for Ex Vivo Engineered Skeletal Muscle

vascular tissue to efficiently deliver nutrients to and eliminate waste products from the engineered tissue. Depending upon the specific application of the functional engineered skeletal muscle, these tissue components will be required to various degrees. Not all components need be incorporated ex vivo. For example, one strategy is to promote vascularization and innervation after implantation of the construct into the host.

The tissue engineer attempts to guide the cellular organization and development in tissue culture by providing the myogenic cells with the appropriate environmental cues (Fig. 28.2). The *in vitro* cell environment includes the proper temperature and atmosphere, the culture media, the substrate chemistry and texture, and fields experienced by the construct, for example, mechanical, electrical, magnetic, and gravitational. The myoblasts express their genes in the context of their environment and developmental history. The working hypothesis is that the optimal environment for engineering functional skeletal muscle ex vivo is to simulate the environment of the developing muscles within the organism. For engineered skeletal muscle intended for force and/or power generation, evaluation of the contractile properties is accomplished with standard techniques of skeletal muscle mechanics, including the measurement of force, power and excitability (Dennis and Kosnik, 2000).

Skeletal myogenesis is the process by which embryonic myoblasts proliferate, organize in

three dimensions, and differentiate to form skeletal muscle tissue de novo. Myogenesis occurs in the fetus during development and can occur *in vitro* if the cells are cultured with the appropriate environmental cues that emulate the fetal environment (Konigsberg, 1963; Turner, 1986; Stockdale and Holtzer, 1961; Strohman et al., 1990). During myogenesis, myoblasts proliferate, fuse into myotubes (Stockdale and Holtzer, 1961), and differentiate into myofibers oriented along the long axis of the muscle. The three-dimensional organization of the myoblasts, fibroblasts, myotubes, and myofibers is coordinated with the development of nervous, connective, and vascular tissue and is critical to the resulting function of the skeletal muscle. Myofibers are stretched during development by the lengthening of the growing bones to which they are attached by tendons. This stimulus plays an important role in the organization of the myofibers relative to each other, the connective tissue, and the bones (Haines, 1932; Weiss, 1932).

Skeletal muscle fibers are highly differentiated and are not mitotically active, yet skeletal muscle has a considerable capacity to regenerate following injury (Carlson and Faulkner, 1983). Regeneration and repair of damaged skeletal muscle is facilitated by myogenic stem cells known as satellite cells, which are located between the sarcolemma and the basal lamina of skeletal muscle fibers (Mauro, 1961). Satellite cells lie in contact with the multinucleated muscle fiber and are

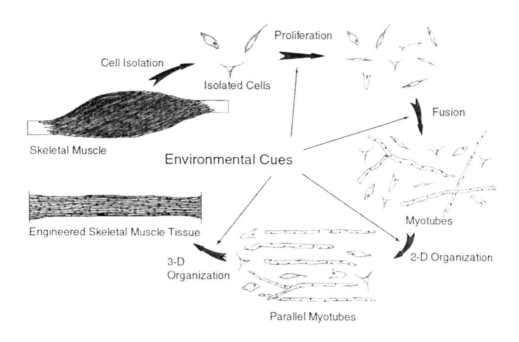

FIGURE 28.2. The tissue engineer isolates myogenic cells from skeletal muscle tissue and uses cues in the culture environment to engineer the cells into a new tissue.

quiescent until activated by overuse, damage, or injury of the myofiber (Bischoff, 1986). The process of repair of a damaged myofiber is similar to that of myogenesis (Snow, 1977a, b), but during repair myoblasts fuse to form new myofibers within the existing basal lamina, while during myogenesis the basal lamina is developed de novo. Once activated, satellite cells divide and give rise to myoblasts. Myoblasts proliferate in the basal lamina tube (Waldeyer, 1865) and then fuse to form the new multinucleated myotube (Stockdale and Holtzer, 1961). During differentiation, myotubes express the proteins that assemble to form the sarcomeres and supporting organelles that allow for contraction. When the nuclei of the myotubes migrate to the periphery of the fiber and are no longer centrally located, the myotube is considered to have differentiated into a myofiber (Carlson and Faulkner, 1983).

Skeletal Muscle Tissue Engineering Components

Embryonic myoblasts and adult skeletal muscle satellite cells are the primary cell sources for skeletal muscle tissue engineering. These cells and their progeny are referred to as myogenic cells in this chapter. Differentiated skeletal muscle fibers are difficult to maintain in culture for long periods of time since they do not retain the ability to regenerate *in vitro* (Lewis and Lewis, 1917; Yaffe, 1968). In contrast, proliferating myogenic cells can be maintained in culture for extended periods (Konigsberg, 1963; Yaffe, 1968). With the appropriate culture conditions, the cells divide, organize, and differentiate into skeletal muscle fibers by a process similar to myogenesis *in vivo* (Dennis and Kosnik, 2000; Strohman et al., 1990; Swasdison and Mayne, 1992; Vandenburgh et al., 1991). Figure 28.3 provides an overview of the processes used to harvest the myogenic cells from skeletal muscle tissue and to manipulate them *in vitro* to engineer new functional skeletal muscle tissue.

Cell Sources

The source of the cells used for skeletal muscle tissue engineering may be dictated by the intended use of the engineered tissue since some cell sources are inappropriate for certain tissue

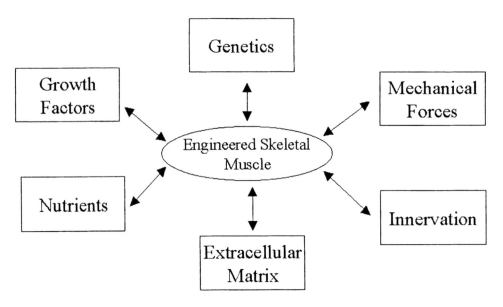

FIGURE 28.3. Factors Influencing Skeletal Muscle Development

engineering applications. When several cell sources are compatible with the intended use of the construct such as for basic science research, myogenic cell lines may be used for their cost effectiveness, ease of use, and uniformity. For skeletal muscle tissue engineering applications in which skeletal muscle constructs will ultimately be implanted into host organisms, the source of the implanted cells obviously will affect the response of the host to the engineered tissue. In the absence of immune suppression, nonautologous cells will be identified by the host as nonself and rejected within several weeks. This may be adequate for short-term applications such as the acute delivery of a therapeutic protein shortly after a myocardial infarct. Tissue engineered skeletal muscle made for implantation from allogeneic immuno-naïve fetal myoblasts may survive longer than those isolated from adults. Ex vivo differentiated myotubes may also survive longer *in vivo* than implanted proliferating myoblasts since the latter express major histocompatibility complex (MHC) Class II molecules while the former do not (Mantegazza et al., 1996). Autologous cells will be required for chronic applications such as performing functional work or long-term protein delivery in which engineered skeletal muscle can be im-

planted into a host organism without a barrier to immune response or the need for immune suppression.

Myogenic Cell Lines

The processes of isolation, purification, and proliferation of primary myoblasts are labor intensive and costly. In addition, primary myoblasts have a limited proliferation potential *in vitro* (Decary et al., 1997) with the exception of murine myogenic cells, which spontaneously immortalize (Hayflick and Moorhead, 1961; Wernig et al., 1995). An alternative for tissue engineering is an established myoblast cell line. Several cell lines have been used to tissue engineer skeletal muscle, including the C2C12 cell line from mouse muscle sarcoma (Okano and Matsuda, 1998a; van Wachem et al., 1996; Vandenburgh et al., 1998a), and the murine G8 cell line (Mulder et al., 1998). Since cell lines do not exhibit all of the same behaviors as primary myogenic cells *in vitro* (Irintchev et al., 1998), their use for tissue engineering studies must be complemented with primary cell studies. For example, muscle constructs engineered from myogenic cell lines exhibited significantly lower excitability and contractility, and prolonged relaxation times after cessation of electrical stimulation compared with constructs engineered from primary adult rodent cells (Dennis et al., 2000).

Myogenic Primary Cells

Three-dimensional skeletal muscle constructs have been engineered from primary cultures of avian (Strohman et al., 1990; Swasdison and Mayne, 1992; Vandenburgh et al., 1991), rodent (Dennis and Kosnik, 2000; Powell et al., 1999; Vandenburgh et al., 1996, 1999) and human (Powell et al., 1999) myogenic cells. The typical method of primary myogenic cell isolation is mechanical dissociation followed by digestion of the tissue with proteases (Konigsberg, 1963; Lagord et al., 1998). In addition to myoblasts, fibroblasts are present in primary cultures of skeletal muscle unless myogenic cells are clonally derived during the culture process (Yaffe, 1968).

Fibroblasts

In the absence of fibroblasts or a supporting scaffold, cell line myogenic cells do not develop into three-dimensional skeletal muscle constructs (Dennis et al., 2000; Shansky et al., 1997). However, myogenic cells from a cell line may be used to form skeletal muscle constructs if the cells are provided with a scaffold (Okano and Matsuda, 1998b). Scaffolds are artificial extracellular matrixes (ECM) that define the initial geometry, structure, and mechanical properties of the engineered skeletal muscle construct, as well as providing support for the myogenic cells during proliferation and fusion. Scaffolds can also be engineered to include specific adhesion molecules and other biochemical signals for the muscle cells. They are typically employed when skeletal muscle constructs are engineered from a myoblast cell line, because the myogenic cells do not produce an adequate ECM to support a cohesive three-dimensional structure in culture (Strohman et al., 1990; Sanderson et al., 1986). Scaffolds used in skeletal muscle tissue engineering include collagen gels (Chromiak et al., 1998; Okano and Matsuda, 1997, 1998a, b; Okano et al., 1997; Vandenburgh et al., 1999), collagen-Matrigel™ gels (Powell et al., 1999; Vandenburgh et al., 1996), alginate hydrogels (Rowley et

al., 1999), and polyurethanes (Mulder et al., 1995, 1998).

Environmental Cues

The tissue engineer uses environmental cues to direct the three-dimensional organization, proliferation, and differentiation of the cells that will form the engineered skeletal muscle. The appropriate environmental cues lead to in changes in gene and protein expression within the cell, which result in changes in cellular metabolism, growth, and organization. The key to controlling the three-dimensional organization, proliferation, and differentiation of the cells is to provide the appropriate cues that will result in the desired outcome for the tissue.

Three general classes of environmental cues have been described: soluble factors, insoluble factors, and cell-cell interactions (Putnam and Mooney, 1996). A fourth class of environmental cues, fields, undoubtedly affect cell growth and differentiation in skeletal muscle. Perhaps one of the most important aspects of the cellular response to environmental cues is the cell's synthesis of multiple cues into an integrated response (Schwartz and Ingber, 1994). Some environmental cues result in different cellular outcomes depending upon the presence or absence of other cues. For instance, myoblasts subjected to mechanical strain decrease cellular α-actin mRNA in the absence of soluble growth factors, but do not change cellular α-actin mRNA in the presence of soluble growth factors (Carson and Booth, 1998). Specific integrin binding to ECM molecules permits proliferation in cultured myoblasts, but only in the presence of serum growth factors (Sastry et al., 1999). Myotubes cultured on a flexible substrate and stretched in serum-free media do not hypertrophy to the same degree as myotubes cultured in the presence of serum (Vandenburgh, 1983). Finally, mechanical stimulation of skeletal myofibers increases their secretion of insulin-like growth factor-1, an important autocrine growth factor for muscle (Perrone et al., 1995). All of the above studies indicate that myogenic cells integrate complex environmental cues and make coordinated responses, even in the presence of conflicting cues.

Soluble Factors

Cell culture medium and its effects on skeletal muscle tissues in culture were the first environmental cues studied (Lewis and Lewis, 1917). In addition to the basic nutrient components necessary for mammalian cell growth (Eagle, 1959), high levels of serum are typically used to allow myogenic cells to proliferate in culture (Yaffe, 1971). Serum-containing media are considered "undefined" because the exact biochemical composition is unknown; they contain numerous growth factors, hormones, and mitogens. Soluble factors such as transforming growth factor beta (TGF-β), fibroblast growth factor (FGF), insulin, and insulin-like growth factor-I (IGF-I) stimulate myogenic cell proliferation and/or differentiation (Allen et al., 1985; Florini et al., 1991; Konigsberg, 1963; Vandenburgh et al., 1991; Yaffe, 1971). More recently, hepatocyte growth factor has been found to be an important factor for muscle satellite cell activation, migration, and proliferation (Sheehan et al., 2000). Serum-free medium has been developed to support muscle cell growth in tissue culture (Allen et al., 1985; Dollenmeier et al., 1981; Ham et al., 1990). In addition, ECM molecules such as fibronectin found in serum allow the adhesion of myogenic cells to the substrate (Chiquet et al., 1979).

Ions are another important component of the tissue culture medium (Cerny and Bandman, 1986; Shainberg et al., 1969). In addition to their role in excitability, they function as second messengers. Since skeletal muscle is an excitable tissue, ions are essential for contraction of the skeletal muscle cells, and for the development and differentiation of the cells that make up the skeletal muscle construct. Myogenic cells can proliferate, but are prohibited from fusing to form multinucleated myotubes in the absence of Ca^{2+} ions (Shainberg et al., 1969). Stretch-induced hypertrophy of skeletal myofibers *in vitro* can be inhibited if the sodium channels that allow ion flow are blocked (Vandenburgh and Kaufman, 1981). Cultured myotubes do not spontaneously contract in the presence of increased potassium ion concentration in the culture media (Bandman and Strohman, 1982) and spontaneous contractions are required for differentiation of myosin heavy chain (MHC) expression from embryonic to neonatal isoforms (Cerny and Bandman, 1986).

Insoluble Factors

ECM molecules are primarily insoluble factors that dramatically affect myogenic cell proliferation, differentiation, and organization, (Adams and Watt, 1993). ECM molecules affect myogenic precursor cells in culture in three distinct ways. The first is by providing the cells with a suitable surface on which to adhere. The second is the role the ECM plays in cell morphology and in the transduction of mechanical signals from the substrate into the cell (Schwartz and Ingber, 1994), where these signals can influence gene expression. The third is to define the tissue shape and material properties (Mooney and Rowley, 1997).

Myogenic cells and adult skeletal muscle fibers have surface receptors for ECM molecules, the most important of which are the integrins (Schwartz, 1992). Myogenic cells exhibit anchorage dependent growth (Stoker et al., 1968) and do not proliferate in the absence of a suitable substrate on which to adhere. They can be cultured on substrates of plastic, glass, or flexible membranes that have been coated with collagen, fibronectin, laminin, Matrigel™, or combinations of these molecules (Chiquet et al., 1979; Turner, 1986). Although avian skeletal muscle cells can be cultured on collagen alone, mammalian skeletal muscle cells typically require the addition of Matrigel™ or laminin in some form in order for them to adhere to substrates (Shansky et al., 1997).

Substrate topography is an important factor for skeletal muscle cell growth. Cultured myogenic cells and myotubes can be aligned with the use of oriented deposits of collagen (Huw and Lawson, 1980) or laminin (Clark et al., 1997), chemically etched grooved surfaces (Acartuk et al., 1999; Clark et al., 1990; Evans et al., 1999), and scratched surfaces (Swasdison and Mayne, 1992). Orienting myotubes in parallel arrays prior to formation of the three-dimensional construct allows for the rapid development of engineered skeletal muscle tissue that is organized into parallel myofibers similar to skeletal muscle *in vivo*. Topographical cues also guide skeletal muscle cell migration, proliferation, and differentiation *in vitro* (Acartuk et al., 1999; Clark et al., 1997; Turner et al., 1983) and can be used as environmental cues to control the developmental path of these cells.

Cell-Cell Interactions

Muscle cell-cell interactions involve cells in physical contact. Cells may also interact with each other by way of one of the other three classes of environmental cues, as when cells secrete mitogens, growth factors, or ECM molecules that affect other cells. Cell-cell contacts are mediated by transmembrane adhesion molecules such as the cadherins that allow cells to bind to and interact with other cells. Cadherins have extracellular domains that adhere to the extracellular domains of cadherin molecules on other cells and have intracellular domains that are continuous with the cytoskeleton (Goichberg and Geiger, 1998). Since the cytoskeleton is in contact with the nucleus, this organization allows for the transduction of extracellular signals directly to the nucleus of the cell, where the signal can effect gene expression (Wang et al., 1993). Myogenic cells interact with myotubes and other myogenic cells via cadherins, and myotube-myotube interactions are mediated by cadherins (Eng et al., 1997). Myogenic cell-myogenic cell adhesions and withdrawal from the cell cycle are mediated by cadherins and are necessary for myogenic cell fusion into myotubes and expression of muscle-specific proteins (Charlton et al., 1997; Eng et al., 1997; Goichberg and Geiger, 1998; Redfield et al., 1997).

Nerve-muscle interactions via innervation are an important type of cell-cell interaction wherein nerve cells interact with myofibers to form specific synaptic structures and to subsequently allow centrally controlled muscle activation. This activation helps guide development and alters gene expression in the myofibers that are activated. For example, in nerve-muscle co-cultures, adult fast MHC are expressed by myotubes only when co-cultured with neurons (Ecob-Prince et al., 1983, 1986). The process of neuromuscular differentiation and the cell-cell interactions that regulate this process have been reviewed (Grinnell, 1995). Despite the importance of neuromuscular cell-cell interactions in skeletal muscle de-

velopment, little *in vitro* tissue engineering research has focused on these interactions.

Fields

Gravitational, electric, fluid flow, and mechanical stress/strain fields affect the proliferation, migration, and differentiation of myogenic precursor cells in culture (Brevet et al., 1976; Dusterhoft and Pette, 1993; Hinkle et al., 1981; Vandenburgh and Kaufman, 1979; Vandenburgh et al., 1991). Time-varying magnetic fields also affect myogenic cells in culture (Dennis et. al, unpublished data). Myogenic cells cultured in the presence of fields are likely affected by the fields themselves in addition to being affected by environmental cues induced by the field. For example, mechanical stain fields may work partially through the fact that ECM molecules have strong piezoelectric material properties (Tanaka, 1999).

Gravitational Fields

In the microgravity of space, skeletal muscle *in vivo* and engineered skeletal muscle tissue ex vivo atrophy relative to control muscles experiencing earth's gravitational field (Edgerton and Roy, 1995; Vandenburgh et al., 1999). The primary cellular mechanism for atrophy is a decrease in normal protein synthesis rates (Stein et al., 1999; Vandenburgh et al., 1999). Since the engineered skeletal muscle constructs in culture did not experience any of the systemic affects that *in vivo* muscles experience in the absence of gravity (e.g., reduced levels of growth hormone in the serum), this study allowed the researchers to look at the intrinsic affects of weightlessness on skeletal muscle without systemic affects. Thus, skeletal muscle wasting experienced by astronauts in the microgravity of space works at least in part by a mechanism specific to the skeletal muscle cells and not to systemic effects. Additionally, skeletal muscle satellite cells cultured in three-dimensional aggregates under simulated microgravity proliferated at lower rates than control cells (Molnar et al., 1997). These results display one ex vivo use of tissue-engineered muscle constructs for basic research and will be extended in the future for screening potential countermeasure agents (Creswick et al., 2000).

Electric Fields

Steady DC electric fields of 36–170 mV/mm delivered via salt-agar bridges caused frog myoblasts cultured in the field to elongate along equipotential lines (Hinkle et al., 1981). DC fields outside of the 36–170 mV/mm range did not align the myoblasts. Chronic electrical stimulation of 4- to 7-day-old cultures of avian myotubes for up to 48 hours resulted in increases in total cellular protein and total MHC relative to controls as long as the electrical stimulation was sufficient to elicit a contractile response from the cultured cells (Brevet et al., 1976). Salt-agar bridges were used to deliver biphasic 10–25 ms pulses for a train duration of 0.6 seconds, repeated every 4 seconds. The authors did not determine whether electrical stimulation itself, the resultant cellular contractions, or some synergistic effect of the two caused the increase in protein, although subthreshold electrical stimulation did not result in protein increases.

The effects of different patterns of electrical stimulation on primary myogenic cells harvested from rats have been investigated in vitro (Naumann and Pette, 1994; Wehrle et al., 1994). Muscles of predominantly slow and fast MHC compositions were harvested from neonatal and adult rats and subjected to 4–8 mA of electrical current in 4-millisecond-long pulses in trains 250 milliseconds long, delivered at 15–100 Hz every 1–100 s. The presence of adult isoforms of MHC in electrically stimulated cultures that are not present in control cultures indicated a shift from developmental isoforms to adult isoforms as the myofibers become differentiated in culture. More frequent electrical stimulation caused shifts toward slow MHC while trains of pulses delivered every 100 seconds resulted in increased adult fast MHC. Although these experiments were carried out on unorganized myofibers in culture, the results suggest that electrical stimulation may be used as an environmental cue for three-dimensional skeletal muscle constructs to enhance differentiation.

Mechanical Strain and Stress Fields

Mechanical stimulation is important for skeletal muscle maintenance and growth whether in vivo or in vitro. Mechanical stimulation can generally be divided into two fundamental components: strain and stress. Strain is a normalized measure of the change in the dimensions of a tissue. Stress is a normalized measure of the force either applied to the tissue externally, or generated within the tissue. In biological terms strain is the percentage elongation of a tissue, and for normal skeletal muscle contraction, is in the 10–15% range. Stress is expressed in terms of force divided by cross-sectional area. In simple elastic materials, stress and strain are related by a proportionality constant, the modulus of elasticity. The relationship between stress and strain is more complex for biological materials, especially in the case of muscle where large stresses may be internally generated without a change in the externally measurable strain. Due to the importance of these two mechanical factors in the environment muscle of cells, it is important to distinguish between the two.

Strain

Mechanical strain is a measure of the change in the dimensions of a tissue. Strain can be uniaxial, biaxial, or triaxial. Of primary interest to the functional tissue engineer are uniaxial and biaxial strain fields. Uniaxial strain fields predominate for tissues under tension or compression along a single axis, such as tendons and most of the muscles of the extremities. Biaxial strain is relevant for sheets of muscles with a radial pattern of fiber distribution, such as the diaphragm. Mechanical strain implies a change in geometry or a relative displacement of elements within a tissue. To be meaningful to the tissue engineer, strain must be normalized by the initial geometry; the change in a dimension is divided by the initial dimension. Thus, for uniaxial length changes, the strain is simply the change in the length divided by the initial length, assuming uniformity along the length of the specimen. By convention, positive values of strain indicate lengthening or stretching, while negative values indicate shortening. Strain is expressed either as a decimal value, or as "% strain," so a strain of 0.10 is equivalent to 10% strain.

Skeletal muscles are strained during development as long bones grow and stretch the attached muscles (Haines, 1932; Weiss, 1932). Mechanical stimulation *in vitro* can simulate the continuous lengthening of muscle, as well as impart intermittent stretch/relaxation cycling, or a complex combination of both events. Whereas continuous stretch resembles the strain experienced by cells attached to long bones during development, intermittent length changes resemble contractile activity that occurs in the fetus soon after nervous innervation of the skeletal muscle.

Continuous passive strain results in hypertrophy of cultured myotubes (Vandenburgh and Karlisch, 1989; Vandenburgh and Kaufman, 1979), and aligns myoblasts and myotubes into parallel arrays *in vitro* (Vandenburgh, 1982). Mechanical stimulation, either passive strain or intermittent stretch-relaxation, is required for long-term maintenance of cultured myofibers without continual loss of protein induced by catabolic glucocorticoid compounds (Chromiak and Vandenburgh, 1992, 1994). Intermittent stretch-relaxation also results in myotube hypertrophy (Vandenburgh et al., 1989). A combination of passive strain and intermittent stretch-relaxation can induce myogenesis and formation of a self-organizing three-dimensional skeletal muscle construct in culture (Vandenburgh et al., 1991).

Stress

Mechanical stress is a normalized measure of the force exerted on an object, divided by the cross-sectional area (CSA) of the object normal to the force vector. Like mechanical strain, stress is normalized so that materials of different geometry can be compared. Distinct from synthetic materials and many biological tissues, skeletal muscle can be activated chemically or electrically to generate considerable mechanical stress on the order of 300 kPa (Gordon et al., 1966a, b). The active mechanical stress generated by cells in culture is a potent driving force in the formation of self-organizing tissues *in vivo* and *in vitro*.

When myogenic cells are co-cultured with fibroblasts, a sheet of multinucleated myotubes, ECM, and fibroblasts form a monolayer covering the substrate surface in the tissue culture dish. Due to a combination of factors, including

the spontaneous contractions of the myotubes, the contractile forces generated by the fibroblasts (Higton and James, 1964; James and Taylor, 1969; Kolodney and Wysolmerski, 1992), and changes in the expression of trans-membrane adhesion molecules in differentiating myotubes, the cell monolayer will eventually detach from the substrate in a process termed *delamination* (Dennis and Kosnik, 2000). Under standard culture conditions, the delamination process proceeds until the monolayer becomes totally detached from the substrate, rolling into a solid spherical body sometimes termed a *myoball* (De la Haba et al., 1975). The myoball contains the ECM, fibroblasts, and myotubes of the original cell monolayer. Typically, the myotubes rapidly necrose as a result of this hypercontraction.

The provision of suitable discrete three-dimensional anchor points in contact with the cell monolayer prior to delamination prevents the complete separation of the monolayer from the dish and subsequent formation of a floating myoball. Suitable anchor materials for avian myogenic cells include stainless steel posts (Swasdison and Mayne, 1992), pins (Strohman et al., 1990), and screen (Vandenburgh et al., 1991). Primary rodent cells do not remain attached directly to metallic anchors, whether coated with adhesion molecules such as laminin or not (Dennis and Kosnik, 2000). Suitable anchor materials for these myogenic cells include acellularized muscle tissue ECM and laminin-coated silk sutures (Dennis and Kosnik, 2000).

The delaminating monolayer of cells is permitted to generate a field of directed mechanical stress by reacting against the anchor points. In so doing, the cell monolayer transforms from a more or less uniform radial stress prior to delamination to one of uniaxial tensile stress, when the delamination from the substrate is complete and the cell monolayer remains attached only to the anchor points. This results in a mechanical stress field generated by the cells themselves in reaction to their environment. Under these conditions of isometric uniaxial stress the cells respond dramatically. Within approximately 72 hours, the surviving myotubes are aligned along the axis defined by the anchor points, and the myotubes and fibroblasts organize into a cohesive three-dimensional cylinder, suspended under tension above the original substrate material between the anchor points (Dennis and Kosnik, 2000; Strohman et al., 1990; Swasdison and Mayne, 1992). The effect of the change in the mechanical stress field is so potent that within days all vestiges of the former organization as a monolayer of cells are lost, giving way to a well-organized, radially symmetric cylindrical structure composed of long bundles of parallel myotubes within an outer layer of fibroblasts.

To permit self-organization of a three-dimensional cylindrical skeletal muscle construct, a minimum of two anchor points are required. The cells respond to a more complex overall stress field by the inclusion of more than two anchor points. The provision of three or four noncolinear anchor points can result in complex, web-like structures of spontaneously contracting muscle (Dennis and Kosnik, 2000), and star-shaped constructs resulted from the arrangement of five or six anchor points in a circular arrangement (Strohman et al., 1990). Similar geometries are also seen for other cell types such as human skin fibroblasts when permitted to reorganize while subjected to oriented mechanical stress and strain fields (Bell et al., 1979).

Skeletal Muscle Tissue Engineering Techniques

Several laboratories have developed tissue engineered muscle constructs *in vitro*. Each of these constructs is designed to provide a specific function. The functions generally include either the production of therapeutic proteins from genetically engineered myogenic cells, drug screening, or the generation of directed force. A brief review of each of these constructs follows.

Organoids

The first force measurements of engineered skeletal muscle were of organoids engineered from primary avian myoblasts and fibroblasts cultured in a mechanical strain field (Vandenburgh et al., 1991). The tissue-organizing strain field included periods of tonic stretching as well as intermittent stretch-relaxation with rest periods. These organoids formed over a period of several weeks without an artificial scaffold from myoblasts and fibroblasts and required a complex computer-controlled mechanical cell stimulator device. The organoids resembled muscle morphologically, having muscle-like cross-sections, fascicular organization, and well-developed sarcomeres. Functionally, organoids generated about 1% of the normalized forces of control skeletal muscle when activated by elevation of extracellular potassium (Dennis and Kosnik, 2000; Vandenburgh et al., 1991). They were developed for space flight studies (Vandenburgh et al., 1999) and are currently used for drug screening applications (Creswick et al., 2000).

BAMS (Bio-Artificial Muscles)

To simplify the process of engineering large numbers of skeletal muscle organoids, the known ability of mesenchynal cells to generate directed forces when cast in collagen gels (Bell et al., 1979) and of skeletal myofibers to survive for long periods in these gels (Vandenburgh et al., 1988) was utilitized (Shansky et al., 1997). Several million primary avian or mammalian myoblasts or mammalian cell lines were suspended in an ice-cold type I collagen solution and cast in silicone rubber molds in the shape of a rodent-sized soleus muscle with artificial tendon attachment sites at each end made of stainless steel wire or Velcro™. The cell-gel mix solidified at 37°C, detached from the molds within 24 hours and formed organized myofibers within 7–10 days. These "bioartificial muscles" or BAMs can be made with human myoblasts (Powell et al., 1999). Using the electrical stimulation method and force measurement protocols of Dennis and Kosnik (2000), active forces of BAMs have been measured (Powell et al., unpublished data). BAMs are designed for use primarily as implantable therapeutic protein delivery devices (Vandenburgh et al., 1996, 1998a).

Myooids

Self-assembling skeletal muscle constructs, termed *myooids* (Dennis and Kosnik, 2000), were developed from primary rodent myoblasts and fibroblasts in the absence of external strain fields in a manner similar to previous work performed with avian muscle cells (Strohman et al., 1990; Swasdison and Mayne, 1992). The myooids used a lower initial cell number than organoids or BAMs, and required 1–3 weeks to develop. Approximately 10 mg of muscle tissue was harvested from either adult or aged rats or mice to produce each myooid. Myooids were also engineered from explicit co-culture of commercially available cell lines of myoblasts (C2C12) and fibroblasts (10T1/2). Myooids generated normalized forces similar to those of organoids, approximately 1% of adult skeletal muscle (Dennis and Kosnik, 2000; Dennis et al., 2000).

Myooids were engineered to be 12 mm long and approximately 0.5 mm in diameter and were spontaneously contractile at approximately 1 Hz. Larger diameter constructs were generally not viable due to the increased diffusion distance from the surface to the center of the construct. Myooids were excited by transverse electrical field stimulation and forces were measured with optical force transducers. Myooids were an order of magnitude less excitable than native skeletal muscle, but similar to neonatal skeletal muscle (Close, 1964) or long-term denervated muscle (unpublished data). They had length-tension, and force-frequency relationships that were qualitatively similar to those of adult skeletal muscle. The excitability and contractility of myooids engineered from cell lines is generally inferior to that of myooids engineered from primary cells (Dennis et al., 2000). They were morphologically similar to skeletal muscle in that they consisted of parallel, unbranched myofibers. Relative to adult skeletal muscle, the lower force output and excitability, prolonged time to peak tension and half-relaxation times indicate that the myooids likely expressed predominantly developmental isoforms of contractile proteins. Under a standard set of culture conditions, myooids will self organize from primary cells from neonatal, adult, or oldest old rodents (Dennis et al., 2000). These constructs have been used primarily for basic science studies of engineered skeletal muscle excitability and contractility (Dennis and Kosnik, 2000; Dennis et al., 2000).

Technical Challenges and Future Directions

One primary limitation of engineered skeletal muscle constructs is that they appear to be developmentally arrested, and do not exhibit the phenotype of adult muscle. This manifests most notably in the very low specific forces generated by all of these constructs, on the order of 1% of control values for adult muscle. A better understanding of the interactions of the various regulating factors outlined in Figure 28.3 on forming adult muscle ex vivo will be required.

To transmit the forces typical in adult muscle, a better interface between the engineered muscle and the anchor points will be required. Thus, engineered tendon and myotendinous junction structures will be necessary to facilitate the ability to generate useful levels of force and power. They will also provide the necessary mechanical interfaces for optimal tissue development. Appropriate cell types for myotendenous formation must be present in the engineered tissue, and mechanical forces generated in or applied to the constructs may be a primary controlling factor in their formation (Banes et al., 1999; Vandenburgh et al., 1991; Woo et al., 1999). Complete ex vivo differentiation to the adult skeletal muscle phenotype for *in vivo* functioning may not be required. Many tissue interfaces for complete adult skeletal muscle phenotype, and optimal force generation may be best formed *in vivo*. These include innervation, vascularization and ECM integration into the implantation site. The primary goal of the tissue engineer should thus be to prepare the muscle construct ex vivo in a manner which allows the most successful acceptance *in vivo*. Since implantation will necessarily require the injury of the host tissue, understanding the potential of the implant "wound" site is mandatory (Martin, 1997). Genetic engineering of the muscle cells may enhance this process. For example, expression of angiogenic factors from foreign gene inserts to stimulate the vascularization of muscle constructs when placed *in vivo* (Lu et al., 2000). Similar processes for stimulating innervation by expressing neurotrophic agents, or enhancing oxygen diffusion by expressing oxygen carriers such as myoglobin, may be required in many cases. Tissue engineering and genetic engineering will thus move forward together for the development of functional skeletal muscle tissues in the future. While simulating the engineering strategy developed by Nature over billions of years of evolution is a reasonable place for the tissue engineer to start, utilizing accelerated processes from our rapidly expanding knowledge in the fields of tissue and genetic engineering will certainly accelerate the formation of the final functional construct.

Another area that will be extremely important for the future in tissue engineering of skeletal muscle will be the development of the appropriate tools for not only engineering the tissue ex vivo, but also for measuring tissue performance and for preimplantation quality control analysis of the final product. Not only must the tools be able to perform in the sterile environment of the tissue culture lab, but they must also be able to allow manufacturing and monitoring of the engineered tissue construct under current Good Manufacturing Process (cGMP) conditions in a cost-effective manner. This will require the combined effort of cell biologists, physiologists, and bioengineers in a multidisciplinary approach to the field of tissue engineering.

Conclusion

In vitro tissue engineering of skeletal muscle involves culturing myogenic cells in an environment that emulates the *in vivo* environment so that the cells proliferate, fuse, organize in three dimensions, and differentiate into functional skeletal muscle. The tissue engineer uses a multitude of *in vitro* environmental cues to direct the proliferation process. The end result will be a skeletal muscle construct that resembles skeletal muscle in both form and function. The construct will be organized like a skeletal muscle, with long multinucleated cells oriented parallel to its long axis, and the construct will be capable of generating useful directed force and power. Such constructs have been developed from avian, rodent, and human primary muscle cells as well as immortalized myogenic cells. Measurements and characterization of the con-

struct's biochemical and contractile functions have begun. Use of these early generation constructs for basic science research, as implantable therapeutic protein delivery devices, and as drug screening constructs are moving forward. Skeletal muscle constructs will likely be implanted into humans as sources of secreted proteins in the near future, and will no doubt one day replace muscle contractile function in patients with functional deficits in force and power generation.

Acknowledgments

The authors thank John Faulkner for editorial comments on an early version of this chapter. This work was supported by grants from the National Institutes of Health (RO1 HL60502, RO1 AG15415) and NASA (NAG2-1205).

References

Acartuk TO, Peel MM, Petrosko, P, LaFromboise W, Johnson PC, DeMilla PA. 1999. Control of attachment, morphology, and proliferation of skeletal myoblasts on silanized glass. *J. Biomed. Mater. Res.* 44:355–370.

Adams JC, Watt FM. 1993. Regulation of development and differentiation by the extracellular matrix. *Development* 117:1183–1198.

Allen RE, Luiten LS, Dodson MV. 1985. Effect of insulin and linoleic acid on satellite cell differentiation. *J. Anim. Sci.* 60:1571–1579.

Bandman E, Strohman RC. 1982. Increased K^+ inhibits spontaneous contractions reduces myosin accumulation in cultured chick myotubes. *J. Cell. Biol.* 93:698–704.

Banes AJ, Horesovsky G, Larson C, Tsuzaki M, Judex S, Archambault J, Zernicke R, Herzog W, Kelley S, Miller L. 1999. Mechanical load stimulates expression of novel genes *in vivo* and *in vitro* in avian flexor tendon cells. Osteoarthritis *Cartilage* 7:141–153.

Bell E, Ivarsson B, Merrill C. 1979. Production of a tissue-like structure by contraction of collagen lattices by human fibroblasts of different proliferative potential *in vitro. Proc. Natl. Acad. Sci. U.S.A.* 76:1274–1278.

Bischoff R. 1986. A satellite cell mitogen from crushed adult muscle. *Dev. Biol.* 115:140–147.

Brevet A, Pinto E, Peacock J, Stockdale FE. 1976. Myosin synthesis increased by electrical stimulation of skeletal muscle cell cultures. *Science* 193: 1152–1154.

Carlson BM, Faulkner JA. 1983. The regeneration of skeletal muscle fibers following injury: a review. *Med. Sci Sports Exerc.* 15:187–198.

Carson JA, Booth FW. 1998. Effect of serum and mechanical stretch on skeletal alpha-actin gene regulation in cultured primary muscle cells. *Am. J. Physiol.* 275:C1438–C1448.

Cerny LC, Bandman E. 1986. Contractile activity is required for the expression of neonatal myosin heavy chain in embryonic chick pectoral muscle cultures. *J. Cell. Biol.* 103:2153–2161.

Charlton CA, Mohler WA, Radice GL, Hynes RO, Blau HM. 1997. Fusion competence of myoblasts rendered genetically null for N-cadherin in culture. *J. Cell Biol.* 138:331–336.

Chiquet M, Puri EC, Turner DC. 1979. Fibronectin mediates attachment of chicken myoblasts to a gelatin-coated substratum. *J. Biol. Chem.* 254:5475–5482.

Chromiak JA, Vandenburgh HH. 1992. Glucocorticoid-induced skeletal muscle atrophy *in vitro* is attenuated by mechanical stimulation. *Am. J. Physiol.* 262:C1471–C1477.

Chromiak JA, Vandenburgh HH. 1994. Mechanical stimulation of skeletal muscle cells mitigates glucocorticoid-induced decreases in prostaglandin production and prostaglandin synthase activity. *J. Cell Physiol.* 159:407–414.

Chromiak JA, Shansky J, Perrone, C, Vandenburgh HH. 1998. Bioreactor perfusion system for the long-term maintenance of tissue-engineered skeletal muscle organoids. *In Vitro Cell Dev. Biol. Anim.* 34:694–703.

Clark P, Coles D, Peckham M. 1997. Preferential adhesion to and survival on patterned laminin organizes myogenesis *in vitro*. *Exp. Cell Res.* 230:275–283.

Clark P, Connolly P, Curtis AS, Dow JA, Wilkinson CD. 1990. Topographical control of cell behaviour: II. Multiple grooved substrata. *Development* 108:635–644.

Close R. 1964. Dynamic properties of fast and slow skeletal muscles of the rat during development. *J. Physiol.* 173:74–95.

Creswick BC, Shansky J, Lee PHU, Wang WY, Vandenburgh HH. 2000. Preliminary studies in support of a space shuttle flight experiment evaluating the ability of rhIGF-1 to attenuate space flight-induced skeletal muscle atrophy. *Gravitational and Space Biology Bulletin.* 14:53 (abstract)

De laHaba G, Kamali HM, Tiede DM. 1975. Myogenesis of avian striated muscle *in vitro:* role of collagen in myofiber formation. *Proc. Natl. Acad. Sci. U.S.A.* 72:2729–2732.

Decary S, Mouly V, Hamida CB, Sautet A, Barbet JP, Butler-Browne GS. 1997. Replicative potential and telomere length in human skeletal muscle: implications for satellite cell-mediated gene therapy. *Hum. Gene Ther.* 8:1429–1438.

Dennis RG, Kosnik PE. 2000. Excitability and isometric contractile properties of mammalian skeletal muscle constructs engineered *in vitro* [In Process Citation]. *In Vitro Cell Dev. Biol. Anim.* 36: 327–335.

Dennis RG, Kosnik PE, Gilbert ME, Faulkner JA. 2000. Excitability and contractility of skeletal muscle engineered from primary cultures and cell lines. *Am. J. Physiol. Cell.* 280:C288–C295.

Dollenmeier P, Turner DC, Eppenberger HM. 1981. Proliferation and differentiation of chick skeletal muscle cells cultured in a chemically defined medium. *Exp. Cell Res.* 135:47–61.

Dusterhoft S, Pette D. 1993. Satellite cells from slow rat muscle express slow myosin under appropriate culture conditions. *Differentiation* 53:25–33.

Eagle H. 1959. Amino acid metabolism in mammalian cell cultures. *Science* 130:432–437.

Edgerton VR, Roy RR. 1996. Neuromuscular adaptations to actual and simulated weightlessness. In *Handbook of Physiology.* S. Churchill, ed. Environmental Physiology, Bethesd, MD: Am. Physiol. Soc. Sect. 4, Vol. III, Chap. 32 p 721–763.

Eng H, Herrenknecht K, Semb H, Starzinski-Powitz A, Ringertz N, Gullberg D. 1997. Effects of divalent cations on M-cadherin expression and distribution during primary rat myogenesis *in vitro*. *Differentiation* 61:169–176.

Evans DJR, Britland S, Wigmore PM. 1999. Differential responses of fetal and neonatal myoblasts to topographical guidance cues *in vitro*. *Dev. Genes Evol.* 209:438–442.

Florini JR, Ewton DZ, Magri KA. 1991. Hormones, growth factors, and myogenic differentiation. *Annu. Rev. Physiol.* 53:201–216.

Goichberg P, Geiger B. 1998. Direct involvement of N-cadherin-mediated signaling in muscle differentiation. *Mol. Biol Cell.* 9:3119–3131.

Gordon AM, Huxley AF, Julian FJ. 1966a. Tension development in highly stretched vertebrate muscle fibres. *J. Physiol. (Lond.)* 184:143–169.

Gordon AM, Huxley AF, Julian FJ. 1966b. The variation in isometric tension with sarcomere length in vertebrate muscle fibres. *J. Physiol. (Lond.)* 184:170–192.

Grinnell AD. 1995. Dynamics of nerve-muscle interaction in developing and mature neuromuscular junctions. *Physiol. Rev.* 75:789–834.

Haines RW. 1932. The laws of muscle and tendon growth. *J. Anat.* 66:578–585.

Ham RG, St. Clair JA, Meyer SD. 1990. Improved media for rapid clonal growth of normal human skeletal muscle satellite cells. *Adv. Exp. Med. Biol.* 280:193–199.

Hayflick L, Moorhead PS. 1961. The serial cultivation of human diploid cell strains. *Exp. Cell Res.* 25:585–621.

Higton DIR, James DW. 1964. The force of contraction of full-thickness wounds of rabbit skin. *Br. J. Surg.* 51:462–466.

Hinkle L, McCaig CD, Robinson KR. 1981. The direction of growth of differentiating neurones and myoblasts from frog embryos in an applied electric field. *J. Physiol, (Lond.)* 314:121–135.

Irintchev A, Rosenblatt JD, Cullen MJ, Zweyer M, Wernig A. 1998. Ectopic skeletal muscles derived from myoblasts implanted under the skin. *J. Cell Sci.* 111(Pt 22):3287–3297.

James DW, Taylor JF. 1969. The stress developed by sheets of chick fibroblasts *in vitro. Exp. Cell Res.* 54:107–110.

Kolodney MS, Wysolmerski RB. 1992. Isometric contraction by fibroblasts and endothelial cells in tissue culture: a quantitative study. *J. Cell Biol.* 117:73–82.

Konigsberg IR. 1963. Clonal analysis of myogenesis. *Science* 140:1273–1284.

Lagord C, Soulet L, Bonavaud S, Bassaglia Y, Rey C, Barlovatz-Meimon G, Gautron J, Martelly I. 1998. Differential myogenicity of satellite cells isolated from extensor digitorum longus (EDL) and soleus rat muscles revealed *in vitro. Cell Tissue Res.* 291: 455–468.

Lewis WH, Lewis MR. 1917. Behavior of cross striated muscle in tissue cultures. *Am J. Anat.* 22:169–194.

Lu Y, Shansky J, Smiley B, Vandenburgh HH. 2001. Recombinant Vascular endothelial growth factor secreted from tissue engineered bioartificial muscles promotes localized angiogenesis. *Circulation* 104: 594–599.

Mantegazza R, Gebbia M, Mora M, Barresi R, Bernasconi P, Baggi F, Cornelio F. 1996. Major histocompatibility complex class II molecule expression on muscle cells is regulated by differentiation: implications for the immunopathogenesis of muscle autoimmune diseases. *J. Neuroimmunol.* 68:53–60.

Martin P. 1997. Wound healing—aiming for perfect skin regeneration. *Science* 276:75–81.

Mauro A. 1961. Satellite cell of skeletal muscle fibers. *J. Biophys. Biochem. Cytol.* 9:493–495.

Molnar G, Schroedl NA, Gonda SR, Hartzell CR. 1997. Skeletal muscle satellite cells cultured in simulated microgravity. *In Vitro Cell Dev. Biol. Anim.* 33:386–391.

Mooney DJ, Rowley JA. 1997. Tissue engineering: Integrating cells and materials to create functional tissue replacements. In: *Controlled Drug Delivery: Challenges and Strategies.* K. Park, ed. American Chemical Society, Washington, DC, pp. 333–346.

Mulder MM, Hitchcock RW, Tresco PA. 1998. Skeletal myogenesis on elastomeric substrates: implications for tissue engineering. *J Biomater. Sci. Polym. Ed.* 9:731–748.

Mulder MM, McElwain JF, Tresco PA. 1995. Three dimensional culture system for myocyte growth and differentiation. *J. Am. Soc. Artif. Intern. Organs* 41:90 (abstract)

Naumann K, Pette D. 1994. Effects of chronic stimulation with different impulse patterns on the expression of myosin isoforms in rat myotube cultures. *Differentiation* 55:203–211.

Okano T, Matsuda T. 1997. Hybrid muscular tissues: preparation of skeletal muscle cell-incorporated collagen gels. *Cell Transplant.* 6:109–118.

Okano T, Matsuda T. 1998a. Muscular tissue engineering: capillary-incorporated hybrid muscular tissues *in vivo* tissue culture. *Cell Transplant.* 7:435–442.

Okano T, Matsuda T. 1998b. Tissue engineered skeletal muscle: preparation of highly dense, highly oriented hybrid muscular tissues. *Cell Transplant.* 7:71–82.

Okano T, Satoh S, Oka T, Matsuda T. 1997. Tissue engineering of skeletal muscle. Highly dense, highly oriented hybrid muscular tissues biomimicking native tissues. *ASAIO J.* 43:M749–M753.

Partridge TA. 1991. Invited review: myoblast transfer: a possible therapy for inherited myopathies? *Muscle Nerve* 14:197–212.

Perrone CE, Fenwick-Smith D, Vandenburgh HH. 1995. Collagen and stretch modulate autocrine secretion of insulin-like growth factor-1 and insulin-like growth factor binding proteins from differentiated skeletal muscle cells. *J. Biol. Chem.* 270: 2099–2106.

Powell C, Shansky J, Del Tatto M, Forman DE, Hennessey J, Sullivan K, Zielinski BA, Vandenburgh HH. 1999. Tissue-engineered human bioartificial muscles expressing a foreign recombinant protein for gene therapy. *Hum. Gene Ther.* 10:565–577.

Putnam AJ, Mooneyy DJ. 1996. Tissue engineering using synthetic extracellular matrices. *Nat. Med.* 2:824–826.

Redfield A, Nieman MT, Knudsen KA. 1997. Cadherins promote skeletal muscle differentiation in three-dimensional cultures. *J. Cell Biol.* 138: 1323–1331.

Rowley JA, Madlambayan G, Mooney DJ. 1999. Alginate hydrogels as synthetic extracellular matrix materials. *Biomaterials* 20:45–53.

Sanderson RD, Fitch JM, Linsenmayer TR, Mayne R. 1986. Fibroblasts promote the formation of a continuous basal lamina during myogenesis *in vitro*. *J. Cell Biol.* 102:740–747.

Sastry SK, Lakonishok M, Wu S, Truong TQ, Huttenlocher A, Turner CE, Horwitz AF. 1999. Quantitative changes in integrin and focal adhesion signaling regulate myoblast cell cycle withdrawal. *J. Cell Biol.* 144:1295–1309.

Schwartz MA. 1992. Transmembrane signaling by integrins. *Trends Cell Biol.* 2:304–308.

Schwartz MA, Ingber DE. 1994. Integrating with integrins. *Mol. Biol Cell.* 5:389–393.

Shainberg A, Yagil G, Yaffe D. 1969. Control of myogenesis *in vitro* by Ca^{2+} concentration in nutritional medium. *Exp. Cell Res.* 58:163–167.

Shansky J, Del Tatto M, Chromiak J, Vandenburgh H. 1997. A simplified method for tissue engineering skeletal muscle organoids *in vitro* [letter]. *In Vitro Cell Dev. Biol. Anim.* 33:659–661.

Sheehan SM, Tatsumi R, Temm-Grove CJ, Allen RE. 2000. HGF is an autocrine growth factor for skeletal muscle satellite cells *in vitro*. *Muscle Nerve* 23:239–245.

Snow MH. 1977a. Myogenic cell formation in regenerating rat skeletal muscle injured by mincing. I. A fine structural study. *Anat. Rec.* 188:181–199.

Snow MH. 1977b. Myogenic cell formation in regenerating rat skeletal muscle injured by mincing. II. An autoradiographic study. *Anat. Rec.* 188:201–217.

Stein TP, Leskiw MJ, Schluter MD, Donaldson MR, Larina I. 1999. Protein kinetics during and after long-duration spaceflight on MIR. *Am. J. Physiol.* 276:E1014–E1021.

Stockdale FE, Holtzer H. 1961. DNA synthesis and myogenesis. *Exp. Cell Res.* 24:508–520.

Stoker M, O'Neill C, Berryman S, Waxman V. 1968. Anchorage and growth regulation in normal and virus-transformed cells. *Int. J. Cancer* 3:683–693.

Strohman RC, Bayne E, Spector D, Obinata T, Micou-Eastwood J, Maniotis A. 1990. Myogenesis and histogenesis of skeletal muscle on flexible membranes *in vitro*. *In Vitro Cell Dev. Biol. Anim.* 26:201–208.

Swasdison S, Mayne R. 1992. Formation of highly organized skeletal muscle fibers *in vitro*. Comparison with muscle development *in vivo*. *J. Cell Sci.* 102(Pt 3):643–652.

Tanaka SM. 1999. A new mechanical stimulator for cultured bone cells using piezoelectric actuator. *J. Biomech.* 32:427–430.

Turner DC. 1986. Cell-cell and cell-matrix interactions in the morphogenesis of skeletal muscle. *Dev. Biol.* 3:205–224.

Turner DC, Lawton J, Dollenmeier P, Ehrismann R, Chiquet M. 1983. Guidance of myogenic cell migration by oriented deposits of fibronectin. *Dev. Biol.* 95:497–504.

van Wachem PB, van Luyn MJ, da Costa ML. 1996. Myoblast seeding in a collagen matrix evaluated *in vitro*. *J. Biomed. Mater. Res.* 30:353–360.

Vandenburgh HH. 1982. Dynamic mechanical orientation of skeletal myofibers *in vitro*. *Dev. Biol.* 93:438–443.

Vandenburgh HH. 1983. Cell shape and growth regulation in skeletal muscle: exogenous versus endogenous factors. *J. Cell Physiol.* 116:363–371.

Vandenburgh HH, Karlisch P. 1989. Longitudinal growth of skeletal myotubes *in vitro* in a new horizontal mechanical cell stimulator. *In Vitro Cell Dev. Biol. Anim.* 25:607–616.

Vandenburgh H, Kaufman S. 1979. *In vitro* model for stretch-induced hypertrophy of skeletal muscle. *Science* 203:265–268.

Vandenburgh HH, Kaufman S. 1981. Stretch-induced growth of skeletal myotubes correlates with activation of the sodium pump. *J. Cell Physiol.* 109:205–214.

Vandenburgh HH, Karlisch P, Farr L. 1988. Maintenance of highly contractile tissue-cultured avian skeletal myotubes in collagen gel. *In Vitro Cell Dev. Biol. Anim.* 24:166–174.

Vandenburgh HH, Hatfaludy S, Karlisch P, Shansky J. 1989. Skeletal muscle growth is stimulated by intermittent stretch-relaxation in tissue culture. *Am. J. Physiol.* 256:C674–C682.

Vandenburgh HH, Swasdison S, Karlisch P. 1991. Computer-aided mechanogenesis of skeletal muscle organs from single cells *in vitro*. *FASEB J.* 5:2860–2867.

Vandenburgh H, Del Tatto M, Shansky J, LeMaire J, Chang A, Payumo F, Lee P, Goodyear A, Raven L. 1996. Tissue-engineered skeletal muscle organoids for reversible gene therapy. *Hum. Gene Ther.* 7:2195–2200.

Vandenburgh H, Del Tatto M, Shansky J, Goldstein L, Russell K, Genes N, Chromiak J, Yamada S. 1998a. Attenuation of skeletal muscle wasting with recombinant human growth hormone secreted from a tissue-engineered bioartificial muscle. *Hum. Gene Ther.* 9:2555–2564.

Vandenburgh HH, Shansky J, Del Tatto M, Chromiak J. 1998b. Organogenesis of skeletal muscle in tissue culture. In: *Methods in Molecular Medicine: Tissue Engineering.* J Morgan, M Yarmush, eds. Humana Press, Tottowa, NJ.

Vandenburgh H, Chromiak J, Shansky J, Del Tatto M, LeMaire J. 1999. Space travel directly induces skeletal muscle atrophy. *FASEB J.* 13:1031–1038.

Waldeyer W. 1865. Uber die Veränderungen der quergestreiften Muskeln bei der Entzündung und dem Typhusprozess, sowie über die Regeneration derselben nach Substanzdefecten. *Virchows Arch. Pathol. Anat. Physiol. Clin. Med.* 34:473–514.

Wang N, Butler JP, Ingber DE. 1993. Mechanotransduction across the cell surface and through the cytoskeleton [see comments]. *Science* 260:1124–1127.

Wehrle U, Dusterhoft S, Pette D. 1994. Effects of chronic electrical stimulation on myosin heavy chain expression in satellite cell cultures derived from rat muscles of different fiber-type composition. *Differentiation* 58:37–46.

Weiss P. 1932. Functional adaptation and the role of ground substances in development. *Am. Naturalist* 67:322–340.

Wernig A, Irintchev A, Lange G. 1995. Functional effects of myoblast implantation into histoincompatible mice with or without immunosuppression. *J. Physiol. (Lond.)* 484(Pt 2):493–504.

Woo SL, Hildebrand K, Watanabe N, Fenwick JA, Papageorgiou CD, Wang JH. 1999. Tissue engineering of ligament and tendon healing. *Clin. Orthop.* S312–S323.

Yaffe D. 1968. Retention of differentiation potentialities during prolonged cultivation of myogenic cells. *Proc. Natl. Acad. Sci U.S.A.* 61:477–483.

Yaffe D. 1971. Developmental changes preceding cell fusion during muscle differentiation *in vitro. Exp. Cell Res.* 66:33–48.

Part VII

Regulatory and Clinical Tissues in Tissue Engineering

29

From Concept Toward the Clinic: Preclinical Evaluation of Tissue-Engineered Constructs

Steven A. Goldstein

Introduction

The purpose of this chapter is to provide a summary of the issues and principles that should be considered in translating and evaluating functional tissue engineering approaches to augment, repair, or replace tissue organ function. It will be assumed that the basic design and scientific validation of the concepts behind the tissue-engineered construct have already been established, as well as the potential for clinical success. The focus of the chapter will be to present the factors that need to be considered in designing preclinical studies that will fulfill regulatory requirements and support commercialization driven validation. It is not the purpose of this chapter to present the policies, specific details, or legal requirements associated with the regulatory process that must be fulfilled in order to bring the tissue-engineered product to market. Instead, it is hoped that this chapter provides a framework that will help to guide investigators to consider and design appropriate experiments and outcome measures to assess the functionality of the tissue-engineered constructs.

Construct Conceptualization and Clinical Demand

The first step toward establishing the necessary experimental paradigm to test a tissue-engineered construct is establishing a clear description of the clinical demand or clinical problem and the associated targeted functional properties needed to solve that clinical problem. Specifically, this characterization should include:

1. The incidence and prevalence of the clinical problem.
2. The demographics of those affected by the problem.
3. Some measure or description of the complexity of the problem.
4. Current treatment options and, in particular, the current standard of care.
5. Timing issues associated with diagnosis, treatment, or follow-up.

The rationale for identifying these features of the clinical problems to be addressed is based on the potential need for establishing appropriate outcome measures and conditions that can simulate the use of the product including caregiver handling, patient comfort, and compliance.

Product Development Parameters

Once the clinical parameters associated with the targeted use of a tissue-engineered construct are established, the process of translating scientific concept into clinical reality involves three major phases; design, manufacturing, and evaluation. Although these three categories of tasks might appear to have a natural sequential order, it is also likely that they may occur in parallel. The reason for the simultaneous approach is based on the recognition that successes and failures in each of these steps will likely influence activity in another step.

The design phase occurs after the initial scientific and clinical concept has been established and involves a clinical, biologic, and engineering optimization process. Based on research findings, this optimization process usually involves the appropriate selection of matrices, cells, and potential biologic factors based on a prescribed functional need of the tissue-engineered product.

The manufacturing phase involves considerations of scale-up with a particular emphasis, in many cases, to design of specialized bioreactors. It is also critical during this phase to consider how the product will be maintained sterile or made sterile and how storage and handling procedures will be prescribed.

The evaluation phase is one of the most visible phases of development. It involves studies to characterize the biologic and structural function and integrity of the constructs, as well as tests for viability, toxicity, and clinical utility. It should be recognized that the results of the evaluation phase may not only provide critical data to support business decisions associated with commercialization of the product, but also support significant marketing efforts through publications or presentations. It should also be noted that current medical, economic, and product commercialization strategies tend to be more successful if thought leaders or clinical champions can be identified to provide its successful use.

Clearly, a key ingredient to this process is establishing the proof of the concept. This typically requires cellular, molecular, or physiologic data to support the technology, which can be either *in vitro* or *in vivo* data. Most importantly will be *in vivo* demonstration of biologic and biomechanical function of the constructs, including the development and utilization of specific biomarkers and functional properties that may also be used during clinical studies. In most cases, animal studies will need to be performed to support functional efficacy.

Regulatory Requirements

The detailed requirements prescribed by the Food and Drug Administration can be found in guidelines and policy documents available from the agency and in federal records. In general, the steps necessary to support regulatory approval include the following.

1. Preclinical safety and efficacy studies in clinically relevant models.
2. The establishment of manufacturing composition, methods, and controls, and measures of reliability associated with the manufacturing processes.
3. Phase I or phase I/II studies utilizing human participants. The design and approval for conducting these studies will be based on an assessment of risk benefit, establishment of necessary and valuable outcome variables and the design of an appropriate adverse event monitoring system. Some tissue-engineered approaches may require dose response or dose escalation studies during these early phase studies. These would likely be more associated with tissue-engineered constructs that include biologic factors such as recombinant proteins, DNA, or other biochemicals that may be delivered by the constructs.
4. Finally, phase III pivotal studies need to be conducted in a way in which statistical testing can be carried out to assess efficacy, complications, and adverse events.

It should be noted that all studies that are conducted prior to clinical evaluation need to be accumulated and provided for regulatory and scientific review. These will include both *in vitro* and *in vivo* studies in many instances. A pivotal preclinical study is typically defined as a study in either large or small animals that utilizes the tissue-engineered construct in a form that mimics how the construct will be utilized in patients. These studies need to consider dose or utilization profiles and also need to establish and utilize appropriate endpoints and outcome measures. These pivotal studies need to be designed to address issues of safety with particular reference to toxicity, bioavailability, and dose response, and, potentially, issues of storage and any unusual handling requirements if necessary.

Design of Preclinical Studies

The most important first step in successful design of a preclinical study is the establishment of an experimental model that simulates the targeted clinical utility of the tissue-engineered construct. An early requirement is the need to consider whether the model should be designed to simulate either a very specific or broad category of clinical application. If a specific or single clinical use is simulated, it may likely result in a regulatory label that might allow approval for only that specific clinical case. On the other hand, a model that can be shown to correspond to a larger class of clinical applications might support regulatory approval with a broader use label. The second important issue associated with model development relates to the compatibility of the selected animal with respect to human physiologic and biologic function. This compatibility might relate to functional demands or biologic response cascades. The model also needs to include the ability to establish quantitative assays. Consideration should be given for the use of assays that could also be used in patient studies. In many cases, functional assays that can be established in an animal model cannot be utilized in human studies. For example, a functional assessment in an animal model might include biomechanical testing of tissue being formed. Clearly, biomechanical testing by destructive means is not feasible in patients. In these cases, efforts should be made toward establishing surrogate measures of the assays that cannot be replicated in human subjects. An example of surrogate measures might be establishing a correlation between noninvasive measure of bone mineral density and bone biomechanical properties. Studies could be performed in either animals or perhaps in cadaver tissue to verify the noninvasive, nondestructive surrogate that could then be utilized in clinical trials. Finally, it is critical that the experimental models will enable a robust statistical assessment. The experiment needs to allow sufficient sample size to be utilized such that an appropriate power will enable statistical testing.

Study Design Example

In an effort to illustrate some of the principles associated with the design of pivotal preclinical studies to evaluate tissue-engineered constructs, specific issues will be discussed as applied toward the evaluation of a functional tissue engineering approach for bone. While many specific biologic, physiologic, or mechanical factors may vary substantially between bone and other tissues, it is hoped that the approach illustrated will serve as a paradigm that can be extrapolated to other tissue constructs.

Clinical Problem Definition

Treatment of complex fracture with a high risk of delayed union and associated loss of bone material subsequent to trauma.

Standard of Care

The standard of care typically involves a surgical approach, to reconstruction with the use of fixation devices and, in most cases, the need for bone graft material to enhance repair. The preferred graft material would be the use of autograft.

Targeted Treatment Regiment

The ideal therapy would involve a definitive treatment at the time of primary surgical approach. The use would be through direct surgical implantation or potentially an injection into the defect site. The ideal material would have a long shelf life, be easily handled in a sterile environment, and add little or no time to the procedure.

The description above provides the clinically based profile for a tissue-engineered construct to augment fracture repair. Without specifying the technology to be utilized, the approach needed to successfully produce a valuable therapeutic project can now be described. Technically, the only assumption that is made is that the tissue-engineered construct may involve cells, scaffolds, and, potentially, biologic factors. Given that there is scientific merit to the underlying technology, the next step would be to establish experimental models to evaluate the efficacy of the technology in a situation that would simulate the expected human clinical environment.

Experimental Model Development

The clinical need described a significant complex fracture with loss of bone. Further evaluation would need to be utilized to determine whether these fractures occur in cortical diaphyseal regions or in more metaphyseal regions. Depending on the review of this issue, an animal model could be developed to try and simulate a typical targeted fracture. The importance of these definitions is illustrated by some of the differences between cortical diaphyseal fractures and metaphyseal fractures. In the mid diaphysis, the mechanical demand is usually very high and the amount of marrow and vascular supply is generally low. In metaphyseal regions, which are rich in trabecular bone, there is a high marrow content and increased vascularity. Metaphyseal regions also tend to have much greater geometric dimension, thereby creating substantially different biomechanical demands than when compared with diaphyseal sites.

The next important decision involves the selection of the animal. Choosing an appropriate animal model depends on the feasibility of creating an appropriate fracture, as well as the overall physiologic characteristics of the animal in comparison to human. Typically, there are cost issues and sample size viability issues that also come into play at this point in program development. For example, human bone is primarily haversion or secondary remodeled bone, while sheep and goats have plexiform bone. These facts are noted to emphasize that investigators should pay close attention to basic morphologic and physiologic aspects of the tissue in proposed models, in comparison to humans. In general, experience has suggested that even animals with morphologic differences in their bone (such as sheep and goats) still demonstrate fracture healing patterns very similar to humans. Scale also plays an important role. If modeling a metaphyseal defect, small rodents may be inappropriate. It is very difficult to apply a surgical model with fixation in the metaphyseal regions of rodents, such as rats. In this case, larger animals, ranging from rabbits to dogs and possibly sheep may be

necessary. Since functional demands and their influences on tissue regeneration and remodeling are also critical, the ambulatory patterns of the selected animal models are important. If a procedure is to be performed on one limb of a dog, for example, experience has shown that dogs can and will often unload the operated limb, thereby reducing functional loading. On the other hand, sheep will always seek to have even load distribution on all four limbs even after surgical procedures. In addition, sheep can be housed in a farm environment, allowing them to ambulate more naturally.

While many examples and specific technical issues can be reviewed, it is the intent to only emphasize the care which should be taken in considering an appropriate animal model.

Development of Outcome Measures

Functional tissue engineering implies that critical outcome variables need to include measures of biomechanical competence, as well as histologic, morphologic, and cellular response. Returning to the bone tissue engineering example, the outcome measures in the animal model should include radiographs, histologic evaluation, and biomechanical tests. Investigators might consider high-resolution techniques in their animal models such as microradiography or even three-dimensional microcomputed tomography. From a histologic perspective, many different assays are available and analysis can be performed at the light microscope or even at the electron microscopic level. It should be recognized, however, that any outcome measures in subsequent human studies would only be able to be done using noninvasive imaging techniques, such as radiographs and, often, computed tomography scans. It is typically unlikely that biopsies or tissue specimens will be available for histologic analysis. As a result, it might be critical to establish the relationship between these noninvasive imaging modalities and biomechanical and histologic competence in the animal model. If correlations can be established between the imaging modalities and biomechanical and histologic variables, then these same correlations might be helpful when only the imaging modalities are available in the human patients. The key point to consider is that even the imaging modalities used in the animal studies should mimic those that might be used in human clinical trials.

Finally, it should be recognized that biomechanical and morphologic studies may be only a portion of the assays needed to evaluate the efficacy and safety of the tissue-engineered construct. If the construct involves biologic factors or cells, there may be the need to perform bioavailability studies that include a sampling of blood or urine. This becomes particularly important if the included factors may cause a systemic increase in specific biochemical responses. Referring to our bone example, these assays may include markers of bone turnover or formation, as well as any systemic biologic factors that are being delivered by a tissue-engineered construct. It should be noted that these same chemistries can be performed on the human patients as well. In fact, many safety studies that will be performed under FDA oversight may require these chemistries to be monitored.

Other Factors to Consider

As noted above, one of the critical issues to consider in the design of the preclinical studies relates to the manufacturing of the tissue-engineered construct that is being evaluated. In order to translate into human clinical trials it would be ideal to utilize a material that is manufactured under the same conditions that are expected to be utilized for the human devices. This includes utilizing scaled dosing regimens, and also assembling the construct using good manufacturing practice guidelines. In addition, it is very typical for many devices to be evaluated in small animals where costs are somewhat minimized prior to large animal studies. Large animal studies are thought to be more representative of human physiology and considered by many to be necessary before clinical trials. Some of these issues should be discussed with experts from the FDA before final experimental designs and studies are planned and implemented.

Summary

In summary, it is hoped that this chapter provided a review of some of the issues and principles that should be considered when attempting to translate tissue-engineered construct design into eventual human trials. Clearly, this chapter could not cover all of the principles or specific issues that need to be considered. However, a general framework and conceptualization has been reviewed that should be applicable for functional tissue engineering approaches to repair or augment any musculoskeletal tissue.

30

Trends in the FDA

Edith Richmond Schwartz

Introduction

Tissue engineering is driven by the need to replace missing or defective tissues and organs with biocompatible functional substitutes. In some instances, the tissue-engineered product will consist solely of a biopolymer that, when placed appropriately at the site of need, has the ability to recruit the host's cells, which will, in time, multiply and differentiate to carry out the needed functions. Alternatively, the tissue-engineered construct may consist of a biopolymer that encompasses viable cells to be delivered as a functional replacement. The cells may be stem or differentiated cells and may or may not be autologous (Schwartz, 1997). Recently, the concept of tissue-engineered constructs has widened to include delivery of genes and proteins for tissue repair. Thus, the biopolymer may serve as a scaffold, a delivery system, and/or a smart material. Furthermore, it may be a homopolymer or heteropolymer, naturally or synthetically derived, and resorbed quickly in the body or be long-lasting *in vivo*.

From the Research Laboratory to the Market Place

Although many university laboratories have been conducting research in tissue engineering since the early 1980s, relatively few resulting products have reached the market place during the past two decades. Several principles underlie the difficult road from basic research to product commercialization. These include the extent of market need; the feasibility of production at reasonable cost; competitive products in development or already on the market; safety and efficacy considerations and reimbursement. Whereas ensuring safety and efficacy is the responsibility of the Food and Drug Administration, reimbursement policy is influenced nationally and internationally both by government agencies as well as private-sector insurers. Particularly in the era of a global economy, reimbursement often is a deciding factor in the development and marketing of tissue-engineered products.

Role of the Food and Drug Administration

The role of the Food and Drug Administration (FDA) is to promote and protect public health by ensuring that biomedical products that enter the market place are safe and effective and remain so during their lifetime of use. In addition to their role in the scientific and clinical review of potential new products, the FDA also carries out facility inspections to ensure that manufacturing is conducted under FDA-approved guidelines and monitors corrective actions when those are necessary (Runner, 2000).

Information about existing FDA rules as well as proposed new FDA rules and guidelines are available on the FDA Website (www.fda.gov), and are published in the Federal Register. Prior to final enactment of new rules, the public is given an opportunity to address and provide an opinion on specific issues that are being considered. In addition, public forums are held at which the public may participate.

Review and approval of biomedical products are under the purview of three FDA Divisions: the Center for Drug Evaluation and Research (CDER), the Center for Devices and Radiological Health (CDRH), and the Center for Biologics Evaluation and Research (CBER). In general, devices are reviewed by CDRH. However, if multiple components, including a drug or biologic are part of a device, CBER or CDER, in addition to CDRH, will partake in the review process. The lead reviewer is selected by discussions between representatives of the participating centers. If dissatisfied with the choice, the applicant organization may appeal the selected lead reviewer to an FDA ombudsman (Feigel, 1999).

Responsibilities of CDRH

CDRH performs independent evaluation of devices for safety and effectiveness and, subsequent to approval, reviews the labeling and advertising to ensure that both the expected performance and risks are accurately described. In addition, personnel from CDRH inspect manufacturing sites to make certain that quality systems are in place so that the public will receive well-designed and well-manufactured products. Sourcing of materials and manufacturing of devices must adhere to FDA standards, whether performed in house or outsourced.

CDRH also is charged with postmarket surveillance. Thus, if a problem is identified with a device that is on the market, CDRH will try to assist the manufacturer in correcting the problem. In serious cases, the device may be recalled and future use restricted.

Instruction on the proper use of a device is the responsibility of the manufacturer and is not a duty of the FDA. In cases of sophisticated implantable devices, in particular, it is vital that the company has an expert team to instruct clinicians on the intended indication and proper use. Deviations from intended use by clinicians are difficult to monitor unless adverse effects are observed and reported.

Legislative Acts that Govern Responsibilities and Actions of CDRH

The basic responsibilities for CDRH were defined in the Medical Device Amendment of 1976. This amendment (1) defined a medical device, (2) gave CDRH the authority to move against violations, (3) established a tiered system for classifying medical devices based on risk, and (4) established standards for marketing claims. In abbreviated form, a device is an instrument, apparatus, implement, implant, or other similar or related article that is intended to affect the structure or any function of the body of man and which does not achieve any of its primary intended purposes through chemical action within or on the body of man and which is not dependent upon being metabolized for the achievement of any of its primary intended purposes (available at: www.verify.fda.gov). Devices are categorized either as class I (least amount of risk), class II (intermediate risk), or class III (highest risk).

Adverse reactions and improper use of some approved devices, including pedicle screws, resulted in the passage of the Safe Medical Devices Act of 1990. This act required CDRH to apply higher scientific standards in the review of new devices prior to marketing. To offset the increase in review time associated in complying with this mandate, Congress increased funding to CDRH in 1992 to permit a 10% increase in staff. At the same time, however, Congress asked that the review process become more efficient and timely.

Recent Changes in the Device Program

The past decade has seen significant improvement in the operations of CDRH. In 1992, the CDRH set out to reduce the time to process applications to market new devices. One of the early initiatives in this reengineering scheme was to focus selectively on high-risk, high-impact products (class III) and reduce allocation of resources on new product applications that fall into class I and are of minimal risk.

For functional tissue engineering, specific initiatives have been instituted under an Inter-Center Tissue Engineering program. Devices considered by this initiative include (1) wound repair and tissue regeneration products, (2) combination products of bone morphogenetic protein, collagen and bone, (3) bone graft substitutes, and (4) encapsulated cells. Applications for FDA review of tissue-engineered devices should define and provide data on material sourcing, cell characterization, the absence of adventitious agents, biomaterial characterization, and methods of sterilization. In addition, clear evidence of product consistency and product stability need be submitted (Runner, 2000).

Investigational Device Exemption

The scientist, developer, manufacturer, and marketer may have several levels of interaction with CDRH in transforming a laboratory finding to a marketable device. Prior to initiating clinical trials with a new device that may pose a significant risk in humans, an Investigational Device Exemption (IDE) must be submitted and approved by the FDA. In recent years, CDRH has encouraged applicants to interact with the agency early in the clinical trial design process so that agreement can be reached in advance on the proposed clinical protocol. This protocol should state the hypothesis for the trial, provide a review of existing information, explain the procedures to be followed, define how the data will be gathered and evaluated, and provide contingency measures to be taken when unanticipated events occur. These early interactions have resulted in a decrease in average IDE approval time from 200 days in 1993 to 70 days in 1997 (Feigal, 2000).

Premarket Approval Application

A Premarket Approval Application (PMA) is required for any new device that is not substantially equivalent to an existing one. Scientific and clinical data are required to support safety and efficacy. Therefore, during the development cycle, studies should be designed toward answering questions that will be presented as claims in the PMA. Prior to initiating studies, the medical claims should be clearly identified, any concerns of participating investigators should be addressed, inclusion and exclusion criteria of the study populations need to be finalized, and the testing site selections approved. To ensure meaningful outcome measurements, continuing discourse with FDA representatives during these processes should be undertaken. The continuing interactions between applicant and CDRH have decreased the average PMA approval time from 26 months in 1993 to 16 months in 1997 (Feigal, 2000).

Premarket Notifications

Major changes have been instituted in application and approval processes for Premarket Notifications, also known as 510(k)s. These applications are required for any device that substantially is equivalent to an already marketed device. Since the implementation of the FDA Modernization Act (FDAMA), the 510(k) paradigm has undergone several significant changes. For example, the Special 510(k) format should be used when applying for approval of a new device that is based on a legally marketed device that has been modified by the manufacturing process, although the fundamental scientific technology or intended use are not changed. A sample template plus essential documents to review in preparation of a Special 510(k) application are available on the FDA Website.

Special 510(k) Case Study

Inregra LifeSciences Corporation, Plainview, New Jersey, markets BioMend® Absorbable Collagen Membrane with indicated use in guided tissue regeneration procedures in periodontal defects to enhance regeneration of the periodontal apparatus. This product is a 510(k) approved device with approval number K924408. The company developed a new device, which is called BioMend Extend™ Absorbable Collagen Membrane. Minor manufacturing modifications altered the physical characteristic so that the resorption rate was changed from 6 to 8 weeks to within 18 weeks. The clinical indication for use was not changed. The company was able to submit a Special 510(k) to gain FDA approval for the new device. Of particular importance to the corporation or any manufacturer that has appropriate use of a Special 510(k) is that the average total time for review of Special 510(k) applications was 29 days in 1999. Therefore, approval for a modified device to market to a new group of users may be obtained in 1 month's time (O'Grady, 2000).

Consensus Standards and the Abbreviated 510(k) Format

An abbreviated 510(k) format may be used to request FDA evaluation for Class II devices where the design and manufacturing process conform to consensus standards. As part of the new 510(k) paradigm issued in March 2000, it is proposed that conformance with recognized consensus standards can provide a reasonable assurance of safety and/or effectiveness for many applicable aspects of medical devices. In the case of 510(k)s, information on conformance with recognized consensus standards may help establish the substantial equivalence of a new device to a legally marketed predicate devices (DHHS/FDA/CDRH, Standards 1998). If a submission includes a declaration of conformity to recognized consensus standards from the party submitting the regulatory application, data relating to the aspects of safety and/or effectiveness covered by the standards will not ordinarily be required in the premarket submission. Use of consensus standards is strictly voluntary for the applicant organization.

Government and private groups including the American National Standard Institute, the American Society for Testing and and Material, and the Institute of Electrical and Electronics Engineers are involved in setting consensus standards. In the international sphere, a Global Harmonization Task Force has study groups to set international standards. All conformance standards referenced in abbreviated 510(k) applica-

tions must have received previous FDA approval. The FDA may conduct a postmarket inspection to confirm the existence of data supporting the declaration of conformity.

Third-Party Reviews and 510(k) Applications

To expedite review times, the FDA began a 2-year pilot program in 1996 to have accredited persons conduct third-party reviews. With the advent of FDAMA in 1997, this policy became an integral part of the class II review process. Accredited persons may not be Federal government employees, must be an independent organization that will not design, manufacture, promote or sell devices; and must operate under accepted professional and ethical business practices (DHHS/FDA/CDRH, Third-Party Reviewers, 1998). There are 13 legally constituted third-party reviewers. Three of these, TUV Product Service, Underwriters Laboratories, and CITECH, have performed 75% of third-party reviews to date. The manufacturer may elect to have a third-party review its application or submit it directly to the FDA for review. If the former route is chosen, the accredited person reviews the 510(k) and communicates the recommendation to the FDA. The FDA then issues a final decision within 30 days. As of June 2000, accredited persons may not review Class III devices or Class II devices that (1) are permanently implantable, (2) are life sustaining or life supporting, or (3) require clinical data in the 510(k). Proposals are in consideration, however, to expand this program to include all Class II devices. Although the average review time for accredited persons is shorter (57 days) compared with FDA (105 days), and all third-party reviews have received FDA approval, thus far only a small percentage (<10%) of applicants have used this available alternative (Rechen, 2000).

Conclusions

A skin-replacement product for patients with severe burns was the first FDA-approved tissue-engineered product. Since that time, tissue-engineered skin replacements have been approved for treatment of diabetic ulcers and reconstructive surgery. Other tissue-engineered products are in development or in clinical trials to serve as substitutes for missing or defective cartilage, bone, blood vessels, cardiovascular tissues, ligaments, tendons, nerve guides, and muscle. The challenges facing scientists, developers, and regulatory agencies are ensuring that the replacement tissues are safe and efficacious. In instances such as bone, where tissue growth is slow, a meaningful surrogate endpoint must be defined to measure efficacy. Further challenges include the design and verification of noninvasive or minimally invasive tests to measure short- and long-term functionality.

As noted above, the FDA is working with other government agencies and private-sector groups to help establish consensus standards for use in the approval process for medical devices. This activity, together with the use of legally constituted third-party reviewers, will continue to expedite the process of bringing new tissue-engineered products to the market place.

References

Feigal Jr DW. 1999. CDRH Top Priorities in 2000. Presentation, Boston, MA, December 1999.

Feigal Jr DW. 2000. CDRH Hot Topics. Presentation, St. Louis Park, MN, July 2002.

O'Grady J. 2000. Special 510(k): Device Modification. Presentation, Washington, DC, July 2000.

Rechen EJ. 2000. Third-Party Review of 510(k)s: An Update Presentation, Washington, DC, July 2000.

Runner S. 2000. Practioners Guide to Tissue Engineering Devices and the FDA: Regulatory Approval. Presentation, Rockville, MD, February 2002.

Schwartz ER. 1997. Tissue Engineering. ATP Focused Program Competition. Washington, DC, Department of Commerce, February 1997.

U.S. Department of Health and Human Services, Food and Drug Administration, Center for Devices and Radiological Health. 1998. The New 510(k) Paradigm: Alternate Approaches to Demonstrating Substantial Equivalence in Premarket Notifications. Rockville, MD.

U.S. Department of Health and Human Services, Food and Drug Administration, Center for Devices and Radiological Health. 1998. Implementation of Third Party Programs Under the FDA Modernization Act of 1997. Rockville, MD.

Index